Surfactants: Fundamentals and Applications in the Petroleum Industry

This book provides an introduction to the nature, occurrence, physical properties, propagation, and uses of surfactants in the petroleum industry. It is aimed principally at scientists and engineers who may encounter or use surfactants, whether in process design, petroleum production, or research and development.

The primary focus is on applications of the principles of colloid and interface science to surfactant applications in the petroleum industry, and includes attention to practical processes and problems. Applications of surfactants in the petroleum industry are of great practical importance and are also quite diverse, since surfactants may be applied to advantage throughout the petroleum production process: in reservoirs, in oil and gas wells, in surface processing operations, and in environmental, health and safety applications. In each case appropriate knowledge and practices determine the economic and technical successes of the industrial process concerned. The book includes a comprehensive glossary, indexed and fully cross-referenced.

In addition to scientists and engineers in the petroleum industry, this book will be of interest to senior undergraduates and graduate students in science and engineering, and to graduate students of surfactant chemistry.

LAURIER L. SCHRAMM is President and CEO at the Petroleum Recovery Institute, and adjunct professor of chemistry at the University of Calgary. Dr. Schramm received his B.Sc. (Hons.) in chemistry from Carleton University in 1976 and Ph.D. in physical and colloid chemistry in 1980 from Dalhousie University, where he studied as a Killam and NRC Scholar. From 1980 to 1988 he held research positions with Syncrude Canada Ltd. in its Edmonton Research Centre. Since 1988 he has held a series of positions, of progressively increasing responsibility, with the Petroleum Recovery Institute.

His research interests have included many aspects of colloid and interface science applied to the petroleum industry, including research into mechanisms of processes for the improved recovery of light, heavy, and bituminous crude oils, such as *in situ* foam, polymer or surfactant flooding, and surface hot water flotation from oil sands. This research has involved the formation and stability of dispersions (foams, emulsions and suspensions) and their flow properties, electrokinetic properties, interfacial properties, phase attachments, and the reactions and interactions of surfactants in solution.

Dr. Schramm has won several national awards for his research, including the Canadian Society for Chemical Engineering – Bayer Award in Industrial Practice and the Natural Sciences and Engineering Research Council of Canada – Conference Board of Canada Award for Best Practices in University–Industry R & D Partnership. He is a Fellow of the Chemical Institute of Canada, a past Director of the Association of the Chemical Profession of Alberta, and a member of the American Chemical Society. He has 100 scientific publications and patents in the open literature and over 220 proprietary research reports for industry. This is his fifth book, following *Emulsions: Fundamentals and Applications in the Petroleum Industry*, *The Language of Colloid and Interface Science*, *Foams: Fundamentals and Applications in the Petroleum Industry*, and *Suspensions: Fundamentals and Applications in the Petroleum Industry*.

Surfactants: Fundamentals and Applications in the Petroleum Industry

Laurier L. Schramm

Petroleum Recovery Institute

CAMBRIDGE UNIVERSITY PRESS
Cambridge, New York, Melbourne, Madrid, Cape Town, Singapore,
São Paulo, Delhi, Dubai, Tokyo, Mexico City

Cambridge University Press
The Edinburgh Building, Cambridge CB2 8RU, UK

Published in the United States of America by Cambridge University Press, New York

www.cambridge.org
Information on this title: www.cambridge.org/9780521157933

© Cambridge University Press 2000

First published 2000
First paperback edition 2010

A catalogue record for this publication is available from the British Library

Library of Congress Cataloguing in Publication Data
Surfactants: fundamentals and application in the petroleum industry / Laurier L. Schramm, editor.
   p.   cm.
   Includes index.
   ISBN 0 521 64067 9
   1. Surface active agents – Industrial applications.   2. Petroleum industry and trade.
   I. Schramm, Laurier Lincoln.
TN871.S76784   2000
665.5—dc21 99-15820 CIP

ISBN 978-0-521-64067-1 Hardback
ISBN 978-0-521-15793-3 Paperback

CONTENTS

v

PREFACE

This book provides an introduction to the nature, occurrence, physical properties, propagation, and uses of surfactants in the petroleum industry. The primary focus is on applications of the principles of colloid and interface science to surfactant applications in the petroleum industry, and includes attention to practical processes and problems. Books available up to now are either principally theoretical (such as the colloid chemistry texts), much more general (like Rosen's *Surfactants and Interfacial Phenomena*, Myers' *Surfactant Science and Technology*, or Mittal's *Solution Chemistry of Surfactants*), or else much narrower in scope (like Smith's *Surfactant Based Mobility Control*). The applications of surfactants in the petroleum industry area are quite diverse and have a great practical importance. The area contains a number of problems of more fundamental interest as well. Surfactants may be applied to advantage in many parts of the petroleum production process: in reservoirs, in oilwells, in surface processing operations, and in environmental, health, and safety applications. In each case appropriate knowledge and practices determine both the economic and technical successes of the industrial process concerned.

In this volume, a wide range of authors' expertise and experiences are brought together to yield the first surfactant book that focuses on the applications of surfactants in the petroleum industry. Taking advantage of a broad range of authors' expertise allows for a variety of surfactant technology application areas to be highlighted in an authoritative manner. The topics chosen serve to illustrate some of the different methodologies that have been successfully applied. Each of the chapters in this book has been critically peer-reviewed and revised to meet a high scientific and editorial standard.

The target audience includes scientists and engineers who may encounter or be able to use surfactants, whether in process design, petroleum production, or in the research and development fields. It does not assume a knowledge of colloid chemistry, the initial emphasis being placed on a review of the basic concepts important to understanding surfactants. As such, it is hoped that the book will be of interest to senior undergraduate and graduate students in science and engineering as well since topics such as this are not normally part of university curricula.

The book provides an introduction to the field in a very applications oriented manner, as the focus of the book is practical rather than theoretical. The first group of chapters (1 to 3) sets out fundamental surfactant principles, including chemistry and uses. Subsequent groups of chapters address examples of industrial practice with Chapters 4–7 aimed at the use of surfactants in reservoir oil recovery processes, Chapters 8–10 covering some oilwell, near-well, and surface uses of surfactants, Chapters 11–13 addressing several environmental, health, and safety applications, and the Glossary containing a comprehensive and fully cross-referenced dictionary of terms in the field.

A recurring theme in the chapters is the use of the fundamental concepts in combination with actual commercial process experiences to illustrate how to approach planned and unplanned surfactant occurrences in petroleum processes. It also completes a natural sequence, serving as a companion volume to my earlier books: *Emulsions: Fundamentals and Applications in the Petroleum Industry*; *Foams: Fundamentals and Applications in the Petroleum Industry*, and *Suspensions: Fundamentals and Applications in the Petroleum Industry*.

Acknowledgments

I thank all the authors who contributed considerable time and effort to their respective chapters. This book was made possible through the support of my family, Ann Marie, Katherine and Victoria who gave me the time needed for the organization, research, and writing. I am also very grateful to Conrad Ayasse for his consistent encouragement and support. Throughout the preparation of this book many valuable suggestions were made by colleagues, the external reviewers of individual chapters, and by the editorial staff of Cambridge University Press, particularly Simon Capelin and Margaret Patterson.

Laurier L. Schramm
Calgary, AB, Canada

SURFACTANT FUNDAMENTALS

Surfactants and Their Solutions: Basic Principles

Laurier L. Schramm[1,2] and D. Gerrard Marangoni[3]

[1] Petroleum Recovery Institute, 100, 3512 – 33rd St. NW, Calgary, AB, Canada T2L 2A6
[2] University of Calgary, Dept. of Chemistry, 2500 University Drive NW, Calgary, AB, Canada T2N 1N4
[3] St. Francis Xavier University, Dept. of Chemistry, PO Box 5000, Antigonish, NS, Canada B2G 2W5

This chapter provides an introduction to the occurrence, properties and importance of surfactants as they relate to the petroleum industry. With an emphasis on the definition of important terms, the importance of surfactants, their micellization and adsorption behaviours, and their interfacial properties are demonstrated. It is shown how surfactants may be applied to alter interfacial properties, promote oil displacement, and stabilize or destabilize dispersions such as foams, emulsions, and suspensions. Understanding and controlling the properties of surfactant-containing solutions and dispersions has considerable practical importance since fluids that must be made to behave in a certain fashion to assist one stage of an oil production process, may require considerable modification in order to assist in another stage.

Introduction

Surfactants are widely used and find a very large number of applications because of their remarkable ability to influence the properties of surfaces and interfaces, as will be discussed below. Some important applications of surfactants in the petroleum industry are shown in Table 1. Surfactants may be applied or encountered at all stages in the petroleum recovery and processing industry, from oilwell drilling, reservoir injection, oilwell production, and surface plant processes, to pipeline and seagoing transportation of petroleum emulsions. This chapter is intended to provide an introduction to the basic principles involved in the occurrence and uses of surfactants in the petroleum industry. Subsequent chapters in this book will go into specific areas in greater detail.

Table 1.   Some Examples of Surfactant Applications in the Petroleum Industry

Gas/Liquid Systems
Producing oilwell and well-head foams
Oil flotation process froth
Distillation and fractionation tower foams
Fuel oil and jet fuel tank (truck) foams
Foam drilling fluid
Foam fracturing fluid
Foam acidizing fluid
Blocking and diverting foams
Gas-mobility control foams

Liquid/Liquid Systems
Emulsion drilling fluids
Enhanced oil recovery *in situ* emulsions
Oil sand flotation process slurry
Oil sand flotation process froths
Well-head emulsions
Heavy oil pipeline emulsions
Fuel oil emulsions
Asphalt emulsion
Oil spill emulsions
Tanker bilge emulsions

Liquid/Solid Systems
Reservoir wettability modifiers
Reservoir fines stabilizers
Tank/vessel sludge dispersants
Drilling mud dispersants

All the petroleum industry's surfactant applications or problems have in common the same basic principles of colloid and interface science. The widespread importance of surfactants in general, and scientific interest in their nature and properties, have precipitated a wealth of published literature on the subject. Good starting points for further basic information are classic books like Rosen's *Surfactants and Interfacial Phenomena* [1] and Myers' *Surfactant Science and Technology* [2], and the many other books on surfactants [3–19]. Most good colloid chemistry texts contain introductory chapters on surfactants. Good starting points are references [20–23], while for much more detailed treatment of advances in specific surfactant-related areas the reader is referred to some of the chapters available in specialist books [24–29]. With regard to the occurrence of related colloidal systems in the petroleum industry, three recent books describe the principles and occurrences of emulsions, foams, and suspensions in the petroleum industry [30–32].

Definition and Classification of Surfactants[4]

Some compounds, like short-chain fatty acids, are amphiphilic or amphipathic, i.e., they have one part that has an affinity for nonpolar media and one part that has an affinity for polar media. These molecules form oriented monolayers at interfaces and show surface activity (i.e., they lower the surface or interfacial tension of the medium in which they are dissolved). In some usage surfactants are defined as molecules capable of associating to form micelles. These compounds are termed surfactants, amphiphiles, surface-active agents, tensides, or, in the very old literature, paraffin-chain salts. The term surfactant is now probably the most commonly used and will be employed in this book. This word has a somewhat unusual origin, it was first created and registered as a trademark by the General Aniline and Film Corp. for their surface-active products.[5] The company later (ca. 1950) released the term to the public domain for others to use [33]. Soaps (fatty acid salts containing at least eight carbon atoms) are surfactants. Detergents are surfactants, or surfactant mixtures, whose solutions have cleaning properties. That is, detergents alter interfacial properties so as to promote removal of a phase from solid surfaces.

The unusual properties of aqueous surfactant solutions can be ascribed to the presence of a hydrophilic head group and a hydrophobic chain (or tail) in the molecule. The polar or ionic head group usually interacts strongly with an aqueous environment, in which case it is solvated via dipole–dipole or ion–dipole interactions. In fact, it is the nature of the polar head group which is used to divide surfactants into different categories, as illustrated in Table 2. In-depth discussions of surfactant structure and chemistry can be found in references [1, 2, 8, 34, 35].

The Hydrophobic Effect and Micelle Formation

In aqueous solution dilute concentrations of surfactant act much as normal electrolytes, but at higher concentrations very different behaviour results. This behaviour is explained in terms of the formation of organized aggregates of large numbers of molecules called micelles, in which the

[4] A glossary of frequently encountered terms in the science and engineering of surfactants is given in the final chapter of this book.

[5] For an example of one of GAF Corp's. early ads promoting their trademarked surfactants, see *Business Week*, March 11, 1950, pp. 42–43.

Table 2. Surfactant Classifications

Class	Examples	Structures
Anionic	Na stearate	$CH_3(CH_2)_{16}COO^-Na^+$
	Na dodecyl sulfate	$CH_3(CH_2)_{11}SO_4^-Na^+$
	Na dodecyl benzene sulfonate	$CH_3(CH_2)_{11}C_6H_4SO_3^-Na^+$
Cationic	Laurylamine hydrochloride	$CH_3(CH_2)_{11}NH_3^+Cl^-$
	Trimethyl dodecylammonium chloride	$C_{12}H_{25}N^+(CH_3)_3Cl^-$
	Cetyl trimethylammonium bromide	$CH_3(CH_2)_{15}N^+(CH_3)_3Br^-$
Nonionic	Polyoxyethylene alcohol	$C_nH_{2n+1}(OCH_2CH_2)_mOH$
	Alkylphenol ethoxylate	$C_9H_{19}{-}C_6H_4{-}(OCH_2CH_2)_nOH$
	Polysorbate 80	$HO(C_2H_4O)_w\ldots(OC_2H_4)_xOH$
	$w + x + y + z = 20,$	$CH(OC_2H_4)_yOH$
	$R = (C_{17}H_{33})COO$	$\vert$
		$CH_2(OC_2H_4)_zR$
	Propylene oxide-modified polymethylsiloxane	$(CH_3)_3SiO((CH_3)_2SiO)_x(CH_3SiO)_ySi(CH_3)_3$
	EO = ethyleneoxy	$\vert$
	PO = propyleneoxy	$CH_2CH_2CH_2O(EO)_m(PO)_nH$
Zwitterionic	Dodecyl betaine	$C_{12}H_{25}N^+(CH_3)_2CH_2COO^-$
	Lauramidopropyl betaine	$C_{11}H_{23}CONH(CH_2)_3N^+(CH_3)_2CH_2COO^-$
	Cocoamido-2-hydroxy-propyl sulfobetaine	$C_nH_{2n+1}CONH(CH_2)_3N^+(CH_3)_2CH_2CH(OH)CH_2SO_3^-$

lipophilic parts of the surfactants associate in the interior of the aggregate leaving hydrophilic parts to face the aqueous medium. An illustration presented by Hiemenz and Rajagopalan [22] is given in Figure 1. The formation of micelles in aqueous solution is generally viewed as a compromise between the tendency for alkyl chains to avoid energetically unfavourable contacts with water, and the desire for the polar parts to maintain contact with the aqueous environment.

A thermodynamic description of the process of micelle formation will include a description of both electrostatic and hydrophobic contributions to the overall Gibbs energy of the system. Hydrocarbons (e.g., dodecane) and water are not miscible; the limited solubility of hydrophobic species in water can be attributed to the hydrophobic effect. The hydrophobic Gibbs energy (or the transfer Gibbs energy) can be defined as the difference between the standard chemical potential of the hydrocarbon solute in water and a hydrocarbon solvent at infinite dilution [36–40]

$$\Delta G_t^\circ = \mu_{HC}^\circ - \mu_{aq}^\circ \tag{1}$$

where μ_{HC}° and μ_{aq}° are the chemical potentials of the hydrocarbon dissolved in the hydrocarbon solvent and water, respectively, and ΔG_t° is

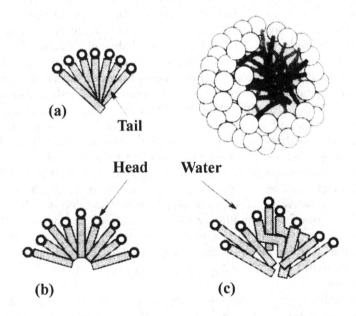

Figure 1. Schematic representation of the structure of an aqueous micelle showing several possibilities: (a) overlapping tails in the centre, (b) water penetrating to the centre, and (c) chains protruding and bending. (From Hiemenz and Rajagopalan [22]. Copyright 1997 Marcel Dekker Inc., New York.)

the Gibbs energy for the process of transferring the hydrocarbon solute from the hydrocarbon solvent to water. In a homologous series of hydrocarbons (e.g., the n-alcohols or the n-alkanes), the value of ΔG_t° generally increases in a regular fashion

$$\Delta G_t^\circ = (a - bn_c)RT \tag{2}$$

where a and b are constants for a particular hydrocarbon series and n_c is the number of carbon atoms in the chain. The transfer Gibbs energy, ΔG_t°, can be divided into entropic and enthalpic contributions

$$\Delta G_t^\circ = \Delta H_t^\circ - T\,\Delta S_t^\circ \tag{3}$$

where ΔH_t° and ΔS_t° are the enthalpy and entropy of transfer, respectively. A significant characteristic of the hydrophobic effect is that the entropy term is dominant, i.e., the transfer of the hydrocarbon solute from the hydrocarbon solvent to water is accompanied by an increase in the Gibbs transfer energy ($\Delta G > 0$) [41]. The decrease in entropy is thought to be the result of the breakdown of the normal hydrogen-bonded structure of water accompanied by the formation of differently structured water, often termed icebergs, around the hydrocarbon chain. The presence of the hydrophobic species promotes an ordering of water molecules in the vicinity of the hydrocarbon chain. To minimize the large entropy effect, the "icebergs" tend to cluster [38], in order to reduce the number of water molecules involved; the "clustering" is enthalpically favoured (i.e., $\Delta H < 0$), but entropically unfavourable. The overall process has the tendency to bring the hydrocarbon molecules together, which is known as the hydrophobic interaction. Molecular interactions, arising from the tendency for the water molecules to regain their normal tetrahedral structure, and the attractive dispersion forces between hydrocarbon chains, act cooperatively to remove the hydrocarbon chain from the water "icebergs", leading to an association of hydrophobic chains.

Due to the presence of the hydrophobic effect, surfactant molecules adsorb at interfaces, even at low surfactant concentrations. As there will be a balance between adsorption and desorption (due to thermal motions), the interfacial condition requires some time to establish. The surface activity of surfactants should therefore be considered a dynamic phenomenon. This can be determined by measuring surface or interfacial tensions versus time for a freshly formed surface, as will be discussed further below.

At a specific, higher, surfactant concentration, known as the critical micelle concentration (cmc), molecular aggregates termed micelles are formed. The cmc is a property of the surfactant and several other factors, since micellization is opposed by thermal and electrostatic forces. A low cmc is favoured by increasing the molecular mass of the lipophilic part of the molecule, lowering the temperature (usually), and adding electrolyte.

Surfactant molar masses range from a few hundreds up to several thousands.

The most commonly held view of a surfactant micelle is not much different than that published by Hartley in 1936 [41, 42] (see Figure 1). At surfactant concentrations slightly above the cmc value, surfactants tend to associate into spherical micelles, of about 50–100 monomers, with a radius similar to that of the length of an extended hydrocarbon chain. The micellar interior, being composed essentially of hydrocarbon chains, has properties closely related to the liquid hydrocarbon.

Critical Micelle Concentration

It is well known that the physico-chemical properties of surfactants vary markedly above and below a specific surfactant concentration, the cmc value [2–9, 13, 14, 17, 35–47]. Below the cmc value, the physico-chemical properties of ionic surfactants like sodium dodecylsulfate, SDS, (e.g., conductivities, electromotive force measurements) resemble those of a strong electrolyte. Above the cmc value, these properties change dramatically, indicating a highly cooperative association process is taking place. In fact, a large number of experimental observations can be summed up in a single statement: almost all physico-chemical properties versus concentration plots for a given surfactant–solvent system will show an abrupt change in slope in a narrow concentration range (the cmc value). This is illustrated by Preston's [48] classic graph, shown in Figure 2.

In terms of micellar models, the cmc value has a precise definition in the pseudo-phase separation model, in which the micelles are treated as a separate phase. The cmc value is defined, in terms of the pseudo-phase model, as the concentration of maximum solubility of the monomer in that particular solvent. The pseudo-phase model has a number of shortcomings; however, the concept of the cmc value, as it is described in terms of this model, is very useful when discussing the association of surfactants into micelles. It is for this reason that the cmc value is, perhaps, the most frequently measured and discussed micellar parameter [39].

Cmc values are important in virtually all of the petroleum industry surfactant applications. For example, a number of improved or enhanced oil recovery processes involve the use of surfactants including micellar, alkali/surfactant/polymer (A/S/P) and gas (hydrocarbon, N_2, CO_2 or steam) flooding. In these processes, surfactant must usually be present at a concentration higher than the cmc because the greatest effect of the surfactant, whether in interfacial tension lowering [30] or in promoting foam stability [31], is achieved when a significant concentration of micelles is present. The cmc is also of interest because at concentrations

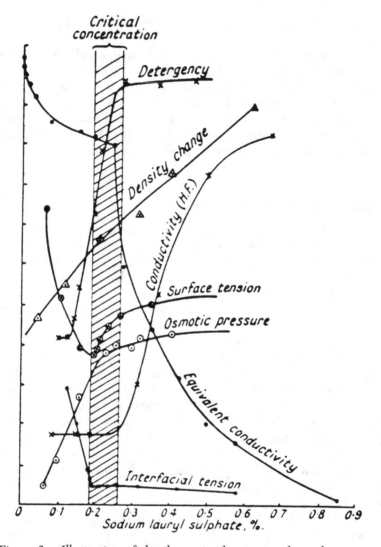

Figure 2.   Illustration of the dramatic changes in physical properties that occur beyond the critical micelle concentration. (From Preston [48]. Copyright 1948 American Chemical Society, Washington.)

above this value the adsorption of surfactant onto reservoir rock surfaces increases very little. That is, the cmc represents the solution concentration of surfactant from which nearly maximum adsorption occurs.

Cmc Measurements.   The general way of obtaining the cmc value of a surfactant micelle is to plot some physico-chemical property of

Table 3.   Some Common Cmc Methods

UV/Vis, IR spectroscopy
Fluorescence spectroscopy
Nuclear magnetic resonance spectroscopy
Electrode potential/conductivity
Voltametry
Scattering techniques
Calorimetry
Surface tension
Foaming

interest versus the surfactant concentration and observe the break in the plot. Table 3 lists the most common cmc methods. Many of these methods have been reviewed by Shinoda [11] and Mukerjee and Mysels [49]. It should be noted that different experimental techniques may give slightly different values for the cmc of a surfactant. However, Mukerjee and Mysels [49], in their vast compilation of cmc values, have noted that the majority of values for a single surfactant (e.g., sodium dodecyl sulfate, or SDS, in the absence of additives) are in good agreement and the outlying values are easily accounted for.

For petroleum industry processes, one tends to have a special interest in the cmc's of practical surfactants that may be anionic, cationic, nonionic or amphoteric. The media are typically high salinity, high hardness electrolyte solutions, and in addition, the cmc values of interest span the full range from ambient laboratory conditions to oil and gas reservoir conditions of temperature and pressure. Irrespective of aiming for process development and optimization under realistic (reservoir) conditions of temperature and pressure, it remains common to determine cmc's experimentally at ambient laboratory conditions and assume that the same hold even at elevated temperatures and pressures. This can be an extremely dangerous assumption.

The nature and limits of applicability of specific methods for determining critical micelle concentrations vary widely. Most methods have been developed for a relatively small set of pure surfactants involving very dilute electrolyte solutions and only ambient temperature and pressure. The determination of cmc at elevated temperature and pressure is experimentally much more difficult than for ambient conditions and comparatively little work has been done in this area. Most high temperature cmc studies have been by conductivity measurements and have therefore been limited to ionic surfactants. For example, cmc's at up to 166 °C have been reported by Evans and Wightman [50]. Some work has been reported using calorimetry, up to 200 °C by Noll [51], and using [19]F

NMR, up to 180 °C by Shinoda et al. [52]. Some work has been reported involving cmc determination by calorimetry (measuring heats of dilution or specific heats). Archer et al. [53] used flow calorimetry to determine the cmc's of several sulfonate surfactants at up to 178 °C. Noll [51] determined cmc's for dodecyltrimethylammonium bromide and commercial surfactants in the temperature range 25–200 °C using flow calorimetry. Surface tension is the classical method for determining cmc's but many surface tension methods are not suitable for use with aqueous solutions at elevated temperatures. Exceptions include the pendant, sessile, and captive drop methods which can be conducted with high-pressure cells [54, 55].

For any of the techniques applied it appears (Archer et al. [53]) that the uncertainties in the experimental cmc determinations increase with increasing temperature because at the same time the surfactant aggregation number decreases and the aggregation distribution increases. That is, the concentration range over which micellization occurs broadens with increasing temperature. Almost all of the elevated temperature cmc studies have involved carefully purified surfactants (not commercial surfactants or their formulations) in pure water or very dilute electrolyte solutions. Conducting cmc determinations at elevated pressure, as well as temperature, is even more difficult and only a few studies have been reported, mostly employing conductivity methods (La Mesa et al. [56]; Sugihara and Mukerjee [57]; Brun et al. [58]; Kaneshina et al. [59]; Hamann [60]) which, again, are unsuitable for nonionic or zwitterionic surfactants and for use where the background electrolyte concentrations are significant.

In the case where one needs to be able to determine cmc's for nonionic or zwitterionic surfactants, in electrolyte solutions that may be very concentrated, and at temperatures and pressures up to those that may be encountered in improved oil recovery operations in petroleum reservoirs, most of the established methods are not practical. One successful approach to this problem has been to use elevated temperature and pressure surface tension measurements involving the captive drop technique [8] although this method is quite time-consuming. Another approach is to use dynamic foam stability measurements. Foaming effectiveness and the ease of foam formation are related to surface tension lowering and to micelle formation, the latter of which promotes foam stability through surface elasticity and other mechanisms [61]. Accordingly, static or dynamic foam height methods generally show that foam height increases with surfactant concentration and then becomes relatively constant at concentrations greater than the cmc (Rosen and Solash [62]; Goette [63]). Using a modified Ross-Miles static foam height apparatus, Kashiwagi [64] determined the cmc of SDS at 40 °C to be 7.08 mM which compared well with values attained by conductivity (7.2 mM) and surface tension (7.2 mM). Rosen and Solash [62] also found that foam production was related to cmc using the Ross-Miles method at 60 °C when they assessed SDS, potassium tetradecyl sulfonate, potassium hexadecyl sulfonate, and sodium hexadecyl sulfate.

Morrison et al. [65] describe a dynamic foam height method for the estimation of cmc's that is suitable for use at high temperatures and pressures. This method is much more rapid than the surface tension method, and is applicable to a wide range of surfactant classes, including both ionic and amphoteric (zwitterionic) surfactants. The method is suitable for the estimation of cmc's, for determining the minimum cmc as a function of temperature, for identifying the temperature at which the minimum cmc occurs, and for determining how cmc's vary with significant temperature and pressure changes. The method has been used to determine the temperature variation of cmc's for a number of commercial foaming surfactants in aqueous solutions, for the derivation of thermodynamic parameters, and to establish useful correlations [55].

Cmc Values. Some typical cmc values for low electrolyte concentrations at room temperature are:

Anionics	10^{-3}–10^{-2} M
Amphoterics	10^{-3}–10^{-1} M
Cationics	10^{-3}–10^{-1} M
Nonionics	10^{-5}–10^{-4} M

Cmc values show little variation with regard to the nature of the charged head group. The main influence appears to come from the charge of the hydrophilic head group. For example, the cmc of dodecyltrimethylammonium chloride (DTAC) is 20 mM, while for a 12 carbon nonionic surfactant, hexaethylene glycol mono-*n*-dodecyl ether ($C_{12}E_6$), the cmc is about 0.09 mM [39, 41, 49]; the cmc for SDS is about 8 mM, while that for disodium 1,2-dodecyldisulfate (1,2-SDDS) is 40 mM [66]. In addition to the relative insensitivity of the cmc value of the surfactant to the nature of the charged head group, cmc's show little dependence on the nature of the counter-ion. It is mainly the valence number of the counter-ion that affects the cmc. As an example, the cmc value for $Cu(DS)_2$ is about 1.2 mM, while the cmc for SDS is about 8 mM [49, 67].

Cmc values often exhibit a weak dependence on both temperature [68–70] and pressure [59, 71], although, as shown in Figure 3, some surfactant cmc's have been observed to increase markedly with temperature above 100 °C [55, 65]. The effects of added substances on the cmc are complicated and interesting, and depend greatly on whether the additive is solubilized in the micelle, or in the intermicellar solution. The addition of electrolytes to ionic surfactant solutions results in a well

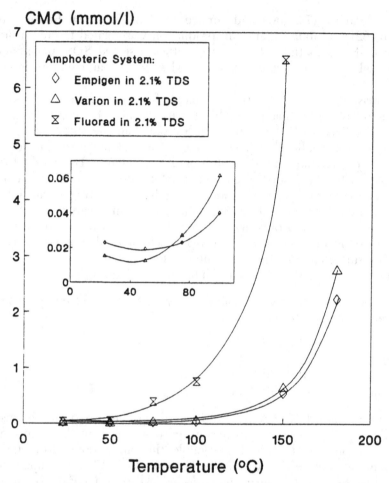

Figure 3.   Temperature variation of the critical micelle concentrations of three amphoteric surfactants in 2.1% total dissolved solids brine solutions. (From Stasiuk and Schramm [55]. Copyright 1996 Academic Press, New York.)

established linear dependence of log (cmc) on the concentration of added salt [72–76]. For nonionic micelles, electrolyte addition has little effect on cmc values. When non-electrolytes are added to the micellar solution, the effects are dependent on the nature of the additive. For polar additives (e.g., *n*-alcohols), the cmc decreases with increasing concentration of alcohol, while the addition of urea to micellar solutions tends to increase the cmc, and may even inhibit micelle formation [77, 78]. Nonpolar additives tend to have little effect on the cmc [79].

The Krafft Point

The solubilities of micelle-forming surfactants show a strong increase above a certain temperature, termed the Krafft point (T_k). This is explained by the fact that the single surfactant molecules have limited solubility whereas the micelles are very soluble. Referring to the illustration from Shinoda [11] in Figure 4, below the Krafft point the solubility of the surfactant is too low for micellization so solubility alone determines the surfactant monomer concentration. As temperature increases the solubility increases until at T_k the cmc is reached. At this temperature a relatively large amount of surfactant can be dispersed in micelles and solubility increases greatly. Above the Krafft point maximum reduction in surface or interfacial tension occurs at the cmc because the cmc then determines the surfactant monomer concentration. Krafft points for a number of surfactants are listed in references [1, 80].

Nonionic surfactants do not exhibit Krafft points. Instead, the solubility of nonionic surfactants decreases with increasing temperature, and these surfactants may begin to lose their surface active properties above a transition temperature referred to as the cloud point. This occurs because above the cloud point a surfactant rich phase of swollen micelles separates, and the transition is usually accompanied by a marked increase in dispersion turbidity.

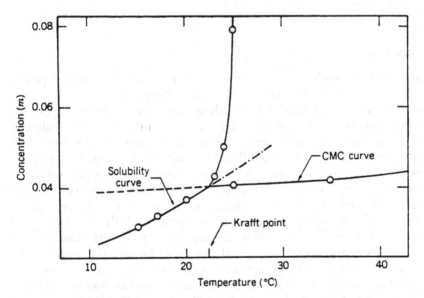

Figure 4.   Example of a "phase behaviour" diagram for a surfactant in aqueous solution, showing the cmc and Krafft points. (From Shinoda et al. [11]. Copyright 1963 Academic Press, New York.)

Table 4.   Typical Methods of Surfactant Analysis

Surfactant Class	Method
Anionic	
alkyl sulfates and sulfonates	Two-phase or surfactant-electrode monitored titration
petroleum and lignin sulfonates	Column or gel permeation chromatography
phosphate esters	Potentiometric titration
sulfosuccinate esters	Gravimetric or titration methods
carboxylates	Potentiometric titration or two-phase titration
Nonionic	
alcohols	NMR or IR spectroscopy
ethoxylated acids	Gas chromatography
alkanolamides	Gas chromatography
ethoxylated amines	HPLC
amine oxides	Potentiometric titration
Cationic	
quaternary ammonium salts	Two-phase or surfactant-electrode monitored titration, or GC or HPLC
Amphoteric	
carboxybetaines	Low pH two-phase titration, gravimetric analysis, or potentiometric titration
sulfobetaines	HPLC

Analysis

Numerous methods have been developed for the quantitative determination of each class of surfactant. The analysis of commercial surfactants is greatly complicated by the fact that these products are mixtures. They are often comprised of a range of molar mass structures of a given structural class, may contain surface-active impurities, are sometimes intentionally formulated to contain several different surfactants, and are often supplied dissolved in mixed organic solvents or complex aqueous salt solutions. Each of these components has the potential to interfere with a given analytical method. Therefore surfactant assays may well have to be preceded by surfactant separation techniques. Both the separation and assay techniques can be highly specific to a given surfactant/solution system. This makes any substantial treatment beyond the scope of the present chapter. Good starting points can be found in the several books on surfactant analysis [81–86]. The characterization and analysis of surfactant demulsifiers is discussed in Chapter 2 of this book. Table 4 shows some typical kinds of analysis methods that are applied to the different surfactant classes.

There are a number of reviews available for surfactants in specific industries [87], and for specific surfactant classes. References [81–90] discuss methods for the determination of anionic surfactants, which are probably the most commonly encountered in the petroleum industry. Most of these latter methods are applicable only to the determination of sulfate- and sulfonate-functional surfactants. Probably the most common analysis method for anionic surfactants is Epton's two-phase titration method [91, 92] or one of its variations [93, 94]. Related, single-phase titrations can be performed and monitored by either surface tension [95] or surfactant-sensitive electrode [84, 85, 96–98] measurements. Gronsveld and Faber [99] discuss adaptation of the titration method to oleic phase samples.

Surfactants and Surface Tension

In two-phase dispersions, a thin intermediate region or boundary, known as the interface, lies between the two phases. The physical properties of the interface can be very important in all kinds of petroleum recovery and processing operations. Whether in a well, a reservoir or a surface processing operation, one tends to encounter large interfacial areas exposed to many kinds of chemical reactions. In addition, many petroleum industry processes involve colloidal dispersions, such as foams, emulsions, and suspensions, all of which contain large interfacial areas; the properties of these interfaces may also play a large role in determining the properties of the dispersions themselves. In fact, even a modest surface energy per unit area can become a considerable total surface energy. Suppose we wish to make a foam by dispersion of gas bubbles into water. For a constant gas volume fraction the total surface area produced increases as the bubble size produced decreases. Since there is a free energy associated with surface area, this increases as well with decreasing bubble size. The energy has to be added to the system to achieve the dispersion of small bubbles. If this amount of energy cannot be provided, say through mechanical energy input, then another alternative is to use surfactant chemistry to lower the interfacial free energy, or interfacial tension. The addition of a small quantity of a surfactant to the water, possibly a few tenths of a percent, would significantly lower the surface tension and significantly lower the amount of mechanical energy needed for foam formation. For examples of this simple calculation for foams and emulsions, see references [61] and [100] respectively.

The origin of surface tension may be visualized by considering the molecules in a liquid. The attractive van der Waals forces between molecules are felt equally by all molecules except those in the interfacial region. This imbalance pulls the latter molecules towards the interior of the liquid. The contracting force at the surface is known as the surface tension. Since the surface has a tendency to contract spontaneously in order to minimize the surface area, bubbles of gas tend to adopt a spherical shape: this reduces the total surface free energy. For emulsions of two immiscible liquids a similar situation applies to the droplets of one of the liquids, except that it may not be so immediately obvious which liquid will form the droplets. There will still be an imbalance of intermolecular force resulting in an interfacial tension, and the interface will adopt a configuration that minimizes the interfacial free energy. Physically, surface tension may be thought of as the sum of the contracting forces acting parallel to the surface or interface. This point of view defines surface or interfacial tension (γ), as the contracting force per unit length around a surface. Another way to think about surface tension is that area expansion of a surface requires energy. Since the work required to expand a surface against contracting forces is equal to the increase in surface free energy accompanying this expansion, surface tension may also be expressed as energy per unit area.

There are many methods available for the measurement of surface and interfacial tensions. Details of these experimental techniques and their limitations are available in several good reviews [101–104]. Table 5 shows some of the methods that are used in petroleum recovery process research. A particular requirement of reservoir oil recovery process research is that measurements be made under actual reservoir conditions of temperature and pressure. The pendant and sessile drop methods are the most commonly used where high temperature/pressure conditions are required. Examples are discussed by McCaffery [105] and DePhilippis et al. [106]. These standard techniques can be difficult to apply to the measurement of extremely low interfacial tensions (<1 to 10 mN/m). For ultra-low tensions two approaches are being used. For moderate temperatures and low pressures the most common method is that of the spinning drop, especially for microemulsion research [107]. For elevated temperatures and pressures a captive drop method has been developed by Schramm et al. [108], which can measure tensions as low as 0.001 mN/m at up to 200 °C and 10,000 psi. In all surface and interfacial tension work it should be appreciated that when solutions, rather than pure liquids, are involved appreciable changes can occur with time at the surfaces and interfaces, so that techniques capable of dynamic measurements tend to be the most useful.

When surfactant molecules adsorb at an interface they provide an expanding force acting against the normal interfacial tension. Thus, surfactants tend to lower interfacial tension. This is illustrated by the general Gibbs adsorption equation for a binary, isothermal system containing excess electrolyte:

$$\Gamma_s = -(1/RT)(d\gamma/d\ln C_s) \qquad (4)$$

Table 5. Surface and Interfacial Tension Methods used in Petroleum Research

Method	Static Values	Dynamic Values	Surface Tension	Interfacial Tension	Contact Angle	High T, P Capability
Capillary rise	✓	≈	✓	✗	✗, need $\theta = 0$	✗
Wilhelmy plate	✓	≈	✓	✗	✓, need to know γ	✗
du Nouy ring	✓	✗	✓	✗	✗, pure liquids only	✗
Drop weight	✓	✗	✓	✓	✗, need $\theta = 0$	✓
Drop volume	✓	✗	✓	✓	✗, need $\theta = 0$	✓
Pendant drop	✓	✓	✓	✓	✗	✓
Sessile drop	✓	✓	✓	✓	✓	✓
Oscillating jet	✓	✓	✓	✗	✗	✗
Spinning drop	✓	≈	✓	✓	✗	✗
Captive drop	✓	✓	✓	✓	✗, forces $\theta = 0$	✓
Maximum bubble pressure	✓	≈	✓	✗	✗	✗
Surface laser light scattering	✓	✓	✓	≈	✗	✓
Tilting plate	✓	≈	✗	✗	✓	✗

where Γ_s is the surface excess of surfactant (mol/cm^2), C_s is the solution concentration of the surfactant (M), and γ may be either surface or interfacial tension (mN/m). This equation can be applied to dilute surfactant solutions where the surface curvature is not great and where the adsorbed film can be considered to be a monolayer. The packing density of surfactant in a monolayer at the interface can be calculated as follows. According to equation 4, the surface excess in a tightly packed monolayer is related to the slope of the linear portion of a plot of surface tension versus the logarithm of solution concentration. From this, the area per adsorbed molecule (a_S) can be calculated from

$$a_S = 1/(N_A \Gamma_s) \tag{5}$$

where N_A is Avogadro's number. Numerous examples are given by Rosen [1].

When surfactants concentrate in an adsorbed monolayer at a surface the interfacial film may take on any of a number of quite different properties which will be discussed in the next several sections. Suitably altered interfacial properties can provide a stabilizing influence in dispersions such as emulsions, foams, and suspensions.

Surface Elasticity

As surfactant adsorbs at an interface the interfacial tension decreases (at least up to the cmc), a phenomenon termed the Gibbs effect. If a surfactant stabilized film undergoes a sudden expansion, the immediately expanded portion of the film must have a lower degree of surfactant adsorption than unexpanded portions because the surface area has increased. This causes an increased local surface tension which produces immediate contraction of the surface. The surface is coupled, by viscous forces, to the underlying liquid layers. Thus, the contraction of the surface induces liquid flow, in the near-surface region, from the low tension region to the high tension region. The transport of bulk liquid due to surface tension gradients is termed the Marangoni effect [27]. In foams, the Gibbs–Marangoni effect provides a resisting force to the thinning of liquid films.

The Gibbs–Marangoni effect only persists until the surfactant adsorption equilibrium is re-established in the surface, a process that may take place within seconds or over a period of hours. For bulk liquids and in thick films this can take place quite quickly, however, in thin films there may not be enough surfactant in the extended surface region to re-establish the equilibrium quickly, requiring diffusion from other parts of the film. The restoring processes are then the movement of surfactant along the interface from a region of low surface tension to one of high surface tension, and the movement of surfactant from the thin film into the now depleted surface region. Thus the Gibbs–Marangoni effect provides a force to counteract film rupture in foams.

Many surfactant solutions show dynamic surface tension behaviour. That is, some time is required to establish the equilibrium surface tension. After the surface area of a solution is suddenly increased or decreased (locally), the adsorbed surfactant layer at the interface requires some time to restore its equilibrium surface concentration by diffusion of surfactant from, or to, the bulk liquid (see Figure 5, [109]). At the same time, since surface tension gradients are now in effect, Gibbs–Marangoni forces act in opposition to the initial disturbance. The dissipation of surface tension gradients, to achieve equilibrium, embodies the interface with a finite elasticity. This explains why some substances that lower surface tension do not stabilize foams [21]; they do not have the required rate of approach to equilibrium after a surface expansion or contraction. In other words, they do not have the requisite surface elasticity.

At equilibrium, the surface elasticity, or surface dilational elasticity, E_G, is defined [21, 110] by

$$E_G = \frac{d\gamma}{d\ln A} \tag{6}$$

where γ is the surface tension and A is the geometric area of the surface. This is related to the compressibility of the surface film, K, by $K = 1/E_G$. E_G is a thermodynamic property, termed the Gibbs surface elasticity. This is the elasticity that is determined by isothermal equilibrium measurements, such as the spreading pressure–area method [21]. E_G occurs in very thin films where the number of molecules is so low that the surfactant cannot restore the equilibrium surface concentration after deformation. An illustration is given in [61].

The elasticity determined from nonequilibrium dynamic measurements depends upon the stresses applied to a particular system, is generally larger in magnitude than E_G, and is termed the Marangoni surface elasticity, E_M [21, 111]. For foams it is this dynamic property that is of most interest. Surface elasticity measures the resistance against creation of surface tension gradients and of the rate at which such gradients disappear once the system is again left to itself [112]. The Marangoni elasticity can be determined experimentally from dynamic surface tension measurements that involve known surface area changes, such as the maximum bubble pressure method [113, 115]. Although such measurements include some contribution from surface dilational viscosity [112, 114] the results are frequently simply referred to in terms of surface elasticities.

Numerous studies have examined the relation between E_G or E_M and foam stability [111, 112, 115]. From low bulk surfactant concentrations,

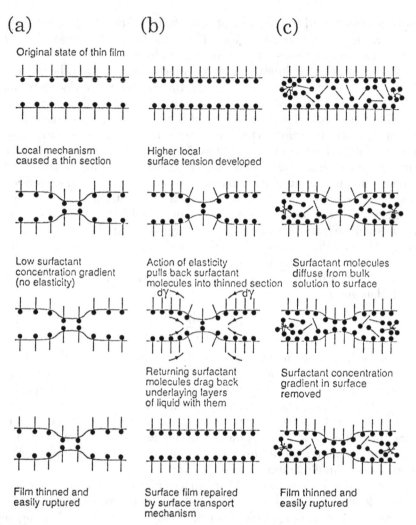

Figure 5. Illustration of the Gibbs–Marangoni effect in a thin liquid film. Reaction of a liquid film to a surface disturbance. (a) Low surfactant concentration yields only low differential tension in film. The thin film is poorly stabilized. (b) Intermediate surfactant concentration yields a strong Gibbs–Marangoni effect which restores the film to its original thickness. The thin film is stabilized. (c) High surfactant concentration (>cmc) yields a differential tension which relaxes too quickly due to diffusion of surfactant. The thinner film is easily ruptured. (From Pugh [109]. Copyright 1996 Elsevier, Amsterdam.)

the Gibbs elasticity increases with an increase in surfactant concentration until a maximum in elasticity is reached, after which the Gibbs elasticity decreases. Surfactant concentrations above the cmc lie well beyond this maximum elasticity region. Lucassen-Reynders [112] cautions that there exists no direct relationship between elasticity and foam stability, although Schramm and Green [113] have found a useful correlation for foams flowing in porous media. Additional factors, such as film thickness and adsorption behaviour, have to be taken into account. Nevertheless, the ability of a surfactant to reduce surface tension and contribute to surface elasticity are among the most important features of foam stabilization. This partially explains why some surfactants will act to promote foaming while others reduce foam stability (foam breakers or defoamers), and still others prevent foam formation in the first place (foam preventatives, foam inhibitors).

Schramm et al. [116] discuss some of the factors that must be considered in the selection of practical foam-forming surfactants for petroleum recovery processes. Kerner [117] describes several hundred different formulations for foam inhibitors and foam breakers.

Surface Rheology

Surface rheology deals with the dynamic behaviour of a surface in response to the stress that is placed on the surface. Two types of viscosities are defined within the interface, a shear viscosity and a dilational viscosity. For a surfactant monolayer, the surface shear viscosity is analogous to the three dimensional shear viscosity: the rate of yielding of a layer of fluid due to an applied shear stress. The surface dilational viscosity expresses the magnitude of the viscous forces during a rate expansion of a surface element. A surfactant monolayer can be expanded or compressed over a wide area range. Thus, the dynamic surface tension experienced during a rate dependent surface expansion is the resultant of the surface dilational viscosity, the surface shear viscosity, and elastic forces. Often, the contributions of shear and/or the dilational viscosities are neglected during stress measurements of surface expansions. Isolating interfacial viscosity effects is rather difficult. The interface is connected to the substrate on either side of it and so are the interfacial viscosities coupled to the bulk viscosities. Therefore, it becomes laborious to determine purely interfacial viscosities without the influence of the surroundings.

A high interfacial viscosity can contribute to emulsion stability by reducing the rate of droplet coalescence [118–121]. This is therefore a property that one may wish to enhance in the formulation of a desirable emulsion. For example, oilfield water-in-oil (W/O) emulsions may be stabilized by the presence of a protective film around the water droplets.

Such a film can be formed from the asphaltene and resin fractions of the crude oil. As drops approach each other the rate of oil film drainage will be determined, in part, by the interfacial viscosity which, if high enough, will significantly retard the final stage of film drainage and possibly even provide a viscoelastic barrier to coalescence. More detailed descriptions are given in references [121–123]. On the other hand, in an enhanced oil recovery process one will generally desire low interfacial viscosity so that once the oil is emulsified and displaced from pores within which it was trapped, the same emulsion drops can later coalesce into an oil bank which can be displaced from the reservoir [30]. Wasan et al. [124] found such a correlation between oil droplet coalescence rate and interfacial viscosity.

As bubbles in a foam approach each other, the thinning of the films between the bubbles, and their resistance to rupture, are thought to be of great importance to the ultimate stability of the foam. Thus, a high interfacial viscosity can promote foam stability by lowering the film drainage rate and retarding the rate of bubble coalescence [125]. Fast draining films may reach their equilibrium film thickness in a matter of seconds or minutes due to low surface viscosity, while slow draining films may require hours due to their high surface viscosity. Bulk viscosity and surface viscosity, thus, do not normally contribute a direct stabilizing force to a foam film, but act as resistances to the thinning and rupture processes. The bulk viscosity will most influence the thinning of thick films, while the surface viscosity will be dominant during the thinning of thin films.

The presence of mixed surfactant adsorption seems to be a factor in obtaining films with very viscous surfaces [27]. For example, in some cases, the addition of a small amount of nonionic surfactant to a solution of anionic surfactant can enhance foam stability due to the formation of a viscous surface layer; possibly a liquid crystalline surface phase in equilibrium with a bulk isotropic solution phase [21, 126]. To the extent that viscosity and surface viscosity influence emulsion and foam stability one would predict that stability would vary according to the effect of temperature on the viscosity. Thus, some petroleum industry processes exhibit serious foaming problems at low process temperatures, which disappear at higher temperatures [21].

Adamson [110] illustrates some techniques for measuring surface shear viscosity. Further details on the principles, measurement and applications are given in references [127–130] for emulsions, and in reference [131] for foams. It should be noted that many experimental studies deal with the interfacial viscosities between bulk phases rather than on droplets or thin films themselves.

Surfactants and Surface Curvature

Surface tension causes a pressure difference to exist across a curved surface, with the greatest pressure being on the inside of a bubble. The pressure difference across an interface between one phase (A), having pressure p_A, and another phase (B), having pressure p_B, for spherical bubbles of radius R, is given by:

$$\Delta p = p_A - p_B = 2\gamma/R \qquad (7)$$

This is the Young–Laplace equation. It illustrates the facts that Δp varies with the radius, and that the pressure inside a bubble exceeds that outside. If the interface had a more complex geometry, then the two principal radii of curvature, R_1 and R_2, would be used,

$$\Delta p = p_A - p_B = \gamma(1/R_1 + 1/R_2) \qquad (8)$$

The Young–Laplace equation forms the basis for some important methods for measuring surface and interfacial tensions, such as the pendant and sessile drop methods, the spinning drop method, and the maximum bubble pressure method [101–103, 107]. Liquid flow in response to the pressure difference expressed by equations 7 or 8 is known as Laplace flow, or capillary flow.

Detergency and the Displacement of Oil. Detergency involves the action of surfactants to alter interfacial properties so as to promote removal of a phase from solid surfaces. Obviously, wetting agents are used, and usually those that rapidly diffuse and adsorb at appropriate interfaces are most effective. In this section we will consider a petroleum industry example.

When a drop of oil in water comes into contact with a solid surface the oil may form a bead on the surface or it may spread and form a film. A liquid having a strong affinity for the solid will seek to maximize its contact (interfacial area) and form a film. A liquid with much weaker affinity may form into a bead. This affinity is termed the wettability. Since there can be degrees of spreading another quantity is needed (see Figure 6, [132]). The contact angle, θ, in an oil–water–solid system is defined as the angle, measured through the aqueous phase, that is formed at the junction of the three phases. Whereas interfacial tension is defined for the boundary between two phases, the contact angle is defined for a three-phase junction.

If the interfacial forces acting along the perimeter of the drop are represented by the interfacial tensions, then an equilibrium force balance equation can be written as,

$$\gamma_{W/O} \cos \theta = \gamma_{S/O} - \gamma_{S/W} \qquad (9)$$

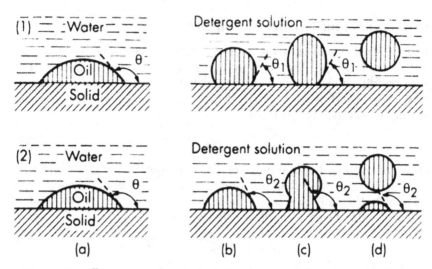

Figure 6.   Illustration of spreading, beading, and the contact angle in a solid/liquid/liquid system. (From Shaw [132]. Copyright 1992 Butterworth–Heinemann, Oxford, UK.)

where the subscripts refer to water, W, oil, O, and solid, S. This is Young's equation. The solid is completely water-wetted if $\theta = 0$ and only partially wetted otherwise. This equation is frequently used to describe wetting phenomena, so two practical points should be remembered. In theory, complete non-wetting by water would mean that $\theta = 180°$ but this is not seen in practice. Also, values of $\theta < 90°$ are often considered to represent "water-wetting" while values of $\theta > 90°$ are considered to represent "non-water-wetting". This is a rather arbitrary assignment based on correlation with the visual appearance of drops on surfaces.

In primary oil recovery from underground reservoirs, the capillary forces described by the Young and Young–Laplace equations are responsible for retaining much of the oil (residual oil) in parts of the pore structure in the rock or sand. It is these same forces that any secondary or enhanced (tertiary) oil recovery process strategies are intended to overcome [26, 29, 30, 133]. The relative oil and water saturations depend upon the distribution of pore sizes in the rock. The capillary pressure, P_c, in a pore is given by,

$$P_c = 2\gamma \cos \theta / R \tag{10}$$

where R is the pore radius, and at some height h above the free water table, P_c is fixed at $\Delta\rho h$ ($\Delta\rho$ is the density difference between the phases). Therefore, as the interfacial tension and contact angle are also fixed, and if the rock is essentially water-wetting (low θ), the smaller pores will tend to

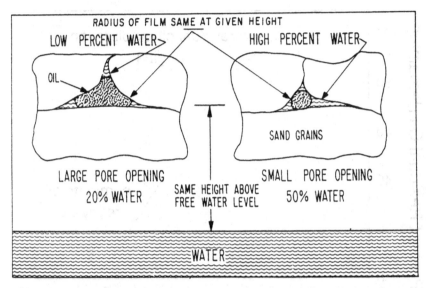

Figure 7. Trapping of oil in water-wet pores of different sizes due to capillary forces. (From Clark [134]. Copyright 1969 Society of Petroleum Engineers, Richardson, TX.)

have more water in them (less oil) than larger pores, as illustrated in Figure 7 [134].

Primary production from an oil reservoir, using only energy inherent in the reservoir, will only recover up to about 15% of the original oil-in-place (OOIP). In secondary oil recovery, flooding the reservoir with water (waterflooding) can produce an additional 15% or so of the oil originally in place. After waterflooding some 70% of the original oil-in-place still remains trapped in the reservoir rock pores. In a water-wet reservoir this residual oil is left in the form of oil ganglia trapped in the smaller pores where the viscous forces of the driving waterflood could not completely overcome the capillary forces holding the oil in place.

In tertiary, or enhanced, oil recovery one generally attempts to reduce the capillary forces restraining the oil and/or alter viscosity of the displacing fluid in order to modify the viscous forces being applied to drive oil out of the pores. The ratio of viscous forces to capillary forces actually correlates well with residual oil saturation and is termed the capillary number. One formulation of the capillary number is:

$$N_c = \eta v/(\gamma\phi) \tag{11}$$

where η and v are the viscosity and velocity of the displacing fluid, γ is the interfacial tension and ϕ is the porosity. A correlation is shown in

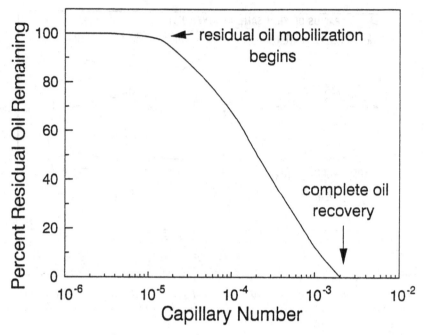

Figure 8.   Correlation between residual oil saturation reduction and the capillary number. (From Taylor and Hawkins [135]. Copyright 1990 Petroleum Recovery Institute, Calgary, AB.)

Figure 8; beginning after even the most efficient waterflooding, when N_c is about 10^{-6} and the residual oil saturation is still around 45% [135].

Suppose that one wished to design a tertiary recovery process so that additional oil would be recovered, reducing the oil saturation to around 25%. A residual oil saturation of 25% requires increasing the capillary number to about 5×10^{-3}. This could be done by raising the viscous forces, i.e., viscosity and velocity, but practical limitations on the size of pumps and the need to avoid inducing fractures in the reservoir prevent one from using these factors to achieve the needed orders of magnitude increase. But, by adding a suitable surfactant to the water one can readily decrease the interfacial tension from say 20 mN/m to 4×10^{-3} mN/m, increasing the capillary number to the desired 5×10^{-3}. Substitution of these interfacial tensions into the above capillary pressure equation shows that with the reduced interfacial tension oil will be recovered from smaller pores down to $R' = 0.0002R$. A more detailed treatment of this topic is given in Chapter 6 of this volume.

In some systems the addition of a fourth component to an oil/water/surfactant system can cause the interfacial tension to drop to near-zero values, ca. 10^{-3} to 10^{-4} mN/m, allowing spontaneous emulsification to very small drop sizes, ca. 10 nm or smaller. The droplets can be so small that they scatter little light and the emulsions appear to be transparent and are termed microemulsions. Unlike coarse emulsions, microemulsions may be thermodynamically stable. Microemulsions can be used in an enhanced oil recovery process. The much lower interfacial tensions produced increase the oil displacement from reservoir rock by increasing the capillary number. The micelles present also help to solubilize the oil droplets, hence this process is sometimes referred to as micellar flooding. The emulsions can be formulated to have moderately high viscosities which help to achieve a more uniform displacement front in the reservoir; this gives improved sweep efficiency, see Figure 9 [*136*]. Thus, there are a

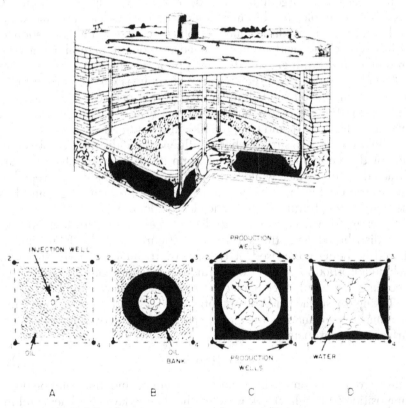

DISPLACEMENT OF OIL IN PETROLEUM RESERVOIRS
BY WATER OR CHEMICAL FLOODING (FIVE-SPOT PATTERN)

Figure 9.   Oil displacement, with good sweep efficiency, in a reservoir. (From Ling et al. [136]. Copyright 1987 Royal Society of Chemistry, Cambridge.)

number of factors that can be adjusted using a microemulsion system for enhanced oil recovery.

Surfactants and Surface Potential

Most substances acquire a surface electric charge when brought into contact with a polar medium such as water. For crude oil/aqueous systems the charge could be due to the ionization of surface acid functionalities. For gas/aqueous systems the charge could be due to the adsorption of surfactant ions. For porous rock or suspensions, the charge could originate from the diffusion of counter-ions away from the surface of a mineral whose internal crystal structure carries an opposite charge due to isomorphic substitution (in clays for example). In a practical petroleum process situation the nature and degree of surface charging is more complicated than in these examples, and surfactant adsorption may cause a surface electric charge to increase, decrease, or not significantly change at all. For example, the bitumen–aqueous interface can become negatively charged in alkaline aqueous solutions due to the ionization of surface carboxylic acid groups, the adsorption of natural surfactants present in the bitumen, and the adsorption of charged mineral solids [139–141]. The degree of such negative charging is very important to the success of in situ oil sands bitumen recovery processes, and surface oil sands separation processes, such as the hot water flotation process (see references [142, 143], and Chapter 10 of this volume).

The presence of a surface charge influences the distribution of nearby ions in the polar medium. Ions of opposite charge (counter-ions) are attracted to the surface while those of like charge (co-ions) are repelled. An electric double layer, which is diffuse because of mixing caused by thermal motion, is thus formed. The electric double layer (EDL) consists of the charged surface and a neutralizing excess of counter-ions over co-ions, distributed near the surface (see Figure 10). The EDL can be viewed as being comprised of two layers, (i) an inner layer that may include adsorbed ions, and (ii) a diffuse layer where ions are distributed according to the influence of electrical forces and thermal motion.

Taking the surface electric potential to be ψ°, and applying the Gouy–Chapman approximation, the electric potential ψ at a distance x from the surface is approximately

$$\psi = \psi^\circ \exp(-\kappa x) \tag{12}$$

Thus ψ depends on surface electric potential and the solution ionic composition (through κ). $1/\kappa$ is called the double layer thickness and for water at 25 °C is given by:

$$\kappa = 3.288\sqrt{I}\,(\mathrm{nm}^{-1}) \tag{13}$$

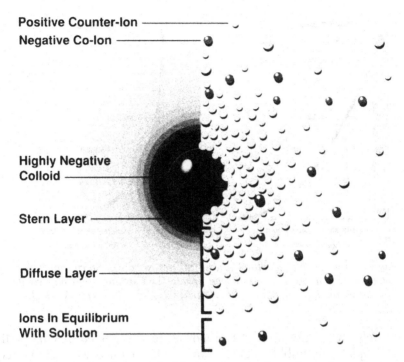

Figure 10. The electric double layer around a charged species in aqueous solution. The left view shows the change in charge density around the charged species. The right view shows the distribution of ions. (Courtesy L.A. Ravina, Zeta-Meter Inc., Staunton, VA.)

where I is the ionic strength, given by $I = (1/2) \Sigma_i c_i z_i^2$, where c_i is concentration of ions and z_i is charge number of ions. For 1:1 electrolyte, $1/\kappa = 1$ nm for $I = 10^{-1}$ M and 10 nm for $I = 10^{-3}$ M.

Also, an inner layer exists because ions are not really point charges and an ion can only approach a surface to the extent allowed by its hydration sphere. The Stern model specifically incorporates a layer of specifically adsorbed ions bounded by a plane known as the Stern plane (see Figures 10 and 11). In this case the potential changes from ψ° at the surface, to $\psi(\delta)$ at the Stern plane, to $\psi = 0$ in bulk solution.

An indication of the surface potential can be obtained through electrokinetic measurements. Electrokinetic motion occurs when the mobile part of the electric double layer is sheared away from the inner layer (charged surface). Of the four types of electrokinetic measurements, electrophoresis, electro-osmosis, streaming potential, and sedimentation potential, the first finds the most use in industrial practise. In electrophoresis, an electric field is applied to a sample causing charged dispersed species, and any attached material or liquid, to move towards the

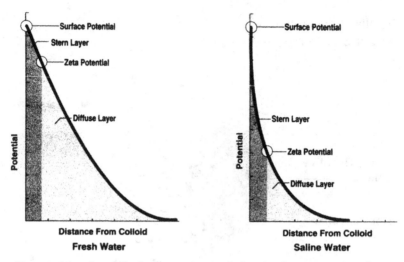

Figure 11. Simplified illustrations of the surface, Stern, and Zeta potentials for a dispersed, charged species in low and high electrolyte concentration aqueous solutions. (Courtesy L.A. Ravina, Zeta-Meter Inc., Staunton, VA.)

oppositely charged electrode. The moving species, which may be parti-cles, droplets, or bubbles, are viewed under a microscope and their electrophoretic velocity is measured at carefully selected planes in the sample cell. The results can be interpreted in terms of the potential (ψ) at the plane of shear, known as the Zeta potential. Since the exact location of the shear plane is generally not known, the Zeta potential is usually taken to be approximately equal to the potential at the Stern plane.

Good descriptions of practical experimental techniques in electro-phoresis and their limitations can be found in references ([144–146]). For the most part, electrophoresis techniques are applied to suspensions and emulsions, but with appropriate cell designs they can sometimes be applied to dispersions of bubbles (e.g., [147–149]). Other electrokinetic techniques, such as the measurement of sedimentation potential, have been used as well. Streaming potential measurements give an indication of the average surface potential in a porous rock, but are strongly influenced by the pore size distribution and may not be sensitive to the contributions of individual mineral constituents [150]. Electrophoresis measurements require crushing of the rock, but have the advantage of being convenient for establishing equilibrium with different solutions and provide information about individual rock components (e.g., clay versus silica components of sandstones) [151]. Figure 12 shows the relative effects of adsorbing anionic, amphoteric or cationic surfactants on different kinds of rock particles (sandstone, limestone and dolomite) dispersed in aqueous solution, as described in reference [152].

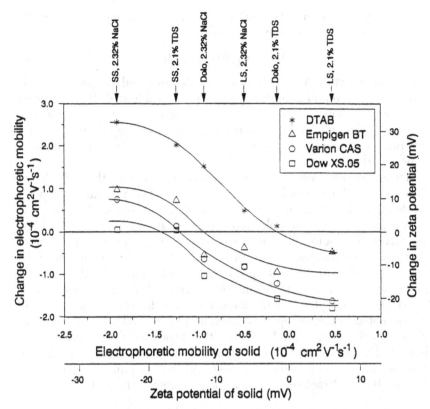

Figure 12.   Relative effects on particle electrophoretic mobility, and Zeta potential, of adsorbing anionic, amphoteric or cationic surfactants. The different kinds of rock particles are sandstone (SS), limestone (LS), and dolomite (Dolo), all dispersed in aqueous solution. (From Mannhardt et al. [152]. Copyright 1992 Elsevier Science Publishers, Amsterdam.)

Surface Potential and Dispersion Stabilization.   Most emulsions, suspensions, and foams are not thermodynamically stable, but may well possess some degree of kinetic stability. Encounters between dispersed species can occur due to Brownian motion, sedimentation, and/or stirring. The stability of the dispersion depends upon how the particles interact when this happens. More details are given in reference [153]. Surfactants are frequently involved in the stabilization of colloidal dispersions of droplets, particles or bubbles by increasing the electrostatic repulsive forces.

In the simplest example of colloid stability the dispersed species would be stabilized entirely by the repulsive forces created when two charged surfaces approach each other and their electric double layers overlap. The repulsive energy, V_R, has the general form,

$$V_R = V_1(\psi) \exp[-\kappa H] \qquad (14)$$

where the function $V_1(\psi)$ depends on the material properties of the particular system and includes the Stern plane potential; H is either the separation distance (emulsions and suspensions) or the film thickness (foams). There is another repulsive force, Born repulsion, which is experienced at very small separation distances where the atomic electron clouds overlap.

The dispersed species will be attracted to each other through electric dipole interactions, which may be due to: (1) two permanent dipoles, (2) dipole–induced dipole, or (3) induced dipole–induced dipole. The latter forces between nonpolar molecules are also called London dispersion forces. Except for quite polar materials, the London dispersion forces are the more significant of the three. Whereas for molecules the force varies inversely with the sixth power of the intermolecular distance, the nature of the variation with distance is somewhat different for dispersions. For dispersed droplets (or particles, etc.) the dispersion forces can be approximated by adding up the attractions between all inter-droplet pairs of molecules. When added this way the dispersion force between two droplets decays less rapidly as a function of separation distance than is the case for individual molecules. For two spheres of radius a, separated by distance H, the attractive energy, V_A, can be approximated by,

$$V_A = -a\left(\sqrt{A_2} - \sqrt{A_1}\right)^2 /12H \qquad (15)$$

when $H \ll a$. Here, A_1 and A_2 are the Hamaker constants of the medium and spheres respectively; they depend on the densities and polarizabilities of the constituent atoms, and are typically about 10^{-20} to 10^{-19} J.

The energy changes that take place when two dispersed species approach each other can be estimated by summing the energies of attraction and repulsion over a range of separation distances, $V = V_A + V_R$. This is known as the DLVO theory, after its originators Derjaguin and Landau (see citations in reference [154], and Verwey and Overbeek [155]).

V_R decreases exponentially with increasing separation distance, and has a range about equal to κ^{-1}, while V_A decreases inversely with increasing separation distance. Figure 13 shows simple attractive and repulsive energy curves, and the total interaction energy curve that results. The shaded areas show that either the attractive van der Waals forces or the repulsive electric double layer forces can predominate at different inter-droplet distances.

The DLVO theory was developed in an attempt to account for the

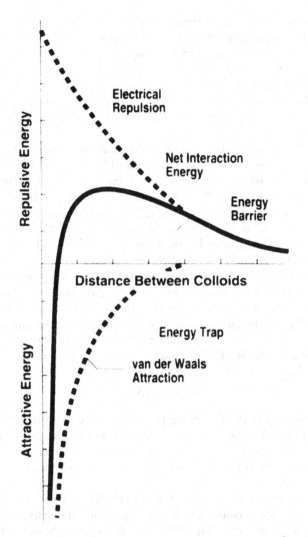

Figure 13.   Example of an attractive energy curve, repulsive energy curve and the total interaction energy curve that results. (Courtesy L.A. Ravina, Zeta-Meter Inc., Staunton, VA.)

observation that colloids coagulate quickly at high electrolyte concentrations, and slowly at low concentrations. The transition from one rate to the other occurs over a very narrow electrolyte concentration range, the midpoint of which is termed the critical coagulation concentration (CCC). Where there is a positive potential energy maximum, a dispersion should be stable if $V \gg kT$, that is, if the energy is large compared to the thermal energy of the particles ($15kT$ is considered unsurmountable). In this case, colliding droplets should rebound without contact, and the emulsion should be stable to aggregation. If, on the other hand, the potential energy maximum is not very great, $V \approx kT$, then slow aggregation should occur. The height of the energy barrier depends on the surface potential, $\psi(\delta)$, and on the range of the repulsive forces, κ^{-1}. The figure shows that an energy minimum can occur at larger interparticle distances. If this is reasonably deep compared to kT, then a loose, easily reversible aggregation should occur.

Examples of interaction energy plots can be found elsewhere for suspensions [32], emulsions [30], and foams [31]. In the case of foams, it is sometimes more helpful to consider the same phenomena in terms of the disjoining pressure. When the two interfaces bounding a foam lamella are electrically charged, the interacting diffuse double layers exert a hydrostatic pressure which acts to keep the interfaces apart. In thin lamellae (film thicknesses on the order of a few hundred nm) the electrostatic, dispersion and steric forces all may be significant and the disjoining pressure concept is frequently employed. The disjoining pressure represents the net pressure difference between the gas phase (bubbles) and the bulk liquid from which the lamellae extend [154], and is the total of electrical, dispersion and steric forces (per unit area) operating across the lamellae (perpendicular to the interfaces). The disjoining pressure, π, may be expressed by taking the derivative of the interaction potential with respect to the film thickness,

$$\pi(t) = -dV/dt \tag{16}$$

A description of how the disjoining pressure can be determined is given in reference [154]. To the extent that the disjoining pressure arises from electrostatic forces, there will be an obvious influence of electrolyte concentration. For very thin films (<100 nm) the disjoining pressure is very important.

It will be apparent that the DLVO calculations can become quite involved, requiring considerable knowledge about the systems of interest. Also, there are some problems. For example, on one hand there will be some distortion of the spherical emulsion droplets as they approach each other and begin to seriously interact, causing a flattening. Also, our view of the validity of the theory is changing as more becomes known about the influence of additional forces such as those due to surface hydration. The DLVO theory nevertheless forms a very useful starting point in attempting to understand complex colloidal systems such as petroleum emulsions. There are empirical "rules of thumb" that can be used to give a first estimate of the degree of colloidal stability that a system is likely to have if the Zeta potentials of the droplets are known.

Many types of colloids tend to adopt a negative surface charge when dispersed in aqueous solutions having ionic concentrations and pH typical of natural waters. For such systems, one rule of thumb stems from observations that the colloidal particles are quite stable when the Zeta potential is about -30 mV or more negative, and quite unstable due to agglomeration when the Zeta potential is between +5 and -5 mV. An expanded set of guidelines, developed for particle suspensions, is given in reference [156]. Such criteria are frequently used to determine optimal dosages of polyvalent metal electrolytes, such as alum, used to effect coagulation in treatment plants.

The transition from stable dispersion to aggregation usually occurs over a fairly small range of electrolyte concentration. This makes it possible to determine aggregation concentrations, often referred to as critical coagulation concentrations (CCC). The Schulze–Hardy rule summarizes the general tendency of the CCC to vary inversely with the sixth power of the counter-ion charge number (for indifferent electrolyte). This relationship follows directly from the DLVO theory when one derives the conditions under which $V = 0$ and $dV/dH = 0$ for high surface potentials. As an illustration, suppose that for a hypothetical emulsion the above equation predicts a CCC of 1.18 M in solutions of sodium chloride. The critical coagulation concentrations in polyvalent metal chlorides would then decrease as follows:

Dissolved Salt	z	CCC (M)
NaCl	1	1.18
$CaCl_2$	2	0.018
$AlCl_3$	3	0.0016

The Schulze–Hardy rule can be applied in the selection of appropriate treatments of injected fluids so as to protect an oil-bearing reservoir from the damage that could otherwise occur due to the release and migration of small, highly charged particles (fines migration) [157]. The particle charges could be due to the nature of the minerals and their interaction with the aqueous environment and/or to the adsorption of ionic surfactants or polymers. The protective action results from compression of the electric double layers and reduction of the Zeta potentials to zero, or near-zero, values.

Surfactant Adsorption in Porous Media

In petroleum recovery [31] and environmental soil remediation processes [158, 159], surfactant adsorption from solution onto solid surfaces most commonly occurs in porous media, either on the walls of pores or throats or else on fine particles in rock pores. This adsorption constitutes a loss of valuable surfactant so it directly affects, and may well dictate, the economics of an oil recovery or remediation process. The adsorption is also of considerable scientific interest because the surfactant can adsorb as individual molecules or as surfactant aggregates of various kinds.

Surfactant adsorption may occur due to electrostatic interaction, van der Waals interaction, hydrogen bonding, and/or solvation and desolvation of adsorbate and adsorbent species. Consider a typical adsorption isotherm for a polar surfactant adsorbing on an oppositely charged surface due to a combination of electrostatic and van der Waals forces. At low surfactant concentration, individual molecules adsorb, with more and more molecules adsorbing as surfactant concentration increases. At some concentration, surfactant aggregation may occur on the surface as hemi-micelles [160–162] in which all the surfactant head groups are towards the surface, and/or admicelles [163–165] in which some of the surfactants are in an opposite orientation to the surface. At higher surfactant concentration, the surface will become covered with a monolayer of surfactant. At still higher surfactant concentration bilayer formation may occur. For more details, see references [18, 166–168].

Figure 14 shows the effect on surface electrokinetic charge of adsorbing increasing amounts of a commercial anionic surfactant onto

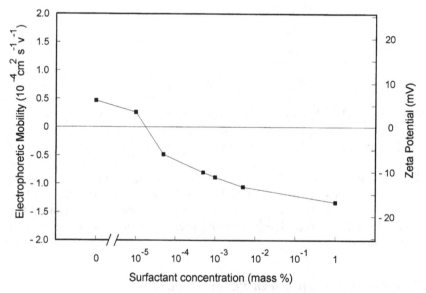

Figure 14.   Illustration of the effect on surface electrokinetic charge of adsorbing increasing amounts of Dow XS84321.05, an anionic surfactant, onto the surfaces of Indiana limestone particles in a high salinity, 2.1% total dissolved solids, brine solution. Plotted from data reported in reference [172].

the surfaces of Indiana limestone particles in a high salinity, 2.1% total dissolved solids, brine solution. In this case, increasing surfactant adsorption first reduces the positive Zeta potential, then causes it to reach zero (the isoelectric point), then makes it increasingly negative.

The extent of adsorption of commercial surfactants developed for use in reservoir recovery processes can vary from near zero to as high as 2.5 mg/g. Surfactant adsorption on rock surfaces is usually measured by either static (batch) or dynamic (coreflood) experiments. The static adsorption method, employing crushed rock samples, is essentially the classical method for determining adsorption isotherms at the aqueous solution/solid interface and involves batch equilibrations of particles in solutions of different initial surfactant concentration. The dynamic coreflood method is more involved but employs a greater solid to liquid ratio and is therefore more sensitive, see references [169–171]. Temperature, brine salinity and hardness, solution pH, rock type, wettability, and the presence of a residual oil phase have all been found to influence the extent of adsorption of different surfactants [116, 152, 172].

Wettability Alteration. Another consequence of surfactant adsorption onto solid surfaces in porous media is that it may alter the wetting properties of the surfaces, which can be an advantage in oil recovery schemes applied to reservoirs of mixed wettability, or where the rock is predominantly oil-wetting. In this case, surfactants may be injected into a reservoir with the intention of having them adsorb onto the rock surfaces in such an orientation that the contact angle is decreased, making the reservoir more water-wetting. A number of studies have shown wettability shifts from oil-wetting towards water-wetting due to surfactant adsorption [173, 174]. This topic is discussed in detail in Chapter 5 of this volume. Another example of surfactant-induced wettability alteration can be found in the treatment of swelling clays, such as montmorillonite, with a cationic surfactant, such as dimethyl di(hydrogenated tallow) ammonium, in order to produce organophilic clay for use in nonaqueous drilling muds [175, 176].

Surfactant Adsorption at the Liquid/Liquid Interface

It was pointed out earlier that surfactant adsorption at liquid interfaces can influence emulsion stability by lowering interfacial tension, increasing surface elasticity, increasing electric double layer repulsion (ionic surfactants), lowering the effective Hamaker constant, and possibly increasing surface viscosity. Surfactant can determine the arrangement of the phases in an emulsion, that is, which phase will form the dispersed versus continuous phase. We will briefly summarize several rules of thumb. A very qualitative rule of emulsion type, Bancroft's rule, states that if a

Table 6. Approximate Surfactant HLB Values[a]

Surfactant	HLB
Oleic acid	1
Sorbitan tristearate (SPAN 65)	2
Sorbitan monooleate (SPAN 80)	4
Diethylene glycol monolaurate	6
Sorbitan monolaurate (SPAN 20)	9
Glycerol monostearate	11
Polyoxyethylene (10) cetyl ether (BRIJ 56)	13
Polyoxyethylene sorbitan monooleate (TWEEN 80)	15
Sodium octadecanoate	18
Sodium dodecanoate	21
Sodium octanoate	23
Dioctyl sodium sulfosuccinate	32
Sodium heptadecyl sulfate	38
Sodium dodecyl sulfate	40
Sodium octyl sulfate	42

[a] Compiled from data in references [2, 23, 181, 184]
SPAN, BRIJ, and TWEEN are trademarks of ICI Americas Inc.

surfactant is most soluble in one of the phases, then more of the agent can be accommodated at the interface if that interface is convex towards that phase, i.e., if that phase is the continuous phase. Very often, mixtures of emulsifying agents are more effective than single components. It is thought that some mixed emulsifiers form a complex at the interface, thus yielding low interfacial tension and a strong interfacial film. A second qualitative rule, the oriented wedge theory, specifies that soaps of monovalent metal cations tend to produce O/W emulsions, while those of polyvalent metal cations will tend to produce W/O emulsions. Because the polyvalent metal cations each coordinate to two surfactant molecules, which are aligned with their polar groups near the metal ion, the hydrocarbon tails adopt a wedge-like orientation. The hydrocarbon tails in a close-packed interfacial layer are most easily accommodated if the oil phase is the continuous phase. It is emphasized that there are exceptions to each of these rules; their utility lies in assisting with initial predictions.

An empirical scale developed for categorizing single-component or mixed (usually nonionic) surfactants is the hydrophile–lipophile balance or HLB scale. This dimensionless scale ranges from 0 to 20 for nonionic surfactants; a low HLB (<9) refers to a lipophilic surfactant (oil soluble) and a high HLB (>11) to a hydrophilic (water soluble) surfactant. Most ionic surfactants have HLB values greater than 20. Some examples of surfactant HLB's are given in Table 6. In general, W/O emulsifiers exhibit

HLB values in the range 3–8 while O/W emulsifiers have HLB values of about 8–18. There exist empirical tables of HLB values required to make emulsions out of various materials [177, 178]. If the value is not known, then lab emulsification tests are required, using a series of emulsifying agents of known HLB values [178]. There are various compilations and equations for determining emulsifier HLB values [177–181].

A limitation of the HLB system is that other factors, such as temperature, are very important as well. Also, the HLB is an indicator of the emulsifying characteristics of an emulsifier but not the efficiency of an emulsifier. Thus, while all emulsifiers having a high HLB will tend to promote O/W emulsions, there will be a considerable variation in the efficiency with which those emulsifiers act for any given system. For example, usually mixtures of surfactants work better than pure compounds of the same HLB.

Just as solubilities of emulsifying agents vary with temperature, so does the HLB, especially for the nonionic surfactants. A surfactant may thus stabilize O/W emulsions at low temperature, but W/O emulsions at some higher temperature. The transition temperature, at which the surfactant changes from stabilizing O/W to W/O emulsions, is known as the phase inversion temperature, PIT. At the PIT, the hydrophilic and oleophilic natures of the surfactant are essentially the same (another term for this is the HLB temperature). As a practical matter, emulsifying agents are chosen so that their PIT is far from the expected storage and use temperatures of the desired emulsions. In one method [182] an emulsifier with a PIT of about 50 °C higher than the storage/use temperature is selected. The emulsion is then prepared at the PIT where very small droplet sizes are most easily created. Next, the emulsion is rapidly cooled to the desired use temperature, where now the coalescence rate will be slow, and a stable emulsion results. Further details can be found in reference [183].

Summary

From the preceding sections it can be seen that surfactants can be extremely important in many facets of the petroleum industry. From a knowledge of some of the basic properties of a surfactant (i.e., the cmc, Krafft point, cloud point, adsorption level and surface or interfacial tension at the cmc), some predictions may be made as to the performance of the surfactant in a given potential oil recovery process. However, it is also clear that there are severe limitations on the extent to which equilibrium micellar properties can be used to predict what are frequently, in practical application, dynamic phenomena. Subsequent chapters in this volume will explore a range of such application areas.

Acknowledgments

The authors thank Susan M. Kutay for valuable advice and resource materials. We gratefully acknowledge the financial and other support provided by the Natural Sciences and Engineering Research Council of Canada and the Petroleum Recovery Institute.

List of Symbols

a	Empirical constant, also used as radius
a_S	Area per adsorbed surfactant molecule
A	Surface area
A_1, A_2	Hamaker constants
b	Empirical constant
c_i	Concentration of ions i
cmc	Critical micelle concentration
C_s	Surfactant concentration in solution
CCC	Critical coagulation concentration
DLVO	Derjaguin, Landau, Verwey, and Overbeek
E_G	Gibbs surface elasticity
E_M	Marangoni surface elasticity
ΔG_t°	Gibbs free energy change for the transfer of a hydrocarbon solute from a hydrocarbon solvent to water
h	Height
H	Separation distance
ΔH_t°	Enthalpy change for the transfer of a hydrocarbon solute from a hydrocarbon solvent to water
HLB	Hydrophile–lipophile balance
I	Solution ionic strength
k	Boltzmann constant
K	Surface film compressibility
n_c	Number of carbon atoms in a hydrocarbon chain
N_A	Avogadro's number
N_c	Capillary number
p_A, p_B	Pressures on each side of an interface
Δp	Pressure difference across an interface
P_c	Capillary pressure
PIT	Phase inversion temperature
R	Radius of a curved surface or interface, also used as the gas constant

R_1, R_2	Principal radii of curvature of a surface or interface
ΔS_t°	Entropy change for the transfer of a hydrocarbon solute from a hydrocarbon solvent to water
t	Fluid film thickness
T	Absolute temperature
T_k	Krafft point
v	Darcy velocity
V	Total potential energy
V_1, V_2	Constants in potential energy equations
V_A	Attractive potential energy
V_R	Repulsive potential energy
x	Distance from a surface or interface
z_i	Charge number of ions i

Greek

Γ_s	Surface excess concentration of surfactant
γ	Surface or interfacial tension
δ	Stern layer thickness
ζ	Zeta potential
η	Viscosity
θ	Contact angle
κ	Debye length (inverse of the double layer thickness)
μ_{HC}°	Chemical potential of a hydrocarbon dissolved in a hydrocarbon solvent
μ_{aq}°	Chemical potential of a hydrocarbon dissolved in water
μ_E	Electrophoretic mobility
π	Disjoining pressure
$\Delta\rho$	Density difference between phases
ϕ	Porosity
ψ	Electric potential
ψ°	Surface electric potential

References

1. Rosen, M.J. *Surfactants and Interfacial Phenomena*, 2nd ed.; Wiley: New York, New York, 1989.
2. Myers, D. *Surfactant Science and Technology*; VCH: New York, 1988.
3. Mittal, K.L., Ed., *Solution Chemistry of Surfactants*; Plenum: New York, 1979, Vols. 1, 2.
4. Mittal, K.L.; Fendler, E.J., Eds., *Solution Behaviour of Surfactants*; Plenum: New York, 1982, Vols. 1, 2.
5. Tadros, Th. F., Ed., *Surfactants*; Academic Press: London, 1984.
6. Mittal, K.L.; Lindman, B., Eds., *Surfactants in Solution*; Plenum: New York, 1984, Vols. 1–3.

7. Mittal, K.L.; Bothorel, P., Eds., *Surfactants in Solution*; Plenum: New York, 1987, Vols. 4–6.
8. Karsa, D.R., Ed., *Industrial Applications of Surfactants*; Royal Soc. Chemistry: London, 1987.
9. Rosen, M.J., Ed., *Surfactants in Emerging Technologies*; Dekker: New York, 1987.
10. Rosen, M.J., Ed., *Structure/Performance Relationships in Surfactants*; American Chemical Society: Washington, 1984.
11. Shinoda, K.; Nakagawa, T.; Tamamushi, B-I.; Isemura, T. *Colloidal Surfactants, Some Physicochemical Properties*; Academic Press: New York, 1963.
12. Friberg, S.E.; Lindman, B., Eds., *Organized Solutions, Surfactants in Science and Technology*; Dekker: New York, 1992.
13. Wasan, D.T.; Ginn, M.E.; Shah, D.O., Eds., *Surfactants in Chemical/ Process Engineering*; Dekker: New York, 1988.
14. Jungermann, E., Ed., *Cationic Surfactants*; Dekker: New York, 1970.
15. DiStasio, J.I., Ed., *Surfactants, Detergents and Sequestrants*; Noyes Data Corp.: Park Ridge, NJ, 1981.
16. Zana, R., Ed., *Surfactant Solutions*; Dekker: New York, 1986.
17. Schick, M.J., Ed., *Nonionic Surfactants: Physical Chemistry*; Dekker: New York, 1987.
18. Sharma, R., Ed., *Surfactant Adsorption and Surface Solubilization*; American Chemical Society: Washington, 1995.
19. Porter, M.R. *Handbook of Surfactants*; Blackie: Glasgow, 1991.
20. Osipow, L.I. *Surface Chemistry Theory and Industrial Applications*; Reinhold: New York, 1962.
21. Ross, S.; Morrison, I.D. *Colloidal Systems and Interfaces*; Wiley: New York, 1988.
22. Hiemenz, P.C.; Rajagopalan, R. *Principles of Colloid and Surface Chemistry*, 3rd ed.; Dekker: New York, 1997.
23. Myers, D. *Surfaces, Interfaces, and Colloids*; VCH: New York, 1991.
24. Smith, D.H., Ed., *Surfactant Based Mobility Control*; American Chemical Society: Washington, 1988.
25. Cahn, A.; Lynn, J.L. In *Kirk-Othmer Encyclopedia of Chemical Technology*, 3rd ed.; Wiley: New York, 1983; Vol. 22, pp 332–432.
26. Morrow, N.R., Ed., *Interfacial Phenomena in Petroleum Recovery*; Dekker: New York, 1991.
27. Clunie, J.S.; Goodman, J.F.; Ingram, B.T. In *Surface and Colloid Science*; Matijevic, E., Ed.; Wiley: New York, 1971; Vol. 3, pp 167–239.
28. Kitchener, J.A. In *Recent Progress in Surface Science*; Danielli, J.F.; Pankhurst, K.G.A.; Riddiford, A.C., Eds.; Academic Press: New York, 1964; Vol. 1, pp 51–93.
29. Shah, D.O.; Schechter, R.S., Eds., *Improved Oil Recovery by Surfactant and Polymer Flooding*; Academic Press: New York, 1977.
30. Schramm, L.L., Ed., *Emulsions, Fundamentals and Applications in the Petroleum Industry*; American Chemical Society: Washington, DC, 1992.
31. Schramm, L.L., Ed., *Foams, Fundamentals and Applications in the Petroleum Industry*; American Chemical Society: Washington, DC, 1994.

32. Schramm, L.L., Ed., *Suspensions, Fundamentals and Applications in the Petroleum Industry*; American Chemical Society: Washington, DC, 1996.
33. Stevens, C.E. In *Kirk-Othmer Encyclopedia of Chemical Technology*, 2nd ed.; Wiley: New York, Vol. 19, 1969, pp 507–593.
34. Lucassen-Reynders, E.H., Ed., *Anionic Surfactants Physical Chemistry of Surfactant Action*; Dekker: New York, 1981.
35. Rieger, M.M. In *Surfactants in Cosmetics*, 2nd ed.; Rieger, M.M.; Rhein, L.D., Eds.; Dekker: New York, 1997, pp 1–28.
36. Israelachvili, J.N; Mitchell, D.J.; Ninham, B.W. *J. Chem. Soc., Faraday Trans. II* **1976**, *72*, 1525.
37. Israelachvili, J.N.; Wennerström, H. *Langmuir* **1990**, *6*, 873.
38. Franks, F. In *Water – A Comprehensive Treatise*; Franks, F., Ed.; Plenum: New York, 1975; Vol. 4, pp 1–93.
39. Lindman, B.; Wennerstrom, H. *Top. Current Chem.* **1980**, *87*, 1.
40. Ben-Naim, A.Y. *Hydrophobic Interactions*; Plenum: New York, 1980.
41. Hartley, G.S. *Aqueous Solutions of Paraffin Chain Salts*; Hermann and Cie: Paris, 1936.
42. Tanford, C. *The Hydrophobic Effect: The Formation of Micelles and Biological Membranes*; 2nd ed.; Wiley: New York, 1980.
43. Harwell, J.H.; Scamehorn, J.F., Eds., *Surfactant Based Separation Processes*; Dekker: New York, 1989.
44. Fendler, J.H.; Fendler, E.H. *Catalysis in Micellar and Macromolecular Systems*; Academic Press: New York, 1975.
45. Mukerjee, P.; Banerjee, K. *J. Phys. Chem.* **1964**, *69*, 45.
46. Stigter, D.; Mysels, K. J. *J. Phys. Chem.* **1955**, *59*, 45.
47. Mittal, K.L., Ed., *Micellization, Solubilization, and Microemulsions*; Plenum: New York, 1977.
48. Preston, W.C. *J. Phys. Colloid Chem.* **1948**, *52*, 84.
49. Mukerjee, P.; Mysels, K.J. *Critical Micelle Concentrations of Aqueous Surfactant Systems*; National Bureau of Standards, NSRDS-NBS 36; U.S. Government Printing Office: Washington, 1971.
50. Evans, D.F.; Wightman, P.J. *J. Colloid Interface Sci.* **1982**, *86*, 515–524.
51. Noll, L.A. *Proceedings, SPE International Symposium on Oilfield Chemistry*, Society of Petroleum Engineers: Richardson, TX, 1991, SPE paper 21032.
52. Shinoda, K.; Kobayashi, M.; Yamaguchi, N. *J. Phys. Chem.* **1987**, *91*, 5292–5294.
53. Archer, D.G.; Albert, H.J.; White, D.E.; Wood, R.H. *J. Colloid Interface Sci.* **1984**, *100*, 68–81.
54. Schramm, L.L.; Fisher, D.B.; Schürch, S.; Cameron, A. *Colloids Surfaces* **1995**, *94*, 145–159.
55. Stasiuk, E.N.; Schramm, L.L. *J. Colloid Interface Sci.* **1996**, *178*, 324–333.
56. La Mesa, C.; Sesta, B.; Bonicelli, M.G.; Ceccaroni, G.F. *Langmuir* **1990**, *6*, 728–731.
57. Sugihara, G.; Mukerjee, P. *J. Phys. Chem.* **1981**, *85*, 1612–1616.
58. Brun,T.S.; Hoiland, H.; Vikingstad, E. *J. Colloid Interface Sci.* **1978**, *63*, 89–96.

59. Kaneshina, S.; Tanaka, M.; Tomida, T.; Matuura, R. *J. Colloid Interface Sci.* **1974**, *48*, 450–460.
60. Hamann, S.D. *J. Phys. Chem.* **1962**, *66*, 1359–1361.
61. Schramm, L.L.; Wassmuth, F. In *Foams, Fundamentals and Applications in the Petroleum Industry*, Schramm, L.L., Ed.; American Chemical Society: Washington, DC, 1994, pp 3–45.
62. Rosen, M.J.; Solash, J. *J. Am Oil Chem. Soc.* **1969**, *46*, 399–402.
63. Goette, E.K. *J. Colloid Interface Sci.* **1949**, *4*, 459–484.
64. Kashiwagi, M. *Bull. Chem. Soc. Jpn.* **1957**, *30*, 193–194.
65. Morrison, C.; Schramm, L.L.; Stasiuk, E.N. *J. Petrol. Sci. Eng.* **1996**, *15*, 91–100.
66. Beck, E.J.; Caplan, J.F.; Comeau, E.K.; Howley, C.V.; Marangoni, D.G. *Can. J. Chem.* **1995**, *73*, 1741–1745.
67. Treiner, C.; Nguyen, D. *J. Phys. Chem.* **1990**, *94*, 2021.
68. Goddard, E.D.; Benson, G.C. *Can. J. Chem.* **1957**, *35*, 1936.
69. Van Os, N.M.; Daane, G.J.; Bolsman, T.A.B.M. *J. Colloid Interface Sci.* **1987**, *115*, 402.
70. Van Os, N.M.; Daane, G.J.; Bolsman, T.A.B.M. *J. Colloid Interface Sci.* **1988**, *123*, 267.
71. Brun, T.S.; Hoiland, H.; Vikingstad, E. *J. Colloid Interface Sci.* **1978**, *63*, 89.
72. Corti, M.; Degiorgio, V. *J. Phys. Chem.* **1981**, *85*, 711.
73. Hayashi, S.; Ikeda, S. *J. Phys. Chem.* **1980**, *84*, 744.
74. Emerson, M.F.; Holtzer, A. *J. Phys. Chem.* **1967**, *71*, 1898.
75. Stigter, D.; Mysels, K.J. *J. Phys. Chem.* **1955**, *59*, 45.
76. Matijevic, E.; Pethica, B.V. *Trans. Faraday Soc.* **1958**, *54*, 587.
77. Singh, H.N.; Swarup, S. *Bull. Chem. Soc. Jpn.* **1978**, *51*, 1534.
78. Abu-Hamidiyyah, M.; Kumari, K. *J. Phys. Chem.* **1990**, *94*, 6445.
79. Bostrom, S.; Backlund, S.; Blokhus, A.M.; Hoiland, H. *J. Colloid Interface Sci.* **1989**, *128*, 169.
80. Meyers, D. In *Surfactants in Cosmetics*, 2nd ed.; Rieger, M.M.; Rhein, L.D., Eds.; Dekker: New York, 1997, pp 29–81.
81. Cross, J., Ed., *Anionic Surfactants – Chemical Analysis*; Dekker: New York, 1977.
82. Cross, J., Ed., *Nonionic Surfactants – Chemical Analysis*; Dekker: New York, 1986.
83. Porter, M.R., Ed., *Recent Developments in the Analysis of Surfactants*; Elsevier: Essex, 1991.
84. Schmitt, T.M. *Analysis of Surfactants*; Dekker: New York, 1992.
85. Cullum, D.C., Ed., *Introduction to Surfactant Analysis*; Blackie: U.K., 1994.
86. Rosen, M.J.; Goldsmith, H.A. *Systematic Analysis of Surface-Active Agents*, 2nd ed., Wiley: New York, 1972.
87. Eldridge, J.M. In *Surfactants in Cosmetics*, 2nd ed.; Rieger, M.M.; Rhein, L.D., Eds.; Dekker: New York, 1997, pp 83–104.
88. Wang, L.K.; Kao, S.F.; Wang, M.H.; Kao, J.F. *Ind. Eng. Chem. Prod. Res. Dev.* **1978**, *17*, 186.

89. Schwartz, A.M.; Perry, J.W.; Berch, J. *Surface-Active Agents and Detergents*, Vol. 2, Kreiger: New York, 1977.
90. Llenado, R.A.; Jamieson, R.A. *Anal. Chem.* **1981**, *53*, 174R.
91. Epton, S.R. *Trans. Faraday Soc.* **1948**, *44*, 226.
92. Epton, S.R. *Nature* **1947**, *160*, 795.
93. Glazer, J.; Smith, T.D. *Nature* **1952**, *169*, 497.
94. Reid, V.W.; Longman, G.F.; Heinerth, E. *Tenside* **1967**, *4*, 292–304.
95. Schramm, L.L.; Smith, R.G.; Stone, J.A. *AOSTRA J. Research* **1984**, 1, 5–14.
96. Birch, B.J.; Clarke, D.E. *Anal. Chim. Acta* **1973**, *67*, 387–393.
97. Vytras, K. *Mikrochimica Acta [Wien]* **1984**, *111*, 139–148.
98. Oei, H.H.Y.; Toro, D.C. *J. Soc. Cosmet. Chem.* **1991**, *42*, 309–316.
99. Gronsveld, J.; Faber, M.J. *Tenside Surf. Det.* **1990**, *27*, 231–232.
100. Schramm, L.L. In *Emulsions, Fundamentals and Applications in the Petroleum Industry*, Schramm, L.L., Ed.; American Chemical Society: Washington, DC, 1992; pp 1–49.
101. Harkins, W.D.; Alexander, A.E. In *Physical Methods of Organic Chemistry*; Weissberger, A., Ed.; Interscience: New York, 1959; pp 757–814.
102. Padday, J.F. In *Surface and Colloid Science*; Matijevic, E., Ed.; Wiley-Interscience: New York, 1969; Vol. 1, pp 101–149.
103. Miller, C.A.; Neogi, P. *Interfacial Phenomena Equilibrium and Dynamic Effects*; Dekker: New York, 1985.
104. Rusanov, A.I.; Prokhorov, V.A. *Interfacial Tensiometry*; Elsevier: Amsterdam, 1996.
105. McCaffery, F.G. *J. Can. Petrol. Technol.* **1972**, *11*, 26.
106. DePhilippis, F.; Budziak, C.; Cheng, P.; Neumann, A.W.; Potoczny, Z.M. *Proceedings, UNITAR/UNDP Int. Conf. Heavy Crude and Tar Sands*, Edmonton, Aug. 7–12 1988, Paper 92.
107. Cayias, J.L.; Schechter, R.S.; Wade, W.H. In *Adsorption at Interfaces*; Mittal, K.L., Ed.; American Chemical Society: Washington, DC, 1975, pp 234–247.
108. Schramm, L.L.; Fisher, D.B.; Schürch, S.; Cameron, A. *Colloids and Surfaces* **1995**, *94*, 145–159.
109. Pugh, R.J. *Adv. Colloid Interface Sci.* **1996**, *64*, 67–142.
110. Adamson, A.W. *Physical Chemistry of Surfaces*, 4th ed. Wiley: New York, 1982.
111. Malysa, K.; Lunkenheimer, K.; Miller, R.; Hartenstein, C. *Colloids and Surfaces* **1981**, 3, 329–338.
112. Lucassen-Reynders, E.H. In *Anionic Surfactants Physical Chemistry of Surfactant Action*, Lucassen-Reynders, E.H., Ed.; Dekker: New York, 1981; pp 173–216.
113. Schramm, L.L.; Green, W.H.F. *Colloid & Polymer Sci.* **1992**, *270*, 694–706.
114. Edwards, D.A.; Wasan, D.T. In *Surfactants in Chemical/Process Engineering*, Wasan, D.T.; Ginn, M.E.; Shah, D.O., Eds.; Dekker: New York, 1988; pp 1–28.
115. Huang, D.D.W.; Nikolov, A.; Wasan, D.T. *Langmuir* **1986**, 2, 672–677.

116. Schramm, L.L.; Mannhardt, K.; Novosad, J.J. In *Proceedings, 14th International Workshop and Symposium, International Energy Agency Collaborative Project on EOR*, Reider, E., Ed.; OMV: Salzburg, Austria, 1993.
117. Kerner, H.T. *Foam Control Agents*; Noyes Data Corp.: Park Ridge, NJ, 1976.
118. Blair, C.M. *Chem. Ind.* **1960**, 5, 538–544.
119. Reisberg, J.; Doscher, T.M. *Prod. Mon.* **1956**, *11*, 43–50.
120. Cairns, R.J.R.; Grist, D.M.; Neustadter, E.L. In *Theory and Practice of Emulsion Technology*; Smith, A.L., Ed.; Academic Press: New York, 1976; pp 135–151.
121. Jones, T.J.; Neustadter, E.L.; Whittingham, K.P. *J. Can. Petrol. Technol.* **1978**, *17*, 100–108.
122. Malhotra, A.K.; Wasan, D.T. In *Thin Liquid Films*, Ivanov, I.B., Ed.; Dekker: New York, 1988; pp 829–890.
123. Cairns, R.J.R.; Grist, D.M.; Neustadter, E.L. In *Theory and Practice of Emulsion Technology*; Smith, A.L., Ed.; Academic Press: New York, 1976; pp 135–151.
124. Wasan, D.T.; Shah, S.M.; Aderangi, N.; Chan, M.S.; McNamara, J.J. *SPE J.* **1978**, *18(6)*, 409–417.
125. Joly, M. In *Recent Progress in Surface Science*; Danielli, J.F.; Pankhurst, K.G.A.; Riddiford, A.C., Eds.; Academic Press: New York, 1964; Vol. 1, pp 1–50.
126. Ross, S. In *Kirk-Othmer Encyclopedia of Chemical Technology*, 3rd ed.; Wiley: New York, 1980; Vol. 11, pp 127–145.
127. Dorshow, R.B.; Swofford, R.L. *Colloids Surfaces* **1990**, *43*, 133–149.
128. Wasan, D.T.; Gupta, L.; Vora, M.K. *AIChE J.* **1971**, *17(6)*, 1287–1295.
129. Flumerfelt, R.W.; Oppenheim, J.P.; Son, J.R. In *Interfacial Phenomena in Enhanced Oil Recovery*; Wasan, D.; Payatakes, A., Eds.; American Institute of Chemical Engineers: New York, 1982; pp 113–126.
130. Goodrich, F.C.; Allen, L.H.; Poskanzer, A. *J. Colloid Interface Sci.* **1975**, *52(2)*, 201–212.
131. Wasan, D.T.; Koczo, K.; Nikolov, A.D. In *Foams, Fundamentals and Applications in the Petroleum Industry*; Schramm, L.L., Ed.; American Chemical Society: Washington, DC, 1994; pp 47–114.
132. Shaw, D.J. *Introduction to Colloid and Surface Chemistry*, 4th ed.; Butterworth-Heinemann: Oxford, UK, 1992.
133. Lake, L.W. *Enhanced Oil Recovery*; Prentice Hall: Englewood Cliffs, NJ, 1989.
134. Clark, N.J. *Elements of Petroleum Reservoirs*, Society of Petroleum Engineers: Richardson, TX, 1969.
135. Taylor, K.T.; Hawkins, B. In *Petroleum Emulsions and Applied Emulsion Technology*, Schramm, L.L., Ed.; Petroleum Recovery Institute: Calgary, AB, 1990.
136. Ling, T.F.; Lee, H.K.; Shah, D.O. In *Industrial Applications of Surfactants*; Royal Society of Chemistry: London, 1987.
137. Takamura, K. *Can. J. Chem. Eng.* **1982**, *60*, 538–545.

138. Takamura, K.; Chow, R.S. *J. Can. Petrol. Technol.* **1983**, *22*, 22–30.
139. Schramm, L.L.; Smith, R.G.; Stone, J.A. *AOSTRA J. Research* **1984**, *1*, 5–14.
140. Schramm, L.L.; Smith, R.G. *Colloids and Surfaces* **1985**, *14*, 67–85.
141. Schramm, L.L.; Smith, R.G. *Can. J. Chem. Eng.* **1987**, *65*, 799–811.
142. Shaw, R.; Czarnecki, J.; Schramm, L.L.; Axelson, D. In *Foams, Fundamentals and Applications in the Petroleum Industry*; Schramm, L.L., Ed.; American Chemical Society: Washington, DC, 1994; pp 423–459.
143. Shaw, R.; Schramm, L.L.; Czarnecki, J. In *Suspensions, Fundamentals and Applications in the Petroleum Industry*; Schramm, L.L., Ed.; American Chemical Society: Washington, DC, 1996; pp 639–675.
144. Hunter, R.J. *Zeta Potential in Colloid Science*; Academic Press: New York, 1981.
145. James, A.M. In *Surface and Colloid Science*; Good, R.J.; Stromberg, R.R., Eds.; Plenum: New York, 1979; Vol. 11, pp 121–186.
146. Riddick, T.M. *Control of Stability Through Zeta Potential*; Zeta Meter Inc.: New York, 1968.
147. Okada, K.; Akagi, Y. *J. Chem. Eng. Japan* **1987**, *20(1)*, 11–15.
148. Yoon, R-H.; Yordan, J.L. *J. Colloid Interface Sci.* **1986**, *113(2)* 430–438.
149. Whybrew, W.E.; Kinzer, G.D.; Gunn, R. *J. Geophys. Res.* **1952**, *57*, 459–471.
150. Kuo, J-F.; Sharma, M.M.; Yen, T.F. *J. Colloid Interface Sci.* **1988**, *126(2)*, 537–546.
151. Schramm, L.L.; Mannhardt, K.; Novosad, J.J. *Colloids and Surfaces* **1991**, *55*, 309–331.
152. Mannhardt, K.; Schramm, L.L.; Novosad, J.J. *Colloids and Surfaces* **1992**, *68*, 37–53.
153. Isaacs, E.I.; Chow, R.S. In *Emulsions, Fundamentals and Applications in the Petroleum Industry*; Schramm, L.L., Ed.; American Chemical Society: Washington, DC, 1992, pp 51–77.
154. Derjaguin, B.V.; Churaev, N.V.; Miller, V.M. *Surface Forces*; Consultants Bureau: New York, 1987.
155. Verwey, E.J.W.; Overbeek, J.Th.G. *Theory of the Stability of Lyophobic Colloids*; Elsevier: New York, 1948.
156. Currie, C.C. In *Foams*; Bikerman, J.J., Ed.; Reinhold: New York, 1953; pp 297–329.
157. Maini, B.; Wassmuth, F.; Schramm, L.L. In *Suspensions, Fundamentals and Applications in the Petroleum Industry*; Schramm, L.L., Ed.; American Chemical Society: Washington, DC, 1996, pp 321–375.
158. West, C.C.; Harwell, J.H. *Environ. Sci. Technol.* **1992**, *26*, 2324–2330.
159. Haggert, G.M.; Bowman, R.S. *Environ. Sci. Technol.* **1994**, *28*, 452–458.
160. Fuerstenau, D.W. *J. Phys. Chem.* **1956**, *60*, 981–985.
161. Somasundaran, P.; Healy, T.W.; Fuerstenau, D.W. *J. Phys. Chem.* **1964**, *68*, 3562–3566.
162. Somasundaran, P.; Fuerstenau, D.W. *J. Phys. Chem.* **1966**, *70*, 90–96.
163. Harwell, J.H.; Hoskins, J.C.; Schecter, R.S.; Wade, W.H. *Langmuir* **1985**, *1*, 251–262.

164. Scamehorn, J.F.; Schechter, R.S.; Wade, W.H. *J. Colloid Interface Sci.* **1982**, *85*, 463–478.
165. Yeskie, M.A.; Harwell, J.H. *J. Phys. Chem.* **1988**, *92*, 2346–2352.
166. Fuerstenau, M.C.; Miller, J.D.; Kuhn, M.C. *Chemistry of Flotation*; Society of Mining Engineers: New York, 1985, pp 177.
167. Cases, J.M.; Villieras, F. *Langmuir* **1992**, *8*, 1251–1264.
168. Somasundaran, P.; Kunjappu, J.T. *Mineral. Metal. Proc.* **1988**, *5*, 68–79.
169. Mannhardt, K.; Novosad, J.J. *Rev. de L'Inst. Français du Petrole* **1988**, *43*, 659–671.
170. Mannhardt, K.; Novosad, J.J. *J. Petrol. Sci. Eng.* **1991**, *5*, 89–103.
171. Mannhardt, K.; Novosad, J.J. *Chem. Eng. Sci.* **1991**, *46*, 75–83.
172. Mannhardt, K.; Schramm, L.L.; Novosad, J.J. *SPE Adv. Technol. Ser.* **1993**, *1(1)*, 212–218.
173. Schramm, L.L.; Mannhardt, K. *J. Petrol. Sci. Eng.* **1996**, *15*, 101–113.
174. Sanchez, J.M.; Hazlett, R.D. *SPE Reservoir Eng.* **1992**, *7(1)*, 91–97.
175. Tatum, J.P. In *Chemicals in the Oil Industry*; Ogden, P.H., Ed.; Royal Society of Chemistry: London, 1988, pp 31–36.
176. Brownson, G.; Peden, J.M. In *Chemicals in the Oil Industry*; Ogden, P.H., Ed.; Royal Society of Chemistry: London, 1983, pp 22–41.
177. *The HLB System*; ICI Americas Inc.: Wilmington, DE, 1976.
178. Courtney, D.L. In *Surfactants in Cosmetics*, 2nd ed.; Rieger, M.M.; Rhein, L.D., Eds.; Dekker: New York, 1997, pp 127–138.
179. Griffin, W.C. In *Kirk-Othmer Encyclopedia of Chemical Technology*, 2nd ed.; Interscience: New York, 1965; Vol. 8, pp 117–154.
180. McCutcheon's *Emulsifiers and Detergents*; MC Publishing Co.: Glen Rock, NJ, 1990; Vol. 1.
181. Griffin, W.C. *J. Soc. Cosmetic Chem.* **1949**, *1*, 311–326.
182. Shinoda, K.; Saito, H. *J. Colloid Interface Sci.* **1969**, *30*, 258–263.
183. Wadle, A.; Tesmann, H.; Leonard, M.; Förster, T. In *Surfactants in Cosmetics*, 2nd ed.; Rieger, M.M.; Rhein, L.D., Eds.; Dekker: New York, 1997, pp 207–224.
184. Little, R.C. *J. Colloid Interface Sci.* **1978**, *65*, 587–588.

RECEIVED for review June 25, 1998. ACCEPTED revised manuscript November 11, 1998.

Characterization of Demulsifiers

R.J. Mikula and V.A. Munoz

CANMET, Advanced Separation Technologies Laboratory, Western Research Centre, Devon, Alberta, Canada

Demulsifiers are a class of surfactants used to destabilize emulsions. This destabilization is achieved by reducing the interfacial tension at the emulsion interface, often by neutralizing the effect of other, naturally occurring surfactants which are stabilizing the emulsion. Demulsifier performance is routinely characterized using simple test procedures developed for use in the field. Because of the complexity of factors determining emulsion stability and, therefore, the effectiveness of any given demulsifier chemical, the wide variety of fundamental, mechanistic approaches to demulsifier selection often give way to empirical methods. A discussion of some of the common demulsifier performance characterization techniques is given along with some empirical methods for demulsifier selection.

Introduction

Several excellent reviews of demulsifier chemistry and properties can be found in the literature [1–5]. For this chapter, the important factors in demulsifier selection and characterization will be discussed, accompanied with specific examples.

Chemical demulsification is commonly used to separate water from heavy oils in order to produce a fluid suitable for pipelining (typically less than 0.5% solids and water). A wide range of chemical demulsifiers are available in order to effect this separation. In order to develop the fundamental understanding necessary to optimize demulsifier selection for a particular emulsion, it should be sufficient, in principle, to obtain a complete chemical and physical characterization of both the emulsion to be separated and the demulsifier to be used.

In practice, however, this is not possible because of the wide range of factors that can affect demulsifier performance. Aside from demulsifier chemistry, factors such as oil type, the presence and wettability of solids, oil viscosity, and the size distribution of the dispersed water phase can all influence demulsifier effectiveness. As a result, demulsifier selection for a particular field operation can still be considered to be an art to some degree.

Oil is produced in combination with water as an emulsion. Some fraction of the water separates easily, while a portion is emulsified and requires some chemical or mechanical processing. Specifications for oil quality for pipeline transportation vary but generally are on the order of less than 0.5% bottom solids and water (BSW), largely determined by water content. The BSW specification for pipeline quality oil is applied using a simple centrifugation test [6, 7].

An extensive variety of chemical demulsifiers are available to enhance resolution of the water-in-oil emulsion that is produced at the wellhead [8–12]. These demulsifiers are simply surfactants that are used to counteract the effect of surfactants naturally present in the wellhead or process emulsions, and which stabilize the water in the oil phase. In the petroleum industry, emulsions of oil in water are known as reverse emulsions. Demulsifiers are also used to destabilize these oil-in-water emulsions. The wide variety of oil types and produced water chemistries in petroleum industry emulsions necessitates an even wider variety of chemical demulsifiers. In addition, production and processing variables require demulsification chemicals tailored to particular process needs.

For example, a long-residence-time settler vessel for demulsification might perform best with a demulsifier which, although slow to reach the interface, results in a high-quality (low water content) oil product. A demulsification process that utilizes centrifuges is better served by a demulsifier that goes to the emulsion interface rapidly (due to the short residence time in the centrifuge). In static testing, the demulsifier, which can rapidly get to the interface often will give a poorer oil quality. In the high gravity environment of the centrifuge, this potential reduction in product oil quality can be overcome. Therefore, from an operations point of view, the same oil quality could be achieved in a settler vessel or in a centrifuge, but each would require a particular demulsifier [2].

The presence of solids further complicates the requirement for an effective demulsifier in that the agent used must ensure that the solid surface is water wet. Various surfactants are more effective at preventing rag layer formation and others are effective over wide concentration ranges (less susceptible to overtreating). Blends of demulsifiers are often employed to satisfy these sometimes conflicting process requirements.

Determining the best demulsifier for resolution of a given water-in-oil emulsion, given a variety of process variables, is not a task that lends itself to solutions based on analysis of the fundamental principles involved. A series of bottle tests are generally performed in order to determine the most effective demulsifier or combination of demulsifiers for a given emulsion. In spite of the difficulties involved, however, several attempts have been made to put demulsifier selection on a solid scientific foundation. Invariably these methods have limited applicability because of the often conflicting effects of the process chemistry and the physical effects of process residence time, emulsion water content, temperature, or pumping conditions [13–15].

Characterization and Selection of Demulsifiers

Background. Demulsifiers are surface-active substances (surfactants) that have the ability to destabilize emulsions. In order to perform, a demulsifier must counteract the emulsifying agent stabilizing the emulsion, and promote aggregation and coalescence of the dispersed phase into large droplets that can settle out of the continuous phase [1–5, 8, 9].

The demulsifier should have a strong attraction to the interface (good surfactant properties) and migrate rapidly through the continuous phase to reach the droplet interface. After concentration at the oil/water interface, the demulsifier counteracts the emulsifying agent and promotes the formation of flocs or aggregates of the dispersed phase. In the flocculated system the emulsifier film is still continuous so the demulsifier must neutralize the emulsifier and facilitate the rupture of the droplet interface film, resulting in coalescence. Ideally this happens rapidly resulting in the separation of the oil and water phases.

Emulsions stabilized by fine particles can be broken up if the wettability of the particles is changed by adding oil- or water-soluble demulsifiers. Iron sulphides, clays, and drilling muds can be made water wet, causing them to leave the interface and be diffused into the water droplets or they can be made oil-wet so that they can be dispersed in the oil. Paraffins and asphaltenes can be dissolved by the demulsifier to make their films less viscous, or crystallization and precipitation can be prevented [1, 3, 16–18].

The role of surfactants in stabilizing emulsions, as well as the relationship between demulsifier structure and performance, has been identified for over 50 years [19]. The classification of surfactants as well as demulsifiers is quite arbitrary, but a commonly used one is based on chemical structure [20, 21]. Chemical types include nonionic, anionic, and cationic. A brief summary of the evolution in demulsifier chemistry over the years and the effective concentration range is presented in Table 1. The development of chemicals which are more surface active has allowed for reductions in the average dosages.

The first anionic surfactants used as demulsifiers are known as soaps and are usually prepared by saponification of natural fatty acid glycerides in alkaline solution [2, 22]. The degree of water solubility is controlled by the length of the alkyl chain ranging from 12 to 18. Multivalent ions such

Table 1.   Summary of Demulsifier Changes in the Petroleum Industry (from Reference 2)

Time Period	Typical Concentration	Chemical Type
1920s	1000 ppm	Soaps, salts of naphthenic acids, aromatic and alkylaromatic sulphonates
1930s	1000 ppm	Petroleum sulphonates, mahogany soaps, oxidized castor oil, and sulphosuccinic acid esters
Since 1935	500 to 1000 ppm	Ethoxylates of fatty acids, fatty alcohols, and alkylphenols
Since 1950	100 ppm	Ethylene oxide/propylene oxide copolymers, p-alkylphenol formaldehyde resins with ethylene/propylene oxides modifications
Since 1965	30 to 50 ppm	Amine oxalkylates
Since 1976	10 to 30 ppm	Oxalkylated, cyclic p-alkylphenol formaldehyde resins, and complex modifications
Since 1986	5 to 20 ppm	Polyesteramines and blends

as calcium and magnesium, commonly present in wellhead fluids, produce marked water insolubility; thus, soaps are not useful in many saline oil field waters.

In the 1930s a number of long-alkyl-chain sulphonates (anionic chemicals), alkyl aryl sulphonates, and sulphates replaced the soaps. Unlike the sulphonates, sulphates are susceptible to hydrolysis, so that pH control is important for sulphate solutions. In dehydration applications the sulphonates exhibit fair to good wetting and water drop performance, some ability to brighten oil, and very little tendency to overtreat, particularly in high-gravity emulsions [1, 2].

The cationic agents are long-chain cations, such as amine salts and quaternary ammonium salts. The amine salts are susceptible to hydrolysis so they are not frequently used. Derivatives of alkyltrimethylammonium salts and alkylpyridium salts are the most common in this group.

The nonionic agents offer advantages regarding compatibility, stability, and efficiency compared to the anionic and cationic agents. They are often divided into those that are relatively water insoluble and those that are quite water soluble. Long-chain fatty acids and their water-insoluble derivatives belong to the first group (fatty alcohols, glyceryl esters, and fatty acid esters).

The ethylene oxide and propylene oxide block copolymers are a class of molecules that are particularly active at the oil/water interface. They became available in the 1940s allowing for the preparation of a wide variety of derivatives including fatty acids, fatty alcohols, and alkylphenol ethoxylates to produce the more water-soluble nonionic agents. These include alkylpolyoxyethylene glycol ethers and alkylphenol (ethylene oxide) ethers. Addition of ethylene oxide and/or propylene oxide to formaldehyde resins and to diamines or higher functional amines yields a variety of modified polymers that perform well at relatively low concentrations. The low molecular weight demulsifiers can be transformed into high-molecular-weight products by reactions with diacids, diepoxides, diisocyanates, and aldehydes [2, 20–23]. This allows for tailoring of demulsifier chemistry to accommodate various oil gravities and surfactant properties, and to adjust surface activity and the rate at which demulsifiers move to the interface.

Demulsifiers synthesized by polycondensation of an ethylene oxide–propylene oxide block copolymer, an oxalkylated fatty amine, and a dicarboxylic acid are known as polyester amines. These demulsifiers have the ability to adhere to natural substances that stabilize emulsions, such as organic materials formed by asphaltenes, oil resins, naphthenic acids, paraffins, and waxes; they also adhere to inorganic particles formed by clays, carbonates, silica, and metallic salts. These properties increase the demulsification efficiency of the polyester amines [2, 5]. The availability of a variety of building blocks allows for the preparation of demulsifiers for specific applications. With this chemical arsenal it is possible to tailor demulsifiers for nearly all problems posed by stable emulsions, including crude oil dehydration and desalting.

A variety of parameters are used to select demulsifiers and predict their performance for given dispersion systems. These include methods that emphasize the demulsifier properties such as the molecular weight, hydrophilic–lipophilic balance (HLB), partition coefficient, relative solubility numbers (RSN), hexane acetone titration (HAT), or preferred alkane carbon number (PACN) [24–27]. Routine characterization of demulsifiers also includes determinations of molecular weight, interfacial tension, infrared spectroscopy, and elemental analysis [5]. Sophisticated methods for the study of surfactant solutions include thermodynamic methods, small-angle neutron scattering, light scattering, rheology, luminescence, nuclear magnetic resonance, spin labels, and chemical relaxation methods [28].

Along with demulsifier characterization, the emulsion system should be characterized as completely as possible. This includes the size distribution of the dispersed phase and the chemistry of the water phase. The composition of the solids and the associated size distribution are also important and can determine emulsion stability and demulsifier performance.

Characterization of the oil phase is also important and involves properties like equivalent carbon number (EACN), acid number, asphaltene content. Other important properties of an emulsified system are due to the synergies between the components. Examples would be interfacial tension and interfacial viscosity, which are properties of the oil–water system. These properties can also be sensitive to the solids present and the nature of the solid surfaces (i.e. whether they are oil or water wet) [1–3, 29–33]. Since the viscosity of the emulsion is affected by both the water content and the droplet size, it can be used to monitor the demulsification process. Microscopic techniques alone or combined with automated image analysis of oil-in-water and water-in-oil emulsions can also provide unique information for the characterization of dispersions and assessment of demulsifier efficiency [34–38].

In practice, the level of characterization required to tailor a demulsifier based on first principles is prohibitive, due mostly to the variety of process variables that impact demulsifier effectiveness. There are many examples of emulsions studied in the laboratory to develop an effective demulsification protocol that resulted in chemical choices that were completely ineffective in the field.

Bottle testing, therefore, is always an essential part of the experimental work prior to pilot or field tests. Work at the Saskatchewan Research Council on characterization of several oilfield emulsions coupled with chemical characterization of commercially available demulsifiers and demulsifier blends showed that physical processes (temperature, pumping, and dispersed water size distribution) were at least as important as the chemical effects associated with demulsifier and oil chemistry in determining demulsifier effectiveness [13–15].

Bottle Testing.   Bottle or jar tests are the most commonly used method for evaluating demulsifier effectiveness or characterizing demulsifier performance. The details of the test procedure vary somewhat depending upon the materials available at a particular oilfield operation. Basically, samples of the process emulsion and the demulsifier to be tested are mixed, and left to separate for a defined period of time. Depending upon the process being mimicked, diluents may be added, the temperature may be controlled, or the sample may be centrifuged. After a defined period of time, the (presumably) separated emulsion is examined for brightness of the oil phase (a bright, shiny, oil phase is indicative of a low water content), clarity of the water phase, sharpness of the interface, and the rate at which the emulsion is resolved into oil and water. The oil phase is evaluated in more detail by dilution with an appropriate solvent and centrifugation to determine the residual water content of the oil phase. Typical pipeline quality oil contains less than 0.5% bottom solids and water as determined from the centrifuge spin test [6, 7]. Sometimes Karl-Fischer water content [39] determinations are used and, in laboratory situations, the water content in the oil phase is determined as a function of distance from the oil/water interface.

Table 2. Bottle Test Data

Height in the separated oil phase (0–25 is closest to the water interface)	BSW (bottom water and solids) in the oil phase		
	Demulsifier 1	Demulsifier 2	Demulsifier 3
75–100	0	0	0
50–75	0.2	0.1	0.2
25–50	0.3	0.2	0.3
0–25	0.7	0.5	0.3
Actual average	0.3	0.2	0.2
Practical average	0.1	0.1	0.1

In field evaluations these detailed determinations are seldom done, sometimes resulting in inappropriate demulsifier selection. This can occur if the oil phase is relatively low in total water but with significantly more water near the oil/water interface compared to the bulk oil phase. An example of this is shown in Table 2.

In this set of bottle test data, the water and solids content (BSW) of the separated oil phase has been evaluated as a function of distance from the separated water. The oil at the surface (75–100) is in all cases water and solids free. The practical average is identical in all cases because during field testing it is often not possible to extract the entire oil phase for BSW testing. It is the most important part of the sample nearest the water phase (often about 10%) that is not analysed. Without detailed BSW data for the oil, these three demulsifiers would be presumed to have identical performance while in fact, demulsifier 1 has a significant amount of the oil which does not meet the 0.5% BSW pipeline specification.

A high water-in-oil content near the oil/water interface in a separation test in the best case can indicate some percentage of off specification oil and in the worst case indicates a propensity for rag layer formation which often results in process upsets. The rag layer is a gel-like emulsion that forms at the interface of the oil and water in a separation vessel. It can be an oil-in-water and/or a water-in-oil dispersion and often shows multiple emulsions. In oil separation vessels, these layers are often allowed to accumulate and are pumped to separate separation processes. Rag layer emulsion separation is one of the most difficult oil–water demulsification problems. When they can be separated at all, they usually are demulsifier intensive and often require elevated temperatures, diluents, or both. This is due to the concentration of emulsion stabilizing components that have built up in the separation vessel where the rag layer accumulates.

Microscopic studies on typical rag layers reveal that the rag layer is formed partly by oil components exhibiting a gel-like structure, along with regions of both oil- and water-dispersed emulsion.

Examples of inorganic solids capable of forming hydrophilic colloidal suspensions are multivalent metal hydroxides (Si, Al, Mg, Fe, etc.), clays (kaolinite, montmorillonite, etc.), and silver halides (AgCl, AgBr, AgI). Organic hydrophilic substances include natural polymers such as polysaccharides (acacia, agar, heparin sodium, pectin, sodium alginate, tragacanth, xanthan gum) and polypeptides (casein, gelatin, protamine sulphate).

The gel formation process in organic nonionic substances requires the existence of dissolved polymers that possess segments in constant Brownian motion. Each polymer chain is encased in a sheath of solvent molecules that solvate its functional groups. In the case of aqueous solutions, water molecules are hydrogen bonded to polar groups such as the carboxylate, hydroxyl, ester, amide, and ether groups. The envelope of water of hydration impedes the chain segments from attracting each other by means of interchain hydrogen bonds and van der Waals forces. Factors reducing the hydration of the macromolecules increase the attracting forces, which establish cross-links between chains, initiating the gelation process.

In the case of ionic substances the stabilization of colloidal suspensions in their sol form is based on electrostatic repulsion and the development of electric double layers. Besides the chemical interaction between the dispersed and continuous phases, gelation can be induced by lowering the temperature of the system and increasing the concentration of the dispersed phase.

In the oil sand and heavy oil industries the components which can play a role in rag layer formation include asphaltenes, oil resins, naphthenic acids, waxes, and oxidised oil components as well as clays, carbonates, silica, iron hydroxides and sulphides: potentially any material that might have an affinity for both the oil and water phases.

In an industrial operation, these components have an opportunity to accumulate at the oil/water interface in the separation vessel, and as the strength of the gel increases, solid components that might otherwise fall to the bottom with the separated water become trapped, further increasing the handling difficulties and separator performance. Figure 1 shows an example of rag layer formation in a BSW spin test where the emulsion is simply mixed with toluene and centrifuged for a period of time. Figure 2 shows a microscopic view of the rag layer showing the complexity of the oil, water, and solids interactions.

Another important parameter in characterizing demulsifier performance is the range of effective concentrations. Usually a demulsifier with poorer performance but a wider range of effective concentrations is better in the field. This is because variations in the water cut in oil field emulsions can result in significant swings in demulsifier concentration on an oil basis and, without a demulsifier that performs well over a range

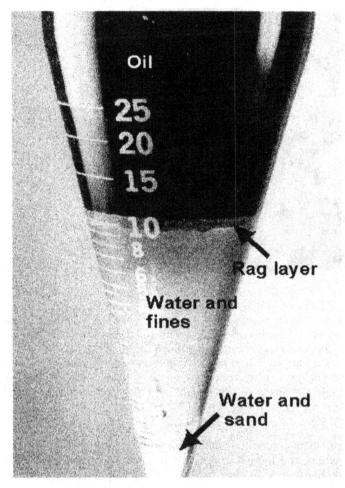

Figure 1. BSW test of an oil emulsion with a high propensity for rag layer formation. Despite the addition of toluene and the centrifugal force applied during the test a clear rag layer was formed at the oil/water interface.

of dosages, overall effectiveness in the field will be reduced. This is illustrated in Figure 3 which shows rag layer formation and water content in the oil phase as a function of demulsifier concentration for a series of jar tests. The oil product water content is only one factor in defining demulsifier performance. Rag layer formation, or overtreatment also needs to be avoided, while optimizing product quality. Figure 4 shows the oil recovery curves which account for loss of oil to rag layer formation. Demulsifier A is probably the better choice because of its wider range of effective concentrations, in spite of the fact that demulsifier B has the

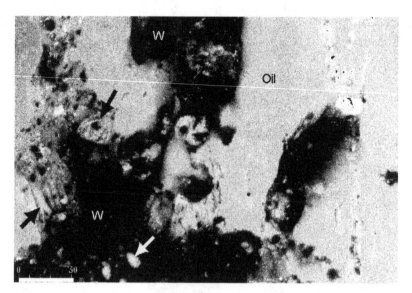

Figure 2.   Confocal micrograph of a sample taken from the rag layer shown in Figure 1. The oil is the bright phase while the water is the dark component. The oil phase shows structures typical of gels (black arrows) which are often found in oxidized oil. This rag layer contained between 20 and 30% oil, which represents a significant potential loss of recovery. The water phase contained dispersed clays forming aggregates and emulsified oil (white arrow). The areas with oil as continuous phase exhibited cavities and structures in which water with dispersed clays was intruded generating a very stable system. Bar = 50 µm.

same performance at about half the addition rate. In the rare field situation where the feed is very consistent, demulsifier B would be the best choice because of the lower addition rate required.

Bottle or jar tests in the field are the only reliable way to characterize demulsifier performance because of the importance of the rate at which the emulsifying components in the oil migrate to the oil/water interface. So-called aged emulsions can be notoriously difficult to separate because the passage of time allows asphaltenes and other naturally occurring surfactants to stabilize the water droplets.

Jar tests are not without their drawbacks in that it is often difficult to reproduce the temperature and pressure conditions encountered in the field. In addition, extrapolation of jar test performance to operational conditions can be subject to serious scale-up problems. Table 3 illustrates the differences in jar test performance which can occur due to wall effects with the glass test containers. The tests were carried out in a variety of glass containers with different volume to surface area ratios. The recoveries are a function of the glass surface area of the container to the

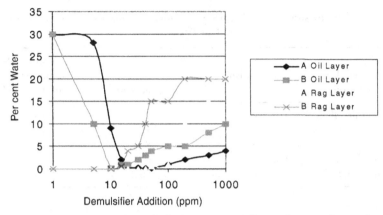

Figure 3. Comparison of the water in oil product and rag layer formation as a function of addition rate for demulsifiers A and B. Rag layer formation at high dosages is characteristic of overtreatment.

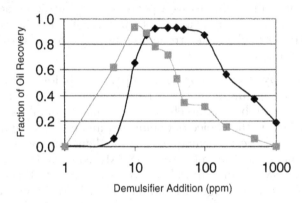

Figure 4. Total oil recovery for demulsifiers A (◆) and B (■) as a function of addition rate. Demulsifier B is effective at a much lower addition rate, but rag layer formation affects the recovery, resulting in a narrow range of effective concentration. Demulsifer A requires a higher addition rate, but is not as susceptible to overtreatment. The consistency of the feed to the process would determine which demulsifier would be most effective. Wide variation in feed properties would make demulsifier A the better choice.

volume of sample. Note that with no demulsifier addition, no separation occurred (0% recovery). In a vessel with a V:S of 10, the emulsion separation (% oil recovery) is very low. In vessel with a V:S of 2.5, the oil recovery is very good. In a vessel with a V:S of 5, an intermediate oil recovery is observed. The test result for mixing at V:S of 5 and separation in a V:S of 2.5 indicates that wall effects can influence separation test

Table 3. Comparison of Oil Recovery (Emulsion Resolution) for a Series of Bottle Tests

Test conditions 100 ppm demulsifier (weight demulsifier/ weight emulsion)	% recovery	Approximate volume : surface of the vessel
20 °C, 50 rpm, 40 min, (2 L)	25.2	10
20 °C, 50 rpm, 5 min, (2 L)	10.3	10
20 °C, hand mixed cylinder, (1 L)	86.7	2.5
20 °C, hand mixed jar, (500 mL)	63.4	5
20 °C, hand mixed jar, cylinder separation	69.2	5, 2.5

results in both the mixing and separation parts of the test. The resolution of this water-in-oil emulsion is significantly enhanced by the interaction of the water and the glass surface. With progressively larger sample-to-container surface ratios, progressively poorer demulsification is achieved. Demulsifier testing is best done in both hydrophilic and hydrophobic containers in order to eliminate erroneous interpretations of results. This is not common practice.

Microscopy. Microscopy is an important tool for characterizing emulsions and evaluating demulsifier performance. A variety of microscopy techniques can be used to characterize complex emulsion systems and therefore help in the choice of demulsifier and in assessing demulsifier performance. These techniques include light microscopy (LM), cryogenic scanning electron microscopy (SEM), confocal laser scanning microscopy (CLSM), and infrared microspectroscopy (IRM) [40–43]. Examples of confocal micrographs and data from IRM are given in this chapter.

CLSM combines some features of LM and scanning electron microscopy (SEM). Like SEM, which scans microscopic entities with an electron beam, CLSM scans the sample with a finely focussed laser beam. The reflected or emitted light (fluorescence) from the specimen is detected by two photomultipliers, digitized, and displayed on a monitor. The main feature of CLSM is that it removes out-of-focus information from the image by means of a spatial filter that consists of an adjustable pinhole (iris) set before the detector. This technique allows for independent imaging of structures with height differences on the order of the wavelength of the light source, thus permitting construction of profiles, three-dimensional images, and quantitative measurements of height.

The CLSM technique can acquire (simultaneously) images in two wavelengths, exciting the fluorescence of some sample components with blue light (488 nm) and detecting the fluorescence image in the green region (514 nm), while also, in a second photomultiplier, other components which can show strong reflection of a longer wavelength (such as 647 nm). Image processing techniques allow one to merge the two images in order to study the association between fluorescent and non-fluorescent sample components. Typically, fluorescence is excited in the organic components and the inorganic components (clays, etc.) are best imaged in the deflectance mode at the longer wavelengths. The use of a pinhole iris and computer reconstruction of only the in-focus information results in an image with a depth of field orders of magnitude better than can be achieved with ordinary light microscopic techniques.

Infrared microspectroscopy (IRM) combines LM with Fourier transform infrared spectroscopy (FT-IR). The IRM technique can obtain infrared spectra of optically distinguishable microscopic structures. In the IRM instrument, the visible light path is coaxial with that of the IR spectrometer, ensuring that the area visually selected is the same area for which the infrared spectrum is being collected. This instrument uses highly polished aluminum mirrors rather than lenses because of adsorption and attenuation of the infrared light that would occur with conventional lenses. Transmittance spectra are acquired for transparent materials, whereas reflectance spectra are acquired for opaque materials. The method is particularly suited to demulsifier and surfactant identification when used as a fingerprint technique in conjunction with a library of commercially available chemicals. In the examples discussed here, rather than assigning the spectral bands to their functional groups, the spectra were compared to a computerized Sadtler library. The spectral data search of the software used in IRM compares the spectral data in the sample with every selected library entry. The comparison is done by using a normalized least squares, dot product algorithm that generates a hit quality index (HQI). A perfect match corresponds to a HQI of zero; however, values lower than 0.5 still provide useful information about the composition of the unknowns. The Sadtler library is divided into groups such as petroleum, surfactants, monomers and polymers, inorganics, and minerals.

Whether or not an emulsion is stabilized by solids will determine the nature of the demulsifier that will be most effective. In addition, the presence of multiple emulsions (water-in-oil-in-water-in-oil, etc.) is often symptomatic of demulsifier overtreatment. Figure 5 shows an oilfield emulsion formed when a free water knockout vessel was contaminated with viscosity reducers from an earlier well fracture. Similar multiple emulsions can result from overtreatment of the produced fluid by demulsifiers in the process.

Interaction of solids at the emulsion interface can also be characterized using microscopy, as can the wettability of the solids. Figures 6 and 7 show two emulsions with nominally the same oil, water, and solids

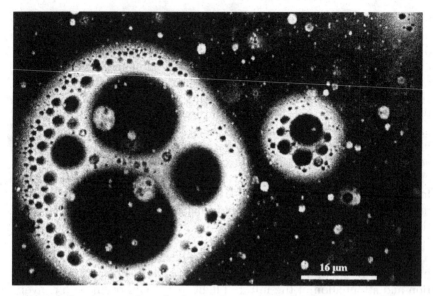

Figure 5.   A typical multiple emulsion resulting from a significant process upset condition. The bright features are the oil phase, the dark areas are water. Some instances of water-in-oil-in-water-in-oil-in-water are visible. Similar multiple emulsions are typical of demulsifier over-treatment conditions.

contents. Figure 6 shows a strong interaction of the solids with the oil phase indicated by the large number of particulates at the oil/water interface. Since certain chemicals are effective in changing the wettability of solids, a demulsifier for that situation would certainly require that type of component in the optimum chemical blend. Figure 7 shows little or no interaction of the solids at the emulsion interface; an optimum demulsifier blend for that emulsion would not likely include wetting agents.

Most of the methods used to measure solids wettability are based on direct determination of the contact angle [44]. A variety of techniques are available for contact angle measurements including the tilting plate, sessile drop, captive bubble, Wilhelmy plate, and capillary rise methods. In general, the data available are for a smooth (usually polished), macroscopic surface of a solid. The material can have different surface properties when it is in powdered form. In the case of fine particles, the material is compressed into a porous plug and the capillary pressure is measured, providing data for calculating the contact angle. Some of the problems associated with this technique are that the packing, surface roughness, particle shape, and porosity play roles that are difficult to correct for in the calculations. Another method is based on the calori-

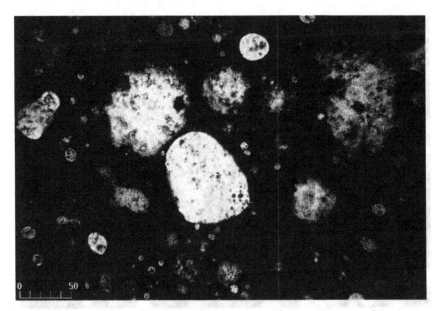

Figure 6. A reverse emulsion showing a significant interaction between solids and the oil phase. A significant percentage of the solids are intruded into the oil droplets (the oil is bright, the water and solids are dark in this image). An effective demulsifier for this fluid will have to include a solids wetting agent.

metric determination of heat evolved upon immersion of the powder in the liquid. The surface area of the powder must then be determined using gas adsorption.

Overall, wettability measurement of small particles is a difficult problem that is further aggravated in the case of heterogeneous surfaces. Some of these problems can result from the presence of patches of different composition in the same particle. It is considered that if these patches are below a critical size of 0.1 mm, the surface is homogeneous regarding its wettability. Several indirect techniques have been developed to measure the surface tension, and thus the wettability of small particles. In these techniques, the surface tensions of the particles are derived from thermodynamic models and include the advancing solidification front or freezing front, sedimentation volume, and particle adhesion techniques [44, 45].

Wettability of solids can also be directly determined from microscopic observations of the immersion or repulsion of the solids by an advancing liquid front at room temperature. The simplicity of the microscopic wettability test allows the use of any non-volatile liquid, such as produced water, deionized water, and oils. As well, the effect of additives such as

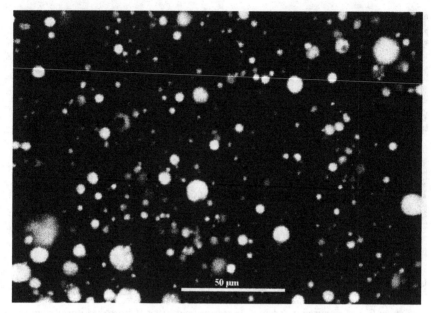

Figure 7. A reverse emulsion with the same oil, solids, and water contents as that shown in Figure 6. In this case, there are no solids at the emulsion interface (the oil is bright, the water and solids are dark in this image).

demulsifiers incorporated in the liquid for treatment of the solids can be readily studied.

Microscopic wettability tests performed at CANMET have demonstrated that solids from oil sands plants (froth, middlings, and tailings) prepared by solvent extraction to remove bitumen and water (Dean Stark) analysis were all oil wetted. It is known, however, that the bulk of the oil sands solids are in fact water wet. Since solvent extraction is the common method for preparation of oil field solids, it raises some questions about the utility of wettability tests on extracted solids using the conventional methods mentioned earlier.

Wettability determinations were performed by depositing the solids in a flat-bottom, $25 \times 9 \times 0.7$ mm ($l \times w \times d$) glass cell with a glass cover 0.17 mm thick. The depth of the cell could be adjusted with spacers so that there was no contact between the glass cover and solids in the cell. The cell was mounted under the light microscope and deionized water or mineral oil (refractive index = 1.5150) was introduced into the cell using a low flow rate by means of a 500-μL syringe. The advance of the liquid front and its contact with the particles were recorded photographically and videotaped. The liquid front appears as a dark line in this optical

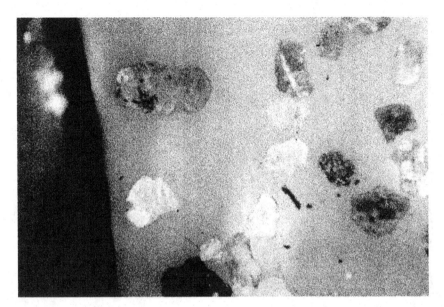

Figure 8. Microscopic wettability test showing the water front before it contacts the particles. The approaching water front is the dark arc at the left of the field of view.

microscope image. Particles that are water wettable will be engulfed by the advancing water front, while hydrophobic particles will be pushed ahead.

Figure 8 shows the solids sample before the water front has contacted the solids. Figure 9 shows the advance of the water front sweeping the particles from the field of view. This demonstrates that the solids are hydrophobic and are not wetted by the moving water front.

Infrared microspectroscopy of the solids in this case showed significant organic components on the particle surfaces (Figure 10). The spectra shown are from a single particle in the microscopic field of view in Figure 9. The example in Figure 10 also shows a spectrum from the Sadtler infrared library of commercial demulsifier and surfactant spectra. Reference to this library is useful in tracing demulsification problems which result from incompatibilities between demulsifiers, corrosion inhibitors or other process aids. In this case it was simply used to confirm the presence of organic components on the normally hydrophyllic mineral surfaces.

The effect of commercial wetting agents on the wettability of the oil wetted particles can be evaluated by treating some of the solids with these reagents and observing their performance. Reagents which do not work result in behaviour similar to that shown in Figures 8 and 9. Wetting

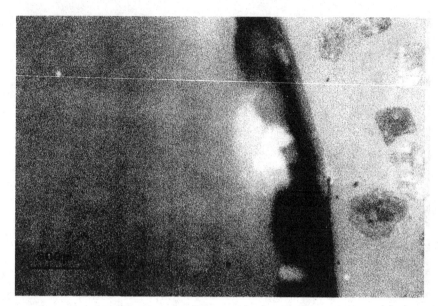

Figure 9.   The same particles as in Figure 8 being pushed ahead of the water front, indicating that they are all hydrophobic. There are no particles in the water on the left side of the field of view.

agents that are most effective will result a complete reversal from oil wetted to water wetted particles. This is shown in Figures 11 and 12.

Figure 11 shows the advance of the water front into the microscopic cell while Figure 12 shows the particles engulfed by the advancing water front, thus demonstrating that the particles are now water wettable due to the demulsifier. The same technique can be used to identify situations where demulsifiers or wetting agents are effective in changing the behaviour of only certain mineral components (clays and quartz but not pyrites for example). This insight into particle behaviour in stable emulsion systems is invaluable in optimizing the choice of demulsifier.

An Empirical Approach to Demulsifier Selection. Research into emulsion fundamentals added greatly to our understanding of the factors that determine emulsion stability and the surface-active chemicals that can be used to manipulate those factors. In spite of these advances, the requirement for blending demulsifiers in order to achieve acceptable field performance means that empirical approaches are often required for demulsifier selection. In fact, complete characterization of emulsion properties, including process residence times, temperatures, and product requirements still only provides guidance in the selection of process demulsifiers. The costs and time involved in achieving the level of characterization required for a fundamental approach can also be

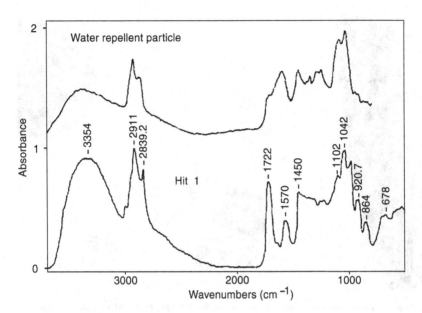

Hit	Library	Name	HQI
1.	CL	CL000159	SUCROSE MONOESTER OF TALL OIL FATTY ACIDS	0.282
2.	BP	BP001337	PHENOLIC RESIN*PAPER LAMINATE	0.305
3.	BP	BP000221	CELLULOSE, CARBOXYMETHYL, SODIUM SALT	0.310
4.	CL	CL000899	HI-FOAM BASE C ANIONIC	0.310
5.	CL	CL001605	AMINO LNO NONIONIC	0.327
6.	CL	CL001788	CLEARATE SO NONIONIC*SOYA LECITHIN	0.332
7.	BP	BP000220	CELLULOSE, CARBOXYMETHYL	0.332
8.	CL	CL000734	LIGNOSOL LC 51.0% ANIONIC*CALCIUM LIGNOSULFONATE	0.343
9.	CL	CL000614	CENTROPHIL C AMPHOTERIC*REFINED LECITHIN	0.343
10.	CL	CL000322	MONAMINE T-100 NONIONIC*TALL OIL ALKANOLAMIDE	0.354

Figure 10.   Infrared microspectroscopy of the hydrophobic particles in Figure 9. The organic coating is indicated by the strong CH_2 stretching peaks at 2800 to 2900 cm^{-1}. The lower spectrum is from the library of commercial demulsifiers and surfactants and represents the best fit to the particle spectra. Ten of the closest spectral matches from the library are listed. The ability to identify commercial additives from the library is useful in determining the extent of incompatibilities between demulsifiers and other process aids (corrosion inhibitors, previously added demulsifiers, etc.). CL: Surfactants Library, BP: Monomers and Polymers Library.

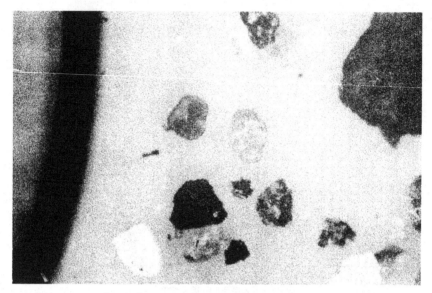

Figure 11.   Oil emulsion processing solids treated with a demulsifier wetting agent, before the water front has contacted the particles. The water front is at the black arc at the left of the field of view.

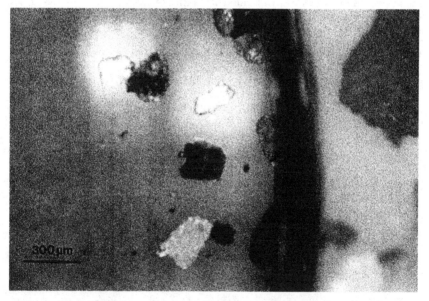

Figure 12.   The same solids as in Figure 11 engulfed by the water front, indicating that the wetting agent has made them hydrophilic. The particles which are water wetted remain in place as the water front advances and can be seen to the left of the advancing water (the dark arc).

prohibitive. Demulsifier vendors generally have suites of chemicals or blends that have been found over the years to be useful in certain oilfields or in certain process demulsification situations. Jar testing carried out on fresh emulsions in the field using these commercially available demulsifiers is often the most cost-effective approach to picking an effective and economical treatment chemical.

Statistical evaluation of demulsifier properties and performance can be useful in reducing the number of jar tests required by targeting chemically similar demulsifier blends. The data used to characterize and group the various demulsifier blends should account for some fundamental demulsifier properties. One method that has been used successfully is the chemical characterization of demulsifiers using carbon 13 nuclear magnetic resonance (C-13 NMR), although many other characteristics could be used.

Principal Component Analysis. Principal component analysis (PCA) is a statistical method that is well established for the quantitative interpretation of large data sets, and it has been particularly useful for characterizing oils [46] and demulsifiers [13–15, 47]. This or any similar statistical method has applications to understanding, at least empirically, the effectiveness of demulsifiers in water in oil emulsion separation. Any large data set of demulsifier properties combined with performance data could be reduced using PCA (48). An example is shown here for C-13 NMR chemical compositional data. The C-13 spectrum has peaks corresponding to the chemical components which make up the demulsifier blend and the area under the peaks is proportional to the amount of component present in the blend. Figures 13 and 14 show a typical example that illustrates the main chemical constituents of the blends including the diluent components, and the corresponding chemical structure.

One of the practical aspects of PCA of NMR data is that the grouping of the demulsifiers into classes can be graphically visualized by the use of a score plot. The scores assigned to the principal components illustrate relationships among the demulsifiers: similar demulsifiers collect together as clusters in the score plot.

In the present case we have the NMR spectra of several demulsifiers where the spectra are described by the peak intensities at 46 specific peak locations. Of these variables, 17 C-13 chemical shifts are from the alkane region, 11 are from the ethylene/propylene oxide region, and 18 are from the aromatic region. These data are represented by a matrix of 198 rows (the demulsifiers) and 46 columns (the peak positions).

The eigenvectors (or principal components) of the data matrix are column matrices with 46 rows corresponding to the importance of each of the 46 NMR peak positions in describing this data set . By multiplying the

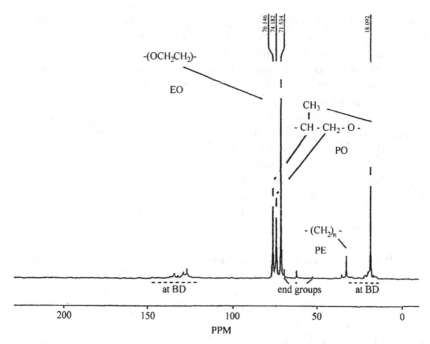

Figure 13.   Carbon 13 nuclear magnetic resonance spectra of a demulsifier blend showing the spectral regions typical of poly-oxyethylene (EO), poly-ethylene (PE), poly-propylene oxide (PO), and xylene diluents (benzene derivatives, BD). The areas under the peaks are proportional to the amount of the component in the blend.

most important eigenvectors or principal components by the row vector representing the NMR peak intensities of a given demulsifier, one can get a "score" for each of the demulsifiers. These scores can be characterized mathematically and clusters or groups of chemically similar demulsifier blends can be found. The number of principal components used to represent the data depends upon the similarities in the various demulsifier blends. In this example, some important points can be illustrated using only the first two principal components since this lends itself to visualization on an x–y plot. The purpose of the PCA is to develop a test which illustrates similarities in chemistry which correlate to performance. The data which is input for the PCA must be related to performance and therefore cannot be either too detailed or too general. It could include any fundamental property of the demulsifier such as molecular weight, HLB, etc.

Figure 15 shows the score plot of the two principal components with four distinct clusters. These clusters represent demulsifier blends with similar chemical compositions. Figure 16 shows the demulsifier performance superimposed on the clusters of similar composition. The corre-

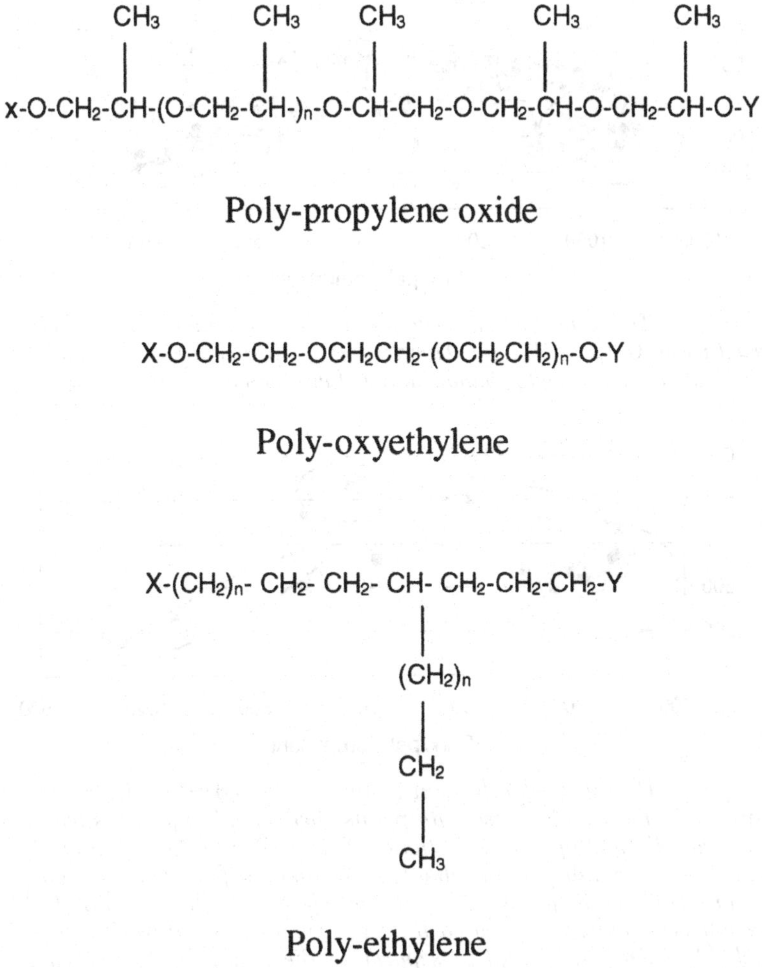

$$CH_3 \quad CH_3 \quad CH_3 \quad CH_3 \quad CH_3$$

$$x\text{-O-CH}_2\text{-CH-(O-CH}_2\text{-CH-)}_n\text{-O-CH-CH}_2\text{-O-CH}_2\text{-CH-O-CH}_2\text{-CH-O-Y}$$

Poly-propylene oxide

$$X\text{-O-CH}_2\text{-CH}_2\text{-OCH}_2\text{CH}_2\text{-(OCH}_2\text{CH}_2)_n\text{-O-Y}$$

Poly-oxyethylene

$$X\text{-(CH}_2)_n\text{- CH}_2\text{- CH}_2\text{- CH- CH}_2\text{-CH}_2\text{-CH}_2\text{-Y}$$

$$(CH_2)_n$$

$$CH_2$$

$$CH_3$$

Poly-ethylene

Figure 14. Chemical structures of the poly-propylene oxide (PO), poly-oxyethylene (EO), and poly-ethylene (PE) components shown in Figure 13.

lation between demulsifier chemistry and performance means that in the field the number of jar tests required can be significantly reduced by testing in detail only those demulsifiers from clusters with similar composition that have good performance. In addition, where good performance is achieved by a few demulsifiers outside the main cluster, commonalities in their properties can be found which could help in understanding the mechanism of demulsification. This approach can be usefully applied to any demulsifier data set in order to minimize the jar testing required for a given oil field.

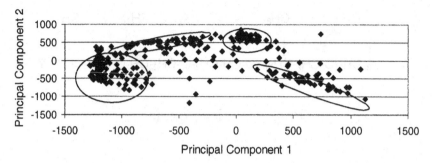

*Figure 15.   Two-dimensional score plot of the 198 demulsifiers charac-
terized using C-13 NMR data. The clusters, or proximity, of demulsifier
scores indicate chemically similar demulsifier blends.*

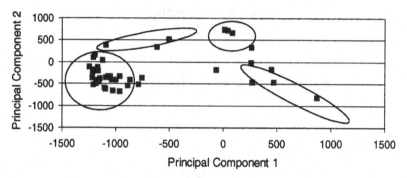

*Figure 16.   Demulsifier performance data superimposed on the clusters
identified in Figure 15. These data points correspond to jar tests where
separated oil quality was less than 0.2% BSW. In field testing of
demulsifiers, the information about which demulsifier chemical types
are most effective helps to reduce the number of jar tests required. A
selection of demulsifiers from each group could be tested to determine
which blends are successful in separating the emulsion. In this case,
chemical types corresponding to the large circle on the left are most
effective. Further optimization would then be carried out only on
chemically similar demulsifiers from the cluster identified in the initial
screening. Some demulsifiers in the other distinctly different chemical
groups are also effective which emphasizes the complexity of the demulsification process.*

Summary

Chemical demulsifiers are a class of surfactants that serve to neutralize
the stabilizing effect of naturally occurring surfactants in oil emulsions. A
good deal of research into the mechanisms of emulsion stability has
allowed demulsifier technology to keep pace with the increasingly
difficult process emulsions now being encountered. Enhanced oil recov-

ery methods either introduce emulsion stabilizing agents (secondary or tertiary oil recovery using chemical injection) to the produced fluids or increase the rate at which natural surfactants can get to and stabilize the oil/water interface (steam flooding, steam assisted gravity drainage). The shift from conventional to heavier oil production also results in emulsions that are more difficult to treat because of the higher concentration of natural surfactants in heavy oils and bitumens. Demulsifier effectiveness has been tailored over the years to deal with these new more difficult emulsions.

The microscopic techniques discussed here represent some of the state of the art in demulsifier performance characterization. Coupled with our appreciation of the fundamental mechanisms by which emulsions form and are stabilized, there is a sense that it should be possible to formulate a demulsifier to efficiently separate any fluid given enough information. This is true on one level in that if it is known that an emulsion is stabilized by solids, then addition of an agent to change the wettability of the solids will destabilize the emulsion. What is not known, is how quickly such a reagent might reach the solids in order to be effective.

In spite of our fundamental appreciation of the role of surfactants in stabilizing and destabilizing emulsions, the choice of an effective demulsifier still depends upon field tests on a selected suite of commercially available chemicals or blends. Field testing on fresh fluids still gives only trends in performance due to the variety of physical factors that impact the rate and extent to which an emulsion can be resolved into its oil and water phases. Selection of the most effective demulsifier in the field also requires some judgement as to the range of demulsifier addition that can be tolerated before an overtreatment situation will occur. Our understanding of demulsification fundamentals gives us an appreciation of the equilibrium factors which determine emulsion stability and demulsifier effectiveness, but the kinetic factors involved are more difficult to predict from even a complete characterization. As a result, increases in our understanding of the science of demulsification have still not completely obviated the "art" of demulsifier selection.

References

1. Smith, V.H.; Arnold, K.E. In *Petroleum Engineering Handbook*, H.B. Bradley, Ed.; 1992, Chapter 19, pp 9–12.
2. Stalss, F.; Bohm, R.; Kupfer, R. *SPE Prod. Eng.* **1991**, *6(3)* 334.
3. Berger, P.D.; Hsu, C.; Arendell, J.P. Society of Petroleum Engineers of AIME, Richardson, TX, USA, SPE 16285, pp 457–464, 1987.
4. Monson, L.T.; Stenzel, R.W. *Colloid Chemistry* vol. VI, Jerome Alexander, Ed.; Rheinhold Publishing Corp., New York, 1946.
5. Van Os, N.M.; Haak, J.R.; Rupert, L.A.M. *Physico-Chemical Properties of*

   Selected Anionic, Cationic and Nonionic Surfactants; Elsevier, New York, 1993.

6. Method for water and sediment in crude oil by centrifuge method (field procedure), *ASTM Designation: D-96* (1983).

7. Method for water and sediment in crude oil by centrifuge method (laboratory procedure), *ASTM Designation: D-1796* (1983).

8. Bessler, D.U. Demulsification of Enhanced Oil Recovery Produced Fluids, *Petrolite Corporation Research & Development*, September 1983.

9. Mukherjee, S.; Kushnick, A.P. In *Oil-Field Chemistry, Enhanced Recovery and Production Simulation*; Borchardt, J.K.; Yen, T.F., Eds.; American Chemical Society: Washington, DC, 1989, pp 364–374.

10. Tambe, D.; Paulis, J.; Sharma, M.M. *J. Colloid Interface Sci.* **1995**, *1(71)*, 463–469.

11. Fuestel, M. *Oil Gas European Magazine* **1995**, *21(4)*, 42–45.

12. Taylor S. *Chemistry and Industry* **1992**, *20*, 770–773.

13. Saskatchewan Research Council, Emulsions Research (1990), Final Technical Report, Publication P-110-186-C-91, April 1990–March 1991.

14. MacConnachie, C.A.; Mikula, R.J.; Kurucz, L.; Scoular, R.J.; Paper No. 39, 5th Saskatchewan Petroleum Conference, Regina, Saskatchewan, 1993.

15. MacConnachie, C.A.; Mikula, R.J. 'Classification of Chemical Demulsifiers with PCA Analysis of ^{13}C NMR Data and Water Dropout Performance Data', CANMET, Western Research Centre, Division Report WRC 92-60(CF), August 1992.

16. Manek, M.B. *Proceedings SPE International Symposium on Oilfield Chemistry*, Society of Petroleum Engineers: Richardson, TX, 1995, SPE paper 28972.

17. Strom-Kristiansen, T.; Lewis, Per Daling, S.; Nordvik, A.B. *Spill Sci. Technol. Bull.*, *2(2/3)*, 133–141.

18. Yang, M.; Stewart, A.C.; Davies, G.A. *Proceedings SPE Annual Technical Conference*, Society of Petroleum Engineers: Richardson, TX, 1996, SPE paper 36617.

19. Zaki, N.; Al-Sabagh, L. *Tenside Surfactants Detergents* **1997**, *34(1)*, 12–17.

20. Sonntag, H.; Strenge, K. *Coagulation and Stability of Disperse Systems.* Halsted Press: New York, 1972.

21. Gennaro, A.R., Ed., *Remington's Pharmaceutical Sciences*, Mack Publishing Co, 17th ed., 1985.

22. Zaki, N.N.; Abdel-Raouf, M.E.; Abdel-Azim, A.A.A. *Monatsh. Chem.* **1996**, *127(12)*, 1239–1245.

23. Amaravathi, M.; Pandey, B.P. *Res. Ind.* **1991**, *36*, 198–202.

24. Makhonin, G.M.; Petrov, A.A.; Borisov, S.I. *Chem. Technol. Fuels Oils* **1983**, *18(7–8)*, 410–413.

25. Binyon, S.J. *Oil Petrochem. Poll.* **1984**, *2(1)*, 57–60.

26. Márquez-Silva, R.L.; Key, S.; Marino, J.; Guzman, C.; Buitriago, S. *Proceedings SPE International Symposium on Oilfield Chemistry*, Society of Petroleum Engineers: Richardson, TX, 1997, SPE paper 37271.

27. Griffin, W.C. *J. Soc. Cosmetic Chem.* **1949**, *1*, 311–315.

28. Zana, Raoul. *Surfactant Solutions. New Methods of Investigations.* Marcel Dekker, Inc.: New York, 1986.
29. Kim, Y.H; Wasan, D.T. *Ind. Eng. Chem. Res.*, **1996** 35(4), 1141–1149.
30. Singh, B.P. *Energy Sources*, **1994**, 16, 377–385.
31. Cooper, D.G.; Zajic, J.E.; Cannel, E.J.; Wood, J.W. *Can. J. Chem. Eng.* **1980**, 58, 576.
32. Krawczyk, M.A.; Wasan, D.T.; Shetty, C.S. *Ind. Eng. Chem. Res.*, **1991**, 30(2), 3367–3375
33. Ivanov, I.B.; Jain, R.K.; Somasundaran, P. *Solution of Chemistry Surfactants*, 2, 817–840.
34. Mikula, R.J.; Munoz, V.A.; Lam, W.W. *Fuel Sci. Technol. International*, **1989**, 7(5–6), 727–749.
35. Mikula, R.J.; Munoz, V.A.; Lam, W.W. *J. Can. Pet. Technol.*, **1989**, 28(6), 29–32.
36. Munoz, V.A.; Mikula, R.J.; Lam, W.W.; Payette, C. *Proc. 20th Microscopical Society of Canada Conference*, Toronto, June 1993.
37. Schramm, L.L., Ed., *Emulsions: Fundamentals and Applications in the Petroleum Industry*, Chapter 3, American Chemical Society: Washington, DC, 1992.
38. Munoz, V.A.; Mikula, R.J. *J. Can. Pet. Technol.* **1997**, 38(10), 36–40.
39. Mitchell, J.; Smith, D.M. *Aquametry Part III*, 2nd ed.; John Wiley & Sons, Inc: New York, 1980.
40. Galopin, R.; Henry, N.F.M. *A Microscopic Study of Opaque Minerals*; W. Heffer and Sons Ltd.: Cambridge, 1972.
41. Mason, C.W. *Handbook of Chemical Microscopy*, Vol. 1; John Wiley and Sons: New York, 1983.
42. Pawley, J.B., Ed., *Handbook of Biological Confocal Microscopy*; Plenum Press: New York, 1990.
43. Messerschmidt, R.G.; Harthcock, M.A., Eds, *Infrared Microspectroscopy: Theory and Applications*; Marcel Dekker, Inc: New York, 1988.
44. Neuman, A.W. *Adv. Colloid Interface Sci.* **1974**, 4, 105–191.
45. Soulard, M.R.; Vargha-Butler, E.I.; Hamza, H.A.; Neuman, A.W. *Chem. Eng. Commun.* **1983**, 21, 329–344.
46. Kvalheim, O.M.; Aksnes, D.W.; Brekke, T.; Eide, M.O.; Sletten, E.; Telnaes, N. *Anal. Chem.* **1985**, 57(14), 2858–2864.
47. Theriault, Y.; Mikula, R.J. 'Demulsifier characterization by ^{13}C NMR spectroscopy and principal component analysis', CANMET, Western Research Centre, Division Report CRL 88–78 (TR), 1988.
48. Luc Massart, D.; Kaufman, L. *The Interpretation of Analytical Data by the Use of Cluster Analysis*; John Wiley and Sons: Toronto, 1983.

RECEIVED for review June 5, 1998. ACCEPTED revised manuscript January 14, 1999.

3

Emulsions and Foams in the Petroleum Industry

Laurier L. Schramm[1,2] and Susan M. Kutay[2]

[1] Petroleum Recovery Institute, 100, 3512 – 33rd St. NW, Calgary, AB, Canada T2L 2A6

[2] University of Calgary, Dept. of Chemistry, 2500 University Drive NW, Calgary, AB, Canada T2N 1N4

Emulsions and foams occur or are created throughout the full range of processes in the petroleum producing industry, including drilling and completion, fracturing and stimulation, reservoir recovery, surface treating, transportation, oil spill and tailings treating, refining and upgrading, and fire fighting. This chapter provides an overview of these examples of surfactants in action.

Introduction

In a petroleum industry context, emulsions comprise a mixture of oil and water in which one of the phases, the dispersed phase, occurs as droplets dispersed within the other, the continuous phase. The droplet diameters are typically of the order of 0.1 to 100 μm, but may be as small as a few nanometres or as large as many hundreds of micrometres. Two types of emulsion are readily distinguished, oil-dispersed-in-water (O/W) and water-dispersed-in-oil (W/O). However, emulsion characterization is not always so simple and it is not unusual to encounter multiple emulsions, O/W/O, W/O/W, and even more complex types [1]. Figure 1 shows an example of a crude oil W/O/W/O emulsion.

Petroleum industry foams comprise a mixture of gas with either oil or water, where the gas phase occurs in the form of bubbles dispersed within the liquid. The bubble diameters are typically on the order of 10 to 1000 μm, but may be as large as several centimetres. Although both aqueous and oleic foams may be encountered, the former are by far the most common. Foams and emulsions may also be encountered simultaneously [2]. Figure 2 shows an example of an aqueous foam with crude oil droplets residing in its Plateau borders.

Petroleum related occurrences of emulsions and foams are widespread, long-standing, and important to industrial productivity.

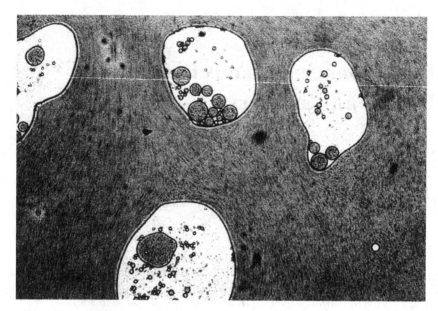

Figure 1.   Example of a petroleum industry water-in-oil-in-water-in-oil (W/O/W/O) emulsion.

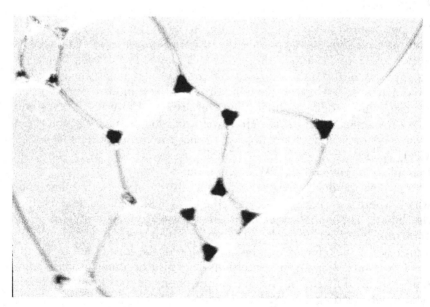

Figure 2.   Example of a petroleum industry foam containing emulsified oil droplets.

Table 1.   Some Desirable and Undesirable Emulsions and Foams

Undesirable	
Well-head emulsions	W/O
Well-head foams	G/O
Fuel oil emulsions	W/O
Oil flotation process froth emulsions	W/O and O/W
Oil flotation process diluted froth emulsions	O/W/O
Oil spill mousse emulsions	W/O
Tanker bilge emulsions	O/W
Distillation and fractionation tower foams	G/O
Fuel oil and jet fuel tank (truck) foams	G/O
Desirable	
Foam drilling fluid	G/W
Foam fracturing fluid	G/W
Foam acidizing fluid	G/W
Producing well-bore foams	G/O
Oil flotation process froths	G/O
Heavy oil pipeline emulsions	O/W
Oil flotation process emulsions	O/W
Emulsion drilling fluid: oil-emulsion mud	O/W
oil-base mud	W/O
Asphalt emulsion	O/W
Enhanced oil recovery *in situ* emulsions	O/W
Fuel-oil emulsion (70% heavy oil)	O/W
Blocking and diverting foams	G/W
Gas-mobility control foams	G/W

Both emulsions and foams may be applied or encountered at all stages in the petroleum recovery and processing industry and both have important properties that may be desirable in some process contexts and undesirable in others (Table 1). This chapter provides an overview of some surfactant and interfacial phenomena applications in the petroleum industry. Further information may be found in the specific references given in the text and elsewhere in this book. A number of other books also provide very useful introductions to the properties, importance, and treatment of emulsions [1, 3–6] and foams [2, 7–9] in the petroleum industry.

Although most emulsions and foams are not thermodynamically stable, in practise they can be quite stable and may resist explicit demulsification, antifoaming and defoaming treatments. Figure 3 shows a magnetic resonance imaging (MRI) slice taken through the centre of an emulsion sample which one of the authors (LLS) had collected from an

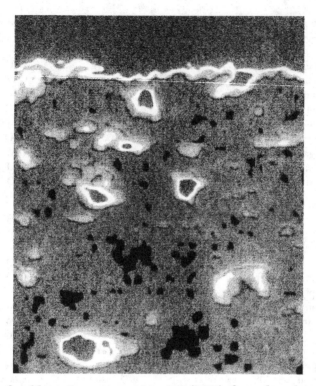

Figure 3.   Magnetic resonance imaging (MRI) slice taken through the centre of an emulsion sample which had been collected from an oil/water separating plant, and then stored for more than 5 years before this image was taken. Aqueous films are shown in white, oil in grey, and solids in black. The larger oil-in-water-in-oil emulsion droplets are about 200 to 250 μm in diameter.

oil/water separating plant, and then stored for more than 5 years before this image was taken. The continuous phase is oil, and one can clearly observe still dispersed oil-in-water-in-oil (O/W/O) and water-in-oil (W/O) droplets. The angular, rather than spherical, shapes suggest the presence of viscoelastic interfacial films surrounding these droplets. Foams can also have long lifetimes. Although the drainage half-life of a typical foam is of the order of tens of minutes, some foams can have much greater stability. In carefully controlled environments, it has been possible to make surfactant-stabilized foam bubbles and films having lifetimes of from months to years [10].

An emulsion or foam can be made by simply mixing oil or gas into water with sufficient mechanical shear. The additional interfacial area created between the two phases is important because, as shown by the

Table 2. Some Factors Involved in Determining the Stability of Foams

Low surface tension (makes it easier to form and maintain large interfacial area)
Gravity drainage (increases the rate of film thinning)
Capillary suction (increases the rate of film thinning)
Surface elasticity (counteracts the effect of surface perturbations)
Bulk viscosity (reduces the rate of film thinning)
Surface viscosity (reduces the rate of film rupture)
Electric double layer repulsion (reduces the rates of film thinning and rupture)
Dispersion force attraction (increases the rates of film thinning and rupture)
Steric repulsion (reduces the rates of film thinning and rupture)

Table 3. Some Factors Involved in Determining the Stability of Emulsions

Low interfacial tension (makes it easier to form and maintain large interfacial area)
Electric double layer repulsion (reduces the rates of aggregation and coalescence)
Surface viscosity (retards coalescence)
Steric repulsion (reduces the rates of aggregation and coalescence)
Small droplet size (may reduce the rate of aggregation)
Small volume of dispersed phase (reduces the rate of aggregation)
Bulk viscosity (reduces the rates of creaming and aggregation)
Small density difference between phases (reduces the rates of creaming and aggregation)
Dispersion force attraction (increases the rates of aggregation and coalescence)

Laplace equation, even a modest interfacial energy per unit area can become a considerable total interfacial energy requirement if many small droplets or bubbles are formed. In practise, the energy requirement is even greater due to the need for droplets and bubbles to deform before being disrupted [*11, 12*]. If this energy requirement cannot be provided, say, by mechanical shear, then another alternative is to use surfactant chemistry to lower the interfacial free energy, or interfacial tension. This can lower the amount of mechanical energy needed for emulsification or foaming by several orders of magnitude. Every meta-stable emulsion or foam that will be encountered in practise contains a stabilizing agent: either surfactant molecules or surface-active fine solids. The stabilizing surfactant makes the emulsion or foam easier to form and may create an interfacial film that helps keep the system from breaking [*1–10*]. Although surfactants and surface and interfacial tensions are very important to the stability of emulsions and foams, there are a considerable number of factors involved in determining the stability of emulsions and foams. Some of these are summarized in Tables 2 and 3. Additional details are given in the books referenced earlier and also in Chapter 1 of this book.

Near-Well Emulsions and Foams

Drilling and Completion.   Both emulsions and foams have
been used as alternatives to suspensions (muds) in drilling fluid formulation. Two kinds of oilwell drilling fluid (or "drilling mud") are emulsion
based: water-continuous and oil-continuous (invert) emulsion drilling
fluids. Here a stable emulsion (usually oil dispersed in water) is used to
maintain hydrostatic pressure in the hole. This is obviously a desirable
kind of emulsion. However, just as with classical suspension drilling muds,
careful formulation is needed in order to minimize fluid loss into the
formation, to cool and lubricate the cutting bit, and to carry drilled rock
cuttings up to the surface. The oils used to make the emulsions were
originally crude oil or diesel oil, but are now more commonly refined
mineral oils [13]. Oil-continuous, or invert, emulsion fluids are typically
stabilized by long chain carboxylate or branched polyamide surfactants.
Borchardt [14] lists a number of other emulsion stabilizers that have been
used. In the case of carboxylate surfactants, the calcium form is often used
to ensure stabilization of the water-in-oil emulsion type (involving the
oriented-wedge mechanism, as is discussed in reference [15]). Organophilic clays have also been used as stabilizing agents. Invert emulsion
fluids provide good rheological and fluid-loss properties, are particularly
useful for high-temperature applications, and can be used to minimize
clay hydration problems in shale formations [13].

Several kinds of foams have been utilized as drilling fluids [16–21].
Figure 4 shows some possible flow regimes corresponding to the use of
air, mist, foam, or liquid as a drilling fluid [22]. Foams have been used to
remove formation brine that has entered a well while air drilling, this is
sometimes called mist drilling because the fluids are injected as a mist,
although the mist changes to foam before returning up the annulus of the
well. Since foams can exhibit a high carrying capacity (viscosity), they can
also be used for sand or scale clean-outs. Foam drilling fluids are now of
much interest for underbalanced, low annular velocity drilling of horizontal wells [23, 24], a method in which the drilling fluid is kept at lower
pressure than the reservoir so that the drilling fluid and cuttings will
neither erode nor penetrate and potentially damage the reservoir. Air,
mist, and foam can also yield superior drilling penetration rates compared
with conventional mud systems. Such foams are typically based on alpha-
olefin sulfonate or alcohol ether sulfate surfactants, having solution
concentrations in the range 0.2 to 2 mass %, and are usually formulated
to have gas contents (foam qualities) in the range 55 to 96% (v/v).
Polymer-thickened foams have also been used for enhanced cuttings
carrying capacity [14, 25]. Foam qualities in the range 95 to 98% (v/v)
tend to provide the best carrying capacities. By carefully selecting the type
of surfactant one can adjust the brine salininty and oil tolerances of the

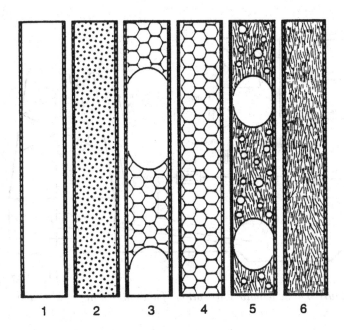

1. Air or gas flow
2. Mist flow
3. Stable foam with gas slugs
4. Stable foam
5. Liquid flow with gas slugs and interspersed bubbles
6. Liquid flow

Figure 4.   Flow regimes corresponding to the use of air, mist, foam, or liquid as a drilling fluid. (From Lorenz [22]. Copyright 1980 Gulf Publishing Co., Houston, TX, USA.)

foam. Polymers, such as guar and xanthan gums, may be added to adjust the foam viscosity, and hence its carrying capacity for cuttings.

Foams intended for use in wells and in several near-well reservoir processes are pre-formed at the surface before injection (Figure 5). One common foam generation method involves simply combining surfactant-containing and gas streams at a high flow velocity and then causing them to experience a sudden pressure drop across, say, a choke or a valve. Another method involves flowing the mixed stream through a foam generating cannister, which may contain screens, steel wool, metal rings or shavings, or glass beads. Both of these methods can produce very high shear rates, which can cause a problem if polymers are incorporated in to the foam. More sophisticated foam generators, that permit some control over the shear forces imparted, have been developed for the generation of polymer-thickened foams [26].

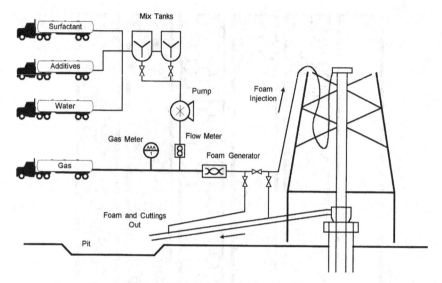

Figure 5.   Schematic illustration of a foam drilling fluid system.

When a foam drilling fluid is brought to the surface, a defoaming strategy is needed to prevent overfoaming of the pit or tank. The traditional approach to defoaming is to add a defoamer such as a polydimethylsiloxane or low molar mass aliphatic alcohol, which will break foam and also act to prevent re-foaming. Borchardt [14] lists a number of defoamers that have been used. Another way this can be handled in the field is to formulate the foam so it is stable at alkaline pH, while in the well, but unstable at acidic pH. In this way the foam can be rapidly destroyed at the surface, the resulting slurry treated to remove the cuttings and other contaminants, then made alkaline again and reused in the drilling operation [27, 28].

Stimulation: Fracturing and Acidizing. Other desirable near-wellbore emulsions and foams are used to increase the injectivity or productivity of wells by fracturing or acidizing. In either case, the goal is to increase flow capacity in the near-well region of a reservoir. Fracturing fluids are injected at high pressure and velocity, through a wellbore, and into a formation at greater than its parting pressure. Fractures (cracks) are created and propagated. Various types of fracturing fluids are available, including water- and oil-based emulsions, and water-, oil-, and alcohol-based foams. Acidizing foams, used to increase the productivity of reservoirs by dissolving fine particles in flow channels, are aqueous foams in which the continuous phase is usually hydrochloric acid (carbonate reservoirs) or hydrofluoric acid (sandstone reservoirs) [29]. Blends of these acids are also used [14]. Borchardt [14] lists a number of other inorganic and organic acids that have been used. Foaming an acidizing fluid will increase its effective viscosity, allowing some mobility control. The propagation of an acidizing fluid can be enhanced by formulating it as an acid-in-oil emulsion, in which case the continuous oil phase acts to reduce contact between the acid and the rock.

Emulsified fracturing fluids are typically very viscous polymer oil-in-water emulsions that may consist of 60–70% liquid hydrocarbon dispersed in 30–40% aqueous solution or gel. The hydrocarbon phase may be diesel fuel, kerosene, or even crude oils and condensates. The aqueous phase may consist of gelled fresh water, a KCl solution or an acid solution. Emulsion fracturing fluids may be applied to oil or gas wells, particularly in low pressure formations susceptible to water blockage, and for bottom-hole temperatures of up to about 150 °C. They can provide excellent fluid loss control, possess good transport properties and can be less damaging to the reservoir than other fluids. However, emulsions are more difficult to prepare and can be more expensive.

Foams have been used as fracturing fluids since the 1970s [30–37]. They were first applied for low pressure reservoirs, but foamed fluids have now been applied to all types of wells: low and high pressure, gas or oil, where it is important to minimize damage. Foam fracturing fluids have been used in liquid sensitive formations to minimize the amount of potentially damaging liquid coming into contact with the reservoir and to permit rapid recovery of the majority of the treatment fluid. Foamed fracturing fluids are typically 60–80% gas (N_2 or CO_2). The liquid phase, water, water/methanol, aqueous gel, or oleic gel, contains surfactants and frequently contains other stabilizers to reduce the likelihood of phase separation. They may be applied to oil or gas wells, particularly in low pressure and water-sensitive formations, and for bottom-hole temperatures of up to about 150 °C. They tend to be less expensive, contribute less liquid contact and less damage to the formation, provide reduced proppant requirements, and have a more rapid recovery and clean-up step compared with other fluids. However, the difficulty of monitoring the rheological properties of these complex systems has caused difficulties with on-site quality control and pressure analysis.

Delayed cross-linking (gelling) nitrogen foams were introduced in the 1980s [38]. Addition of cross-linkers increases the viscosity of the fracturing fluid. Various additives that have been used include polymers (guar gum and guar derivatives) and cross-linking agents (aluminate, borate, titanate, and zirconate). Cross-linked foams can be used to more easily place proppant in a formation compared with the use of a non-cross-linked polymer-foam. Nitrogen, being inert, allows foams to be formulated with many types of cross-linkers. The addition of cross-linkers to CO_2 foams has allowed their application to deeper, hotter reservoirs [39]. Quite thin filtercakes (0.10 mm) are deposited with cross-linked foams [40], but since the residue is cross-linked, it is much more difficult to remove than is the case with non-cross-linked polymers. Even though the cost of foam treatments is typically 10 to 20% greater than non-foamed gel stimulation treatments, the rapid fluid recovery and mini-mized damage to the reservoir has made foamed fluid treatment pro-cesses attractive. Chambers provides a useful review of the applications of foam stimulation fluids [41].

Reservoir Occurrences of Emulsions and Foams

Primary and Secondary Production. Emulsions are com-monly produced at the wellhead during primary and secondary (water-flood) oil production. For these processes the emulsification has not usually been attributed to formation in reservoirs, but rather to formation in, or at the face of, the wellbore itself [42]. However, at least in the case of heavy oil production, laboratory [43] and field [44, 45] results suggest that water-in-oil emulsions can be formed in the reservoir itself during water and steam flooding. Energy is needed for emulsification, partly because of the increased surface area that is created in forming small droplets and partly because deformation of large drops is needed before smaller drops can pinch off. The type of emulsion that will be formed is influenced by the critical Weber number [11, 12]. The Weber number, We, is given by:

$$We = (\eta_1 \dot{\gamma} R)/\gamma_{12}$$

where η_1 is the viscosity of the continuous phase, $\dot{\gamma}$ is the shear rate, R is the droplet radius, and γ_{12} is the interfacial tension. Figure 6 shows that for a given viscosity ratio, η_2/η_1, between the dispersed (η_2) and contin-uous (η_1) phases, reducing the interfacial tension increases the Weber number, lowering the energy needed to cause droplet breakup. The figure also shows that for a given flowing system in a heavy oil reservoir, the viscosity ratio will be smaller, and an emulsion easier to form, if it is a water-in-oil emulsion rather than an oil-in-water emulsion.

During primary production, the pressure is greater in the reservoir at the locations from which oil is being drained and lower near and in the wellbore. As oil moves toward a producing well and then into the bottom of the well, the reduced pressure it experiences can cause dissolved gas to be released. When this happens to a light oil, the gas normally separates from the oil. In the case of some heavy oils, however, the gas remains dispersed in the oil as an *in situ* oil foam [46]. In petroleum industry terminology this is called foamy-oil production, and can be associated with increased primary oil production compared to what would be expected from non-foamy-oil production. It is thought that the formation of foamy-oil delays the formation of a continuous gas phase (increases the

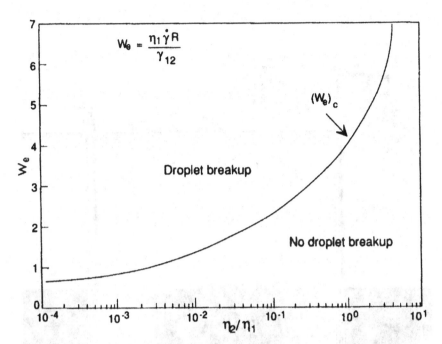

Figure 6. The critical Weber number: for a given viscosity ratio, reducing the interfacial tension increases the Weber number, lowering the energy needed to cause droplet breakup. (From [11]. Copyright 1992 American Chemical Society, Washington, DC, USA.)

trapped gas saturation) and contributes a natural pressure-maintenance function [46].

Secondary and Tertiary Production.   *Chemical Flooding.* In oilfields, after the primary and secondary cycles of oil recovery, chemicals may be injected to drive out additional oil in an enhanced oil recovery process, which may involve creating *in situ* emulsions in the reservoir. Figure 7 shows a reservoir schematic with such a chemical flood.

In a petroleum reservoir the relative oil and water saturations depend upon the distribution of pore sizes in the rock. The capillary pressure, or pressure difference across an oil/water interface spanning a pore, is given by:

$$P_c = 2\gamma \cos \theta / r$$

where γ is the oil/water interfacial tension, θ is the contact angle, measured through the water phase at the point of oil/water/rock contact, and r is the effective pore radius. In a water-wet reservoir the water will have been imbibed most strongly into the smallest radius pores, while the

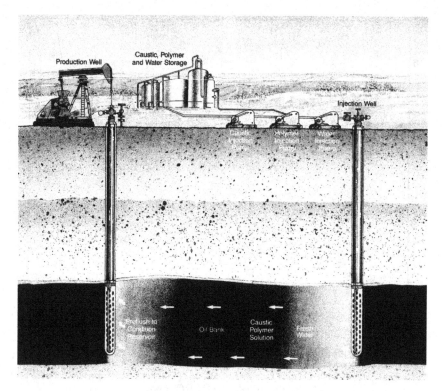

Figure 7.   Illustration of a reservoir enhanced oil recovery process, which may involve creating in situ emulsions or foams. (Courtesy Alberta Oil Sands Technology and Research Authority, Edmonton, AB, Canada.)

largest pores will retain high oil contents. As water is injected during a secondary recovery process the applied water pressure increases and the larger pores will imbibe more water, displacing oil which may be recovered at producing wells. There is a practical limit to the extent that the applied pressure can be changed by pumping water into a reservoir however, so that after waterflooding some residual oil will still be left in the form of oil ganglia trapped in the larger pores where the viscous forces of the driving waterflood could not completely overcome the capillary forces holding the oil in place.

The ratio of viscous forces to capillary forces correlates well with residual oil saturation and is termed the capillary number. One formulation of the capillary number is:

$$N_c = \eta v / (\gamma \phi)$$

where η and v are the viscosity and velocity of the displacing fluid. The functional form of the correlation is illustrated in Figure 8 in Chapter 1 of this book. During waterflooding N_c is about 10^{-6} and at the end of the waterflood the residual oil saturation is still around 45%. In order to recover the remaining oil one must increase the capillary number. This could be done by raising the viscous forces, i.e. viscosity and velocity, but in practise does not achieve the desired orders of magnitude increase.

Chemical flooding involves the injection of a surfactant solution which can cause the oil/aqueous interfacial tension to drop from about 30 mN/m to near-zero values, on the order of 10^{-3} to 10^{-4} mN/m, allowing spontaneous or nearly spontaneous emulsification and displacement of the oil [47, 48]. Sharma [49] has reviewed the kinds of surfactants used for enhanced oil recovery processes. The exact kind of emulsion formed can be quite variable, ranging from fine macroemulsions, as in alkali/surfactant/polymer flooding [50], to microemulsions [51, 52]. Microvisualization studies suggest that with such low interfacial tensions, multiple emulsions may form, even under the low flow rates that would be produced in a reservoir. Figure 8 shows an example of multiphase flow in an etched glass micromodel wherein crude oil is being displaced by an alkali/surfactant/polymer solution at low flow rate (advance rate of about 2 m/day). Even at such a low flow rate, the displacement and tortuous flow have combined to produce both water-in-oil emulsion (top of the pores) and water-in-oil-in-water multiple emulsion (lower regions of the pores). Details of the chemical formulation are given in reference [53].

Microemulsion/Micellar Flooding. Microemulsions are stable emulsions of hydrocarbons and water in the presence of surfactants and co-surfactants. They are characterized by spontaneous formation, ultra-low interfacial tension, and thermodynamic stability. The wide-spread

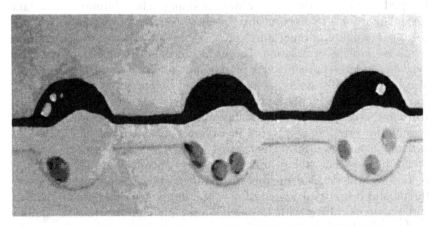

Figure 8. Videomicrographic image of multiple emulsions formed during low tension flooding of oil in a microvisual cell. The droplets in the upper field of view are W/O while the lower droplets are W/O/W.

Table 4.   Components of Microemulsions for EOR

Component	Composition
Oil	Crude oil or white oil
Brine	Formation water or water from sea, lakes, and rivers with variable saline conditions and concentrations (mg/L to g/L)
Chemicals	Primary surfactant (e.g., petroleum sulfonate) Co-surfactant/co-solvent (e.g., C_3 to C_5 alcohol) Polymer (e.g., xanthan) Alkaline agents (e.g., sodium carbonate) Bactericides (e.g., formaldehyde) Sacrificial adsorption agents

interest in microemulsions and use in industrial applications are based mainly on their high solubilization capacity for both hydrophilic and lipophilic compounds, their large interfacial areas and on the ultra-low interfacial tensions achieved when they coexist with excess aqueous and oil phases. The properties of microemulsions have been extensively reviewed elsewhere [54–60]. The ultra-low interfacial tension achieved in microemulsion systems has application in several phenomena involved in oil recovery as well as in other extraction processes (e.g., soil decontamination and detergency).

As with alkali/surfactant/polymer formulations, microemulsions are injected into reservoirs as part of enhanced oil recovery (EOR) processes which use interfacial tension lowering to mobilize the residual oil left trapped in the reservoirs after waterflooding. The optimum surfactant formulation for a microemulsion system is dependent on many variables (e.g., pH, salinity, temperature, etc.). Table 4 lists some of the components in a typical formulation. The surfactants and co-surfactants must be available in large amounts at a reasonable cost. In addition, they should also be chemically stable, brine soluble and compatible with the other formulation components. Common surfactants used are petroleum sulfonates and ethoxylated alcohol sulfates [49, 50]. The degree of interfacial tension lowering depends on the phase behaviour of the oil/brine/surfactant mixture. Surfactants are generally used at concentrations much higher than their critical micelle concentration (cmc). Phase behaviour will depend on the surfactant partition coefficient between the oil and brine. The advent of new and more cost-effective surfactants and polymers, along with improved reservoir characterization, should lead to substantial design improvements.

Although producing a more efficient oil displacement than alkali/surfactant/polymer flooding, microemulsion flooding has developed slowly so far because of its complex technology and higher costs. Nevertheless, numerous field pilot tests have been reported, primarily using previously waterflooded reservoirs [61, 62]. Many field experiments have failed or have displayed poor performances because of inadequate well patterns, poor knowledge of reservoir characteristics, or degradation of chemicals, leading to loss of mobility control. Some pilot tests, with better reservoir characterization and properly designed chemicals, have been reported to be technically successful with recoveries in the order of 50% of the oil at the start of the flood, recovering two-thirds of the residual oil [63, 64]. Specific chapters on chemical flooding for reservoir oil recovery (Chapter 6) and for environmental soil remediation (Chapter 11) appear elsewhere in this book.

Macroemulsion Flooding. Some emerging applications involve the possible use of macroemulsions, as opposed to the microemulsions discussed in the previous section. These emulsions would be injected or produced *in situ* in order either for blocking and diverting, or for improved mobility control. Broz et al. [65] and French et al. [66] have proposed the use of oil-in-water emulsions for blocking and diverting of injected steam. For mobility control, there is some evidence to suggest that the *in situ* formation of heavy water-in-oil emulsions, such as sometimes happens during cyclic steam stimulation of heavy oil reservoirs, can improve the oil mobility, and hence recovery, in water-wet reservoirs [67]. This apparently only occurs for certain conditions of emulsion properties, flow conditions, and rock wettability, because the improved oil mobility in the reservoir has to occur despite the fact that the bulk phase emulsion viscosity would be greater than that of the oil alone.

Sarma and Maini [68] have suggested that emulsified solvent flooding might be a viable alternative to hydrocarbon or CO_2 miscible displacement of heavy oil from thin reservoirs. In this case one would be able to both improve sweep efficiency (reduce fingering) and significantly reduce the amount of solvent needed for a flood by injecting it in the form of a solvent-in-water emulsion.

Foam Injection Processes. Foams can be injected in to a reservoir for mobility control or for blocking and diverting. The foam can thus act to reduce the effects of:

- poor mobility ratio between injected and reservoir fluids
- other causes of poor areal sweep efficiency
- poor vertical sweep efficiency
- non-oil-saturated or thief zones
- reservoir heterogeneities

For example, major problems occur in gas flooding methods due to the displacing agent's high mobility and low density compared with those of reservoir fluids. In such cases, channelling (fingering) and gravity override both reduce the sweep efficiency, contribute to early breakthrough of injected fluid, and therefore reduce the amount of oil recovered. Injecting the gas as a foam can counteract these kinds of problems because the foam lowers the gas mobility in the swept and/or higher permeability parts of the formation and diverts at least some of the displacing medium (gas) into other parts of the formation that were previously unswept or underswept. It is from these latter areas that the additional oil is recovered. Since foam mobility tends to be reduced disproportionately more in higher permeability zones, improvement in both vertical and horizontal sweep efficiency can be achieved. Suitable foams can be formulated for injection with air/nitrogen [69–72], natural gas [72–76], carbon dioxide [72, 77, 78], or steam [72, 79–81].

A major challenge is the proper selection of foam-forming surfactants, and there have probably been several hundred papers published in the past 35 years on appropriate foam characteristics. Some of the characteristics thought to be necessary for a foaming agent to be effective (including cost-effective) in porous media under reservoir conditions are as follows [82]:

- good solubility in the brine at surface and reservoir conditions
- good thermal stability under reservoir conditions
- low adsorption onto the reservoir rock
- low partitioning into the crude oil phase
- strong ability to promote and stabilize foam lamellae
- strong ability of the foam to reduce gas mobility in porous media
- good tolerance of the foam to interaction with crude oil in porous media

These requirements can severely limit the number of surfactant candidates. For example, the desired process might be the hydrocarbon gas miscible foam flooding of high salinity (ca. 100,000 to 300,000 ppm), high hardness (ca. 5000 to 25,000 ppm), moderately high temperature (to 130 °C) reservoirs such as are found in Western Canada. In one study, from an initial set of 157 commercially available foaming surfactants recommended by suppliers around the world, solubilities were determined [83] at different salinity, hardness, pH and temperature conditions and it was found that only nine had sufficient solubility and thermal stability under such conditions. The most salinity and hardness tolerant surfactants were mostly betaines and sulfobetaines. Borchardt [14] lists a number of foaming surfactants that have been used in less demanding environments.

Foaming capability relates to both foam formation and foam persistence. Relevant are such factors as surface tension lowering, surface elasticity, surface viscosity and disjoining pressure [84]. Surface tension lowering is necessary but not sufficient. Some combination of surface elasticity, surface viscosity and disjoining pressure is needed, but the specific requirements for an effective foam in porous media remain elusive, partly because little relevant information is available and partly because what information there is appears to be somewhat conflicting. For example, both direct [85] and inverse [86] correlations have been found between surface elasticity and gas mobility reduction in foam floods. Overall, it is generally found that the effectiveness of foams in porous media is not reliably predicted based on bulk physical properties or on bulk foam measurements. Instead, it tends to be more useful to study the foaming properties in porous media at various laboratory scales: micro, meso and macro.

Micro-scale experiments involve the microscopic observation of flowing foams in etched-glass micromodels. Here the pore dimensions are typically on the order of hundreds of micrometres. Such experiments provide valuable and rapidly obtainable qualitative information about foam behaviour in constrained media under a variety of experimental conditions, including the presence of a residual oil saturation [82, 87, 96].

Meso-scale experiments involve conducting foam floods in samples of porous rock, which may be reservoir core samples or quarried sandstones and carbonates, the quarried samples being more reproducible. The overall rock dimension here is of the order of 10 cm. These meso-scale foam floods allow the determination of gas mobility reduction by foams under widely varying conditions [1]. The mobility reduction factor (MRF) is the ratio of pressure drops across a core resulting from the simultaneous flow of gas and liquid in the presence and absence of surfactant in the liquid phase. Mobility reduction factors achieved depend on many factors [82, 83, 88, 89] including:

- the nature of the surfactant
- the composition of the brine
- the composition of the gas
- the nature of the porous medium
- the foam quality
- the foam texture
- the foam flow rate
- the temperature and pressure

Macro-scale experiments involve a special apparatus that allows foam floods to be performed at reservoir conditions of temperature and pressure in an integral two metre length of rock sample, that is, in porous media samples one order of magnitude longer than the meso-scale. In addition to being a first step in scale-up, this allows the study of dynamic foam behaviour that would be impossible in short core samples.

The conduct and interpretation of such macro-scale experiments are illustrated in Chapter 7 of this book.

The economics of foam flooding are determined to a large degree by the amount of surfactant required to generate and propagate a foam. Surfactant loss through partitioning into the crude oil phase and through adsorption on the rock surfaces cannot be completely eliminated, and these are therefore important (but undesirable) mechanisms of surfactant loss. Surfactant loss through partitioning into the crude oil phase can be responsible for surfactant losses of as much as 30%. However, for the very hydrophilic surfactants chosen for many foam flooding applications, the partitioning into crude oil is very nearly zero. More serious are the results of a number of systematic studies of the adsorption properties of surfactants suitable for foam flooding, e.g., [90–92]. These have shown that effective foaming surfactants may exhibit adsorption levels from near zero up to quite high levels on the order of 2.5 mg/g, depending, not only on the nature of the surfactant, but also on factors such as temperature, brine salinity and hardness, rock type, wettability, and the presence of a residual oil phase. These factors can lead to vastly different distances of foam propagation in a reservoir, so that selection of a surfactant formulation with acceptable adsorption levels at reservoir conditions is crucial. Surfactant adsorption is discussed in more detail in Chapter 4 of this book.

Results from field testing have suggested that foams may achieve lower gas mobility reductions than anticipated due to the defoaming action of residual crude oil [93], which has led to an interest in the formulation of oil tolerant foams. Although crude oils tend to act as defoamers, microvisual and coreflood studies have shown [94–96] that foams actually exhibit a wide range of sensitivities to the presence of crude oils. Temperature, brine salinity and hardness, and the nature of the crude oil phase have all been found to influence the oil tolerance of a given foam and many attempts have been made to correlate foam–oil sensitivity with physical parameters, e.g., [94–100]. These have met with varying degrees of success [87, 101]. Overall, it is clearly possible to make foams that are reasonably stable in the presence of light through heavy crude oils [96], using either relatively pure foaming agents (usually quite expensive), or else with specially formulated mixtures, which can be cost-competitive with traditional foaming agents (e.g., [102, 103]). Some of these foams, intended for mobility control, can even improve microscopic displacement, by emulsifying oil into droplets that are small enough to permit their passage inside the foam's lamellar structure, and thus contribute an incremental oil recovery [96, 103].

The foregoing summarizes much of what can be gained from laboratory testing in terms of surfactant selection. Necessary subsequent steps in the evaluation of foam effectiveness include reservoir simulations and field tests. Reservoir foam applications may involve slug injection, in which foaming surfactant solution is injected into the gas stream at the wellhead over a period of a few hours, semi-continuous injection, in which surfactant solution is injected at intervals over a period of a day or so, and continuous injection, in which surfactant solution is injected continuously for months or even years. The application mode is chosen depending upon a number of factors, including the reservoir characteristics, the prior production history, and a project's specific economics. Currently, most attention is being paid to near-well applications of foams. Some examples are given in references [72, 74–76]. Disadvantages of deeper penetration and full-field foam application include surfactant costs (especially due to replenishing surfactant lost due to adsorption on reservoir rock) and the fact that it can take a considerable amount of time (months) to build up an appreciable flow resistance in deep reservoir applications.

Polymer Thickened Foams. Polymer enhanced (thickened) foams have also found increasing use in the petroleum industry. Incorporating polymers into foaming solutions affects the solution properties primarily by increasing the liquid phase viscosity, which enhances foam stability by decreasing the rate of drainage and reducing the rate of interbubble gas diffusion (e.g., [104–111]). There is a small but growing literature on the development of polymer thickened foams in terms of both fundamental and applied work (e.g., [106–108, 112, 113]).

Several uses of polymer thickened foams for reservoir recovery were patented in the late 1960s and early 1970s (e.g., [106, 114]), and in 1974 Minssieux showed that the addition of a polymer to a foaming solution could diminish the loss of foam viscosity due to foam degradation in the presence of oil [115]. Studies [108, 113] have also shown that polymer thickened foams can possess extremely high effective viscosities in both bulk and in sand-pack porous media flow tests. These studies also showed that the effective polymer-foam viscosities can be comparable to the viscosities of the polymer solutions (no gas) used in formulating these foams. This suggests that polymer thickened foams, with their enhanced viscosities and stability, could be effective mobility control agents [105, 114, 116]. Otherwise, polymer thickened foams can be formulated using the same range of types of gases, surfactants, and other additives as is the case for conventional surfactant-stabilized foams. A range of polymer additives have been tested, including polyacrylamide, polyvinyl alcohol, polyvinylpyrrolidone, and xanthan biopolymers. In addition to changing foam quality and texture, the effective viscosities of polymer thickened foams can also be adjusted by varying the polymer concentration and molar mass. In general, polymer thickened foams are shear thinning.

Polymer thickened foams to which time-delayed cross-linking agents have been added, gelling foams, can be used to improve the efficiency of oil displacement by blocking swept zones and by diverting fluids into underswept zones in reservoirs containing large permeability variations and/or fractures [117–119]. Once gelled, these foams can function in similar fashion to conventional gels, but with only a small fraction of the pore space being occupied by gelled liquid. Figure 9 shows photomicrographs of gelled-foam lamellae in a Berea sandstone core, showing films, rods, and intermediate structures.

Micro-Foams (Colloidal Gas Aphrons). The terms microfoam and colloidal gas aphrons refer to a dispersion of aggregates of very small foam bubbles in aqueous solution. The latter term was coined by Sebba [120–122] in the 1970s. They can be created by dispersing gas into surfactant solution under conditions of very high shear. An apparatus for this purpose is described in references [121, 123]. The concept is that, under the right conditions of turbulent wave breakup, one can create a dispersion of very small gas bubbles, each surrounded by a bimolecular film of stabilizing surfactant molecules (Sebba termed this film a soapy shell). Under ambient conditions the bubble diameters are typically in the range 50 to 300 μm. Figure 10 shows an example of such a micro-foam, generated with an apparatus modelled after that of Sebba. There is some evidence that such micro-foams tend to be more stable than comparable foams that do not contain the bimolecular film structure [120, 121, 123]. Other claims for special properties have been made [124–130], but are less well supported, or even conflicting within the literature.

Two kinds of applications relating to petroleum recovery have been reported in the literature, micro-foam flushing for soil remediation [126–128] and micro-foam injection for reservoir oil recovery [129, 130]. Despite the fact that these papers make conflicting claims regarding the physical properties of these foams, and, although their results should be interpreted with caution pending additional independent studies, these papers provide interesting reading and suggest that micro-foams may well find useful application in reservoir oil recovery processes.

Emulsions and Foams in Surface Operations

An emulsion that was useful in the reservoir may be, or may become, an undesirable type of emulsion (W/O) when produced at the wellhead. Pipeline and refinery specifications place severe limitations on the water, solids, and salt contents of oil they will accept, in order to avoid corrosion, catalyst poisoning, and process upset problems. Foaming, as well, can cause surface handling and refinery upset problems.

Oilfield Emulsions and Foams and their Treatment. A typical W/O petroleum emulsion from a production well might contain

60 to 70% water. Some of this, the free water, will readily settle out. The rest (bottom settlings) requires some kind of specific emulsion treatment. The specific kind of treatment required can be highly variable. It is often said that each oilfield location produces a unique kind of emulsion requiring a custom treatment approach. Some of the complexity and variability of oilfield emulsions can be appreciated if one considers that crude oils consist of, at least, a range of hydrocarbons (alkanes, naphthenes, and aromatics) as well as phenols, carboxylic acids and metals. There may be a significant fraction of sulphur and nitrogen compounds present as well. The carbon numbers of all these components range from one (methane) through 50 or more (asphaltenes). Some of these components can form films at oil surfaces while others are surface active. In addition, due to the wide range of possible compositions, crude oils can exhibit a wide range of viscosities and densities (so much so that these properties are used to distinguish light, heavy and bituminous crude oils [131]).

In many surface separation processes there will occur three distinct phases or process streams, an oil product stream, which may contain emulsified water, an aqueous tailings stream, which may contain emulsified oil, and an interface or "rag layer" emulsion stream, which may contain emulsified oil and/or water. The interface emulsion layer may build up to a certain level in a process, continuously reform and break in the separator and never cause operational problems. On the other hand, the interface emulsion layer may build to such an extent that it requires removal and treatment. Knowledge of the nature of the dispersed phase will be required to determine an effective treatment. Figure 3 illustrated the simultaneous presence of W/O and O/W/O emulsion. Mikula shows (Figure 1 in reference [132]) a photomicrograph of a quite stable interface emulsion (rag layer emulsion) in which one can clearly observe the simultaneous occurrences of both O/W and W/O emulsions in different regions of the same sample.

The first step in systematic emulsion breaking is to characterize the emulsion in terms of its nature (O/W, W/O, or multiple), the number and nature of immiscible phases, the presence of a protective interfacial film around the droplets, and the sensitivity of the emulsifiers. In oilfield W/O emulsions, a stabilizing interfacial film can be formed from the asphaltene and resin fractions of the crude oil. This causes special problems because if the films are viscoelastic then a mechanical barrier to coalescence exists, which may be quite intractable and yield a high degree of emulsion stability. More detailed descriptions are given in references [133–135]. Based on an emulsion characterization, a chemical addition could be made to neutralize the effect of the emulsifier, followed by mechanical means to complete the phase separation.

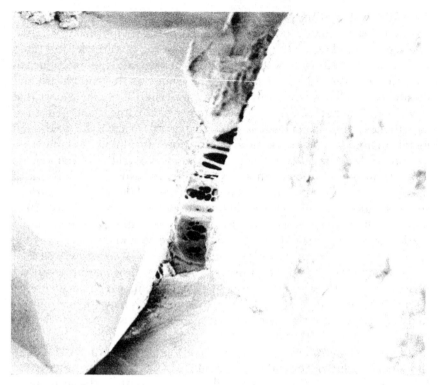

Figure 9(a)

Figure 9.   Photomicrographs of gelled-foam lamellae in a Berea sandstone core, taken using a low energy scanning electron microscope (SEM). Image (a) shows a view down a pore surrounded by rock grains. Image (b) is a magnification of the upper centre region.

Demulsifying agents are designed to reduce emulsion stability by displacing or destroying the effectiveness of the original stabilizing agents at the interfaces. Examples of the primary active agents in commercial demulsifiers include ethoxylated (cross-linked or uncross-linked) propylene oxide/ethylene oxide polymers or alkylphenol resins. These products are formulated to provide specific properties including hydrophile–lipophile balance (HLB), solubility, rate of diffusion into the interface, and effectiveness at destabilizing the interface [6, 136]. Demulsifiers are usually added to the continuous phase, within which they must then diffuse to the interface and disrupt the stabilizing interfacial film. The demulsifier should usually be added far enough upstream to permit these actions to take place, and for droplet coalescence to occur, before the emulsion reaches a separating vessel.

A variety of physical methods are used in emulsion breaking. These

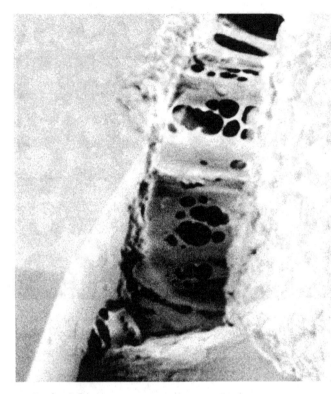

Figure 9(b).

are all designed to accelerate coagulation and coalescence. For example, oilfield W/O emulsions may be treated by some or all of settling, heating, electrical dehydration, chemical treatment, centrifugation and filtration. The mechanical methods, such as centrifuging or filtering, rely on increasing the collision rate of droplets and applying an additional force driving coalescence. An increase in temperature will increase thermal motions to enhance the collision rate and also reduce viscosities (including interfacial viscosity), thus increasing the likelihood of coalescence. In the extremes, very high temperatures will cause dehydration due to evaporation, while freeze–thaw cycles will break some emulsions. Electrical methods may involve electrophoresis of oil droplets, causing them to collide, to break O/W emulsions. With W/O emulsions, the mechanism involves deformation of water droplets, since these are essentially nonconducting emulsions. Here the electric field causes an increase in the droplet area, disrupting the interfacial film. Increased droplet contacts increase the coalescence rate, breaking the emulsion. More details on the application of these methods in large-scale continuous processes are given elsewhere [*137, 138*].

Figure 10. Photomicrograph of a freshly generated micro-foam (colloidal gas aphron). The bubble diameters are in the range 30 to 150 μm.

The produced water in an oilfield typically contains emulsified oil at levels of a few hundred to a few thousand mg/L [*139*]. The oil content usually must be greatly reduced in order to reuse this water for reinjection (<20 mg/L), steam generation (<1 mg/L), local irrigation (<5 mg/L), or ocean disposal (<25 mg/L) [*139*]. The emulsified oil is usually separated by some combination of skim tanks, filters, induced gas flotation (IGF) cells, centrifuges, and hydrocyclones. Two kinds of surfactants are added to the IGF cells, flotation aids and demulsifiers, typically at concentrations in the range 1 to 10 mg/L. Further details can be found in references [*138, 139*]. Oilfield produced water may also foam, which can cause problems in handling and in gas separation. This is usually dealt with by adding antifoaming or defoaming chemicals such as silicones or polyglycol esters at levels in the range 1 to 30 mg/L.

When oil nears and enters the annulus of a production well, it experiences a decreased system pressure, dissolved gas may come out of solution, and the oil may foam. In conventional oil production this foam is thought to be detrimental to oil production rates so antifoaming or defoaming agents are sometimes placed in the wells [*140*]. In the non-thermal production of heavy oil, however, foaming of the oil is thought to improve production, as discussed in an earlier section.

Transportation Emulsions. Some emulsions are made to reduce viscosity so that an oil can be made to flow. Emulsions of asphalt, a semi-solid variety of bitumen dispersed in water, are formulated to be both less viscous than the original asphalt and stable so that they can be transported and handled. In application, the emulsion should shear thin and break to form a suitable water-repelling roadway coating material. Another example of emulsions that are formulated for lower viscosity with good stability are those made from heavy oils and intended for economic pipeline transportation over large distances. Here again the emulsions should be stable for transport but will need to be broken at the end of the pipeline. It is desirable for the dispersion to possess poor stability under static conditions to permit easy separation of the oil and water. In addition, the oil that has undergone separation is often re-emulsified for further treatment/application.

Oil Sand Processing. The large oil sands surface-mining and processing operations involve a number of kinds of emulsions and foams in a variety of process steps. Here, bitumen is separated from the sand matrix, in large tumblers, and forms an oil-in-water (O/W) emulsion containing not just oil and water, but also dispersed solids and gas. The emulsified oil is further separated from the solids by a flotation process which produces an oleic foam termed bituminous froth, which may be either gas dispersed in the oil (primary or secondary flotation) or the reverse, gas dispersed in water (secondary or tertiary flotation). In this case of bituminous froths, the foams contain not just oil and gas, but also emulsified water and some dispersed solids. The froth has to be broken in order to permit pumping and subsequent removal of entrained water and solids before the bitumen can be upgraded to synthetic crude oil. This is facilitated by deaeration and dilution with naphtha. The diluted froth contains multiple emulsion types including tenacious multiple emulsions [15] which complicate the downstream separation processes. These aspects are discussed in Chapter 10 of this book and are reviewed in references [141, 142].

Oil Spills and Tailings. Emulsions may be discharged to or created in tailings ponds, such as in the tailings ponds created by surface processing of mined oil sands (see Chapter 10 of this book). Oil is produced at off-shore drill sites in the form of oil-in-water emulsions (containing reservoir water) which may have to be transported to an on-shore processing centre, at which the primary emulsion is separated into its components and the oil is often re-emulsified (fresh water) for other applications.

Oil spills at sea can cause significant environmental damage. The most straightforward method is to contain the spill and remove it mechanically, but this is not always feasible in practise. Often chemical treatment agents must be incorporated into the clean-up procedures. Following the actual spill of an oil onto the sea, a slick is formed which spreads out from the source with a rate that depends on density, interfacial tension, viscosity, and the nature and degree of emulsification that has occurred [143]. With sufficient wind and wave energy advection and turbulence can cause an O/W emulsion to be formed, which helps disperse oil into the water column and away from sensitive shorelines. Droplets may rise but usually become weathered, lose their lighter components to evaporation, and eventually settle out. Otherwise, the oil may pick up water to form a water-in-oil "mousse" emulsion, probably stabilized by asphaltenes and/ or natural surfactants [144–147].

These W/O mousse emulsions can contain abnormally high water contents (>80%) without inverting. Such high dispersed phase volume fraction emulsions have very high viscosities. This complicates the use of dispersants and/or demulsifiers. As their common name implies, the mousse emulsions not only have viscosities that are much higher than the original crude oil but can become semi-solid. With increasing time after a spill these emulsions weather (evaporation of lighter components, wind and wave effects, and photo-oxidation of remaining components) making the emulsions more stable, more solid-like, and considerably more difficult to handle. The presence of mechanically strong films [147, 148] makes it hard to get demulsifiers, which are usually sprayed from vessels or aircraft, into these emulsions (Figure 11). Oil spill dispersants are therefore more effective if applied while the spill is still fresh, before mousse emulsion can form [149].

A fairly large number of demulsifying/dispersing formulations have been created for application to marine oil spills [143, 150, 151]. These are usually formulated to have a tendency to promote oil-in-water emulsion formation, and tend to be moderately hydrophilic, having hydrophile–lipophile balances (HLB's) in the range 10 to 12 [150, 152]. They are usually formulated in a solvent that will be miscible with the spilled oil [153]. Such values can be obtained, for example with an appropriate blend of Span® and Tween® surfactants. Some surfactant dispersants include sulfosuccinates, sorbitan esters of oleic or lauric acid, polyethylene glycol esters of oleic acid, and ethoxylated fatty alcohols. Determining effectiveness of a chemical agent is a complex issue because it is a function of the oil type, composition, the amount of oil present and how long it has weathered. This has been a major stumbling block in the development of a universal treatment agent. More hydrophilic surfactants than those used for oil spills may be effective in treating/breaking emulsions that have been recovered in skimmers or tanks where the water solubility of the agent is not as important an issue. Figure 12 illustrates the addition and action of a dispersant [153]. Dispersant application is influenced by factors such as oil spill slick thickness, degree of oil weathering that has

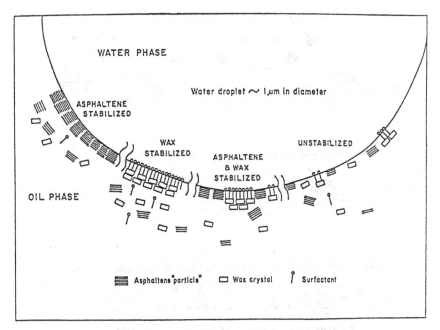

Figure 11.   Illustration of possible structures in the interface in a water-in-oil mousse emulsion. Some possibilities are, from left to right, stabilized by asphaltenes, by surfactants and waxes, by both, and unstabilized. (From Mackay [147]. Copyright 1987 Environment Canada, Ottawa, ON.)

already taken place, and the prevailing weather and sea conditions. For instance, surfactants that are effective in Arctic waters will not behave the same in more temperate conditions. Further details are presented in Chapter 12 of this book.

Surface-washing agents remove the oil from solid surfaces such as shorelines through a different mechanism, detergency (see Chapter 1 of this book). The longer the oil remains on the shoreline/beach the more difficult it is to remove, therefore response time is critical. Good surface-washing agents are poor dispersants and vice versa. Low dispersant effectiveness is a benefit for surface-washing applications because the oil is to be recovered, not dispersed. The post-treatment of recovered oil can be another complex issue. Depending on the condition of the recovered oil, the oil may have to undergo various treatment procedures. For example, before oil can be sent back to a refinery it must be de-watered and free of suspended solids. Very viscous recovered oils may be too difficult to treat, necessitating either incineration or land-fill disposal.

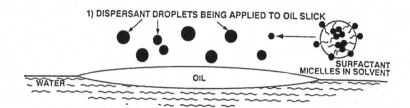

1) DISPERSANT DROPLETS BEING APPLIED TO OIL SLICK

SURFACTANT MICELLES IN SOLVENT

WATER   OIL

2) DISPERSANT DROPLETS COALESCE WITH OIL SLICK AND DIFFUSE AS LENSES INTO OIL

WATER   OIL

3) SOLVENT DELIVERS SURFACTANT THROUGHOUT OIL AND TO OIL-WATER INTERFACE

WATER   OIL

SOLVENT DENSITY PROMOTES SPREADING AT INTERFACE

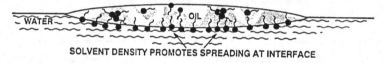

4) DISPERSANT-ENRICHED OIL READILY DISPERSES INTO DROPLETS

WATER   OIL

FINE OIL DROPLETS BREAK AWAY

Figure 12.   Illustration of the mechanism of oil spill dispersion. (From Fiocco et al. [153]. Copyright 1995 American Society for Testing and Materials, Philadelphia, PA.)

Downstream Occurrences

Refineries and Upgraders.   Wellhead or other surface-produced emulsions may have to be broken and reformulated as new emulsions more suitable for transportation by pipeline to an upgrader or refinery. At the upgrader or refinery, the new emulsion will have to be broken and the water removed, which otherwise would cause operating problems. At any of the emulsion breaking stages the presence of solid particles and film-forming components from the crude oil or bitumen can make this very difficult.

Some agents will act to reduce the foam stability of a system (termed foam breakers or defoamers) while others can prevent foam formation in the first place (foam preventatives or foam inhibitors). There are many such agents, Kerner [154] describes several hundred different formulations for foam inhibitors and foam breakers. In all cases the cause of the reduced foam stability can be traced to various kinds of changes in the nature of the interface.

The addition to a foaming system of any soluble substance, that can become incorporated by co-solubilization or by replacement of the original surfactants into the interface, may decrease dynamic foam stability if the substance acts against the formerly present stabilizing factors. Some branched, reasonably high molecular mass alcohols can be used for this purpose. Not being very soluble in water, they tend to be adsorbed at the gas/liquid interface, displacing foam promoting surfactant and breaking or inhibiting foam. Alternatively, a foam can be destroyed by adding a chemical that actually reacts with the foam-promoting agent(s). Foams may also be destroyed or inhibited by the addition of certain insoluble substances.

Petroleum emulsions have been used to prevent the formation of foams, or destroy foams already generated, in various industrial processes [155]. The rapid spreading of drops of low surface tension oil over lamellae ruptures them by providing weak spots [156]. Polydimethylsiloxanes are frequently used as practical antifoaming agents because they are insoluble in aqueous media (and some oils), have low surface tension and are not overly volatile. They are usually formulated as emulsions for aqueous foam inhibiting so that they will readily become mixed with the aqueous phase of the foam. A review of refinery foam occurrences and treatment is given by Lewis and Minyard [157].

Finally, many kinds of foams pose difficult problems wherever they may occur. In surface emulsion treaters (e.g., oil–water separators) and in refineries (e.g., distillation towers), the occurrence of foams is generally undesirable and any such foams will have to be broken [2].

Emulsions for Paving Roads.   Asphalt emulsions are used to produce a smooth, water-repellant surface in road paving. First, an asphalt oil-in-water emulsion is formulated which has sufficiently low viscosity to be easy to handle and apply, and which has sufficient stability to survive transportation, brief periods of storage, and the application process itself. After application the emulsion needs to break quickly. An additional advantage of the emulsion over asphalt alone lies in its ability to be applied to wet gravel or rock [3, 158].

The asphalt emulsions are usually stabilized either by natural naphthenic surfactants released by treatment with alkali (for a somewhat similar situation, see also Chapter 10 in this book), or else by the addition of anionic or cationic surfactant [158]. When stabilized by cationic surfactant, the positive charge may facilitate binding of the asphalt droplets to the gravel or rock surfaces [3].

Fire Fighting Foam Systems. Fire fighting foams were first introduced in the early 1900s, and complete fire fighting foam systems were in use by the military by the 1940s [159]. Some of the history of this development, and a description of the early formulations of suitable foaming agents, are given by Perri [159]. Fire fighting foams function by a combination of the following [160, 161].

- blanketing the burning fuel surface and smothering the fire
- suppressing and preventing air from mixing with flammable vapours
- separating flames from the fuel's surface
- cooling the fuel and its surface by the action of the water in the foam

Some of these are illustrated in Figure 13. The foam is created by mechanically mixing air with a concentrated solution of surfactant in water. These foams are often formulated to contain fluorocarbon surfactants, sometimes blended with hydrocarbon surfactants and/or polymers.

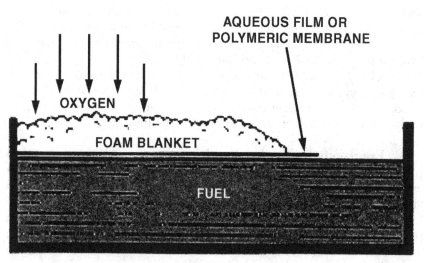

Figure 13.   Some functions of a fire fighting foam: blanketing the burning fuel surface, smothering the fire, preventing air from mixing with flammable vapours, and separating flames from the fuel's surface. (Reliable Fire Equipment Company, Alsip, IL.)

Foams that can be effective on hydrocarbon fuel fires are typically characterized as protein (hydrolyzed protein surfactants), fluoroprotein (hydrolyzed protein and perfluorinated surfactants), aqueous film-forming (AFFF, blend of perfluorinated surfactants), alcohol resisting aqueous film-forming (AFFF-AR), high expansion, or alcohol (polar-solvent) foams [160]. The practical formulations may contain numerous other additives to control freezing, viscosity, bacterial degradation, oxidation, corrosion, and so on [161]. The most commonly used foams for fire fighting contain 75–97% air; these are known as "low expansion" foams. For a fire fighting foam to be effective it must possess the following attributes [162]:

- resistant to large electrolyte concentrations (e.g., sea water)
- insignificant toxicity and biodegradability
- long term storage stability
- undamaged by inadvertent freezing–thawing cycles
- freeze-resistant for cold climates

Foam selection criteria include classical properties like static half-lives, but also properties like expansion and fire extinguishing performance. Corrie [161] describes a range of laboratory evaluation methods. Fire fighting foams may be found in any of the many industrial operations involving the transportation, processing, or handling of flammable petroleum liquids, including refineries and offshore production platforms.

Acknowledgments

The authors thank Jana Vander Kloet for valuable advice and resource materials. We gratefully acknowledge the financial and other support provided by the Natural Sciences and Engineering Research Council of Canada and the Petroleum Recovery Institute.

References

1. Schramm, L.L., Ed., *Emulsions, Fundamentals and Applications in the Petroleum Industry*; American Chemical Society: Washington, DC, 1992.
2. Schramm, L.L., Ed., *Foams, Fundamentals and Applications in the Petroleum Industry*; American Chemical Society: Washington, DC, 1994.
3. Becher, P. *Emulsions: Theory and Practice*; Reinhold: New York, 1965.
4. Sumner, C.G. *Clayton's The Theory of Emulsions and Their Technical Treatment*; 5th ed.; Blakiston Co. Inc.: New York, 1954.
5. Becher, P., Ed., *Encyclopedia of Emulsion Technology*; Vols. 1–3, Dekker: New York, 1983, 1985, 1988.

6. Lissant, K.J. *Demulsification, Industrial Applications*; Dekker: New York, 1983.
7. Bikerman, J.J. *Foams, Theory and Industrial Applications*; Reinhold: New York, 1953.
8. Akers, R.J., Ed., *Foams*; Academic Press: New York, 1976.
9. Bikerman, J.J. *Foams*; Springer-Verlag: New York, 1973.
10. Isenberg, C. *The Science of Soap Films and Soap Bubbles*; Tieto: Clevedon, England, 1978.
11. Isaacs, E.E.; Chow, R.S. In *Emulsions, Fundamentals and Applications in the Petroleum Industry*; Schramm, L.L., Ed.; American Chemical Society: Washington, DC, 1992, pp 51–77.
12. Walstra, P.; Smulders, I. In *Food Colloids*; Royal Society of Chemistry: London, England, 1997, pp 367–381.
13. Jones, T.G.J.; Hughes, T.L. In *Foams, Fundamentals and Applications in the Petroleum Industry*; Schramm, L.L., Ed.; American Chemical Society: Washington, DC, 1994, pp 463–564.
14. Borchardt, J.K. In *Oil-Field Chemistry*; Borchardt, J.K.; Yen, T.F., Eds.; American Chemical Society: Washington, DC, 1989, pp 3–54.
15. Schramm, L.L. In *Emulsions, Fundamentals and Applications in the Petroleum Industry*; Schramm, L.L., Ed.; American Chemical Society: Washington, DC, 1992, pp 1–49.
16. Anderson, G.W. In *Proceedings, Rotary Drilling Conference of IADC*, March 9–12, 1976.
17. Bentsen, N.W.; Veny, J.N. *J. Petrol. Technol.* **1976** (October), 1237–1240.
18. Hutchinson, S.O.; Anderson, G.W. *Oil and Gas J.* **1972** (May), 74–79.
19. Beyer, A.H.; Milhone, R.S.; Foote, R.W. In *Proceedings, 47th Annual Fall Meeting*, Society of Petroleum Engineers: Richardson, TX, 1972, paper SPE 3986.
20. Anderson, G.W. *World Oil* **1971** (Sept.), 39–42.
21. Anderson, G.W.; Harrison, T.F.; Hutchinson, S.O. *The Drilling Contractor* **1966** (May–June), 44–47.
22. Lorenz, H. *World Oil* **1980** (June), 187–193.
23. Rovig, J. *APPEA J.* **1996**, 557–561.
24. Biesman, T.; Emeh, V. In *Proceedings, 1st Underbalanced Drilling International Conference*, 1995.
25. Russell, B.A. In *Proceedings, SPE/IADC Drilling Conference*, Society of Petroleum Engineers: Richardson, TX, 1993, SPE/DOE paper 25766.
26. Southwell, G.P. U.S. Patent 5 356 565, October 18, 1994.
27. Thomas, T.R. U.S. Patent 5 385 206, January 31, 1995.
28. Thomas, T.R. U.S. Patent 5 591 701, January 7, 1997.
29. Gdanski, R.; Behenna, R. In *Proceedings, Field Application of Foams for Oil Production Symposium*; Olsen, D.K.; Sarathi, P.S., Eds.; U.S. Dept. Of Energy: Bartlesville, OK, 1993, Paper FS7, pp 163–172.
30. Bentsen, N.W.; Veny, J.N. *J. Petrol. Technol.* **1976** (October), 1237–1240.
31. Anderson, G.W. In *Proceedings, Rotary Drilling Conference of IADC*, March 9–12, 1976.

32. Hutchinson, S.O.; Anderson, G.W. *Oil and Gas J.* **1972** (May), 74–79.
33. Beyer, A.H.; Millhone, R.S.; Foote, R.W. In *Proceedings, 47th Annual Fall Meeting,* Society of Petroleum Engineers: Richardson, TX, 1972, paper SPE 3986.
34. Harms, W.M. In *Oil-Field Chemistry*; Borchardt, J.K.; Yen, T.F., Eds.; American Chemical Society: Washington, DC, 1989, pp 55–100.
35. Phillips, A.M.; Mack, D.J. US Patent 5 002 125, 1991.
36. Valko, P. et al. In *Proceedings, Symposium on Formation Damage Control,* Society of Petroleum Engineers: Richardson, TX, 1992, paper SPE 23778.
37. Harris, P.C. In *Proceedings, International Meeting on Petroleum Engineering,* Society of Petroleum Engineers: Richardson, TX, 1992, paper SPE 22394.
38. Watkins, E.K.; Wendorff, C.L.; Ainley, B.R. In *Proceedings, Annual Technical Conference of SPE;* Society of Petroleum Engineers: San Franscisco, CA, 1983, paper SPE 12027.
39. Harris, P.C. In *Proceedings, Annual Technical Conference of SPE;* Society of Petroleum Engineers: New Orleans, LA, 1990, paper SPE 20642.
40. Harris, P.C. In *Proceedings, International Meeting on Petroleum Engineering,* Society of Petroleum Engineers: Beijing, China, 1983, paper SPE 22394.
41. Chambers, D.J. In *Foams, Fundamentals and Applications in the Petroleum Industry*; Schramm, L.L., Ed.; American Chemical Society: Washington, DC, 1994, pp 355–404.
42. Dow, D.B. *Oil-Field Emulsions*; US Bureau of Mines Bulletin 250, 1926.
43. Chung, K.H.; Butler, R.M. *J. Can. Petrol. Technol.* **1989**, *28*, 48–55.
44. Vittoratos, E. *J. Can. Petrol. Technol.* **1991**, *30*, 82–86.
45. Vittoratos, E. In *Supplementary Notes for Petroleum Emulsions and Applied Emulsion Technology*; Schramm, L.L., Ed.; Petroleum Recovery Institute: Calgary, AB, 1992.
46. Maini, B.B.; Sarma, H. In *Foams, Fundamentals and Applications in the Petroleum Industry*; Schramm, L.L., Ed.; American Chemical Society: Washington, DC, 1994, pp 405–420.
47. Poettmann, F.H. In *Improved Oil Recovery*; Interstate Compact Commission: Oklahoma City, OK, 1983; pp 173–250.
48. Lake, L.W. *Enhanced Oil Recovery*; Prentice Hall: Englewood Cliffs, NJ, 1989.
49. Sharma, M.K. In *Particle Technology and Surface Phenomena in Minerals and Petroleum*; Sharma, M.K.; Sharma, G.D., Eds.; Plenum: New York, 1991, pp 199–222.
50. Taylor, K.; Hawkins, B. In *Emulsions, Fundamentals and Applications in the Petroleum Industry*; Schramm, L.L., Ed.; American Chemical Society: Washington, DC, 1992; pp 263–293.
51. Prince, L.M., Ed., *Microemulsions Theory and Practice*; Academic Press: New York, 1977.
52. Neogi, P. In *Microemulsions: Structure and Dynamics*; Friberg, S.E.; Bothorel, P., Eds.; CRC Press: Boca Raton, FL, 1987; pp 197–212.

53. Nasr-El-Din, H.A.; Green, K.A.; Schramm, L.L. *Rev. Inst. Franç. du Pétrole* **1994**, *49*, 359–377.
54. Robb, I.D., Ed., *Microemulsions*; Plenum: New York, 1977.
55. Overbeek, J.T.G.; de Bruy, P.L.; Verhoeckx, F. In *Surfactants*; Tadros, Th.F., Ed.; Academic Press: New York, 1984, pp 111–131.
56. Tadros, Th.F. In *Surfactants in Solution*; Mittal, K.L.; Lindman, B., Eds.; Plenum: New York, 1984, pp 1501–1532.
57. Shah, D.O., Ed., *Macro- and Microemulsions: Theory and Applications*; American Chemical Society: Washington, DC; 1985.
58. Robinson, B.H. *Nature* **1986**, *320*, 309.
59. Friberg, S.E.; Bothorel, P., Eds., *Microemulsions: Structure and Dynamics*; CRC Press: Boca Raton, FL, 1987.
60. Leung, R.; Jeng Hou, M.; Shah, D.O. In *Surfactants in Chemical/Process Engineering*; Wasan, D.T.; Ginn, M.E.; Shah, D.O., Eds.; Marcel Dekker: New York, 1988, pp 315–367.
61. Thomas, S.; Farouq Ali, S.M. *J. Can. Pet. Technol.* **1992**, *31*, 53–60.
62. Moritis, G. *Oil Gas J.* **1992**, *4*, 51–71.
63. Chapotin, D.; Lomer, J.F.; Putz, A. In *Proceedings, SPE/DOE 5th Symposium on EOR*: Society of Petroleum Engineers: Richardson, Texas, 1986, paper SPE/DOE 14955.
64. Reppert, T.R.; Bragg, J.R.; Wilkinson, J.R.; Snow, T.M.; Maer Jr., N.K.; Gale, W.W. In *Proceedings, 7th Symposium on Enhanced Oil Recovery*; Society of Petroleum Engineers: Richardson, Texas, 1990, paper SPE/DOE 20219.
65. Broz, J.S.; French, T.R.; Corroll, H.B. In *Proceedings, 3rd UNITAR/UNDP International Conference Heavy Crude and Tar Sands*; United Nations Institute for Training and Research: New York, 1985.
66. French, T.R.; Broz, J.S.; Lorenz, P.B.; Bertus, K.M. In *Proceedings, 56th Calif. Regional Meeting*: Society of Petroleum Engineers: Richardson, Texas, 1986, paper SPE 15052.
67. Woo, R.; Jackson, C.; Maini, B.B., unpublished results, Petroleum Recovery Institute, Calgary, AB, Canada, 1992.
68. Sarma, H.K.; Maini, B.B., unpublished results, Petroleum Recovery Institute, Calgary, AB, Canada, 1992.
69. Holm, L.W. *J. Petrol. Technol.* **1970**, *22*, 1499–1506.
70. Kuehne, D.L.; Ehman, D.I.; Emanuel, A.S. In *Proceedings, SPE/DOE Symposium on EOR*, Society of Petroleum Engineers: Richardson, TX, 1988, SPE/DOE paper 17381.
71. Thach, S.; Miller, K.C.; Lai, Q.J.; Sanders, G.S.; Styler, J.W.; Lane, R.H. In *Proceedings, SPE Annual Technical Conference*, Society of Petroleum Engineers: Richardson, TX, 1996, SPE paper 36616.
72. Olsen, D.K.; Sarathi, P.S., Eds., In *Proceedings, Field Application of Foams for Oil Production Symposium*, U.S. Department of Energy: Bartlesville, OK, 1993.
73. Liu, P.C.; Besserer, G.J. In *Proceedings, SPE 63rd Annual Technical Conf.*, Society of Petroleum Engineers: Richardson, TX, 1988, SPE paper 18080.
74. Aarra, M.G.; Skauge, A. In *Proceedings, SPE 69th Annual Technical*

Conf., Society of Petroleum Engineers: Richardson, TX, 1994, SPE paper 28599.

75. Aarra, M.G.; Skauge, A.; Sognesand, S.; Stenhaug, M. *Petrol. Geoscience* **1994**, *2*, 125–132.

76. Svorstøl, I.; Blaker, T.; Tham, M.J.; Hjellen, A. In *Proceedings, 9th European Symposium on Improved Oil Recovery*, The Hague, The Netherlands, 1997, paper 001.

77. Holm, L.W.; Garrison, W.H. *SPE Res. Eng.* **1988**, *3*, 112–118.

78. Heller, J.P. In *Foams, Fundamentals and Applications in the Petroleum Industry*; Schramm, L.L., Ed.; American Chemical Society: Washington, DC, 1994, pp 201–234.

79. Cooke, R.W. In *Proceedings, SPE International Thermal Operations Symposium*, Society of Petroleum Engineers: Richardson, TX, 1991, SPE paper 21531.

80. Patzek, T.W.; Koinis, M.T. In *Proceedings, SPE/DOE Symposium on EOR*, Society of Petroleum Engineers: Richardson, TX, 1988, SPE/DOE paper 17380.

81. Isaacs, E.E.; Ivory, J.; Green, M.K. In *Foams, Fundamentals and Applications in the Petroleum Industry*; Schramm, L.L., Ed.; American Chemical Society: Washington, DC, 1994, pp 235–258.

82. Schramm, L.L.; Mannhardt, K.; Novosad, J.J. In *Proceedings, 14th International Workshop and Symposium, International Energy Agency Collaborative Project on Enhanced Oil Recovery*; Rieder, E., Ed.; OMV Energie: Vienna, Austria; 1993, paper 18.

83. Novosad, J.J.; Ionescu, E.F. In *Proceedings, CIM Annual Technical Meeting*, Canadian Inst. Mining, Metallurgical and Petroleum Engineers: Calgary, AB, 1987, paper CIM 87–38–80.

84. Schramm, L.L.; Wassmuth, F. In *Foams, Fundamentals and Applications in the Petroleum Industry*; Schramm, L.L., Ed.; American Chemical Society: Washington, DC, 1994, pp 3–45.

85. Huang, D.D.W.; Nikolov, A.; Wasan, D.T. *Langmuir* **1986**, *2*, 672–677.

86. Schramm, L.L.; Green, W.H.F. *Colloids & Surfaces* **1995**, *94*, 13–28.

87. Schramm, L.L. In *Foams, Fundamentals and Applications in the Petroleum Industry*; Schramm, L.L. Ed.; American Chemical Society: Washington, DC, 1994, pp 165–197.

88. Mannhardt, K.; Novosad, J.J.; Schramm, L.L. *J. Petrol. Sci. Eng.* **1996**, *14*, 183–195.

89. Schramm, L.L.; Mannhardt, K. *J. Petrol. Sci. Eng.* **1996**, *15*, 101–113.

90. Mannhardt, K.; Schramm, L.L.; Novosad, J.J. *Colloids and Surfaces* **1992**, *68*, 37–53.

91. Mannhardt, K.; Novosad, J.J. In *Foams, Fundamentals and Applications in the Petroleum Industry*; Schramm, L.L. Ed.; American Chemical Society: Washington, DC, 1994, pp 259–316.

92. Novosad, J.J. In *Chemicals in the Oil Industry: Developments and Applications*; Ogden, P.H. Ed.; Special Publication 97, Royal Society of Chemistry: Cambridge, 1991, pp 159–173.

93. Chad, J.; Matsalla, P.; Novosad, J.J. In *Proceedings, CIM Annual*

Technical Meeting, Canadian Inst. Mining, Metallurgical and Petroleum Engineers: Calgary, AB, 1988, paper CIM 88-39-40.

94. Schramm, L.L.; Novosad, J.J. *Colloids and Surfaces* **1990**, *46*, 21–43.

95. Schramm, L.L.; Turta, A.; Novosad, J.J. *SPE Reservoir Engineering* **1993**, *8*, 201–206.

96. Schramm, L.L.; Novosad, J.J. *J. Petrol. Sci. & Eng.* **1992**, *7*, 77–90.

97. Manlowe, D.J.; Radke, C.J. *SPE Reservoir Engineering* **1990**, *5*, 495–502.

98. Hanssen, J.E.; Dalland, M. In *Proceedings, SPE/DOE 7th. EOR Symposium*, Society of Petroleum Engineers: Richardson, TX, 1990, paper SPE/DOE 20193.

99. Raterman, K.T. In *Proceedings, 64th Annual Technical Conference*, Society of Petroleum Engineers: Richardson, TX, 1989, paper SPE 19692.

100. Kristiansen, T.S.; Holt, T. In *Proceedings, 12th International Workshop and Symposium, IEA Collaborative Project on EOR*, AEA: Winfrith, UK, 1991.

101. Mannhardt, K.; Novosad, J.; Schramm, L.L. In *Proceedings, SPE/DOE Improved Oil Recovery Symposium*, Society of Petroleum Engineers: Richardson, TX; 1998, paper SPE 39681.

102. Rendall, W.A.; Ayasse, C.; Novosad, J.J. U.S. Patent 5 074 358, December 24, 1991.

103. Schramm, L.L.; Ayasse, C.; Mannhardt, K.; Novosad, J.J. U.S. Patent 5 060 727, October 29, 1991.

104. Rand, P.B. US Patent 4 442 018, 1984.

105. Patton, J.T.; Kuntamukkula, M.S.; Holbrook, S.T. *Polymer Preprints* **1981**, *22*, 46–48.

106. Dauben, D.L. U.S. Patent 3 530 940, 1970.

107. Ram Sarma, D.S.H.S.; Pandit, J.; Khilar, K.C., *J. Colloid Interface Sci.* **1988**, *124*, 339–348.

108. Sydansk, R.D. In *Proceedings, International Symposium on Oilfield Chemistry*, Society of Petroleum Engineers: Richardson, TX, 1993, SPE paper 25168.

109. Lionti-Addad, S.; di Meglio, J-M. *Langmuir* **1992**, *8*, 324–327.

110. Pradhan, M.S.; Sarma, D.S.H.R.; Khilar, K.C. *J. Colloid Interface Sci.* **1990**, *139*, 519–526.

111. Pradhan, M.S.; Khilar, K.C. *J. Colloid Interface Sci.* **1994**, *168*, 333–338.

112. Friberg, S.E.; Fang, J-H. *J. Colloid Interface Sci.* **1987**, *118*, 543–552.

113. Sydansk, R.D. In *Proceedings, International Symposium Oilfield Chemistry*, Society of Petroleum Engineers: Richardson, TX, 1993, SPE paper 25175.

114. Bernard, G.G.; Holm, L.W. U.S. Patent 3 393 738, 1968.

115. Minssieux, L. *J. Petrol. Technol.* **1974** (January), 100–108.

116. Zhukov, I.N.; Polozova, T.I.; Shatava, O.S. *Kolloidn. Zh.* **1987**, *49*, 758–762.

117. Hazlett, R.D.; Strom, E.T. U.S. Patent 4 844 163, July 4, 1989.

118. Miller, M.J.; Fogler, H.S. In *Proceedings, Annual Technical Conference*,

Society of Petroleum Engineers: Richardson, TX, 1992, SPE paper 24662.
119. Sydansk, R.D. U.S. Patent 5 105 884, April 21, 1992.
120. Sebba, F. *J. Colloid Interface Sci.* **1971**, *35*, 643–646.
121. Sebba, F. *Foams and Biliquid Foams – Aphrons*; Wiley: New York, 1987.
122. Sebba, F. *Chemistry and Industry* **1984**, *10*, 367–372.
123. Sebba, F. *Chemistry and Industry* **1985**, *4*, 91–93.
124. Roy, D.; Valsaraj, K.T.; Tamayo, A. *Sep. Sci. Technol.* **1992**, *27*, 1555–1568.
125. Roy, D.; Valsaraj, K.T.; Constant, W.D.; Darji, M. *J. Hazardous Mat.* **1994**, *38*, 127–144.
126. Roy, D.; Kommalapati, R.R.; Valsaraj, K.T.; Constant, W.D. *Water Res.* **1995**, *29*, 589–595.
127. Roy, D.; Kongara, S.; Valsaraj, K.T. *J. Hazardous Mat.* **1995**, *42*, 247–263.
128. Enzien, M.V.; Michelsen, D.L.; Peters, R.W.; Bouillard, J.X.; Frank, J.R. In *Proceedings, 3rd International Symposium In-Situ and On-Site Reclamation*, San Diego, CA, 1995, pp 503–509.
129. Mirzadjanzade, A.K.; Ametov, I.M.; Bogopolsky, A.O.; Kristensen, R.; Anziryaev, U.N.; Klyshnikov, S.V.; Mamedzade, A.M.; Salatov, T.S. In *Proceedings, 7th European IOR Symposium*, Moscow, Russia, 1993, pp 469–473.
130. Mirzadjanzade, A.K.; Ametov, I.M.; Bokserman, A.A.; Filippov, V.P. In *Proceedings, 7th European IOR Symposium*, Moscow, Russia, 1993, pp 27–29.
131. Meyer, R.F.; Wynn, J.C.; Olson, J.C., Eds., *The Future of Heavy Crude and Tar Sands*; UNITAR: New York, NY, 1982.
132. Mikula, R.J. In *Emulsions, Fundamentals and Applications in the Petroleum Industry*; Schramm, L.L., Ed.; American Chemical Society: Washington, DC, 1992, pp 79–129.
133. Malhotra, A.K.; Wasan, D.T. In *Thin Liquid Films*; Ivanov, I.B., Ed.; Dekker: New York, 1988, pp 829–890.
134. Cairns, R.J.R.; Grist, D.M.; Neustadter, E.L. In *Theory and Practice of Emulsion Technology*; Smith, A.L., Ed.; Academic Press: New York, 1976, pp 135–151.
135. Jones, T.J.; Neustadter, E.L.; Whittingham, K.P. *J. Can. Petrol. Technol.* **1978**, *17*, 100–108.
136. Grace, R. In *Emulsions, Fundamentals and Applications in the Petroleum Industry*; Schramm, L.L., Ed.; American Chemical Society: Washington, DC, 1992, pp 313–339.
137. Scoular, R.J. In *Supplementary Notes for Petroleum Emulsions and Applied Emulsion Technology*; Schramm, L.L., Ed.; Petroleum Recovery Institute: Calgary, AB, 1992.
138. Leopold, G. In *Emulsions, Fundamentals and Applications in the Petroleum Industry*; Schramm, L.L., Ed.; American Chemical Society: Washington, DC, 1992, pp 342–383.
139. Zaidi, A.; Constable, T. *Produced Water Treatment Design Manual*; Wastewater Technology Centre: Burlington, ON, 1994.

140. Bikerman, J.J. *Foams*; Springer-Verlag: New York, 1973, pp 250–251.
141. Shaw, R.C.; Czarnecki, J.; Schramm, L.L.; Axelson, D. In *Foams, Fundamentals and Applications in the Petroleum Industry*; Schramm, L.L., Ed.; American Chemical Society: Washington, DC, 1994, pp 423–459.
142. Shaw, R.C.; Schramm, L.L.; Czarnecki, J. In *Suspensions, Fundamentals and Applications in the Petroleum Industry*; Schramm, L.L., Ed.; American Chemical Society: Washington, DC, 1996, pp 639–675.
143. *Using Oil Spill Dispersants on the Sea*; National Research Council, National Academy Press: Washington, DC, 1989.
144. Bobra, M.A. In *Proceedings of the International Oil Spill Conference*; American Petroleum Institute: Washington, DC, 1991, pp 483–488.
145. Bobra, M.A. *A Study of Water-in-Oil Emulsification*; Manuscript Report EE-132, Environment Canada: Ottawa, ON, 1992.
146. Canevari, G.P. In *Proceedings of the Oil Spill Conference*; API Publication 4452, American Petroleum Institute: Washington, DC, 1987; pp 293–296.
147. Mackay, D. *Formation and Stability of Water-in-Oil Emulsions*; Manuscript Report EE-93, Environment Canada: Ottawa, ON, 1987.
148. Urdahl, O.; Sjöblom, J. *J. Dispersion Sci. Technol.* **1995**, *16*, 557–574.
149. Cormack, D.; Lynch, W.J.; Dowsett, B.D. *Oil Chem. Pollut.* **1986/87**, *3*, 87–103.
150. Fingas, M.F. *Spill Technology Newsletter* **1994**, *19*, 1–10.
151. Bocard, C.; Gatellier, C. In *Proceedings of the Oil Spill Conference*; API Publication 4452, American Petroleum Institute: Washington, DC, 1981, pp 601–607.
152. Fingas, M.F.; Duval, W.; Stevenson, G. *The Basics of Oil Spill Cleanup*; Government Publishing Centre, Ottawa, Ontario, 1979.
153. Fiocco, R.J.; Lessard, R.R.; Canevari, G.P.; Beckeer, K.W.; Daling, P.S. In *The Use of Chemicals in Oil Spill Response*; Lane, P., Ed.; American Society for Testing and Materials: Philadelphia, PA, 1995, pp 299–309.
154. Kerner, H.T. *Foam Control Agents*; Noyes Data Corp.: Park Ridge, NJ, 1976.
155. Currie, C.C. In *Foams*; Bikerman, J.J., Ed.; Reinhold: New York, 1953, pp 297–329.
156. Kitchener, J.A. In *Recent Progress in Surface Science*; Danielli, J.F.; Pankhurst, K.G.A.; Riddiford, A.C., Eds.; Academic Press: New York, 1964, Vol. 1, pp 51–93.
157. Lewis, V.E.; Minyard, W.F. In *Foams, Fundamentals and Applications in the Petroleum Industry*; Schramm, L.L., Ed.; American Chemical Society: Washington, DC, 1994, pp 461–483.
158. *The Asphalt Handbook*; The Asphalt Institute: College Park, MD, 1965.
159. Perri, J.M. In *Foams: Theory and Industrial Applications*; Bikerman, J.J., Ed.; Reinhold: New York, 1953, pp 189–242.
160. Nguyen, H.M. *Not All Firefighting Foam Systems Are Created Equal!*; U.S. Coast Guard Marine Safety Center: Washington, DC, 1998; also on http://www.bts.gov/smart/cat/foams.html.

161. Corrie, J.G. In *Foams*; Akers, R.J., Ed.; Academic: New York, 1976, pp 195–215.
162. Briggs, T. In *Foams: Theory, Measurements, and Applications*; Prud'homme, R.K.; Khan, S.A., Eds.; Marcel Dekker: New York, 1996, pp 465–509.

RECEIVED for review August 21, 1998. ACCEPTED revised manuscript September 21, 1998.

SURFACTANTS IN POROUS MEDIA

4

Surfactant Adsorption in Porous Media

Laura L. Wesson and Jeffrey H. Harwell

The University of Oklahoma, Norman, OK, USA

An overview of some of the significant findings of surfactant adsorption research is presented. Subjects include the importance of surfactant adsorption in petroleum applications, some history of surfactant adsorption research, the mechanisms which have been proposed to explain observed adsorption behavior, and a review of several significant surfactant adsorption studies. The emphasis of this review is understanding the mechanisms of surfactant adsorption as they relate to applications of surfactants in petroleum processes.

Introduction

Surfactants have a variety of applications in the petroleum industry, and surfactant adsorption is a consideration in any application where surfactants come in contact with a solid surface. In enhanced or improved oil recovery (EOR or IOR) surfactants can be used in classic micellar/polymer (surfactant) flooding, alkaline/surfactant/polymer (ASP) flooding or in foams for mobility control or blocking and diverting. Surfactants can act in several ways to enhance oil production: by reducing the interfacial tension between oil trapped in small capillary pores and the water surrounding those pores, thus allowing the oil to be mobilized; by solubilizing oil (some micellar systems); by forming emulsions of oil and water (alkaline methods); by changing the wettability of the oil reservoir (alkaline methods) or by simply enhancing the mobility of the oil [1]. In selecting a suitable surfactant for any EOR application, one of the criteria for economic success is minimizing surfactant loss to adsorption. Factors affecting surfactant adsorption include temperature, pH, salinity, type of surfactant and types of solids found in the reservoir. Usually the only factor which can be manipulated for EOR is the type of surfactant to be used; the other factors being determined by reservoir conditions.

When an oil reservoir is first produced, forces such as overburden pressure and evolution of gases dissolved in the reservoir oil cause spontaneous production of oil because of the pressure gradient between the interior of the reservoir and the production well. This spontaneous production is commonly referred to as primary recovery. Following the completion of the primary recovery phase, 60 to 80% of the oil originally in the reservoir commonly remains in the formation. Production has ceased because a pressure gradient no longer exists to mobilize the oil. Secondary recovery consists of water-flooding to displace the remaining oil from the injection to the production well.

Nevertheless, a point is soon reached where the amount of oil produced by water-flooding is insufficient to justify the operating costs of the project. At this time it is common for 30 to 60% of the original reservoir oil to remain in the formation. The oil is trapped in the pores of the rock by capillary forces arising from the high oil/water interfacial tension. Additional water injected into the formation simply bypasses the trapped oil droplets on its way to the production well, following the path of least resistance to the flow.

It has long been known that surfactants lower oil/water interfacial tensions, thus reducing capillary forces such as those trapping the remaining oil. This raises the possibility of releasing trapped oil droplets by injecting surfactants into the reservoir. Early demonstrations of the technical feasibility of enhanced oil recovery by surfactant flooding (sometimes referred to as micellar or chemical flooding) were done in the laboratory by Novosad et al. in 1982 [2] and in field tests by Lake and Pope in 1979 [3] and by Holm in 1982 [4]. In addition to the technical feasibility, economic feasibility must also be determined; however, the economic feasibility depends on a complex of factors such as oil prices, international economies, and the cost of the surfactants. Generally, the cost of the surfactant is the single most expensive item in the cost of a chemical flood. These costs include both the initial investment in purchasing the surfactant, as well as the cost of replacing surfactant which has been lost to adsorption. It is frequently found that the amount of surfactant adsorbed accounts for most of the cost of the surfactant. Since these surfactants are synthesized from petroleum, their costs will rise at least as fast as that of the oil they are used to produce. So simply waiting for oil prices to increase will not necessarily make EOR economically feasible. The oil produced by a chemical flood must then be sufficient to replace the oil used for the surfactant (unless some means of recovering the surfactant from the reservoir is feasible), to pay for the price of producing the surfactant from the oil, to pay for all the additional engineering, equipment and operating costs during the several years the flood is occurring, and to provide a reasonable return on investment. All of these demands must be satisfied in a volatile oil market in which oil prices may fluctuate between the beginning of a surfactant flood and the time the tertiary oil is finally produced. Producing more barrels of oil for each kilogram of surfactant injected into the reservoir is a technological problem that has direct bearing on the economics of the process. Understanding and controlling the amount of surfactant adsorbed directly affects project economics. The following is a sample calculation which illustrates just how substantial the costs associated with losing surfactant to adsorption can be.

An area one acre (4047 m^2) by 3 meters deep is to be swept with a surfactant solution. Core samples reveal that the subsurface is approximately 70% solid material having a density of 2.5 g/cc. Thus approximately 2.12×10^{10} grams of solid material are available for adsorption of the surfactant. If the specific area of this solid were 0.5 m^2/g then the surface area of the solid would be 1.06×10^{10} m^2. Assuming surfactant adsorption reaches bilayer coverage at a density of 1 molecule per 0.5 nm^2 of available surface area (typical for an ionic surfactant), then approximately 3.5×10^4 moles of surfactant would be adsorbed onto the solid. If the surfactant had an average molecular weight of 500 g/mol this would result in 1.76×10^4 kilograms of surfactant being adsorbed. Assuming a purchase price of $2/kg for this surfactant, then the resultant loss to adsorption would be $34,300. If this surfactant were being used to produce oil worth $18/bbl then 1960 barrels would have to be produced just to compensate for the adsorbed surfactant. Looking at this situation in terms of EOR, if there is 50% residual saturation and EOR is expected to remove 50% of the residual or 5727 barrels of oil, then the cost of the surfactant loss to adsorption would account for approximately one-third of the total value of the oil recovered by EOR. Obviously, it is critical to the economic success of an EOR project that adsorption be minimized in the design of the project; to do so requires an understanding of surfactant adsorption mechanisms.

In the first part of this chapter, reviews of the background research on surfactant adsorption and the mechanisms involved in surfactant adsorption are presented. In the second part of the chapter, several pertinent experimental studies are presented which illustrate the mechanisms of surfactant adsorption in various systems. As already stated, there are multiple factors which affect adsorption These factors will now be presented, beginning with the characteristics of the solid materials commonly used in adsorption studies.

Solid Surface Chemistry

Many surfactants adsorb onto a solid due, in a large part, to the electrostatic interactions between charged sites on the solid surface and the charged headgroups of the ionic surfactants. The adsorption of nonionic surfactants is discussed later in this chapter. The structures of several solids used in adsorption research and the electrical properties associated with them are discussed in this section. Most mineral surfaces in reservoirs can be assumed to be charged.

Types of Solids. There have been a variety of solids used in surfactant adsorption research. These solids have included "ideal" reservoir materials such as alumina (Al_2O_3) and silica (SiO_2), and "real" materials such as kaolinite clays, river alluvium, and sandstones.

There are several crystalline phases of alumina arising from the different configurations possible for the aluminum and oxygen ions. The surface charge on alumina in contact with a surfactant solution arises indirectly from the crystal structure of the alumina. The most commonly used alumina in adsorption studies has been α-alumina or corundum which has a rhombohedral crystal structure comprising a hexagonal close-packed array of oxygen ions with aluminum ions on two-thirds of the octahedral sites (5). The other two forms of alumina are the η-phase, which has a cubic structure, and the θ-phase, which has a monoclinic structure.

Crystalline silica can exist as quartz, cristobalite, and tridymite, with quartz being the form most commonly used in adsorption studies. Many studies also use amorphous silicon oxides. The quartz crystal consists of silica tetrahedra with the silicon ions located in the center and the oxygen ions located at the corners. The tetrahedra are arranged to form interlinked helical chains [5]. The different forms of quartz are distinguished by the differences between the angles formed by the Si—O—Si bond, with the α-form being the most common.

Kaolinite is a clay mineral with the chemical formula: $Al_2(OH)_4Si_2O_5$ [5] or $(OH)_8Si_4Al_4O_{10}$ [6]. The basic unit of kaolinite consists of a single silica tetrahedral sheet and a single alumina octahedral sheet such that the oxygen atoms at the tips of the silica tetrahedrons and one of the oxygen atoms of the alumina octahedral sheet form a common layer.

In illustrating the mechanisms of surfactant adsorption we will discuss adsorption on a river alluvium. River alluvium from the Canadian River in Cleveland County, Oklahoma has been used in several recent adsorption studies [7–9]. Palmer et al. [7] profiled the alluvium and found that it consisted of 91% sand, 2% silt and 7% clay.

An adsorption medium often considered typical of reservoir solids is sandstone. Sandstone is an agglomeration of individual minerals, but the primary component is usually quartz. Other minerals comprising sandstone include chert, feldspar, mica, illite, kaolinite and calcium carbonate. A common type of sandstone used in adsorption research is Berea sandstone [10, 11].

Other solids used in surfactant adsorption research include rutile (TiO_2) [12–15], carbonates, and graphite [16–19]. Studies with carbonates

have included purified calcium and magnesium carbonate [20] and
Indiana limestone [11] and Baker dolomite [11, 21].

With the exception of graphite, the common characteristic of the
solids used in adsorption research is the capacity of the surface of these
solids to have an electrical surface charge. This capacity arises from the
interaction between the oxygen atoms in the structure and water
molecules. It is especially significant to note that under typical reservoir
conditions carbonates and sandstones have opposite charges.

**Electrical Characteristics and the Electrical Double
Layer.**   Electrical surface charges arise from charge imbalances due to
imperfections in the crystal structure and preferential adsorption of
counter or potential determining ions [22, 23]. At low surfactant concentrations the surface charge largely determines the surfactant adsorption.
However, as the surfactant concentration increases other factors such as
the tendency of the surfactant to aggregate, become significant.

Imperfections in the crystal structure include isomorphous replacement of ions within the crystal lattice, broken bonds, dislocations, and
lattice defects [24]. Ion replacement leads to a charge imbalance within
the lattice resulting in a charged surface. A common substitution is the
replacement of silicon atoms in kaolinite by aluminum atoms.

When a surface is fractured, bonds between layers, such as the
alumina–silica layers in kaolinite or the metal–oxygen bonds in alumina,
can be broken, leaving ions with unsatisfied valence conditions. The
resulting charge can be either negative or positive depending on the type
of bond broken. A related source of surface charge is the partial
dissolution of the solid surface by water. This also leaves surface ions
with unsatisfied valences.

Lattice defects are holes within the lattice due to missing ions. The
missing ions leave the lattice with unbalanced charges. Charge imbalances can also arise in crystal structures due to dislocations. There are two
types of dislocations. In the screw dislocation a section of a crystal is
skewed one atom spacing. In the edge dislocation an extra plane of atoms
has been inserted into a section of a crystal. The charge imbalances arise
at the sites of the dislocations.

Charge imbalances and broken bonds are accommodated by chemical
adsorption of water by the solid surface. The chemically adsorbed water
molecule forms an amphoteric site on the surface. Deprotonation of the
group leaves a negative charge on the surface. Protonation of the
amphoteric group leads to a positive charge on the surface. This charging
mechanism makes the surface charge highly dependent on the pH of the
contacting solution.

Electrical Double Layer.   One of the earliest theories proposed
for explaining interactions of charged particles at the solid/liquid interface

was the electrical double layer theory developed to describe the formation of a charge on a mercury electrode surface. In early studies of surfactant adsorption on minerals, the surfactant concentrations were low enough that there were no interactions between surfactant monomers on the solid surface. The simple electrical double layer model developed for mercury electrodes was adequate for describing the adsorption behavior. When the surfactant concentrations increased to levels where surfactants began to interact with one another at the surface, this theory could no longer describe the adsorption behavior. The addition of higher electrolyte concentrations also affected the ability of this theory to describe adsorption behavior. The equations given below provide an introduction to the kinds of interactions which must be considered in describing adsorption behavior. In addition this theory serves as a starting point for many of the more complex models of surfactant adsorption. Some discussion of current models is provided in the next section.

The adsorption of counterions or potential determining ions at relatively low concentrations can be addressed by the concept of the electrical double layer which develops in response to a charge on the mineral surface. An electrical potential exists across an interface when there is an unequal distribution of charges across that interface. This unequal distribution results in each side of the interface acquiring net charges of opposite sign.

The idea of the electrical double layer was proposed by Helmholtz in 1879, and modified by Stern in 1924 [25]. In the Stern modification the counterions in the solution, opposite in charge relative to the surface, were divided into two layers: (1) a layer of ions adsorbed close to the surface (generally referred to as the Stern layer) and (2) a diffuse layer of counterions sometimes referred to the Gouy layer. As shown in Figure 1, the potential decreases rapidly within the Stern layer (δ) and more gradually within the diffuse layer (d). The net charge in the Stern layer plus the Gouy layer is equal and opposite in sign to the surface charge. For minerals the surface charge is primarily controlled by the pH and the nature of the mineral.

The diffuse layer charge, σ_d, which extends out from the plane δ seen in Figure 1, can be described by the following:

$$\sigma_d = -\sqrt{\frac{2\varepsilon kT}{\pi}}\sqrt{n_0}\sinh\left(\frac{Ze\Psi_\delta}{2kT}\right) \qquad (1)$$

where ε is the dielectric constant of water, k is Boltzmann's constant, T is the absolute temperature, Z is the valence charge including the sign of the adsorbing ion, e is the elemental charge, Ψ_δ is the electrical potential at the plane a distance δ from the surface (the Stern plane) and n_0 is the

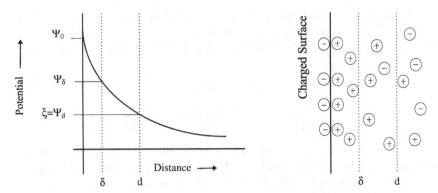

Figure 1.   Gouy–Chapman model of the electrical double layer and the potential distribution where δ is the Stern plane within which counterions are adsorbed close to the surface and d is the diffuse layer of counterions.

number of ions/cc in the bulk phase (where the electrical potential is zero).

The adsorption of counterions at the plane δ from the surface can be described by the Gouy–Chapman equation as:

$$\Gamma_\delta = 2rC \exp\left(\frac{-W_\delta}{kT}\right) \tag{2}$$

where Γ_δ is the adsorption density, r is the radius of the adsorbed ion, C is the concentration of ions in the bulk, and W_δ is the work required to bring ions from the bulk solution to the plane δ and is comprised of electrostatic and interaction terms:

$$W_\delta = Ze\Psi_\delta - \phi \tag{3}$$

In equation 3 $Ze\Psi_\delta$ is the electrical work of bringing the ion into the Stern plane, and ϕ is the free energy change associated with the partial removal of the alkyl chain for a surfactant from the water phase.

Assuming for an alkyl chain of n carbon atoms,

$$\frac{\phi}{kT} = \frac{n\phi'}{kT} \tag{4}$$

where ϕ' is the interaction energy per CH_2 group between adjacent chains of adsorbed surfactant molecules. This interaction begins when the bulk surfactant concentration reaches the hemimicelle concentration, which is the concentration at which the first surfactant aggregates form on the solid surface. Details of the concept of the hemimicelle are presented in the next section of this chapter.

In applying these equations to surfactant adsorption research for surfactant concentrations greater than the hemimicelle concentration, Somasundaran et al. [26] put equation 2 into logarithmic form and differentiated to give:

$$\frac{d \ln \Gamma_\delta}{d \ln C} = 1 - \frac{d\left(\frac{Ze\Psi_\delta}{kT}\right)}{d \ln C} - \frac{\phi'}{kT} \frac{dn}{d \ln C} \tag{5}$$

where n is the number of carbon atoms and $dn/d \ln C$ indicates that the effective number of carbon atoms that can be removed totally from the aqueous environment by chain–chain association increases as the surface coverage increases [23].

The concept of the electrical double layer works well in describing the behavior of simple ions like Na^+ or Cl^- or single ions like surfactants when a nonelectrostatic term is added to the adsorption potential. For potential determining ions such as H^+ and OH^- however, the "site-binding" model is frequently used. This model is used to describe the development of a surface charge at a mineral/solution interface. It requires knowing the reactions responsible for surface charge development and the potential–charge relationships at the interface. It also limits the concentration of the surface species to the total number of sites available on the surface. These interactions are specific for individual systems [27–29].

The adsorption of H^+ and OH^- at the surface affects the charge on the surface of the solid. The charge on the surface can be negative, positive or neutral. The neutral condition is referred to as the point of zero charge or pzc. The pzc is the pH at which the net charge on the surface is zero. At a pH value above or below the pzc the surface is negatively or positively charged, respectively. In the case of alumina with a pzc of approximately 9 [22], the surface is positively charged at a pH less than 9 and negatively charged for pH values greater than 9. For silica the pzc is 2–3 [22], so the surface is negative above pH 3. For kaolinite the pzc is approximately 4.5 [24]. If adsorption is desirable then the surfactant and surface should have opposite charges. If adsorption is undesirable, which is the case for applications such as EOR, then it may be advantageous to have a surfactant with the same charge as the solid. Surfactants will still adsorb on like charged surfaces, however, especially at high concentrations (above the CMC) and in the presence of multivalent counterions.

Mechanisms of Surfactant Adsorption

Single Surfactant Systems.   Surfactant adsorption at the solid/liquid interface has been studied for several decades. Much of the early work reported in the literature was based on selecting the most effective surfactant for purifying ores by flotation. These studies focused on determining the interactions which bring about adsorption and determining the structures of the surface aggregates formed. Currently there is general agreement on the interactions which bring about adsorption, but there is still much discussion concerning the structure of the surfactant surface aggregates.

Some of the experimental techniques employed in these studies have included determining the change in surfactant concentration in the bulk solution upon adsorption, zeta potential measurements, and probe techniques (electron spin resonance and fluorescence). Attempts to describe the adsorption behavior exhibited in the adsorption isotherms has led to the development of several mathematical models [26, 30–33]. To date, none of the models are capable of fully accounting for all of the phenomena which affect surfactant adsorption without introducing ad hoc assumptions and adjustable parameters, but they have offered some interesting insights.

Most adsorption studies have employed the surfactant depletion method with the results being presented as isotherms which are simply plots of the amount of surfactant adsorbed per gram of solid or per surface area of solid versus the equilibrium surfactant concentration at a constant temperature. These plots can be constructed using log–log, linear–log or linear–linear scales with the most common choice being the log–log scale. Koopal [34] presents a discussion of the advantages and disadvantages for the different scales. The log–log scale can be used to obtain information over wide ranges of adsorption and surfactant concentrations, and the plots generally have abrupt changes in slope with increasing surfactant concentration. A typical four-region isotherm constructed on a log–log scale for a monoisomeric anionic surfactant is shown in Figure 2. The reasons for the changes in slope are discussed below.

Not all log–log isotherms seen in the literature consist of four regions. Some of the earliest adsorption studies were conducted by de Bruyn, 1955, Shinoda, 1963, and Jaycock and Ottewill, 1963 using surfactant concentrations well below the CMC, and the reported isotherms consisted of only two regions [26]. Further studies showed that at higher surfactant concentrations log–log isotherms exhibited three distinct regions [35] and at still higher concentrations four regions [36, 37]. It is important to note that the exact shape of the isotherm will depend on several factors including the type of surfactant, the charge on the surface, and the presence or absence of additional compounds including electrolytes, co-surfactants, hydrotropes or alcohols.

The mechanisms driving surfactant adsorption are generally discussed in terms of the four-region isotherms. At low surfactant concentrations, designated as region I (see Figure 2), the adsorption behavior can usually

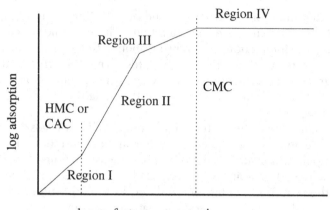

Figure 2.   Typical four-region adsorption isotherm for a monoisomeric surfactant.

be described by Henry's Law, i.e. linear with a slope of one. This is also the region where the simple Stern/Gouy double layer model is appropriate. Early work by de Bruyn [38] and Gaudin [39, 40] determined that in this region surfactant monomers adsorbed as individual ions with no interaction between the adsorbed molecules. This conclusion was based on the zeta potential measurements of quartz/dodecylammonium chloride systems at low surfactant concentrations being nearly identical to the zeta potential measurements of quartz/sodium chloride systems [40]. Today, it is known that the surface–surfactant interaction depends on the type of surfactant. For nonionic surfactants the interactions involve hydrogen bonding between surface hydrogens and proton acceptors in the head groups and hydrophobic bonding between the hydrocarbon tails of surfactants and the surface. Hydrophobic bonding is explained in Chapter 1 of this book. Scamehorn et al. [30] and Harwell et al. [31] showed that a tail–surface interaction involving adsorbed monomers affects the value of the Henry's Law coefficient. For ionic surfactants there are electrostatic interactions between the head groups of the surfactants and charged sites on the surface. This electrostatic attraction is typically described in terms of the interaction of the charged surfactant ion with the electrical double layer of the solid.

The mechanism dominating adsorption in region II was described in 1955 by Gaudin and Fuerstenau [40] as being due to the association of the adsorbed surfactants into patches at the solid/liquid interface. These associations were attributed to tail–tail interactions, which are the same hydrophobic interactions by which micelle formation is described today.

The region I/region II break corresponds, therefore, to the surfactant concentration at which the first surfactant aggregates form on the surface.

This concentration is referred to as the hemimicelle concentration (HMC) [35] or as the critical admicelle concentration (CAC) [31]. This aggregate formation can be viewed as a two-dimensional phase transition occurring on the highest energy patches on the solid surface [30]. The CAC/HMC varies with surfactant chain length and branching in the same manner as CMC varies with these parameters [41]. If the system contains ionic surfactants, the addition of an electrolyte will decrease the CAC in the same manner that electrolytes reduce the critical micelle concentration (CMC) [42]. A note of practical application: in systems with added electrolyte care must be taken to avoid precipitation of the surfactant by the electrolyte [43]. The presence of a precipitate is easily hidden by the solid material upon which adsorption is supposed to be occurring, and the decrease in surfactant concentration due to precipitation could be interpreted as greater surfactant adsorption than what is actually occurring. Familiarity with the precipitation phase boundaries of the surfactant for a given electrolyte or preliminary precipitation analyses using the surfactant and electrolyte concentrations of interest may eliminate this error.

In region III a decrease in the slope relative to the slope in region II is seen. There have been several theories proposed to explain this change. Somasundaran et al. [26, 35] attributed this change in slope to the surfactant ions having filled all of the surface sites by the end of region II with further adsorption being due to association between first and second layer hydrocarbon chains in region III. The observed change in slope was also attributed to a reversal in surface charge due to the adsorbed surfactant ions. Scamehorn et al. [30] proposed that bilayer formation began in region II and continued into region III but at a different rate. This can also be viewed as adsorption taking place on the least energetic patches on the surface in region III.

Region IV or plateau adsorption generally begins at or near the critical micelle concentration (CMC) and is characterized by little or no increase in adsorption with increasing surfactant concentration. In this region micelles exist in the bulk solution and act as a chemical potential sink for any additional surfactant added to the system. Most researchers agree that the surfactant aggregates have a bilayer structure when the solution concentration exceeds the CMC. The total adsorption above the CMC may still be substantially less than complete a bilayer, however, and depends strongly on surface charge and, therefore, pH.

There have been several surfactant structures proposed in attempts to describe the adsorption isotherm. Two of them have been mentioned already, the hemimicelle and local bilayer or admicelle. In 1955, Gaudin and Fuerstenau [40] introduced the term hemimicelle to describe the adsorption behavior they had observed in region II. Hemimicelles can be described as aggregates of adsorbed surfactant molecules in which the surfactant monomers are arranged in a single layer with the head groups facing the solid surface. This structure was proposed to explain the increased adhesion of bubbles to the surface of minerals in region II of the adsorption isotherm. The bilayer structure consists of surfactant monomers arranged such that the head groups of the first layer are facing the surface and those of the second layer face the surrounding solution. The tail groups of the two layers interact in the same manner as they do in micelles. The bilayer structure was first proposed in the 1940s [40]. In 1985 the term admicelle was introduced and was used to describe surfactant surface aggregates which were bilayered in structure and which had formed without an intermediate hemimicelle structure existing at a lower surfactant concentration [31]. Such structures almost certainly dominate at the CMC when the total surface coverage is well below complete bilayer coverage.

Some additional structures which have been proposed are the surface micelles proposed by Gao [44], and the hemicylinders and cylinders proposed by Manne et al. [17, 18]. Surface micelles are aggregates described as spheres with only one surfactant monomer adhering to the solid surface. The hemicylinder and cylinder structures were based on atomic force microscopy (AFM) images of several surfactant systems involving adsorption of cetyltrimethylammonium bromide on pyrolitic graphite [17], tetradecyltrimethylammonium bromide ($C_{14}TAB$) on silica, $C_{14}TAB$ and didodecyldimethylammonium bromide on mica [18], and $C_{14}TAB$, hexadecyltrimethylammonium hydroxide, and sodium dodecyl-sulfate on gold [45]. These cylindrical structures are arranged such that the head groups of the surfactants are facing outward.

As the study of surfactant adsorption has evolved, the debate over the exact structure of the adsorbed surfactant aggregates has become more confused rather than becoming clarified. Until the recent advent of AFM studies most of the debate had focused on monolayer (hemimicelle) and bilayer (admicelle) structures. Current literature indicates that many researchers are beginning to believe that the structure of the adsorbed surfactant depends on the system being studied. In 1989 Somasundaran et al. [46] introduced the term solloid to describe any surfactant aggregates at the solid/liquid interface without attempting to define its morphology. Despite the uncertainty or at least the complexity of the structure of adsorbed surfactant aggregates, it is clear that micelle-like aggregates form spontaneously at concentrations well below the bulk CMC. Also, a complete bilayer is formed at the maximum adsorption of surfactants adsorbing onto surfaces of opposite charge.

Mixed Surfactant Systems.   Most surfactant systems used in the petroleum industry are comprised of more than one surfactant. The similarities and differences between pure component and mixed surfactant systems have been presented by Harwell and Scamehorn [47]. As adsorption behavior from single surfactant systems mirrors the behavior of micelle formation in solution, so too does adsorption from mixed surfactant systems mirror mixed micelle behavior. Since there is no interaction between surfactant molecules in region I, adsorption in this region for a mixed surfactant system is driven by the same interactions as for single surfactant systems, and the surfactants adsorbing from a mixture will behave as their pure components; but as the surfactant concentrations increase, the position of the region I/region II break may shift relative to the break in the single component adsorption isotherms. Adsorption from a mixture may fall in region II when the adsorption of either pure component would still be in region I. This behavior is exactly analogous to the lowering of the CMC in mixed surfactant systems.

The position of the remainder of the isotherm (regions II, III and IV) relative to the adsorption isotherms of the pure component will depend on the types and amounts of surfactants in the mixture. When surfactants of similar head groups are mixed, the adsorption of the mixture will vary monotonically between the adsorptions of the pure components. This is the same as the CMC of the mixture varying monotonically with the mole fraction of each component. If the mixture exhibits negative deviations from ideal mixing behavior such as when ionic and nonionic surfactants are mixed, then the CAC will also exhibit negative behavior. That is the mixture CAC will be lower than either of the CAC's of the pure components. If anionic and cationic surfactants are mixed then deviations more negative than those seen for ionic–nonionic systems will be seen. This results in CAC's that will again be lower than that of either of the pure components adsorptions. To summarize: mixtures exhibiting non-ideal behavior can produce the same surface coverage but with lower total surfactant concentrations relative to the pure component systems.

Experimental Studies

Surfactant adsorption research covers many disciplines. Theoretical studies include attempts to create models capable of accounting for every facet of surfactant adsorption including determining the structure of the adsorbed aggregates and determining the mechanisms driving the adsorption process. Practical studies focus on evaluating surfactant systems suitable for applications like ore flotation, improved oil recovery, in situ and ex situ soil remediation (a field which has its origins in EOR), cleaning applications, surfactant based separation processes, and wetting. Often the results obtained from a study have both theoretical and practical applications.

The literature review in this section is intended as an introduction to the types of research that have been done; and is divided into the following sections: general adsorption studies with an emphasis on those results which were significant in furthering the basic understanding of surfactant adsorption; applied studies with an emphasis on EOR and related fields; and recent studies involving gemini surfactants, a new class of high performance surfactants.

Fundamental Adsorption Studies. This section presents a few of the studies which were fundamental to understanding the mechanisms of surfactant adsorption and several recent studies which have served to expand our basic knowledge.

Cationic Surfactant onto Quartz. In the early 1950s attempts were made to move the ore flotation process from an art to a science. These attempts were driven by the more complex ores being mined and the recognized need for a systematic approach to selecting a suitable collector (surfactant) for a given ore. The purpose of the collector was to promote adhesion of ore fines to bubbles sparged into a slurry of ore. The early studies focused on quartz using dodecylammonium acetate concentrations which spanned what are now termed regions I and II adsorption. Figure 3 is the isotherm which Gaudin and Bloecher [48] obtained. They noted that the observed change in slope in the isotherm occurred slightly below the bulk critical micelle concentration (CMC), and that the adsorption was reversible. They also found that the amount adsorbed in a flotation process which resulted in almost complete recovery of the oxide was under 5% of the amount required for monolayer coverage.

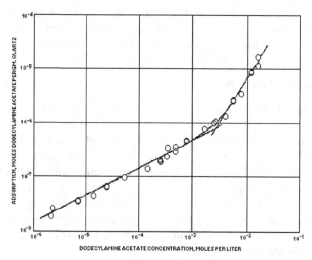

Figure 3.   Two-region adsorption isotherm of dodecylamine on quartz [48].

Based on zeta potential measurements of dodecylammonium chloride adsorption onto quartz [40], it was proposed that the observed change in the adsorption behavior was due to the association of the adsorbed surfactants into patches at the solid/liquid interface. It was hypothesized that these aggregates of adsorbed surfactant formed for the same reasons surfactant monomers associate to form micelles in solution, and the term hemimicelle was introduced. The aggregates were proposed to be "half" micelles on the surface because the surface now spontaneously dewet to allow bubble attachment. Zeta potential measurements showed that the surface potential changed from negative to positive in systems containing multivalent ions. This change in potential was also observed in systems containing dodecylammonium ions. Based on these observations it was proposed that the association of adsorbed ammonium ions acted as multivalent cations.

Continuing research led to the application of electrical double layer theory to describe surfactant adsorption. This theory was applied to sodium dodecyl sulfonate adsorption onto alumina [35, 37]. Region III adsorption was observed and attributed to the surfactant ions having filled all of the first layer sites by the end of region II with further adsorption being due to association between of the first layer hydrocarbon chains and second layer hydrocarbon chains. The region II/III transition was attributed to reversal of the surface charge leading to repulsion of the surfactant ions from the surface in region III thus reducing the electrical component of the adsorption potential. For a time after these studies, researchers tended to focus on the electrical interactions between the surfactant ions and the surface while other features, such as patchwise adsorption and the presence of bilayers, were overlooked.

Cationic Surfactants onto Silica.   The effect of surfactant types on the adsorption mechanism was illustrated by a study of dodecylpyridinium chloride (DPC) and cetylpyridinium chloride (CPC) onto silica [49]. Due to the more hydrophobic nature of the silica surface relative to rutile and alumina, both the head and tail groups of DPC and CPC can interact with the solid surface. Adsorptions were conducted with various pH values and salt concentrations. The resulting isotherms, on a logarithmic scale, consisted of four regions, but the shape of the isotherms depended on the potassium chloride concentration and the pH. Figure 4 illustrates the differences in these isotherms relative to the typical isotherm shown in Figure 2 for varying pH values. Adsorption of DPC onto the commercial silica Aerosil was also conducted at two salt concentrations (0.001 M and 0.1 M) For both salt concentrations used, the slopes of the region I adsorptions were approximately one, and the region IV plateau adsorption was only slightly higher for the higher salt system. For the low salt system region II appeared as a pseudoplateau.

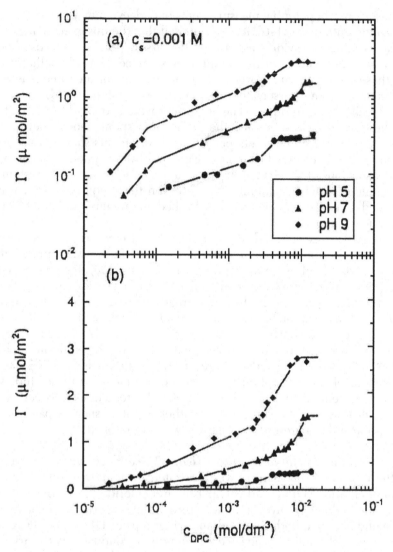

Figure 4. *Adsorption isotherm of DPC onto Aerosil OX50 [49].*

This behavior was attributed to both the head and tail groups adsorbing onto the surface and thus inhibiting adsorption. For the high salt concentration there was no pseudoplateau but a steep increase in the slope as is typically seen for surfactant adsorption on mineral oxides. This steep increase in the slope was attributed to the salt ions being able to screen the head group repulsion. As in micelle formation this screening allows the head groups to approach each other more closely and facilitate surfactant aggregation. Similar results were obtained for tetramethylammonium bromide (TMAB) and cetyltrimethylammonium bromide (CTAB) [50].

Anionic Surfactant onto Alumina and Kaolinite. A study involving the adsorption behavior of isomerically pure alkylbenzene sulfonates onto alumina and kaolinite and the development of a predictive patchwise adsorption model to explain the observed isotherm examined the underlying forces causing surfactant adsorption and provided information which could aid in minimizing the loss of surfactant to adsorption in EOR [30]. The resulting isotherms did not exhibit apparent adsorption maxima or minima that had been seen in previous studies. The presence of maxima and minima in earlier studies was attributed to using surfactants which were not monoisomerically pure and to the interactions which occur between the isomers in mixed micelles during adsorption. This is analogous to the minima observed in surface tension curves for mixed surfactant systems. The agreement between the theoretical calculations and experimental data was good for both mineral oxide systems except for the region just prior to the plateau region for the kaolinite isotherms. The structures predicted by the patchwise adsorption model were unassociated molecules in region I, hemimicelles then mixtures of hemimicelles and bilayers in region II, and bilayers in regions III and IV.

A second aspect of this work was comparing the plateau adsorption on alumina at varying pH values to bilayer values calculated from adsorption densities for monolayers. The monolayer values were estimated to range from 1.94 to 2.87 molecules/100 $Å^2$. These estimates were determined from surface tension data, film pressure studies on sodium dodecyl sulfonate, and sulfonate head group densities for cubic packing. When the plateau adsorption values were compared with adsorption densities for bilayer coverage calculated from the monolayer values it was observed that the plateau adsorption values fell in between the values calculated for bilayer coverage below pH 7 but fell below the bilayer range for adsorption above pH 7, thus indicating the formation of bilayers at pH values far below the pzc of approximately 9. As the pH approaches the pzc it is only natural that there would not be complete bilayer coverage since the charge on the surface is becoming less positive.

Mixture of Anionic Surfactants onto Alumina. Most EOR surfactants are mixtures of isomers, but these mixtures are too complex for application of basic theory. In contrast, the effectiveness of ideal solution theory in explaining region II adsorption for binary mixtures of anionic surfactants has been demonstrated [51]. These controlled isomeric mixtures allow application of the ideal solution theory. The application of this theory utilized a reduced adsorption equation for mixtures of anionic surfactants [52]. The parameters for this reduced equation were obtained from the individual adsorption isotherms for sodium octyl sulfate (C_8SO_4), sodium decyl sulfate $(C_{10}SO_4)$ and sodium dodecyl sulfate $(C_{12}SO_4)$ onto α-alumina at 30 °C. The alumina had a surface area of 160 m^2/g and the pH was adjusted to produce an equilibrium pH of 8.4. This pH results in a positive charge to the alumina surface leading to high adsorption of the anionic surfactants. Figure 5 illustrates the agreement between ideal solution theory and experimental

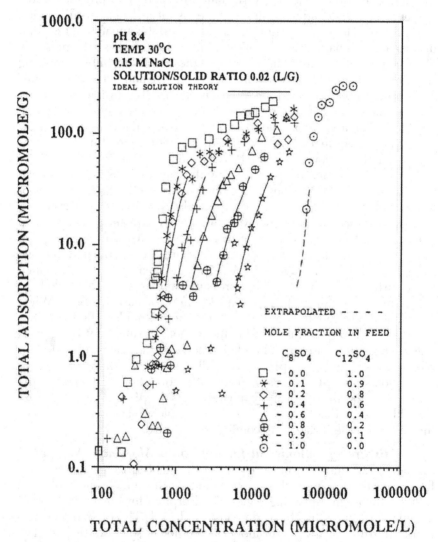

Figure 5.   Mixed adsorption isotherms of C_8SO_4 and $C_{12}SO_4$ onto α-alumina [51].

data for a binary mixture of $C_8SO_4/C_{12}SO_4$. Agreement was also demonstrated for a binary mixture of $C_{10}SO_4$ and $C_{12}SO_4$ on γ-alumina [53]. Regular solution theory has been shown to describe adsorption of mixtures of anionics and nonionics [54]. One important observation in these mixture studies is the reinforcement of the view that micelle formation and mixed micelle formation play a central role in the behavior of such systems, as proposed earlier by Trogus et al. [55]. Another important conclusion is that mixed admicelle and hemimicelle formation is very similar to that of mixed micelle formation.

Cationic Surfactants onto Porous Silicas.   Remarkably, while many of the studies of surfactant adsorption have been on porous materials, little attention has been paid to the effect of pore size on the isotherms. Indeed, all of the models of surfactant adsorption that have been developed ignore the effect of pore structure on the electrical double layer, treating the surface as a plane. Recently, the influence of pore size on the adsorption of cationic surfactants onto porous and nonporous silicas was examined in a study using the cationic surfactants hexadecylpyridinium chloride and dimethylbenzyltetradecylammonium chloride (TBzCl) [56]. The porous silicas were Sorbsil C30 from Rhone-Poulenc with an average pore volume of 0.6 ml/g and Sipernat 50S from Degussa-France with a pore volume of 0.003 ml/g. The corresponding BET surface areas are 700 and 450 m^2/g, respectively. It might be expected that the higher surface area silica would have the highest plateau adsorption, but this was not the case. For TBzCl in 0.01 mol/L NaCl the maximum adsorption on the Sorbsil C30 was 5.5×10^{-6} mol/g and on Sipernat 50S it was 9×10^{-6} mol/g. This behavior was attributed to the Sipernat 50S having large pore diameters while the Sorbsil had small diameters. This behavior had been seen in a previous study [57] using a nonionic surfactant, Triton X-100 adsorbed onto various silicas with well characterized pore radii. Again, as the BET surface areas increased, the pore radii decreased and the plateau adsorption decreased.

Applied Adsorption Studies.   This section deals primarily with the application of surfactant adsorption to EOR processes and related fields. From early work involving the formation of optimum microemulsions (three phase or Winsor Type III systems) to the current use of surfactants in other tertiary processes such as foam, CO_2, steam, and alkaline floods, surfactant adsorption has always played a significant role in surfactant selection. The following are just a few of the many articles that have involved studies of surfactant adsorption. These articles range from having surfactant adsorption as the primary topic to those in which adsorption is but one facet of the work being presented.

Anionic Surfactants onto Kaolinite and Illite.   In the investigation of the adsorption of sodium dodecylbenzenesulfonate (SDBS) and sodium dodecyl sulfate (SDS) onto asphalt covered kaolinite and illite surfaces, Siffert et al. [58] observed Langmuir type I isotherms for SDS adsorption onto Na^+ kaolinite and Na^+ illite while the SDBS exhibited a maximum in adsorption with a decrease beginning near the CMC. Adsorption maxima were observed near the CMC for both surfactants in the Ca^{2+} kaolinite and Ca^{2+} illite systems. The adsorption behavior was explained as precipitation of the calcium salt of the surfactants (an idea supported by other studies), and the interaction of the aromatic ring in SDBS with the asphalt. This interaction favors desorption of the asphalt rather than adsorption of the SDBS. The amount of asphalt desorbed by SDBS was twice that desorbed by SDS. Other explanations for adsorption maxima include mixed micelle formation [55] and electrostatic repulsion of micelles from the bilayer covered surface [59].

Anionic Surfactant onto Kaolinite.   The adsorption of a petroleum sulfonate surfactant, TRS 10-80, onto Na-kaolinite was conducted in batch experiments at low-to-medium salinity and under conditions in which liquid-crystal suspensions formed in alcohol-containing brines [60]. TRS 10-80 was described as not being very brine-soluble. The adsorption studies were conducted at 30 °C with pH values ranging from 7 to 13. The alcohol used was 2-butanol and its concentration was held constant at 30 g/l.

The adsorption of systems containing NaCl alone and NaCl-Na$_2$CO$_3$ were markedly different. For NaCl systems (26.2 and 21 g/l NaCl) the adsorption isotherms are marked by maxima of approximately 55 and 40 g/l, respectively. The maxima for both systems occurred at an equilibrium sulfonate concentration slightly below 5 g/l. In contrast the NaCl-Na$_2$CO$_3$ adsorption plateaued at approximately 10 mg/l.

Other findings presented in this study include (1) the observation that sodium hydroxide, while producing a higher pH than sodium carbonate (12.2 versus 11.3), did not decrease the adsorption as effectively as sodium carbonate; (2) the substitution of SO_4^{2-} ions for Cl^- ions at constant ionic strength strongly diminishes sulfonate adsorption; (3) adding sodium silicates to the NaCl brines was said to give adsorption results similar to those with Na$_2$CO$_3$ (no data was presented) while systems containing phosphates gave adsorptions of less than 5 mg/g, and in some systems negative adsorption was observed.

Some of the conclusions presented were that, for these systems, the pH dependent part of adsorption is small, the decrease in the adsorption was correlated with the lowering of the sulfonate activity, and sodium carbonate reduces sulfonate adsorption more than sodium hydroxide.

Anionic Blends onto Sand and Clay. Following a successful enhanced oil recovery demonstration using a surfactant blend in a foam flood, research was conducted to examine the fate of the blends in core studies [61]. The surfactant blend was composed of alpha olefin sulfonates (AOS's) and the DOWFAX® surfactant 3B2. This surfactant is a disulfonated alkyldiphenyloxide (DPOS). This line of surfactants is discussed in more detail in the third part of this section.

It was pointed out that the use of AOS's was desirable for EOR applications due to their low cost, but that they tend to precipitate in the presence of such cations as calcium and magnesium. The DPOS surfactant, on the other hand, does not tend to precipitate in the presence of the cations because of the disulfonate anion.

Solubility experiments indicated that a 50:50 blend of the surfactants was soluble in 90,000 ppm Ca^{2+}; sufficient for most conditions encountered in oil reservoirs. Adsorption studies on sand indicated that the pure surfactants had maximum adsorptions of approximately 150 and 50 μg/g for AOS and DPOS respectively, while the 50:50 blend had a maximum adsorption of about 75 μg/g. This reduced adsorption for the disulfonate is consistent with the role of surface charge in surfactant adsorption mechanism.

In static studies using the clay montmorillonite, surfactant adsorption as a function of blend composition was examined. It was found that when the blend consisted of more than 30% DPOS, total adsorption was suppressed, again consistent with reduction of adsorption when charge repulsion between surface and surfactant is increased.

Column studies were conducted on sand using each of the pure surfactants, and a 50:50 blend of the surfactants in a 5% (weight to volume) sodium chloride solution. In each of the three cases the surfactant solution was injected in 1/4 pore volume slugs and the effluent continuously monitored. The DPOS was the least adsorbed, the AOS the most adsorbed, and the degree of adsorption of the blend fell between the two pure surfactant adsorptions, but was still much less than the AOS adsorption.

Conclusions concerning the adsorption work presented were that the blend provided increased calcium tolerance and losses of surfactant due to precipitation by calcium and adsorption onto reservoir rocks can be reduced by the presence of the disulfonate.

Cationic and Anionic Surfactants onto Carbonates. The adsorption onto several carbonates of the cationic surfactants, cetylpyridinium chloride (CPC) and dodecyl pyridinium chloride (DPC) were compared to the adsorption of the anionic surfactant sodium dodecyl sulfate (SDS) [20]. It was expected that cationic surfactants would exhibit adsorption lower than anionic surfactants on carbonate minerals, which tend to be positively charged. The carbonate solids were a synthetic calcite ($CaCO_3$) and natural dolomite ($CaMgCO_3$).

The surface charges on these carbonate minerals were attributed to preferential dissolution of lattice ions, Mg^{2+}, Ca^{2+} and CO_3^{2-} and the adsorption of H^+ or OH^- which may act as potential determining ions for carbonates. Therefore, the adsorption was conducted using various concentrations of $MgCl_2$, $CaCl_2$, Na_2CO_3, but no attempt was made to regulate the pH, although the pH was measured. The average pH value of the $MgCl_2$ and $CaCl_2$ systems was approximately 8.0, and the average pH of the Na_2CO_3 systems was approximately 10.0. The pzc of the calcite was 9.2 and the pzc of the dolomite was 7.4 [62].

The adsorption behaviors of the CPC and DPC on the two carbonates were markedly different. DPC exhibited region I (linear) adsorption over the entire range of surfactant concentrations examined, up to the CMC, with the exception of the DPC/calcite/$MgCl_2$ system. For this system there was approximately zero adsorption until the equilibrium surfactant concentration reached approximately 4500 μmolar. At this concentration the adsorption was measured as a negative value. Negative adsorption was explained as the repelling of the like-charged surfactant from the surface and the subsequent concentration of surfactant in the region of the solution from which the analyte sample was collected. The negative adsorptions seen for the CPC/calcite systems are shown in Figure 6.

CPC with no electrolyte present had nearly constant adsorption (0.05–0.1 μmole/gram) on calcite. Adsorptions from 0.05 M $MgCl_2$ onto calcite and from 0.05 M $CaCl_2$ onto calcite resulted in nearly constant adsorptions at approximately 0.05 μmole/gram until the equilibrium surfactant concentration approached 200 μmolar, then the adsorption became negative. In contrast, adsorption isotherms for the dolomite system were more like the traditional isotherm shown in Figure 2. For the system with no additional electrolyte the adsorption values ranged from approximately 0.01 to 2.0 μmole/gram. Adsorption values from 0.05 M $MgCl_2$ ranged from 0.02 to 0.35 μmole/gram. While the values from 0.05 M $CaCl_2$ ranged from 0.2 to 0.4 μmole/gram.

The adsorption isotherms of anionic SDS on the carbonates indicated typical surfactant adsorption behavior with the plateau adsorption occurring at 9–10 μmole/gram for the system containing no additional electrolyte, and in the $MgCl_2$ solutions on both carbonates. The maximum adsorption for both carbonate systems containing Na_2CO_3 was approximately 4 μmole/gram while the $CaCl_2$ systems were approximately 5 μmole/gram.

The conclusions reached by the authors were that the addition of lattice ions from the solid can enhance the adsorption of the anionic surfactant while reducing the adsorption of a cationic surfactant by

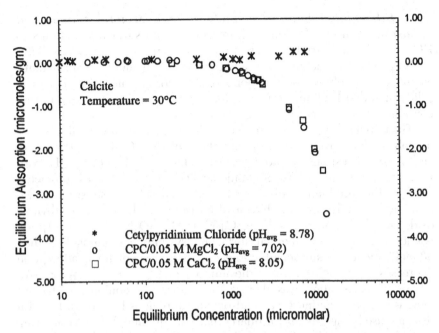

Figure 6. *Negative adsorption exhibited by CPC onto synthetic calcite* [20].

directly affecting the surface charge. The enhancement of the anionic surfactant adsorption arises from the decrease in the electrostatic repulsion between the head groups of the adsorbed surfactant molecules due to the addition of the divalent cations. This effect would not, of course, be observed in the cationic surfactant systems.

Further, just as in micelle formation, the addition of counterions can reduce the repulsion between the head groups of the anionic surfactants by compressing the electrical double layer between them. This compression acts to increase the adsorption. This increase in adsorption was not observed for addition of divalent cations to the cationic surfactants systems, however.

The authors proposed that a reduction in surfactant losses for EOR in carbonate reservoirs would be possible by using a cationic surfactant with an appropriate concentration of added multivalent electrolyte where the cations were also lattice ions for the mineral. In addition to lower adsorption losses the cationic surfactant offer good corrosion inhibiting capabilities and antibacterial properties. Unfortunately, cationic surfactants are more expensive than anionic surfactants; however, to the author's knowledge the economics of the proposed application have never been examined.

Ethoxylated Sulfate Surfactants onto Mineral Oxides and Sandstone Cores. Various features of anionic surfactant systems in EOR have been illustrated in a series of studies using ethoxylated sulfates as the primary surfactants with additives which included co-surfactants, alcohols, electrolytes, polyethylene oxide and polymers [63–70]. The solids included kaolinite, quartz, sandstone cores, Berea cores, and oil containing reservoir cores.

The initial study [63] examined the adsorption of commercial mixtures of polyethylene oxide nonyl-phenolether sulfates, C_9-Ph-$(EO)_x$-SO_3Na and their corresponding nonionic surfactants, C_9-Ph-$(EO)_x$-OH with $x = 2$, 4, 5.5, 6, and 9. The nonionic surfactant is present as unreacted feed in the production of the sulfated material. These studies also examined the adsorption behavior of an isomerically pure polyethylene oxide nonyl-phenolether sulfate, C_9-Ph-$(EO)_4$-SO_3Na. The adsorption isotherms for the commercial surfactant systems containing less than 30 mol% anionic surfactant indicated that the plateau adsorption decreases, on a mole basis, as the number of EO-groups increases. An additional observation was that as the amount of nonionic surfactant increases there is an increase in adsorption in regions II through IV, with the increase being greater for quartz than for kaolinite. When comparing the adsorption of the isomerically pure sulfate with sulfate/nonionic mixtures, it was observed that both anionic and nonionic surfactants adsorbed onto kaolinite and quartz, but adsorption of the nonionic was approximately 50% greater on both solids. Since the quartz is negatively charge, this is again consistent with our understanding of the central role of electrostatics in surfactant adsorption.

A later study [66] focused on the nonequilibrium adsorption of C_9-Ph-$(EO)_6$-SO_3Na, 88 mol% sulfonate and 12 mol% unconverted nonionic surfactant, with a polymer, xanthan, onto oil-containing sandstone cores from the North Sea. Addition of the polymer reduced the surfactant adsorption by 80% relative to adsorption without xanthan, yet there was no complex formation between the surfactant and the xanthan. This study reflects one of the current trends of using systems containing surfactant–polymer mixtures and emphasizes the need for system specific adsorption studies in EOR applications.

A more recent study [70] examined the effects of the polymer on surfactant adsorption in a low tension polymer water flood (LTPWF). The surfactant was alkylpropoxyethoxy sulfate, C_{12-15}-$(PO)_4$-$(EO)_2$-$OSO_3^-Na^+$, and the polymers were xanthan and a copolymer of acrylamide and sodium 2-acrylamido-2-methylpropane sulfonate (AN 125 from Floerger). The solid materials were sandstone cores from a North Sea oil reservoir, Berea, and Bentheim cores. For these systems the xanthan caused a 20% reduction in the adsorption of the surfactant. It was also observed that surfactant adsorption appeared to increase as the water wettability decreased under LTPWF conditions and also as the salinity of the brine increased. No mechanism was presented to explain the effect of wettability on surfactant adsorption; however, it was proposed that the polar components in the crude oil were adsorbed onto the solid surface and would not be displaced under the LTPWF conditions. The effect of the brine was explained as a reduction in lateral electrostatic repulsion among adsorbed surfactant ions which causes a closer packing of the adsorbed molecules, thus facilitating formation of admicelles.

Mixed Anionic Surfactants onto East Vacuum Grayburg–San Andres Unit (EVGSAU) and Baker Dolomite Cores. The surfactant CHASER[TM] CD1045 (Chevron Chemical Co.) described only as a mixed surfactant was adsorbed onto EVGSAU and Baker dolomite cores as part of a study examining CO_2-foam in mobility control (21). The adsorption portion of this study was conducted at room temperature and atmospheric conditions.

On Baker dolomite cores the results of four studies were reported. The differences between the studies were the presence or absence of additional electrolytes (4% brine) and the porosity of the cores. Comparing the adsorption of the surfactant from distilled water versus from the 4% brine solution for cores with similar porosity indicates that the adsorption from the distilled water was slightly less than the adsorption from the brine solution. For example, for cores with porosities averaging 18.8% with equilibrium surfactant concentrations of approximately 2180 ppm, the adsorption was 3200 lb/acre-ft for the distilled water system and 3577 lb/acre-ft for the 4% brine system. The adsorption was described as "reasonably Langmurian", meaning that the slope decreased as adsorption increased. For the distilled water systems, the adsorption appears to be just beginning to plateau at the maximum of the surfactant concentration range studied. The brine systems exhibit an adsorption plateau of approximately 3500 lb/acre-ft. The differences in the adsorption behavior between the distilled water and brine systems was attributed to the brine shifting the surface charge of calcite towards less negative or even positive values. Any anionic surfactant would tend to adsorb to a greater extent under increased electrolyte concentrations.

The adsorption behavior of the CHASER[TM] CD1045 onto EVGSAU cores was similar to the behavior seen for the dolomite cores. For the distilled water it appears that the adsorption is just beginning to plateau, but greater surfactant concentrations would have to be studied in order to confirm this. For the 4% brine system the adsorption plateaus at slightly greater than 2000 lb/acre-ft which is less than the adsorption seen on the dolomite cores.

In a similar study [71] using Chaser[TM] CD-1045 for CO_2-foam applications, the adsorption of the surfactant onto Baker dolomite was determined. The dolomite used in this study was similar in porosity to the previous study discussed but the studies were conducted using synthetic South Camden Unit (SCU). The average surfactant adsorption was approximately 420 lb/acre-ft which is considerably lower than that reported for the East Vacuum Grayburg–San Andres Unit. The composition of the brine was not provided in this report, but it may be at least part of the reason for the great difference between the two adsorption studies conducted on Baker dolomite.

There are two additional types of chemical flooding systems that involve surfactants which are briefly mentioned here. One of these systems utilizes surfactant–polymer mixtures. One such study was presented by Osterloh et al. [72] which examined anionic PO/EO surfactant microemulsions containing polyethylene glycol additives adsorbed onto clay. The second type of chemical flood involves the use of sodium bicarbonate. The aim of the research was to demonstrate that the effectiveness of sodium bicarbonate in oil recovery could be enhanced with the addition of surfactant. The surfactant adsorption was conducted in batch studies using kaolinite and Berea sandstone [73]. It was determined that the presence of a low concentration of surfactant was effective in maintaining the alkalinity even after long exposures to reservoir minerals. Also, the presence of the sodium bicarbonate is capable of reducing surfactant adsorption.

Adsorption of Gemini Surfactants.   Gemini surfactants, characterized by two hydrophilic groups and at least two hydrophobic groups, have attracted a significant amount of recent attention because of several unique properties. A few of these properties are low CMC values, low adsorption at the air/water interface, and closer packing of the hydrophobic groups [74]. This section examines two studies, one directly applied to EOR and the second involving a suite of gemini isomers for their possible use in remediation applications.

Anionic Surfactant Blend and Amphoteric Surfactants onto Berea Sandstone, Indiana Limestone, Baker Dolomite, and Quartz.   The first study to be presented examined the adsorption behavior of two amphoteric surfactants, a betaine (Empigen BT) and a sulfobetaine (Varion CAS); and a 50:50 blend of a C_{10} diphenyl ether disulfonate (DOWFAX® 3B2), and a C_{14-16} α-olefin sulfonate [11]. The anionic surfactant blend was designated as DOW XS84321.05. The C_{10} diphenyl ether disulfonate surfactant is one isomer in a suite of surfactants which differ in their degree of alkylation and sulfonation and in their chain lengths. This suite consists of monoalkyl disulfonates (MADS), dialkyl disulfonates (DADS), monoalkyl monosulfonates (MAMS), and

Table 1.   Plateau Adsorption Values (mg/g) from Adsorption Isotherms[a]

| Rock Type | Brine | Surfactant Types[b] | | |
		Anionic	Betaine	Sulfobetaine
Sandstone	2.1% TDS	0.11	1.31	1.30
Sandstone	10.5% TDS	0.26	1.12	—
Sandstone	2.32% NaCl	0.03	0.94	1.29
AGSCO Quartz	2.1% TDS	0.15	0.23	—
AGSCO Quartz	10.5% TDS	0.15	0.09	—
Indiana Limestone	2.1% TDS	0.37	0.31	0.21
Indiana Limestone	10.5% TDS	0.30	0.33	—
Indiana Limestone	2.32% NaCl	0.21	0.12	0.19
Baker Dolomite	2.1% TDS	0.13	0.38	0.32
Baker Dolomite	2.32% NaCl	0.12	0.37	0.34

[a] Taken from Table 5 of Mannhardt et al. [11]
[b] Anionic surfactant (DOW XS84321.05); betaine (Empigen BT); sulfobetaine (Varion CAS)

dialkyl monosulfonates (DAMS). DOWFAX® 3B2 is a mixture of C10 MADS and C10 DADS. The general structure of these surfactants is shown in Figure 7.

The adsorption studies were conducted on core samples of Berea sandstone, Indiana limestone, Baker dolomite, and quartz sand from three brines (a sodium chloride solution of 2.32% and two synthetic reservoir brines with total dissolved solids of 2.1 and 10.5%). Conclusions were based on the maximum or plateau adsorption values obtained, and these values are shown in Table 1.

As shown in Table 1, the anionic surfactant blend gave the lowest adsorption onto sandstone and onto the dolomite for all three of the brine conditions examined. While onto quartz, the adsorption of the anionic surfactant remained constant for both of the synthetic reservoir brines

$$SO_3Na \qquad (SO_3Na)$$

$$(R) \qquad\qquad R$$

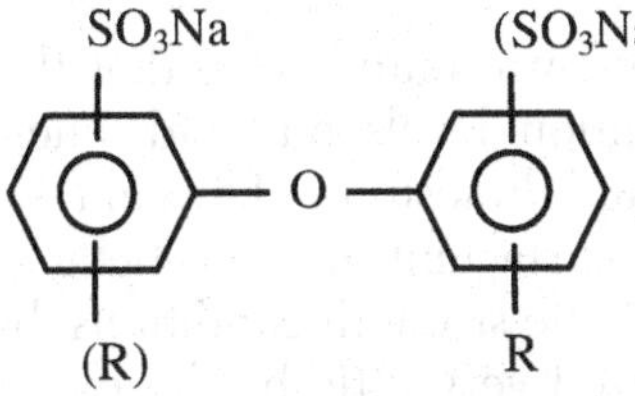

Figure 7.   Structure of the DOWFAX® gemini surfactant where R are alkyl chains of C6, C10, C12, or C16.

tested with amounts in between those seen for the betaine surfactant. For limestone the adsorption of the anionic and the betaine were approximately the same in the two synthetic reservoir brines, but both were greater than the adsorption of the sulfobetaine. For adsorption onto limestone from the brine solution the anionic adsorption was greater than the betaine and slightly greater than that of the sulfobetaine.

Upon examining the adsorption data from the reservoir brines, there was no consistent pattern seen in the adsorption behavior. Based on this it was concluded that the tendency of surfactant adsorption to increase with increasing salt concentration is minor and that the trends in adsorption can be explained solely on the basis of the interaction of the charge on the surfactant with the solid surface charges.

The surface charge on the solid depended, in part, on the brine solution surrounding the surface. Electrophoretic mobilities were determined in the 2.1% TDS and the 2.32% NaCl brines, and at pH 7 the following trend was observed:

> Berea clays < quartz < dolomite < limestone
> most negative   least negative

The Berea sandstone had been split into clay and quartz fractions, but the Berea whole rock was still more negative relative to the other core listed for this study. Even though the trend was the same for both brines, the divalent cations in the 2.1% TDS brine produced less negatively charged surfaces than did the NaCl brine. This behavior was attributed to adsorption of these ions into the Stern layer or, in the case of carbonates, to preferential dissolution of CO_3^{2-} over Ca^{2+} or Mg^{2+} in the presence of excess divalent cations in the aqueous phase. It was also noted that adsorption of metal hydroxide ions or mineral transformation reactions at the solid surface may play a role.

A detailed discussion was presented on the relationship between surfactant adsorption and the solid surface charge. For the anionic surfactant, as expected, as the surface became increasingly positive the adsorption increased. This increase in positive charge occurred either when the rock type was changed in the order shown above while keeping the brine fixed or when divalent cations were added to the brine for a fixed rock type.

Some of the conclusions presented were that the anionic surfactant blend exhibits low adsorption levels on sandstone, but adsorbs more strongly onto dolomite and limestone. Divalent cations increase the adsorption of the anionic surfactant and the betaine on sandstone and limestone under constant ionic strength conditions, but the adsorption of the sulfobetaine was affected very little by the presence of the divalent cations. Increasing the total dissolved solids at constant ionic composition gave mixed results, increasing the adsorption for some surfactant/rock

Table 2.   Gemini Surfactant Description

Surfactant	Avg. MW (g/mol)	CMC (M)[a]
C10 MAMS	423	3.53×10^{-4}
C10 MADS	523	1.40×10^{-4}
C10 DADS	617	1.33×10^{-4}
C12 MADS	551	1.30×10^{-4}
C16 MADS	600	2.53×10^{-4}

[a] Reported by Dow Chemical and determined at room temperature and native electrolyte conditions

combinations, but decreasing it for others (see Table 1). In terms of adsorption levels, the anionic surfactant appears to be the best choice of the systems studied for applications in sandstone and dolomite reservoirs. In limestone reservoirs, the sulfobetaine would be best, particularly in the presence of hardness ions. Finally, the trends in the adsorption of the anionic surfactant appear consistent with the electrostatic mechanisms.

Anionic Surfactants onto Canadian River Alluvium (CRA) and Alumina. The second study on gemini surfactants to be discussed was conducted in order to determine strategies for designing gemini surfactants in order to minimize adsorption. The adsorption studies were conducted on Canadian River alluvium (CRA) [9] and on alumina at room temperature. CRA is primarily a sand and is expected to behave similarly to sandstone cores. The anionic gemini surfactants were supplied by DOW and used as received. The alkyl groups used in the CRA and alumina studies were linear and included alkyl chain lengths of C6, C10, C12, and C16. The DAMS components and the C12 and C16 DADS were not studied due to their low water solubilities.

Adsorption onto Canadian River Alluvium (CRA). The adsorption studies done on CRA were part of a larger study focused on examining the behavior of the gemini surfactants in soil remediation processes [9]. For the adsorption onto CRA, the surfactants studied consisted of C10 and C16 MADS, C10 DADS, and C10 MAMS.

Prior to use, the CRA was crushed and sieved. The BET (N_2) surface area was determined to be 4.63 m^2/g with an average pore diameter of 55.52 Å. Five grams of soil were used with 0.1 ml of a calcium chloride solution (0.005 M) added and allowed to dry. Previous research had shown the calcium chloride to be necessary to get the soil fines to separate from the bulk solution. When the soil was dry, 25 ml of each surfactant solution (1/5 to 10 times the CMC) was added. The CMC values for the individual components are given in Table 2. The samples were placed on a finger-tip shaker for 24 hours then centrifuged for 20 minutes. The

amount of adsorption was determined by calculating the change in
surfactant concentration in the bulk solution. The equilibrium surfactant
concentrations were determined by HPLC with a UV detector set at
254 nm and methanol as the mobile phase. Prior to analysis on the HPLC,
all samples were passed through a 0.2 μm syringe filter to remove any
suspended soil particles.

As shown in Figure 8, the greatest adsorption is shown by the C10
MAMS component (4.91 mg/g). The MAMS surfactant is an isomeric
variation of the theme of this suite of gemini surfactants which has lost the
defining structure of the geminis. The higher adsorption of the C10
MAMS is then attributed to the monosulfonated component having less
electrostatic and steric hindrances than the disulfonated gemini components. The maximum adsorption of the C10 MADS was 0.6 mg/g,
considerably lower than the 4.91 mg/g seen for the C10 MAMS. The
more hydrophobic nature of the MAMS relative to the MADS component
arises from the absence of the second sulfonate group. This greater
degree of hydrophobicity is the source of the increase in adsorption.

Figure 9 depicts the adsorption isotherms of the three C10 components. For the reasons stated above, the monosulfonate had the greatest
adsorption (4.91 mg/g). While between the two disulfonated components
(MADS and DADS), the dialkyl component had the greatest amount of
adsorption (2.25 versus 0.6 mg/g). This is due to the greater hydrophobicity of the dialkyl component. The difference seen in the maximum
adsorption of the MAMS and DADS components, 4.91 and 2.25 mg/g,
respectively, is due to both steric hindrances caused by the second alkyl

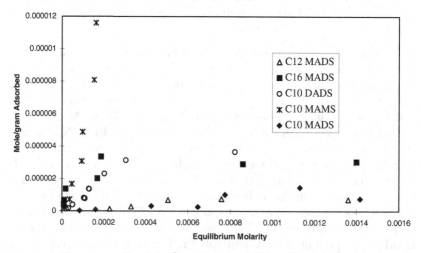

Figure 8.   Gemini adsorption onto CRA, all components.

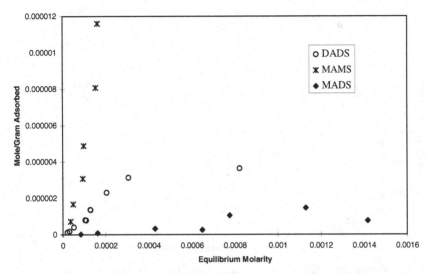

Figure 9.   Gemini adsorption onto CRA, C10 components.

group and the more ionic nature of the DADS component due to the second sulfonate group.

The adsorption isotherms of the C10 and C16 MADS components are shown in Figure 10. It can be seen that the adsorption increases at lower surfactant concentration with increasing chain length, from C10 to C16 (0.6 to 1.83 mg/g). This increase is attributed to increasing hydrophobicity with increasing chain length.

Adsorption studies have been conducted with other gemini surfactants on CRA. Rouse et al. [8] studied the adsorption of DOWFAX® 8390, a commercially available C16 surfactant and found the maximum adsorption of 4.3 mg/g while the maximum adsorption of the C16 MADS component was 1.8 mg/g. The commercial product contains approximately 35 weight percent active component which is approximately 80 wt.% monoalkylated and 20 wt.% dialkylated. The increase in the dialkylated component in the commercial mixture is responsible for its greater maximum adsorption compared to the C16 MADS.

The adsorption behavior of the C10 DADS relative to the C16 MADS is worth special note. Comparing initial concentrations, it is seen that at lower surfactant concentrations the C16 has the greatest amount of adsorption; however, as the plateau region is approached the C10 adsorption exceeds that of the C16. Since they are both disulfonates this behavior can be attributed to the difference in chain lengths and the degree of alkylation, both of which are directly related to the hydro-phobicity of the surfactants. At the lower surfactant concentrations the

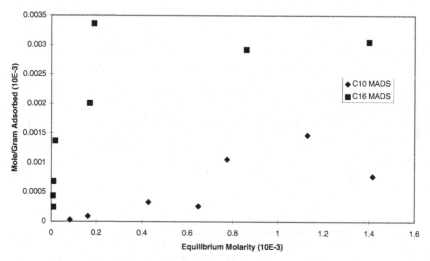

Figure 10.   Gemini adsorption onto CRA, MADS components.

longer chain length dominates the adsorption, but at higher concentrations the dialkylation dominates.

Adsorption onto Alumina. The goal of the adsorption studies on alumina was to study the adsorption of the gemini surfactants on an oppositely charged substrate, leading to high adsorption and bilayer formation . The adsorptions of the C10, C12, and C16 MADS and the C10 DADS components were studied.

The alumina used was manufactured by LaRoche and has a BET (N_2) surface area of 301.83 m^2/g and an average pore diameter of 105.08 Å. Due to the high surface area, only 0.05 grams of alumina and 30 ml of surfactant solution were used. These quantities allowed enough surfactant to remain in the bulk solution at equilibrium to be analyzed. The surfactant feed concentrations ranged from approximately 1/5 to 10 times the CMC with NaCl concentrations of 0.15 M for the MADS solutions and 0.09 M for the DADS solutions. The pH of the alumina/ surfactant solutions was measured and adjusted to values ranging from 2.3 to 3.5 using sulfuric acid. The pH was allowed to equilibrate without further adjustment. The vials were placed on a table-top type shaker for 24 hours. The alumina in the adsorption solutions was allowed to settle prior to analysis. Centrifuging was found not to be effective in separating the alumina from the surfactant solutions. The equilibrium pH was then measured. The equilibrium surfactant concentrations were determined by HPLC using methanol as the mobile phase with a UV detector with the wavelength varying from 254 to 264 nm depending on the surfactant.

Prior to injection into the HPLC the solutions were passed through a 0.2 μm syringe filter to remove any suspended alumina.

Of the MADS components the order of increasing adsorption was C10, C12, and C16. This is in agreement with adsorption studies of sodium alkylsulfonates onto alumina conducted by Wakamatsu and Fuerstenau [41]. From the isotherms shown in Figure 11, the behavior of the C16 and C12 MADS are very similar, noticeably the sharp increase in adsorption just prior to the plateau region. The C10 MADS component has a more gradual increase with no sharp break before apparently plateauing. The third, fourth and eighth data points of the C10 MADS isotherm were at lower feed pH's than the remaining points on the curve. The effect of varying feed pH's on C12 and C16 MADS was not as noticeable as on the C10 MADS.

As seen in Figure 11, the adsorption isotherms of C10 MADS and C10 DADS overlap over the entire isotherm. This was not expected. Generally, the dialkyl component would be expected to have the greater adsorption due to the greater hydrophobicity. Since this behavior was not seen in the adsorption isotherms on the CRA, there must be a difference in the interaction of the surfactants and the alumina. Even though there is a greater hydrophobicity for the dialkyl component, there are also greater steric constraints for aggregate formation. On the more porous alumina it is possible that the pore structure was such that the second alkyl group prevented adsorption of the dialkyl component in some pores, while no such constraint existed for the monoalkyl component. There were no significant variations in the feed or equilibrium pH values, and the difference in the salinity was 0.06 M. A lower salinity was

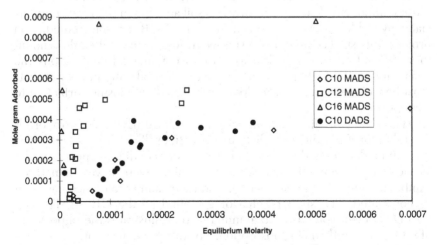

Figure 11. Gemini adsorption onto alumina, all components.

Table 3.   Maximum Adsorptions of Gemini Surfactants

Surfactant	CRA		Alumina	
C10 MAMS	4.91 mg/g	$(1.16 \times 10^{-5}$ mol/g$)$	—	
C10 MADS	0.6	(1.15×10^{-6})	236.5 mg/g	$(4.52 \times 10^{-4}$ mol/g$)$
C10 DADS	2.25	(3.66×10^{-6})	237.0	(3.86×10^{-4})
C12 MADS	—		299.1	(5.43×10^{-4})
C16 MADS	1.83	(3.05×10^{-6})	527.5	(8.79×10^{-4})

required for the DADS component due to the tendency of these solutions to form insoluble materials.

In general the gemini components adsorbed as expected on both surfaces. As expected the higher surface area, oppositely charged alumina had significantly greater adsorption than that on the CRA (see Table 3 for the actual values). The monosulfonates showed the greatest amount of adsorption due to the lack of electrostatic and steric hindrances. In the series of MADS components there was increasing adsorption with increasing chain length due to increasing hydrophobicity.

There were several observations made concerning the effect pH had on the adsorption on alumina. Adsorption of the DADS component onto alumina was very sensitive to variations in the feed pH. This was exhibited by the several points that did not fit smoothly on the isotherm (see Figure 11). It was also noted that for any of the surfactants if the feed pH was at or below 2.5 or 2.6 the final pH was usually below 3.0, but if the feed pH approached 3.0 the final pH would approach 4.0.

In conclusion, the adsorption studies on the CRA indicated that the purer components had significantly less soil adsorption than the commercially available DOWFAX surfactants. In EOR or soil remediation projects this smaller amount of surfactant loss to the soil could amount to significant cost savings, but as with all commercial applications the economics of the higher costs associated with producing purer products must be weighed against profit losses associated with losing surfactant to adsorption.

While the gemini surfactants discussed above have many properties favorable for EOR and remediation applications, like all surfactants their use will be determined by their behavior under conditions specific to the application. In general their adsorption values were lower than those exhibited by many other surfactants and their salinity tolerance has been demonstrated. The similar behavior exhibited by the C10 MADS and C10 DADS components onto alumina further illustrates the importance of pore size in influencing adsorption. For this particular suite of surfactants the components which are not "true gemini" surfactants tend to have the more favorable adsorption properties, monoalkyl versus dialkyl components.

Summary

Surfactants are used extensively in enhanced oil recovery. Applications include micellar floods or flooding in conjunction with polymers, alkalis, steam or carbon dioxide. Another application is the generation of foams for mobility control or blocking and diverting. For each of these applications care must be taken in selecting the surfactants. Surfactants tend to be a major portion of the costs associated with EOR, and losing surfactant to adsorption leads to substantial economic losses.

Surfactant adsorption depends on many factors. Factors discussed in this chapter include the electrical nature of the solid surface, pH of the system, and the structure of the surfactant. For most of the solids in the various studies reviewed, the charge on the surface is determined in large part by the pH of the system. Adsorption is enhanced in those systems in which the solid surface and the surfactant have opposite charges, and the greater the surface charge the greater the surfactant adsorption. Higher surface area solids tend to have increased adsorption, but pore size can also affect the degree of adsorption. Care must be taken to avoid confusing precipitation of the surfactant for adsorption; hence, familiarity with the solubility of a surfactant in the presence of counterions is necessary.

The studies reviewed for this chapter are examples of the types of studies which have been done on surfactant adsorption. These studies included those upon which fundamental theories and models have been developed as well as those used to develop practical applications of surfactants in the field of enhanced oil recovery.

Acknowledgment

We would like to express our appreciation to Lisa Quencer of the DOW Chemical Company for all of her help in the gemini surfactant research conducted on Canadian River alluvium and alumina.

References

1. Taber, J.J.; Martin, F.D.; Seright, R.S. *SPERE* Aug. **1997**, 199–205.
2. Novosad, J.; Maini, B.; Batycky, J. *JAOCS* **1982**, 59, 833.
3. Lake, L.W.; Pope, G.W. *Pet. Eng. Int.* **1979**, 51, 38.
4. Holm, L.W. *Proceedings of the 57th Annual Fall Technical Conference of*

SPE; Society of Petroleum Engineers: New Orleans, LA, 1982, paper SPE 11196.

5. Greenwood, N.N.; Earnshaw. A. *Chemistry of the Elements, 2nd ed.*; Butterworth-Heineman: Boston, 1997; pp 242–4, 352.
6. Grim, R. *Clay Mineralogy*; McGraw-Hill: New York, 1968; p 59.
7. Palmer, C.; Sabatini, D.A.; Harwell, J.H. In *Transport and Remediation of Subsurface Contaminant*; Sabatini, D.A.; Knox, R.C. Eds.; American Chemical Society: Washington DC, 1992; pp 169–81.
8. Rouse, J.D.; Sabatini, D.A.; Brown, R.E.; Harwell, J.H. *Water Env. Res.* **1996**, *68*, 162–8.
9. Sabatini, D.A.; Harwell, J.H.; Deshpande, S.; Wesson, L.; Wade, D. *Water Research*, submitted for publication.
10. Novosad, J.J. *SPEJ* **Dec. 1982**, 962–70.
11. Mannhardt, K.; Schramm, L.L.; Novosad, J.J. *Colloids Surfaces* **1992**, *68*, 37–53.
12. Böhmer, M.R.; Koopal, L.K. *Langmuir* **1992**, *8*, 2649–59.
13. Böhmer, M.R.; Koopal, L.K. *Langmuir* **1992**, *8*, 2660–5.
14. Koopal, L.K.; Lee, E.M.; Böhmer, M.R. *J. Colloid Interface Sci.* **1995**, *170*, 85–97.
15. Lee, E.M.; Koopal, L.K. *J. Colloid Interface Sci.* **1996**, *177*, 478–89.
16. Zhu, B-Y.; Gu, T.; *Colloids Surfaces* **1990**, *46*, 339–45.
17. Manne, S.; Cleveland, J.P.; Gaub, H.E.; Stucky, G.D.; Hansma, P.K. *Langmuir* **1994**, *10*, 4409–13.
18. Manne, S.; Gaub, H. *Science* **1995**, *270*, 1480–2.
19. Krishnakumar, S.; Somasundaran, P. *Colloids Surfaces* **1996**, *117*, 227–33.
20. Tabatabai, A.; Gonzalez, M.V.; Harwell, J.H; Scamehorn, J.F. *SPERE* **1993**, 117–22.
21. Tsau, J-S.; Heller, J.P.; Moradi-Araghi, A.; Zornes, D. R.; Kuehne, D.L. *SPE/DOE Ninth Symposium on Improved Oil Recovery*: Tulsa, OK, 1994, paper SPE/DOE 27785.
22. Fuerstenau, D.W. *Pure Appl. Chem.* **1970**, *24*, 135–64.
23. van Olphen, H. *An Introduction to Clay Colloid Chemistry*; Interscience: New York, 1963; pp 17–20.
24. Fernandez, M.E. Master's Thesis, University of Texas at Austin, 1978.
25. Rosen, M.J. *Surfactants and Interfacial Phenomena*; Wiley, New York, 1989, 2nd ed.; pp 35–8.
26. Somasundaran, P.; Healy, T.W.; Fuerstenau, D.W. *J. Phys. Chem.* **1964**, *68*, 3562–6.
27. Davis, J.A.; James, R.O.; Leckie, J.O. *J. Colloid Interface Sci.* **1978**, *63*, 480–99.
28. James, R.O.; Parks, G.A. In *Surface and Colloid Science*; Matijevic, E., Ed.; Plenum Press: New York, 1982, Vol. 12; pp 119–216.
29. Hankins, N.P.; O'Haver, J.H.; Harwell, J.H. *Ind. Eng. Chem. Res.* **1996**, *35*, 2844–55.
30. Scamehorn, J.F.; Schecter, R.S.; Wade, W.H. *J. Colloid Interface Sci.* **1982**, *85*, 463–78.
31. Harwell, J.H.; Hoskins, J.C.; Schecter, R.S.; Wade, W.H. *Langmuir* **1985**, *1*, 251–62.

32. Zhu, B-Y.; Gu, T. *J. Chem. Soc., Faraday Trans.* **1989**, *85*, 3813–7.
33. Böhmer, M.R.; Koopal, L.K. *Langmuir* **1992**, *8*, 1594–602.
34. Böhmer, M.R.; Koopal, L.K. *Langmuir* **1992**, *8*, 2649–59.
35. Somasundaran, P.; Fuerstenau, D.W. *J. Phys. Chem.* **1966**, *70*, 90–6.
36. Scamehorn, J.F.; Schecter, R.S.; Wade, W.H. *J. Colloid Interface Sci.* **1982**, *85*, 479–93.
37. Chandar, P.; Somasundaran, P.; Turro, N.J. *J. Colloid Interface Sci.* **1987**, *117*, 31–46.
38. de Bruyn, P.L. *Trans AIME* **1955**, *202*, 291–6.
39. Gaudin, A.M.; Morrow, J.G. *Trans AIME* **1954**, *199*, 1196–202.
40. Gaudin, A.M.; Fuerstenau, D.W. *Trans. AIME* **1955**, *202*, 958–62.
41. Wakamatsu, T.; Fuerstenau, D.W. In *Adsorption From Aqueous Solution*; Weber, W. J.; Matijevic E., Eds.; American Chemical Society: Washington, DC, 1968; pp 161–72.
42. Bitting, D.; Harwell, J.H. *Langmuir* **1987**, *3*, 500–11.
43. Amante, J.C.; Scamehorn, J.F.; Harwell, J.H. *J. Colloid Interface Sci.* **1991**, *144*, 243–53.
44. Gao, Y.; Du, J.; Gu, T. *J. Chem. Soc., Faraday Trans.* **1987**, *83(8)*, 2671.
45. Jaschke, M.; Butt, H.-J.; Gaub, H.E.; Manne, S. *Langmuir* **1997**, *13*, 1381–4.
46. Somasundaran, P.; Kunjappu, J.T. *Colloids Surfaces* **1989**, *37*, 245–68.
47. Harwell, J.H.; Scamehorn, J.H. In *Mixed Surfactant Systems*; Ogino, K.; Abe, M., Eds.; Marcel Dekker: New York; pp 263–81.
48. Gaudin, A.M.; Bloecher, F.W. *Trans AIME* **1950**, *187*, 499–505.
49. Goulob, T.P.; Koopal, L.K. *Langmuir* **1997**, *13*, 673–81.
50. Goulob, T.P.; Koopal, L.K.; Bijsterbosch, B.H.; Sidorova, M.P. *Langmuir* **1996**, *12*, 3188–94.
51. Lopata, J.J.; Harwell, J.H.; Scamehorn, J.F. In *Surfactant-Based Mobility Control; Progress in Miscible-Flood Enhanced Oil Recovery*; Smith, D.H., Ed.; American Chemical Society: Washington, DC, 1988; pp 205–19.
52. Scamehorn, J.F.; Schecter, R.S.; Wade, W.H. *JAOCS* **1983**, *60*, 1345–9.
53. Roberts, B.L.; Scamehorn, J.F.; Harwell, J.H. In *Phenomena in Mixed Surfactant Systems*; Scamehorn, J.F., Ed; American Chemical Society: Washington, DC, 1986; p 200.
54. Lopata, J.J. Master's Thesis, University of Oklahoma, Norman, OK, 1987.
55. Trogus, F.J.; Schecter, R.S.; Wade, W.H. *J. Colloid Interface Sci.* **1979**, *70*, 293–305.
56. Treiner, C.; Montocone, V. In *Surfactant Adsorption and Surface Solubilization*; Sharma, R., Ed.; American Chemical Society: Washington, DC, 1995; pp 36–48.
57. Giordano, F.; Desnoyel, R.; Rouquerol, J. *Colloids Surfaces A* **1993**, *71*, 293–8.
58. Siffert, B.; Jada, A.; Wersinger, E. *Colloids Surfaces* **1992**, *69*, 45–51.
59. Ananthapadmanabhan, K.P.; Somasundaran, P. *Colloids Surfaces* **1983**, *7*, 105–14.
60. Bavière, M.; Ruaux, E.; Defives, D. *SPE International Symposium on Oilfield Chemistry*: Anaheim, CA, 1991, paper SPE 21031.
61. Dawe, B.; Oswald, T. *J. Canadian Pet. Tech.* **1991**, *30*, 133–7.

62. Gonzalez, M. Master's Thesis, University of Oklahoma, Norman, OK, 1989.
63. Austad, T.; Løvereide, T.; Olsvik, K.; Rolfsvåg, T.A.; Staurland, G. J. Petrol. Sci. Eng. **1991**, 6, 107–24.
64. Austad, T.; Bjørkum, P.A.; Rolfsvåg, T.A. J. Petrol. Sci. Eng. **1991**, 6, 125–35.
65. Austad, T.; Bjørkum, P.A.; Rolfsvåg, T.A.; Øysæd, K.B. J. Petrol. Sci. Eng. **1991**, 6, 137–48.
66. Austad, T.; Rørvik, O.; Rolfsvåg, T.A.; Øysæd, K.B. J. Petrol. Sci. Eng. **1992**, 6, 265–76.
67. Austad, T.; Fjelde, I.; Rolfsvåg, T.A. J. Petrol. Sci. Eng. **1992**, 6, 277–87.
68. Austad, T.; Fjelde, I.; Veggeland, K. J. Petrol. Sci. Eng. **1994**, 12, 1–8.
69. Fjelde, I.; Austad, T.; Milter, J. J. Petrol. Sci. Eng. **1995**, 13, 193–201.
70. Austad, T.; Ekrann, S.; Fjelde, I.; Taugbøl, K. Colloids Surfaces **1997**, 127, 69–82.
71. Contracts for Field Projects and Supporting Research on Enhanced Oil Recovery, U.S. Department of Energy, Progress Review No. 86, Contract No. DE-FG22-94BC14991, 1996.
72. Osterloh, W.T.; Jante, M.J., Jr. SPE/DOE Eighth Symposium on Enhanced Oil Recovery: Tulsa, OK, 1992, paper SPE/DOE 24151.
73. Peru, D.A.; Lorenz, P.B. SPE Res. Eng. **1990**, 5(3), 327–32.
74. Rosen, M.J.; Tracy, D.J. J. Surf. Detergents **1998**, 1, 547–54.

RECEIVED for review July 7, 1998; ACCEPTED revised manuscript January 6, 1999.

5

Surfactant Induced Wettability Alteration in Porous Media

Eugene A. Spinler and Bernard A. Baldwin

Phillips Petroleum Company, Phillips Research Center, Bartlesville, OK, USA

It is the wettability of the reservoir rock that controls the distribution of oil and water and affects their movement through pore spaces. Understanding wettability in porous media is, by itself, a difficult problem. Controlling it to modify the behavior of reservoir rock presents a more complex problem. Surfactants provide a tool that can transform the wettability of the porous rock. There are numerous methodologies and practices for studying and measuring wettability and its modification. The interactions of surfactants with reservoir materials to alter wettability are highly dependent upon the pore surface composition and pore structure as well as the characteristics of the surfactants. Wettability alteration of the porous rock from surfactants can affect drilling, well completion, well stimulation, secondary or tertiary oil production and environmental clean-up.

Role of Wettability Alteration

Wettability alteration of porous reservoir rock with surfactants is one means to improve the flow and distribution of fluids in a reservoir. However, much remains to be learned regarding how surfactants interact with the rock minerals and organics found both within the pores and on the pore surfaces of reservoir rock. It is the objective of this chapter to provide an initial understanding and review of the following:

1. Theoretical aspects of wettability
2. Methods for measuring wettability
3. Surfactant induced wettability alteration
4. Laboratory and field studies

   Background. Surfactants have been introduced into oil reservoirs to increase oil recovery, minimize adverse mobility ratios, clean plugged wellbores and improve drilling. In the first two processes, the surfactant is selected to either alter the interfacial tension between injected and in-situ fluids and/or alter the wettability of the reservoir rock. The former changes wettability by altering the fluid(s) while the latter alters the chemical nature of the immobile surface. Both processes are capable of increasing hydrocarbon production, however, most studies have focused on the reduction of interfacial tension. Understanding the role surfactant-altered wettability plays in hydrocarbon recovery may be of equal, or greater, importance. The latter two processes are designed to either remedy a production problem or minimize negative interactions between the reservoir and the drilling procedure. Without careful selection of surfactant, the latter two may seriously alter the reservoir wettability and permeability near the production wellbore causing catastrophic reduction in hydrocarbon production and alter recovered core so that it no longer accurately represents the reservoir. Thus, surfactants can play a very important role, both positively and negatively, in oil production.

Theoretical Aspects for Wettability

Fundamental Equations of Wettability.   This section provides the basic equations and concepts that are needed to understand and permit a practical discussion of wettability. Rigorous definitions and mathematics can be found in the references. Wettability, in this chapter, describes the interaction between fluids and the rock surface, i.e. whether a surface prefers to be in contact with oil or water. Although some general terminology will be used, wetting will generally be described with water as the wetting fluid and oil as the second fluid phase.

When two fluids, mutually immiscible with each other, both contact a solid surface, the less wetting fluid will retreat from contact with the solid while the stronger wetting fluid will be attracted to the surface. At the point of intersection between the two fluid phases and the solid surface, a contact angle is produced. If the fluids are not moving and their interaction with the surface and each other is thermodynamically stable, the three phase contact angle that forms is the result of the mechanical equilibrium of the three interfacial tensions (free energy per unit area). Young's equation [1] denotes the equilibrium relationship

$$\gamma_{SO} - \gamma_{SW} = \gamma_{WO} \cos \theta \tag{1}$$

γ_{SO} is the interfacial tension between a solid and oil, γ_{SW} is the interfacial tension between a solid and water and γ_{WO} is the interfacial tension between water and oil. In an oil/water/solid system, the contact angle, θ, is customarily measured through the water phase (Figure 1). This contact

Water

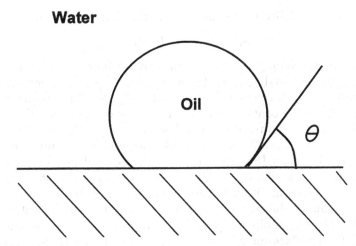

Figure 1. View of the contact angle, θ, as measured through the water phase.

angle provides a direct macroscopic measurement of wettability on flat surfaces under the appropriate experimental conditions.

Contact angles can be either static or dynamic. Static contact angles are formed with a surface under no applied force other than gravity, i.e. surface is perpendicular to the gravitational field and neither the fluids nor the surface(s) are being displaced. Dynamic contact angles are distorted from the static case by an applied force and can be advancing or receding. Advancing contact angles are formed at the front of the encroaching wetting phase. Likewise, receding contact angles are formed at the front of the encroaching non-wetting phase.

The right side of equation 1 is known as the adhesion tension, τ_{SWO}, for water and oil in contact with a solid:

$$\tau_{SWO} = \gamma_{WO} \cos \theta \qquad (2)$$

Measuring the adhesion tension provides a means to determine the contact angle for smooth clean surfaces. This is discussed under wettability measurement by the Welhelmy method.

Regardless of which fluid(s) wets the surface, any spontaneous spreading that occurs decreases the free energy of the total system. The change in free energy is called the spreading coefficient, S_{WS}, for water and a solid, and in terms of the three interfacial tensions can be written as

$$S_{WS} = \gamma_{SO} - \gamma_{WO} - \gamma_{SW} \qquad (3)$$

Spontaneous spreading of water only occurs when S_{WS} is zero or greater. A change in the interfacial tension term of any component produces an alteration of wettability. Combining equation 1 with 3 results in

$$S_{WS} = \gamma_{WO}(\cos \theta - 1) \tag{4}$$

Spontaneous spreading of water occurs when water has a "nonfinite" contact angle (i.e. less than zero). In this case water spreads spontaneously on the surface, which can hold a uniform film of water that is stable at any thickness. When the contact angle is finite, and is less than 90°, the water "wets" but does not spread; it forms a sessile drop with a large area of contact on the surface. When the contact angle is finite, and is greater than 90° but less than 180°, the water does not spread and does not wet either; it forms a sessile drop with a very small area of contact on the surface. The discussion is analogous for oil spreading, but one has to be careful to maintain the contact angle as measured through the water.

To describe wettability in a porous reservoir rock requires inclusion of both the fluid surface interaction and curvature of pore walls. Both are responsible for the capillary rise seen in porous media. The fundamental equation of capillarity is given by the equation of Young and Laplace [2]

$$\Delta P = \gamma_{WO}(1/r_1 + 1/r_2) \tag{5}$$

The pressure difference, ΔP, between the two fluid phases that causes capillary rise is a function of the interfacial tension (surface free energy) and the mutually perpendicular radii of curvature, r_1 and r_2, for the interface between the two fluids. The pressure difference is known as the capillary pressure, P_c, and is defined as the pressure in the oil phase, P_O, less the pressure in the water phase, P_W. If the porous medium is regarded as a bundle of capillaries with an average radius, $\check{r}$, equation 5 can be expressed as

$$P_c = 2\gamma_{WO} \cos \theta/\check{r} \tag{6}$$

When liquid penetrates a single capillary of radius r, the length of flow l, in time t, for fluid of viscosity μ, is given by the Washburn [3, 4] equation

$$l^2 = (\gamma_{WO}rt \cos \theta)/(2\mu) \tag{7}$$

To apply to a porous medium, defined as bundle of capillaries, the equation can be written as

$$l^2 = ((c\check{r})\gamma_{WO}t \cos \theta)/(2\mu) \tag{8}$$

where c is a constant to allow for randomly oriented capillaries.

Forces Associated with Wettability. Molecules in a thin film reside in a different environment than those in a bulk phase. The fundamental forces that produce the interfacial tensions at surfaces are intermolecular. When the distances separating interfaces are small these forces interact proportionally to the distance of separation. The interaction at these distances can be described by a force per unit area known as the disjoining pressure, $\Pi(h)$, where h is the thickness of the thin film. When the disjoining pressure is positive, the interfaces repel each other. If the disjoining pressure is negative, the interfaces attract each other. The disjoining pressure can be approximated by the sum of three force components, electrostatic (Πe), structural (Πs), and van der Waals (Πm) [5].

$$\Pi(h) = \Pi e(h) + \Pi s(h) + \Pi m(h) \qquad (9)$$

The descriptions of these components are as follows:

- $\Pi e(h)$ – Ionic molecules in water give rise to double layers at the fluid–fluid and at the fluid–solid interfaces that consist of a compact surface charged layer and a diffuse layer of counterions (see Figure 2). As a film thins, the counterions of the two double layers approach each other and repulsion of these double layers takes place. This repulsion stabilizes the film by preventing further thinning and creates the electrostatic component of the disjoining pressure.

- $\Pi s(h)$ – Forces that interact at molecular dimensions and depend upon the solvent structure constitute the structural portion of the disjoining pressure. Dipole–dipole interaction is the attraction of the positive end of one polar molecule for the negative end of another polar molecule. The hydrogen bonding of water molecules are an example of dipole–dipole interaction. Non-polar molecules such as hydrocarbons do not dissolve in water because the attraction between adjacent water molecules is greater than that between water and hydrocarbons. This results in an enhanced

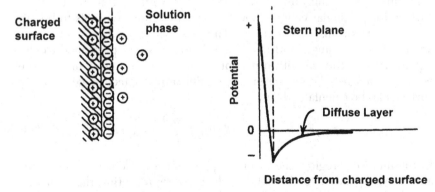

Figure 2. The electric double layer showing a compact surface charged layer and a diffuse layer of counterions. (Reprinted by permission from reference 38. Copyright 1989.)

attraction, hydrophobic bonding, between two particles when the solvent–particle interaction is weaker than the solvent–solvent interaction. Consequently, if two hydrophobic surfaces are immersed in water, the interfacial energy (interfacial tension) of the water–hydrocarbon interface is reduced as the hydrophobic surfaces are brought together.

- $\Pi m(h)$ – London–van der Waals forces, also known as the "dispersion" forces, arise from the mobility of electrons in a molecule such that molecules are attracted to or repulsed from each other by the polarizable electron cloud. The interaction between molecules varies with separation and provides for the van der Waals component of the disjoining pressure.

The Young–Laplace equation (equation 5) describes the force balance in terms of capillary pressure for two fluid phases in contact with each other and a surface. If one of the phases is present as a thin film, the equilibrium relationship that accounts for the thin film is the augmented Young–Laplace equation [6–8],

$$P_c = \Pi(h) + 2H\gamma_{WO} \qquad (10)$$

where H is the mean curvature of the film interfaces. In the meniscus region where the distance between interfaces is large, $\Pi(h)$ is zero and equation 10 reverts to equation 5. When the solid interface is flat and the interfaces are parallel, H becomes zero and P_c becomes equal to $\Pi(h)$.

An observable contact angle may exist for a meniscus even when one of the fluid phases completely wets a solid substrate with a thin film (see Figure 3). The thickness of the film can be less than the wavelength of visible light (4000 to 7000 angstroms). Melrose [9] estimated the bilayer film thickness of water to be on the order of 0.5 nanometers. Zorin [10] calculated the equilibrium thickness of the wetting film on a quartz surface for particular cationic and anionic surfactant solutions to be 100 and 770 angstroms, respectively. The observable contact angle is an apparent contact angle and under conditions that are presumed to give equilibrium is the equilibrium contact angle [7, 8] (used in previous equations above). This apparent or equilibrium contact angle on a flat surface can be calculated from

$$\gamma_{WO}\cos\theta = \gamma_{WO} + P_c h_O + \int_{h_O}^{\infty} \Pi(h)\,\partial h \qquad (11)$$

In the meniscus region, the macroscopic contact angle can be measured and is extrapolated to the solid surface by representing the film as a membrane of zero thickness. Although derived for a flat surface, equation 11 applies to curved surfaces if the film is thin relative to the curvature of the surface.

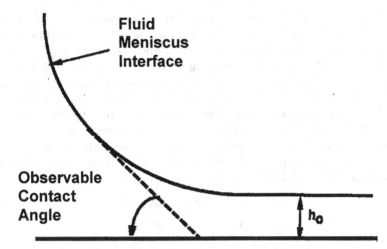

Figure 3.   *An observable contact angle may exist for a meniscus even when one of the fluid phases completely wets a solid substrate with a thin film of thickness* h_O.

Summary.   Most of the above fundamental equations have been developed for simple surfaces and form the basis by which wettability can be studied. Some researchers [11] have endeavored to extend these relationships to more complex surfaces. The addition of surfactant to a system can, to a limited extent, use the above equations to describe wettability alteration. As will be seen further below, the intricacy of the problem to alter wettability by surfactants extends well beyond contact angle and the pore shape complexity.

Wettability Measurement Methods

This section briefly describes the more customary measurement methods that can be used to study surfactant induced wettability alteration of porous and non-porous media. The information is given to make the reader aware of the significant differences between the various wettability measurement methods and the various quantities that they measure. The reader is referred to the references for more detailed information about any given method. Anderson [12] has produced a good literature review of many of the various wettability methods.

Amott–Harvey.   The wettability test devised by Amott [13] and its modification, the Amott–Harvey Relative Displacement Index (RDI) [14] are the most common quantitative measures of wettability employed for porous media by the oil industry. It relies on measurements of the saturation changes produced by spontaneous imbibition for both water,

ΔSw_S, and oil, ΔSo_S, compared to the maximum saturation change by forced imbibition of these same fluids, ΔSw_F and ΔSo_F, respectively, in the porous rock sample. Spontaneous imbibition occurs by displacement of the non-wetting fluid by a wetting fluid via capillary forces. Forced displacement (sometimes referred to as drainage, forced imbibition or viscous displacement) occurs via an external pressure gradient applied to the fluids to overcome capillary forces resisting further displacement of the fluid. Figure 4 illustrates those saturation changes in relation to the pressure required for fluid displacement. The Amott formulation consists of two terms, one defined as the water index (WI) and a second defined as an oil index (OI) as follows:

$$WI = \Delta Sw_S/(\Delta Sw_S + \Delta Sw_F) \qquad (12)$$

$$OI = \Delta So_S/(\Delta So_S + \Delta So_F) \qquad (13)$$

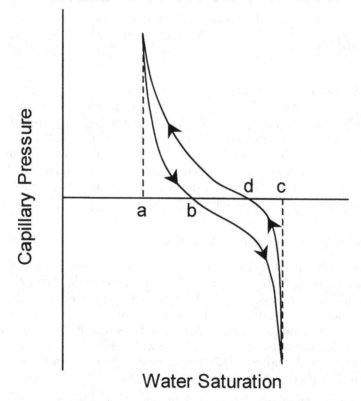

Water Saturation

Figure 4. The relationship of saturation states used for the Amott wettability indices or Amott–Harvey Relative Displacement Index to capillary pressure. ΔSw_S is the saturation change from a to b. ΔSw_F is the saturation change from b to c. ΔSo_S is the saturation change from c to d. ΔSo_F is the saturation change from d to a.

These indices vary from 0 to 1 for neutral to strongly wet, respectively.

The Amott–Harvey methodology combines these indices into a single expression, the Relative Displacement Index, RDI, by

$$RDI = WI - OI \tag{14}$$

RDI varies from 1 to -1, for highly water-wet and highly oil-wet porous media, respectively. Neutral wettability has an RDI equal to 0.

The main shortcoming, reported by Anderson [12] for the Amott methods, is their insensitivity near neutral wettability. In addition, the RDI relates relative volumes of imbibition, but one also needs to look at these individual volumes to obtain a better understanding of fluid displacement as represented by this index. More recently, Ma [15] reported that the Amott test also does not discriminate adequately at strongly water-wet conditions and proposed an imbibition rate method for wettability (see below). Other shortcomings include variations in laboratory procedures. The temperature and length of time employed for the spontaneous imbibition and the pressure used during the forced displacement cycle are often modified to match specific field parameters or for ease of measurement in the laboratory. Consequently, the quantitative values that are obtained can differ from one laboratory to another for reasons that are not related just to the wettability of the rock sample.

This imbibition method is dependent on both the chemical properties of the rock surface and the geometry of the pore network, particularly the pore throat size and shape. Since pore structure can affect the measurement results, the reduction of the interfacial tension between pore fluids by surfactants can affect fluid movement through the pore structure. This method, therefore, is not strictly a measure of surface wettability, but is a measure of the efficiency of fluid displacement that is normally most strongly affected by the wettability state of the pore surfaces. Within this framework of understanding, wettability alteration of the pore surfaces using surfactants can be studied with this measurement method.

USBM. The United States Bureau of Mines (USBM) method, described by Donaldson [16], is another commonly used method for determining the wettability of porous rocks. Although similar in overall practice to the Amott–Harvey method, it was based on the free energy change of water per unit of pore space, δF, accompanying a change of water saturation, δSw.

$$\delta F = -P_c \delta Sw \tag{15}$$

By relating the free energy change to that produced by forcibly displacing water (called secondary drainage) to that produced by forcibly displacing oil (called forced imbibition) into the same porous media, a measure of wetting is obtained. The data is normally obtained in a series of steps using conventional centrifuge techniques.

The integration of equation 15 is used to obtain the area under the individual curves (A_1 for secondary drainage and A_2 for forced imbibition). A limit of final saturation or capillary pressure must be chosen to provide consistent results. Wettability is defined as

$$W = \log(A_1/A_2) \tag{16}$$

Figures 5a to 5c show the USBM method applied, respectively, to water-wet, oil-wet and neutral-wet porous rocks. The wettability for each rock was determined using equation 15. Three distinct categories of wettability are readily apparent from these plots: water-wet, intermediate wettability, and oil-wet.

The USBM method is similar to the Amott methods in that it is affected

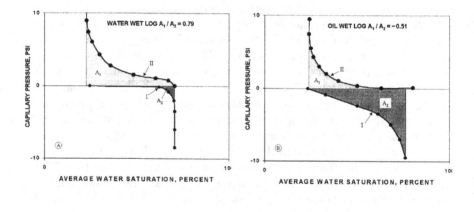

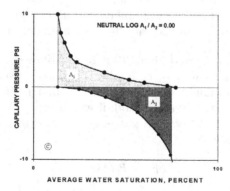

Figure 5.   The areas under the forced imbibition and forced drainage are the basis for the USBM wettability method. (Reproduced by permission from reference 16. Copyright 1969.)

by both the surface interaction and the pore network geometry. Likewise, with similar caveats, wettability alteration of the pore surfaces using surfactants can be studied with this measurement method.

Spontaneous Imbibition Index. The Spontaneous Imbibition Index (SII), as a quantifiable measure of wettability to one fluid component, was defined by Spinler [17] as the ratio of measured spontaneous water imbibition to highly water-wet spontaneous imbibition, ΔSw_{S-WW}:

$$SII = \Delta Sw_S / \Delta Sw_{S-WW} \qquad (17)$$

The relationship of SII to the Amott WI can be seen in Figure 6. This index is the same as the Amott WI at both neutral and highly water-wet conditions, 0 and 1, respectively. The denominator in equation 17 is normally estimated using a correlation obtained from conventional WI tests and can be formulated in terms of porosity, φ, and initial water saturation, Swi, both raised to powers.

$$\Delta Sw_{S-WW} = \varphi^a (1 - Swi)^b \qquad (18)$$

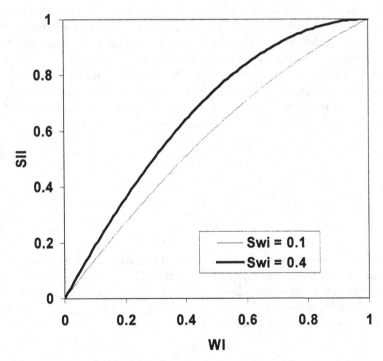

Figure 6. SII varies from the Amott WI with initial water saturation. SII is less sensitive to wettability at highly water-wet conditions.

Thus, SII is sometimes expressed as

$$SII = (Swf - Swi)/(\varphi^a(1 - Swi)^b) \qquad (19)$$

The above expression has been used not only for laboratory measurements, but to estimate apparent wettability from well log measurements for waterflooded reservoirs in which the displacement process was dominated by capillary forces [17].

Although SII has only been defined in terms of water imbibition, an analogous definition could be made for oil imbibition. Many of the problems that apply to the Amott indices also apply to SII, but the denominator of SII, once determined for a specific rock type, simplifies the determination of wettability to measurements of spontaneous imbibition.

Like the other imbibition methods, SII is dependent on both surface chemistry and pore network geometry. It works best for rocks that imbibe only one fluid, but could be modified to a form like the Amott–Harvey RDI. It also has the potential, by its formulation, to evaluate wettability alteration by surfactants in reservoirs.

Imbibition Rate.   Spontaneous imbibition rate has long been considered only a qualitative measure of wettability because the rate is dependent upon fluid and rock properties in addition to wettability. Ma [15], however, defined a method of quantifying wettability from the rate of spontaneous imbibition using the area under the imbibition curve as a measure of the work of displacement that results from the decrease in surface free energy. Wettability is defined as the ratio of pseudo-work of spontaneous imbibition, W, to the pseudo-work of spontaneous imbibition for strongly water-wet imbibition, W_{S-WW}:

$$W_R = \frac{W}{W_{S-WW}} \qquad (20)$$

where

$$W = \int_{S_{WI}}^{S_{WF}} \frac{\delta S_W}{t_D^b} \qquad (21)$$

for both the numerator and denominator of equation 20. The dimensionless time, t_D, in the denominator of equation 21 is raised to a power and used to normalize the dependency of the spontaneous imbibition rate on fluid and rock properties. The form of t_D used by Ma was based on the work of Rapoport [18] and Mattax et al. [19] as

$$t_D = At \sqrt{\frac{k}{\varphi} \frac{\gamma}{\mu_W} \frac{1}{L_C^2}} \qquad (22)$$

where A is a constant, t is imbibition time, k is permeability, φ is porosity, γ is interfacial tension, μ_W is viscosity and L_C is a characteristic length.

This index ranges from 0 to 1 for neutral to highly wet, and has a similar form for both water-wet and oil-wet systems. It has the advantage of being able to discriminate at strongly wet conditions. Figure 7 shows that until the contact angle exceeds about 50° on rough surfaces, the spontaneous imbibition endpoint ($P_c = 0$) does not vary significantly. In this case, indices such as the Amott and SII could not distinguish between contact angles less than about 50°. Since the imbibition rate technique uses the area under the imbibition curve and this area is proportional to the magnitude of P_c, this technique should be able to discern these different wettabilities. Otherwise, the same constraints apply to this technique that applied to the other imbibition techniques.

Contact Angle.   Precise measurements of fluid/surface interaction are made by determining the angle that forms between a drop of fluid and a surface [20, 21]. Wetting is noted by very low contact angles,

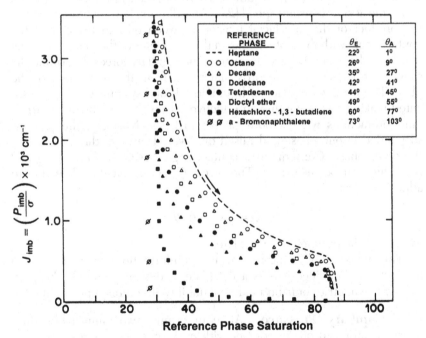

Figure 7.   This illustrates the effect of contact angle on imbibition capillary pressures for air/liquid in Teflon cores. θ_E is the contact angle on a smooth flat plate and θ_A is the advancing contact angle on a roughened Teflon surface. J_{imb} is the displacement curvature; P_{imb} is the imbibition capillary pressure; σ is the surface tension. (Reproduced by permission from reference 103. Copyright 1978.)

neutrality is observed at a contact angle near 90° and non-wetting is indicated by a high contact angle ($>90°$). Quantitative observation of the contact angle requires a very flat surface. Thus, this measurement is often considered more of a basic research method than a practical tool for analysis of porous rock. However, this technique is valuable because it provides a method for determining surface/fluid interaction, independent of pore throat geometry.

Contact angle is generally determined by optically observing the angle of the air/liquid/solid interface, or in the case of two fluids the liquid/liquid/solid interface. The actual measurement can be performed visually or by mathematically analyzing a digital image. The technique is extremely sensitive to surface contamination and requires diligence to obtain accurate values. When properly performed the analysis is highly repeatable and accurately describes the surface/fluid interaction.

Wilhelmy Plate.   The dynamic Wilhelmy plate provides a direct method of measuring adhesion tension, τ_{SWO}, from the force exerted when a plate is passed perpendicularly through an interface such as oil and brine at slow constant speed [22–25]. Adhesion tension (equation 2) is the product of the interfacial tension between oil and brine, and the contact angle with the plate acting on the perimeter of the plate in contact with the oil–water interface. Buoyancy and capillary forces determine the shape of force–distance records that result from the movement of the plate. The wetting behavior of liquid/liquid/solid systems can be investigated making this technique an effective means for investigating surfactant interactions with smooth planar surfaces. Normally the plate is suspended in one phase and raised or lowered through the interface to the other phase. Consequently, contact angle hysteresis (advancing and receding) can be measured. The force, Q, acting on the plate due to adhesion tension is:

$$Q = \gamma_{OW} p \cos \theta \tag{23}$$

where p is the perimeter of the plate.

The Wilhelmy plate method measures information about the surface interactions, but since it uses a flat plate, it does not provide the pore network geometric contribution for wettability alteration of porous media.

Capillary Pressure.   At equilibrium, two immiscible fluid phases (water and oil) in contact with each other in a porous material will distribute themselves in such a manner to minimize the free energy of the total system. This distribution is a function of saturation history, surface wettability for each fluid, pore structure, interfacial tension, fluid densities, and fluid height. The pressures within the water and oil phases reflect the distribution of fluids in a porous medium and consequently the free energy of the system. Capillary pressure, as previously defined, at any given height is the pressure difference between the oil and water pressure in the porous medium

$$P_c = P_O - P_W \qquad (24)$$

In porous media, by definition, the capillary pressure is positive when water is the more wetting phase and, accordingly, the capillary pressure is negative when oil is the more wetting phase. Capillary pressure becomes zero at the oil–water interface, also called the free fluid level that would exist outside of the porous medium.

Where surfactant is introduced to the reservoir via some carrying fluid, the relevant capillary pressure curves are from imbibition, both spontaneous and forced. Porous plate/membrane desaturation [26], flow or centrifuge effluent production [27, 28] and direct measurement of saturation in the porous media [29] are capable of measuring the complete imbibition curve. When used with reservoir-like fluids, the results from these methods reflect the wettability state of the porous rock. When surfactant is introduced, these methods should also be able to see the impact of wettability alteration. Two of these methods, the porous plate/membrane and the direct measurement of saturation methods require no modeling to be accurate for laboratory use.

Cryomicroscopy. Cryomicroscopy is a method that looks directly at pore surfaces and internal fluids to evaluate wettability [30]. Porous rock samples containing oil and water are frozen quickly to cryogenic temperatures and then viewed with conventional scanning electron microscope (SEM) techniques. The quick freezing process locks the pore fluids in place without expansion or movement. A cryogenic cold stage is added to the SEM to keep the sample cold. When coupled with X-ray spectroscopy analysis, it is possible to analyze pore wall mineralogy and geometry, and to differentiate brine from oil in the images of fluid distribution in pore spaces (see Figure 8).

Nuclear Magnetic Resonance (NMR). The surface sensitive nature of proton NMR relaxation provides a technique for qualitative wettability measurement in porous media. NMR measures the behavior of the magnetic dipoles of hydrogen nuclei in the presence of an applied magnetic field. Proton relaxation rates are strongly dependent upon the pore size and surface properties (wettability) of the pore wall. Water molecules near a water-wet surface will relax faster than near an oil-wet surface (and vice versa for oil). Most efforts to utilize NMR for core wettability have not demonstrated the capability to provide quantitative measures of wettability. However, Howard et al. [31] used relaxation time populations generated by a non-linear optimization technique to distin- guish between water and hydrocarbon phases in chalk at various water

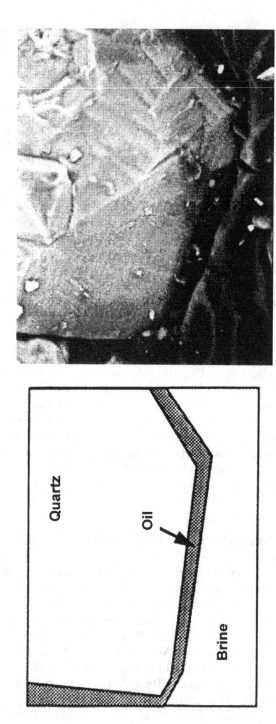

Figure 8. Magnification of the lower part of a quartz grain that was altered to be oil-wet. An oil film can be seen to surround the quartz grain. (Reproduced from reference 30. Copyright 1991.)

saturations. Shifts of these relaxation time populations were interpreted as indicators of pore-wall wettability. A comparison made with SII measurements indicated some promise for this technique.

Proton relaxation is strongly affected by paramagnetic impurities that are found in some porous rocks that can limit the application of this technique.

Other. Some of the other methods that are sometimes used to obtain a quantitative or qualitative measure of wettability include: immersion microcalorimetry [32], environmental scanning electron microscopy [33], capillary penetration [34], dye adsorption [35], pore surface analysis [36] and flotation [37].

Summary. The Amott, USBM, Spontaneous Imbibition Index, imbibition rate, and capillary pressure are all displacement methods applicable to porous media and the possible evaluation of wettability alteration by surfactants. However, these methods must be complemented by more fundamental studies using contact angles or adhesion studies (Wilhelmy), etc. to meld the understanding of surface interactions with the macroscopic displacement of fluids. To comprehend how a surfactant alters the contact angle on a flat surface provides only part of the information to predict how the surfactant will interact in porous media. To measure only the fluid displacement in porous media provides little information on surface interactions. NMR and/or cryomicroscopy could help span this gap. Cryomicroscopy can directly look at pore surfaces, but for the moment, it is difficult and time consuming to use. Both techniques provide more of a qualitative measure of wettability than quantitative, but they are tools that can complement and help bridge between more fundamental measurements and quantitative displacement methods.

Surfactant Induced Wettability Alteration

A number of factors affect the interaction of surfactants with the solid surface of porous rock and consequently affect wettability. Some of the more obvious items include:

1. Surfactant structure
2. Surfactant concentration
3. Kinetics
4. Pore surface composition
5. Other factors
   a. Surfactant stability
   b. Co-surfactants
   c. Electrolytes and pH
   d. Temperature

    e. Pore structure/surface roughness
    f. Reservoir structure

The interplay between these various factors is complex and often requires experimental measurement under as realistic conditions as possible to appropriately determine the impact of surfactant on wettability. It is the migration to, and the adsorption of, the surfactant at the fluid and solid interfaces along with the orientation and density of the adsorbed surfactant molecules that modifies the fluid–surface interfacial tension/wettability. Surfactant adsorption at an interface is a necessary, but not a sufficient condition for wettability alteration. Although details of adsorption will be covered in Chapter 4, this section includes a brief treatise on it with the other known variables that can affect wettability modification with surfactants.

Surfactant Structure.   Surfactants in this section are defined as amphipolar or amphipathic molecules composed of a hydrophilic head and a hydrophobic tail group. A detailed description of surfactants and surfactant structure can be found in Chapter 1. Surfactants are generally classified according to their hydrophilic head group. Common classifications are:

1. Anionic – negative charge
2. Cationic – positive charge
3. Amphoteric – charge changes with pH (sometimes grouped with Zwitterionics)
4. Zwitterionic – both positive and negative charges
5. Nonionic – no charge

The hydrophobic tail of the surfactant can consist of a hydrocarbon, perfluoroalkyl or polysiloxane group. The structure of the tail group such as its length or branching can significantly alter the physical behavior of the surfactant.

The dual nature of surfactants produces a strong affinity for interfaces between immiscible fluids such as oil and water or fluid/solid interfaces. The concentration of surfactant at an interface minimizes the free energy of the total system. The surfactant, by adsorbing at a fluid/solid interface, reduces interfacial tension and modifies the ability of water or oil to wet the solid surface. A surfactant that orients itself on a surface such that the surfactant molecules have the hydrophobic tail groups away from the surface or along the surface will decrease water-wetting and increase oil-wetting. Likewise, the orientation of a surfactant with the head group away from the surface can make the surface more water-wet.

In addition to the dual nature of surfactant molecules, specific structural characteristics can increase or decrease the packing of surfactant molecules and consequently influence the modification of surface wettability. Rosen [38] made some general observations of modifications to hydrophobic tail groups that can be applied to the water/solid interface as follows:

1. Increasing its length can
   a. improve packing of the surfactant molecules at an interface provided the head group permits it, and
   b. increase the tendency of the surfactant to absorb at an interface.
2. Introducing branching or undersaturation can cause looser packing of surfactant molecules at the interface.
3. The presence of an aromatic nucleus can
   a. cause looser packing of surfactant molecules at the interface, and
   b. increase adsorption of the surfactant onto polar surfaces.
4. The presence of a polyoxypropylene chain can increase surfactant adsorption onto polar surfaces.
5. The replacement of the hydrophobic tail by a perfluoroalkyl group could create a surface that is neither water nor hydrocarbon wet.

The literature cited next does not include experimental work on reservoir-like materials, but the general nature of the observations should be indicative of the effect that the structural modifications to the hydrophobic tail group has on adsorption:

- Varadaraj et al. [39] used Wilhelmy plate measurements to study dynamic contact angles of Guerbet sulfate (branched hydrophobic tail) and monodisperse ethoxy sulfate surfactants (linear hydrophobic tail) on the Teflon–water–air interface. Comparison of C16 linear with C16 Guerbet surfactants revealed that hydrocarbon chain branching decreases the advancing and receding angles by about 30°, representing increased water-wetting effectiveness. This change was attributed to an increased structural rigidity of the branched hydrophobic tail group as well as an increased area of coverage.
- Varadaraj et al. [34] also evaluated the influence of the composition and structure of the hydrophobic tail group on wettability alteration of hydrophobic sand packs using capillary penetration wetting techniques. The composition of the surfactant tail group and its branching exerted a significant influence on wetting. The rate of wetting at half saturation was observed to order as ethoxylates > sulfates > ethoxysulfates. However, wetting effectiveness was observed as sulfates > ethoxylates based on maximum amount of water imbibed. Branching of the tail group was observed to increase the wetting rate and effectiveness.
- Schechter et al. [40] compared the adsorption of alkyl benzene

sulfonates (anionic) on alumina for varying lengths of hydrophobic tail groups. Adsorption below the critical micelle concentration (CMC) increased with increasing tail length. For the plateau level, at and above the CMC, increasing tail length produced a small decrease in adsorption.

Experiments such as these show that it is important to match the surfactant to the specific surface of interest to achieve optimum wettability alteration.

Surfactant Concentration.   In solutions below the CMC, the surfactant in solution consists of monomers. Above the CMC, surfactant micelles are formed and the monomer concentration remains relatively constant with increasing surfactant concentration. Surface coverage by surfactant molecules, however, varies in a more complex manner with concentration. Chapter 4 of this book provides a detailed description of surfactant adsorption isotherms.

A typical ionic surfactant adsorption isotherm for an oppositely charged substrate is illustrated in Figure 9. A number of researchers [41,

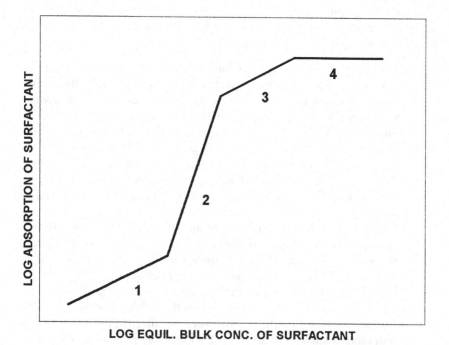

Figure 9.   A typical ionic surfactant adsorption isotherm for an oppositely charged substrate. The regions where adsorption behavior changes are numbered and described accordingly in the text.

42] have described the change in surfactant adsorption behavior as follows:

- Region 1 corresponds to low surface coverage by individual surfactant molecules with an absence of surfactant aggregate formation on the surface.
- Region 2 indicates the formation of surfactant aggregates (called admicelles or hemimicelles) which produces the sharp increase in the slope of the isotherm.
- Region 3 represents a sufficient accumulation of surfactant aggregates with the decrease in the slope of the isotherm caused by a change in the sign of the surface charge to that of the oppositely charged surfactant, resulting in electrostatic repulsion of further ionic surfactant molecules.
- Region 4 begins at the CMC and is described as completion of bilayer coverage of the surface.

The adsorption of ionic surfactants on a like-charged substrate is less understood, but can occur via hydrogen bonding or dispersive forces [43]. In the experimental work reported by Alveskog et al. [44], the wettability alteration of Berea sandstone, containing negatively charged mineral surfaces, with an anionic surfactant appeared consistent with the stated mechanisms. Details are provided in the Laboratory and Field Studies section of this chapter.

Nonionic surfactants are described as having Langmuir type adsorption isotherms [43] on charged substrates, with the surfactant at low concentrations lying prone on the surface and at higher concentrations with the hydrophobic group displaced from the surface. For concentrations at or above the CMC, either a monolayer or a bilayer may form. If only a single monolayer is adsorbed and the hydrophobic end of the surfactant is outward from the surface or lying along the surface, this would render the surface more oil-wet. If a bilayer is adsorbed, the hydrophilic end of the surfactant would tend to render the surface water-wet with a reversed ionic surface charge for an ionic surfactant. A reversal in surface charge was measured by Zorin et al. [10] for a quartz surface as a function of a cationic surfactant concentration (cetyltrimethyl-ammonium bromide) (see Table 1).

Rosen [43] also describes the surfactant adsorption isotherm for non-polar, hydrophobic substrates as Langmuir type. At low concentrations, the surfactant orientation occurs with the hydrophobic tail group close or parallel to the surface and the hydrophilic head group towards the water. As more surfactant is adsorbed, the orientations of the surfactant molecules become more perpendicular to the surface until the CMC, where surface saturation is achieved. The different orientations of the

Table 1

CTAB Concentration (mole/l)	Charge Sign of Quartz Surface
0	negative
10^{-6}	negative
10^{-5}	negative
10^{-4}	positive
5×10^{-4}	positive

Reprinted with permission from reference 10.
Copyright 1992

surfactant molecules would make the surface change from oil-wet at low concentrations to water-wet at the CMC and higher concentrations.

Complicating the typical behavior of surfactants with increasing concentration, as described above, is the reported work by McGuiggan et al. [45], in which different adsorption procedures can produce different adsorption/wetting characteristics. Sequential adsorption by several exposures to small amounts of dihexadecyl dimethylammonium acetate on mica surfaces produced an intercalated monolayer exhibiting low adhesion; whereas direct adsorption by a single exposure to a higher surfactant concentration produced a hydrophobic monolayer of high adhesion. The low adhesion of the intercalated monolayer was a result of head groups pointing both into and away from the mica surface. In such a configuration, the positively charged head groups reduce interfacial energy by electrostatic interaction with the mica or the aqueous environment. Conversely, with a hydrophobic monolayer where the surfactant molecules are oriented in the same direction, the hydrophobic tails create a high energy interface when interacting with the water-wet mica surface or the aqueous solution. An intercalated monolayer has a lower water/hydrophobe interfacial energy, but at the cost of an increase in the mica–surfactant interfacial energy.

Kinetics.   Kinetics also play a role in wettability alteration of porous media by surfactants. In reservoirs, surfactants must be transported through the pore networks by an injected fluid phase, usually water or oil. The ability to alter wettability is related to surfactant diffusion rates and adsorption rates. Surfactants must diffuse through the bulk fluid phase to the meniscus interface and the fluid–pore interface. Surfactants must also interact and adsorb on the pore surfaces. If the diffusion rate or the adsorption rate for surfactant is slow relative to the creation of new water–rock interfaces, because of the water displacement rate, the wettability of the pore surfaces may vary with time. These types of nonequilibrium effects are not well understood and may control the displacement of oil in porous rock. Experimental and theoretical work illustrates the importance of understanding surfactant kinetics:

- Chesters et al. [46] have addressed the transport of low concentrations of surfactant in a theoretical framework. A hydrodynamic model of steady wetting was extended to include the effect of a nonionic surfactant. This model indicated that under certain conditions surfactant may concentrate at the contact line between a surface and a meniscus to significantly reduce the contact angle.

- Damania et al. [47] used the Wilhemy technique to study the processes of diffusing anionic and cationic surfactants on quartz surfaces. The surface tension response for populating a newly formed meniscus surface (on contacting a liquid–air interface) with anionic surfactant decreased for the slow diffusion process with minimal adsorption on the similarly charged quartz surface. The surface tension response for the cationic surfactant, however, showed an oscillating behavior indicating a varying concentration of surfactant at the meniscus as scavenging by adsorption on the quartz surface competed with populating the newly formed meniscus surface for the slow diffusing surfactant.

- Similar erratic behavior was found by Princen et al. [48], who performed an experimental study of instabilities that occur during wetting. These instabilities manifested themselves as stick–slip phenomena, where an advancing fluid of surface-active molecules (trioctylamine) moved in a non-uniform manner onto hydrophyllic quartz surfaces. Depending upon the velocity of the moving surface, the instabilities could be periodic or random. It was postulated that this effect was caused by the diffusion and adsorption of molecules in front of the liquid edge, increasing the contact angle.

- Dynamic capillary pressures and interfacial tensions using surfactants (cetyltrimethylammonium bromide, a cationic surfactant and ethoxylated isononylphenol, a nonionic surfactant) were studied by Churaev [49] and Churaev et al. [50]. The rates of spontaneous and forced displacement of aqueous surfactant solutions by oils were measured in thin quartz as a function of pressure drop and surfactant concentration. The surface of an advancing or receding meniscus was observed to become depleted of surfactant by adsorption or concentrated by desorption. Thus, capillary pressures, contact angles, and interfacial tensions become rate dependent as a result of mass exchange processes between the moving meniscus and a wetting film on capillary walls. A model representation of the mass transfer of surfactant molecules as shown in Figure 10 illustrates the nature of the problem.

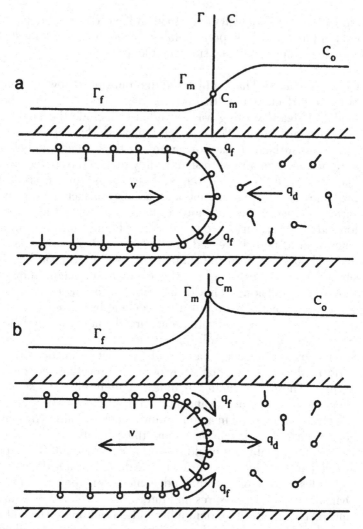

Figure 10.   This schematic illustrates the mass transfer of surfactant molecules between a receding meniscus (a), and advancing meniscus (b), and a wetting film on the capillary wall. Γ is surfactant adsorption, C is surfactant concentration, q is the flux of surfactant between the meniscus and into the fluid or onto the wetting film and v is velocity. Subscript f refers to the wetting film, m refers to the meniscus, o refers to the bulk solution and d refers to diffusion. (Reprinted by permission from reference 50. Copyright 1996.)

Table 2

Mineral	Isoelectric Point
Quartz	<2.0 to 3.7
Kaolinite	<1 to 5
Calcite	none/4 to 10.1
Dolomite	<6 to 7

Reprinted with permission from reference 54.
Copyright 1996

Pore Surface Composition. The bulk matrices of most reservoir formations are composed of mixtures of quartz, clays, limestone and dolomite. Based on the normally water-wet character of these rock matrix components for most formations, it could be concluded that most reservoirs are water-wet. However, Treiber et al. [51] reported that laboratory evaluations of 55 different oil producing reservoirs from various areas of the world showed a different picture with 66% of the lab reservoir core as oil-wet with a contact angle from 105 to 180° and 7% as intermediate-wet with a contact angle from 75 to 105°. Taber [52], using Morrow's [53] definition of intermediate wettability as systems that do not spontaneously imbibe, an advancing contact angle of 62° to a receding contact angle of 133°, changed this allocation to 27% of the lab reservoir core as oil-wet and 47% of the lab reservoir core as intermediate-wet. The factor that masked the water-wet character of bulk rock composition was the presence of films on the pore surfaces that resulted from the deposition of crude oil components and/or original organic films.

The surface potential of the reservoir bulk matrix materials cited above depends upon pH of the water phase in contact with the mineral. The pH for which the Zeta potential on a surface is zero is the isoelectric point. These surfaces are positively charged at pH less than the isoelectric point and negatively charged at pH greater than the isoelectric point. Table 2 contains isoelectric points for the common reservoir materials assembled by Maini et al. [54]. Ionic surfactants therefore can be attracted or repulsed from a surface depending upon pH.

Under most reservoir conditions, quartz grains are considered as having a negative surface charge and carbonate grains as having a positive surface charge. Clay materials, because of their structure, can have a different charge on the surface of the clay crystal than on the edge of the crystal, but typically have a negative surface charge and a positive edge charge [55].

• Some clay minerals can also coat a pore rock surface and change its wettability behavior. Trantham et al. [56] noted that the oil-wet character of the North Burbank reservoir was due to a coating of chamosite clay, that covered an estimated 70% of the quartz surface.

- Fassi-Fihri et al. [30] used cryomicroscopy on Brent reservoir sandstone to study the placement of oil. Illite was found associated with brine and the reservoir oil had an affinity for kaolinite. Quartz and feldspar were noted to be generally in contact with the brine. Wettability was heterogeneous on the pore scale. The field was considered to have mixed wettability since in laboratory tests both oil and brine would spontaneously imbibe into the rock.

Reservoir mineral surfaces tend to be contaminated by hydrocarbon components. Crude oil components may adsorb on these high-energy mineral surfaces by a variety of mechanisms [57] including precipitation, and polar, acid/base and ion-binding interactions. Furthermore, the latter two processes require the presence of brine. Yan et al. [58] induced wettability alteration of Berea sandstone plugs, even at relatively high water saturations, by precipitation of asphaltene fractions. Xie [59] used the Wilhelmy plate method to measure the wettability alteration and found that at higher temperatures, the more asphaltic crude oils resulted in higher contact angles. Clementz [60] demonstrated that asphaltenes adsorb on clays at low water content. Radke et al. [11] used the concept of disjoining and critical capillary pressures to describe the collapse of water films on pore walls with negative curvature to provide sites for asphaltene adsorption from crude oils. Given geological time and the higher temperatures that often exist in reservoirs, it is likely that, by diffusion through thin water films and other mechanisms, all reservoir rock surfaces have some degree of hydrocarbon contamination.

Some limestone rocks may even retain original organic films since limestone has an organic origin. Chalk is composed of the skeletal remains of numerous nanofossils such as coccoliths. Associated with chalks are thin (few hundred angstroms) organic films [61, 62] that appear well preserved despite the depositional environment and age.

The result of the varied composition for reservoir pore surfaces is that surfactants can interact with these surfaces in a variety of ways on a pore level and their consequential orientation can alter wettability. The ionic surfactants will be attracted to or repelled from charged surfaces, non-ionic surfactants may adsorb on negatively charged surfaces via hydrogen bonding, and the hydrophobic tail group regardless of the head group may adsorb by dispersion or hydrogen bonding to hydrophobic organics.

Not only are the pore surfaces complex in the reservoir, but it is difficult to obtain pore surfaces for laboratory studies like those found in the reservoir. The problem is caused by a number of factors including:

1. Contamination from drilling and coring.
2. Changes in temperature and pressure from reservoir conditions

that can cause precipitation of components from the brine, deposition of crude components and/or dissolution of surface materials.

3. Improper storage that can result in evaporation of water and the lighter crude components, changes in clays and the oxidation of surfaces.

4. Restoration process in the laboratory that includes solvent extraction and resaturation with non-reservoir-like fluids.

Such artifacts affect the laboratory response of the core materials to surfactants and can complicate the evaluation and selection of surfactants for use in the reservoir.

Other Factors. There are a number of other factors that can potentially affect wetting. Their effects on wetting, however, are little studied, particularly in porous media. They are mentioned here to make the reader aware of them.

1. Surfactants may spend long periods of time being transported in the reservoir before interacting to alter the wettability of pore surfaces. The surfactant must maintain its chemical structure and interfacial properties during that time. The long term stability of surfactants at elevated temperatures in an appropriate brine can be monitored in the laboratory from cloud point and interfacial tension measurements [63].

2. Co-surfactants or impurities can interact with the surfactant in solution and at interfaces to improve or worsen the behavioral characteristics of the surfactant. Most commercial surfactants are not pure blends and are often shipped as a complex mixture of surfactants and alcohols to improve their handling characteristics. Some alcohols, when mixed with surfactants, are known to synergistically reduce interfacial tension between fluids. They could also interact with the surfactant at the solid–liquid boundary. Binary mixtures of surfactants such as anionic and nonionic have been investigated to reduce surfactant adsorption [64], but the effect on wettability was not determined. Mixtures of surfactants and co-surfactants will usually separate chromatographicaly [65] in the porous rock as the stronger adsorbing components are more rapidly depleted resulting in a compositional change of the mixture as it propagates through the reservoir.

3. Electrolytes and pH can affect both the surfactant and the rock pore surface interaction with the surfactant. Complicating the situation for an oil reservoir, is the presence of pH and ions for the injected surfactant solution that are not in equilibrium with the reservoir rock or connate brine. Electrolytes are known to affect

the solubility of surfactants [66]. Schramm et al. [67] provided a broad study of the effect of brine composition, ionic strength and pH in the range of most reservoirs on the Zeta potential for Berea sandstone, Indiana limestone and Baker Dolomite. The presence of organic films on pore surfaces may dampen or have other significant effects on such surface interactions. Bitting et al. [68] investigated the effect of counterion type and concentration on adsorption for various salts of dodecyl sulfate (anionic) on alumina. Counterion binding induced aggregate formation by an anionic surfactant on a like-charged surface with the monovalent counterions between the adsorbed surfactant aggregate and the mineral surface. The dependence of the advancing and receding contact angle on pH (approximately $2 < pH < 12$) for sodium dodecyl sulfate (anionic) and cetyltrimethylammonium (cationic) surfactants were measured by Zorin et al. [10] in 0.1 normal NaCl brine on polished quartz surfaces. Both the advancing and receding contact angles were different and decreased with increasing pH. No hysteresis was observed with the anionic surfactant. The cationic surfactant experienced substantial hysteresis with both the advancing and receding contact angles increasing for intermediate pH ($4 < pH < 10$).

4. Temperature can also alter wettability by affecting either the surfactant or the surfactant–surface adsorption characteristics. Ziegler et al. [69] reported that the adsorption of a nonionic (nonylphenoxypolyethanol) decreased with temperature increase for low concentrations, whereas the opposite was true for high concentrations. Noll et al. [66] reported adsorption calorimetry results that indicated an increase in temperature decreased adsorption for sodium dodecylsulfate (anionic) and decyltrimethylammonium bromide (cationic) surfactants regardless of surface wettability. Similar results were reported for nonionic commercial surfactant (Triton^{TM} X-100) except for adsorption on an oil-wet surface. These trends were consistent with an increase in adsorption associated with conditions that caused a decrease in surfactant solubility in solution.

5. Pore surface roughness may affect the apparent wettability induced by surfactants in the same way that surface roughness can affect equilibrium or advancing and receding contact angles. Figure 11 shows how the apparent equilibrium contact angle, θ_R, can appear to be something other than the true equilibrium contact angle, θ_{TRUE}. The roughness effect on contact angle is sometimes represented by $\cos \theta_R = R \cos \theta_{TRUE}$ [70]. If θ_{TRUE} is greater than 90°, it is increased by surface roughness; if θ_{TRUE} is less than 90°, it is decreased by surface roughness. Some of the

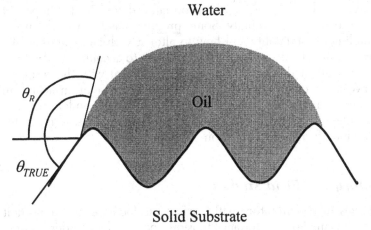

Figure 11.   The upper drawing shows the misleading effect of roughness on the equilibrium contact angle for a water-wet substrate. The lower drawing shows the same effect for an oil-wet substrate.

hysteresis effects seen in advancing and receding contact angles have been associated with rough surfaces [71, 72].

6. Large reservoir features can influence wettability alteration of a reservoir by surfactants. The sweep efficiency of a reservoir can influence the quantity of surfactant that contacts the reservoir rock matrix. Wettability alteration may be limited by the permeability contrast and the displacement mechanism. For a highly fractured reservoir with capillary dominated oil displacement, only the water

that enters a matrix block by spontaneous imbibition will be able to transport surfactant to the matrix. At lower surfactant concentrations, there may be insufficient surfactant to satisfy the matrix adsorption, but sufficient to alter the wettability of the portions of pore surfaces first encountered or that have a higher affinity for the surfactant. For viscous dominated oil displacement, water is continuously displaced through the reservoir matrix, carrying surfactant. Most of the injected surfactant is transported through the higher permeability parts of the reservoir. For lower concentrations of surfactant, the quantity of surfactant may eventually be sufficient to satisfy matrix adsorption with ample pore volumes of surfactant containing injected water. In either case, the adsorption of surfactant is controlled by the injected surfactant concentration, its dilution by formation water and the volume of water displaced through the matrix.

Summary.   The interplay between the static and dynamic factors is very complex and often requires experimental measurement under conditions as realistic as possible to appropriately determine the impact of a given surfactant on wettability. Some progress has been made in the understanding of wettability and how to alter it with surfactants in the reservoir. Ultimately, it may require a field test to account for all of the variables that affect the interaction of the surfactant with the reservoir rock. However, by comprehending the probable behavior of surfactants in the laboratory, and within the framework of conceptual models, one can likely reduce the effort required to select a suitable surfactant for wettability modification of a reservoir.

Laboratory and Field Studies

Most studies in the literature with surfactants focus on the important aspect of reducing liquid interfacial tensions, nonetheless, understanding the role that surfactants can have to alter wettability is of equal importance. As an example, some researchers [73] have reported that maximum oil recovery occurs near neutral wettability. In actuality, the optimum wettability condition for maximum oil recovery depends upon numerous factors and can vary from reservoir to reservoir [74]. Surfactants provide an opportunity to modify reservoir wettability for maximum secondary or tertiary oil recovery. Other opportunities exist to improve drilling, etc. This section provides laboratory studies and field examples of wettability alteration by surfactants in porous media.

Capillary Pressure and Relative Permeability.   Surprisingly, no specific experimental studies were found on the effect of surfactants on capillary pressure or relative permeability (K_r). These parameters are essential to mechanistic numerical modeling of oil displacement. There are studies [53, 75] that show the effect of wettability on P_c and K_r, but these are not surfactant generated and can not be concluded to be representative. Nevertheless, since it is known that both P_c and K_r are affected by wettability, surfactants are expected to have a significant effect on these properties.

Imbibition Tests. Alveskog et al. [44] measured the influence of surfactant concentration for n-dodecyl-ortho-xylene-sulfonate (anionic) on Berea sandstone plugs. Since the quartz surfaces of Berea sandstones are normally negatively charged, adsorption of the like-charged surfactant should be low and wettability alteration at a minimum. However, as surfactant concentration was increased, the wettability as measured by the Amott–Harvey method changed from strongly water-wet to weakly oil-wet in a very narrow range, 0.005 to 0.008 weight % surfactant. This wettability reversal occurred at a surfactant adsorption level that was 10% of the adsorption at the CMC and remained weakly oil-wet for surfactant concentrations well above the CMC. The tabulated data included by Alveskog indicated substantial spontaneous imbibition of water, 50 to 70% of final water saturation quantity after forced imbibition, in addition to spontaneous oil imbibition throughout the surfactant concentration range for which the Berea plugs were labeled weakly oil-wet (see Figure 12). This suggests that the plugs had a substantial number of continuous water-wet pathways in addition to continuous oil-wet pathways, otherwise spontaneous imbibition of both oil and water would not occur. Surfactant adsorption would be expected to primarily take place through hydrogen bonding or dispersive forces for the quartz Berea surfaces. If this is the case, it is possibile that the surfactant hydrophobe covered some of the quartz surfaces creating oil-wet pathways and the surfactant head group provided water-wet pathways. Another possibility was that clays or other than quartz surfaces were present and have affected the results.

Austad et al. [76] conducted imbibition tests in outcrop chalk that had been aged in crude oil to achieve desired wettabilities. Chalk character-ized as nearly oil-wet because of a slow spontaneous imbibition rate, saw a dramatic increase in the countercurrent imbibition rate and in oil recovery when the imbibition water contained 1 weight % dodecyltri-methylammonium bromide surfactant (cationic). This was attributed to a change in wettability to a more water-wet state. The increase in the countercurrent imbibition rate indicated that the loss of capillary forces expected for the decrease in interfacial tension from the surfactant was overcome by the increase in capillary forces from the wettability change.

Foams. Foams are employed to reduce the mobility ratio of injected gas to in-situ fluids in a reservoir. This reduction provides a

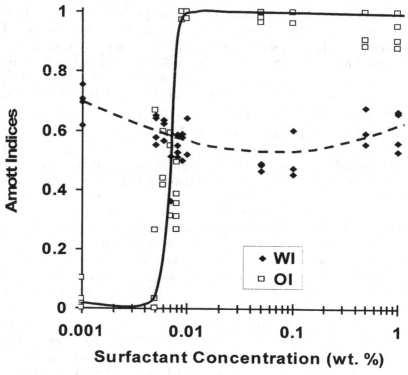

Figure 12. Alteration of wettability of Berea plugs with an anionic surfactant created continuous oil paths as well as water paths in the rock as evidenced by the high indices.

better sweep efficiency and consequently increased recovery of hydrocarbons. A foam is produced by mixing surfactants with water and injected gas. Limited studies [77, 78] have been made where attention to wettability alteration by the surfactants in foams have been addressed. Foams, made with similar surfactant concentrations, were reported equally effective at reducing gas permeability in both oil-wet and water-wet porous media. The ability to form a stable foam in an oil-wet porous medium resulted from the alteration of the initially oil-wet surface to water-wet by surfactant adsorption. Wettability alteration of the oil-wet surface was evidenced by a dramatic shift in liquid relative permeability when surfactant was present in the aqueous phase. The liquid relative permeability curve of the oil-wet porous medium, in the presence of surfactant, essentially matched that of the porous medium when it was water-wet. Similar shifts in liquid relative permeability have not been observed for foam flow in strongly water-wet porous media. The adsorption of surfactants from foams was substantial for some surfactants [79].

Since surfactant adsorption can alter rock surface wettability, it is possible that a surfactant could change a water-wet surface to oil-wet and break the foam. Such foam effects on porous media surfaces must be considered in the design of the foam.

Carbon Dioxide Flood.   Smith et al. [80] studied the impact of wettability on tertiary oil recovery by carbon dioxide flooding after a secondary waterflood. It was reported that oil recovery could be improved by the wettability alteration of reservoir rock surfaces using surfactants. In this study, water-wet sandstone rock surfaces were modified by treatment with solutions of surfactants to neutral and even moderately oil-wet states. The laboratory results indicated that maximum tertiary oil recovery, after waterflood, by carbon dioxide flooding increased as the wettability of the sandstone decreased from highly water-wet to a neutral-wet or a slightly oil-wet surface.

Microbial Enhanced Oil Recovery (MEOR).   The use of in-situ or injected microbes to increase oil production is very attractive because of low cost, environmental concerns and distribution of the microbes throughout the reservoir. The dominant mechanism(s) involved in MEOR apparently has not been positively identified. It is known that microbe growth produces gases, acids, surfactants and polymers. The gases can increase oil production by viscosity reduction and expulsion of oil by free gas [81, 82]. Acids are believed to increase permeability by dissolving the constrictions in the pore network [76, 83]. Polymers and the biomass of cells can plug the larger pores and provide a better flow pattern [81, 82]. Microbes produce a variety of biosurfactants, with the type and amount dependent on both the specific strain of microbes and growth conditions [84, 85]. Marsh et al. [86] showed that generation of biosurfactant was essential for increased oil recovery by comparing flooding with whole microbe fluids, cell-free fluid and fluid from a microbe which did not produce surfactant. They concluded that the production of acids, solvents and gases alone could not explain the observed increases in oil recovery by Bacillus strain JF-2.

Researchers tend to clump all biosurfactants as interfacial tension reducers [81, 83, 84], however, their measured interfacial tensions are often at least two-orders of magnitude greater than required for increased oil production by interfacial tension reduction. This discrepancy is generally explained by suggesting that in the rock the biosurfactants are locally concentrated or that the presence of co-surfactants, such as bioalcohols, produces a lower IFT than measured in the laboratory. Bala et al. [87] observed that MEOR with a wide range of oils was about the same for all oils. Furthermore, they noted that there was no correlation between IFT reduction and MEOR. This suggests that an alternate explanation might be a change in rock wettability produced by the biosurfactants. To achieve a change in wettability would not require the concentration of biosurfactant generally associated with substantial IFT reduction. The change in surface wettability could also explain the generally observed increase of oil permeability during MEOR. One could even speculate that the produced bioacids may react with crude oil components to form surfactants. This area of MEOR seems potentially fruitful for further work because of low operating cost and the placement of the surfactant producing microbes throughout the reservoir. The latter eliminates the significant loss of surfactant through adsorption, which often occurs for conventional surfactant injection into wells.

Drilling Mud.   Drilling muds and completion fluids often contain surfactants to either suspend the components in the additive package or affect the drilling process. For example, cationic and anionic surfactants are included in oil-based muds to wet the surface of the cuttings and facilitate their removal from the wellbore by floatation. Thomas [88] showed that drilling fluids could change the wettability of core. Sharma et al. [89] identified the major contributor for the wettability changes caused by drilling muds as the surfactants. They found that oil-based muds had the most profound change on the wettability of sandstone and carbonate outcrop rocks. Water-based muds, even ones containing surfactant, had only a minimal effect. Highly water-wet samples were more affected by the oil-based muds than the less water-wet samples. The altered wettability ranged from slightly less water-wet to oil-wet. For the muds tested, the original water-wet state of the rock was largely restored with cleaning. Menezes et al. [7] found that most oil-based mud components changed the wettability of pure quartz. McDonald et al. [90] investigated oil-based mud emulsifiers and actual mud filtrates containing surfactants that changed the wettability of sandstone rock. Sanner et al. [91] reported that oil-based mud filtrate reversed the wettability of both sandstone and carbonate rock. This altered wettability reduced the effective oil permeability in low permeability rock and increased oil permeability in high permeability rock, by reducing water saturation in high permeability rock. Most of the wettability alterations caused by drilling muds are inadvertent. The surfactant was added to produce some property other than wettability alteration. However, Christensen et al. [92] described a water-based drilling mud additive with a strongly lipophilic surfactant that was included to decrease shale sloughing and borehole instability while drilling. This was accomplished by reducing the water-wettability of the rock and preventing water from interacting with the formation and causing the damage.

Surfactant Enhanced Oil Recovery.   Most traditional surfactant floods were conducted to minimize the interfacial tension between the oil and water. In water-wet reservoirs, oil is generally trapped in pore spaces by capillary forces and can be present as discontinuous oil ganglia following a waterflood. In these cases, by reducing interfacial tension, the capillary forces are reduced and oil recovery may be improved. In oil-wet or mixed-wet conditions, a substantial portion of the oil is trapped by being attached to pore surfaces. Although a reduction in interfacial tension will help mobilize the oil, a change in wettability of the pore surfaces to a more water-wet state will release oil from the surfaces and consequently can improve oil recovery.

A study [93], which included laboratory evaluation and a field test in Santa Fe Energy Company's Torrance Field, reported increased oil production when the reservoir wettability was altered from oil-wet to water-wet using a synthetic surfactant. These surfactants, called thin film spreading agents (TFSA), such as alkoxylated nonylphenol resins, displace asphaltene molecules from both oil-water interfaces and mineral surfaces. The laboratory studies on demulsification, wettability alteration, and oil recovery efficiency indicated that TFSA molecules recovered incremental oil by coalescing near wellbore emulsions, rendering oil-wet reservoir rock surfaces water-wet, thus increasing oil permeability and improving areal sweep efficiency. Over an 18 month period in a 36-acre waterflood pilot, a 0.1 pore volume bank containing 239 mg TFSA/kg of rock was injected into an irregular pattern of one injector surrounded by nine producing wells. An estimated 8150 ± 850 bbl of additional oil was recovered.

A similar test in the Basal Tannehill reservoir, Shackleford County Texas [94], was not as conclusive as the Torrance pilot, due to the short time period of stabilized oil production prior to TFSA injection. However, an estimated 1510 barrels of additional incremental oil was recovered.

These field responses to wettability alteration appear to corroborate the laboratory work that showed increased oil recovery with wettability alteration under idealistic laboratory conditions. However, there were insufficient follow up studies to confirm that wettability alteration was the primary reason for additional recovery. Additional field studies such as coring of the pilot area could have helped to evaluate the effectiveness of wettability alteration.

Alkaline Floods. Alkaline floods, typically using sodium hydroxide, generate surface active products by an in-situ chemical reaction between the injected alkali and the organic acids of the crude. Four possible mechanisms [95] are responsible for the recovery of oil by alkaline floods: (1) emulsification and entrainment, (2) emulsification and entrapment, (3) wettability reversal from oil-wet to water-wet, and (4) wettability reversal from water-wet to oil-wet. One example in the literature of wettability alteration by alkali [96] was reported for an offshore field in the Gulf of Mexico that had a low recovery factor from primary production. The wettability of this reservoir was found, using the

Amott method, to be preferentially oil-wet, RDI = −0.82. Laboratory work was undertaken to determine the feasibility of injecting alkaline solutions to improve oil recovery. These experiments were designed to produce surfactants in-situ. The surfactants would both lower the interfacial tension and react with the reservoir rock surface to modify the wettability of the porous media. The experimental work considered the injection of seawater and sodium hydroxide mixtures into cores. The experimental results show that the oil recovery was higher than 50% when the alkaline solution was injected. The conclusion was that surfactant produced by alkaline injection altered the rock wettability from oil-wet to intermediate-wet, increasing oil recovery. One precaution with alkaline flooding is that the range of reactions and the change in pH can cause unexpected variation in oil recovery if the reservoir and fluids are not well characterized.

Cementation.   The alteration of wettability by surfactants can be applicable to well cementation. Carriay [97] describes a method whereby a water-oil microemulsion was used as a spacer to reverse the wettability on the walls of the borehole by eliminating the oil film. This wettability alteration made the surface water-wet and permitted the cement, whose continuous phase is water, to adhere more strongly to the formation. An amphoteric surfactant with a co-surfactant was found to be most suitable.

Well Stimulation.   The injection of water and production of oil are often limited by the near wellbore permeability of the fluids. One method of altering the oil or water relative permeability is to change the wettability.

One interesting way proposed to improve relative permeability was a patent obtained by Penny [98] on the use of anionic alkyl polyamine and anionic perfluoro compounds as reservoir wettability-altering agents. It was claimed that these highly surface active compounds reduce both the oil-wet and water-wet tendencies of reservoir carbonate rock. In an injection well, it was claimed that these agents increase water permeability by both reducing the residual oil saturation and the interaction of the water with the reservoir surface. The increase in water permeability means that more water can be injected with the same equipment or that it costs less to inject the same amount of water. In a production well, this non-wetting property allows the hydrocarbons to displace water, leaving a lower water saturation and an increased flow of hydrocarbons through capillaries and flow channels in the formation to the production wellbore. The authors and reviewers have some reservations about some of the claims in this patent, but the concept deserves due consideration.

Micellar acidizing solvents [99] are sometime used in removing skin damage and near-wellbore plugging problems. These micellar systems are produced by combining a surfactant, co-surfactant, and a high molecular weight, water-insoluble, polar organic solvent that is added to an acidic solution. They tend to exhibit the same types of properties as mutual solvents, but do so through an entirely different scheme of chemistry. Like most mutual solvents, the micellar acidizing solvent is totally miscible in acid, but exhibits the properties of a much stronger surfactant, and the solvency of a much stronger organic solvent. The mixture is melded into a clear, single-phase micellar system. This micellar system has the properties of a water-wetting agent, surfactant, dispersing agent, and an organic solvent in a totally miscible treating solution. In a producing well, the relative permeability to oil is increased by changing the surface to water-wet. In an injection well, removal of residual oil around the wellbore and water-wetting the surface promotes single-phase flow for the injected water.

Soil Remediation. Surfactants have been used to remove hazardous waste contamination from soil. Kimball [100] has applied the technology developed for enhanced oil recovery to in-place remediation of soil containing hydrophobic substances. It was found that surfactants (sandoxylate sx-408, naxchem k, and inprove) were effective at removing hexadecane, *o*-cresol and phenanthrene by desorption of the contaminants through flushing of soil. This desorption method indicated that recovery of contaminants resulted from a change in surface wetting with the introduction of surfactant.

Researchers at New Mexico Institute of Mining and Technology have developed a method of treating natural zeolites with cationic surfactants to produce a sorbent for non-polar organics and inorganic oxyanions while causing minimal decrease in the zeolite's sorption of transition metals [101]. The water-wet zeolites, after modification with surfactants, absorbed organic hydrocarbons and chlorinated hydrocarbons from aqueous solutions. They also removed Pb^{2+} from solution with or without the surfactant treatment. The surfactant treatment changed the wettability of the zeolite surface to promote hydrocarbon removal without affecting the internal structure that removed the transition metal ions.

One major constraint for soil and aquifer remediation is that surfactants, if left behind, must not impose an environmental threat. Environmental issues [102] concerned with the transport of surface active compounds through the subsurface must be addressed as part of the advancement of this technology.

Conclusion

This chapter provides an overview of much of the current methodology and practices for studying wettability and its alteration by surfactants in porous rock. The cited references should provide the reader with a starting point for a more comprehensive review of any particular aspect. The complex nature of the subject can be overwhelming. The reader should have noted that there are few studies that directly address wettability alteration by surfactants in reservoir rock. The authors contend that this is probably a result of a poor understanding of wettability and surfactants on reservoir rock, both in the laboratory and in the reservoir, as well as a continuing negative backlash from numerous failed efforts to develop a cost effective surfactant flood process in the 1970s. However, surfactants do have their purpose, as seen in the previous section on field studies. Some challenges in this area are to better merge the theoretical understanding of wettability and surfactants in porous media and to present the results in forms understandable to the reservoir engineers and others who ultimately apply the results.

List of Symbols

A	constant
A_1	area under secondary drainage curve
A_2	area under forced imbibition curve
C	surfactant concentration
CMC	critical micelle concentration
h	film thickness
h_0	thin film thickness
H	mean curvature of the film interfaces
J_{imb}	displacement curvature
k	permeability
K_r	relative permeability
l	length of flow
L_C	characteristic length
OI	Amott oil index
p	perimeter
P_c	capillary pressure
P_{imb}	imbibition capillary pressure
P_O	pressure in oil phase
P_W	pressure in water phase
Q	force from adhesion tension
R	ratio of apparent to true equilibrium contact angle
r	single capillary radius
$\check{r}$	average capillary radius
r_1	radius of curvature perpendicular to r_2
r_2	radius of curvature perpendicular to r_1
RDI	Amott–Harvey Relative Displacement Index
SII	Spontaneous Imbibition Index

S_w	water saturation
Swf	final water saturation after spontaneous imbibition
Swi	initial water saturation before imbibition
S_{WS}	spreading coefficient
t	imbibition time
t_D	dimensionless time
W	pseudo-work of spontaneous imbibition
WI	Amott water index
W_R	ratio of pseudo-works of spontaneous imbibition
W_{S-WW}	pseudo-work of spontaneous imbibition for strongly water-wet imbibition

Greek

γ	interfacial tension
γ_{SO}	interfacial tension between a solid and oil
γ_{SW}	interfacial tension between a solid and water
γ_{WO}	interfacial tension between water and oil
Γ	surfactant adsorption
δF	free energy change of water per unit of pore space
ΔP	pressure difference
ΔSo_F	forced imbibition oil saturation change
ΔSo_S	spontaneous imbibition oil saturation change
ΔSw	water saturation change
ΔSw_F	forced imbibition water saturation change
ΔSw_S	spontaneous imbibition water saturation change
ΔSw_{S-WW}	highly water-wet spontaneous imbibition water saturation change
θ	contact angle measured through water
θ_A	advancing contact angle
θ_E	contact angle
μ	viscosity
μ_W	water viscosity
$\Pi(h)$	disjoining pressure
$\Pi e(h)$	electrostatic component of the disjoining pressure
$\Pi m(h)$	van der Waals component of the disjoining pressure
$\Pi s(h)$	structural component of the disjoining pressure
σ	surface tension
τ_{SWO}	adhesion tension
φ	porosity

References

1. Adamson, A.W. *Physical Chemistry of Surfaces*; Wiley-Interscience: New York, 1976; p 341.

2. Willhite, G.P. *Waterflooding*; SPE Textbook Series: Richardson, TX, 1986; p 10.
3. Washburn, E.W. *Phys. Rev.* **1927**, *1*, 273.
4. Davies, J.T.; Rideal, E.K. *Interfacial Phenomena*; Academic Press: New York, 1961; p 423.
5. Derjaguin, B.V. *Surface Forces and Surfactant Systems*; Kilian, H.G.; Lagaly, G., Eds.; Progress in Colloid & Polymer Science, Vol. 74: New York, 1987; pp 17–18.
6. Hirasaki, G.J. *Proceedings of the SPE/DOE Enhanced Oil Recovery Symposium;* Society of Petroleum Engineers, Richardson, TX, 1988; paper SPE/DOE 17367.
7. Menezes, J.L.; Yan, J.; Sharma, M.M. *Proceedings of SPE International Symposium on Oilfield Chemistry*; Society of Petroleum Engineers, Richardson, TX, 1989; paper SPE 18460.
8. Mohanty, K.K.; Davis, H.T.; Scriven, L.E. In *Surface Phenomena in Enhanced Oil Recovery*; Shah, D.O., Ed.; Plenum Press: New York, 1981; pp 596–599.
9. Melrose, J.C. *Proceedings of the 57th Annual Fall Technical Conference and Exhibition of SPE;* Society of Petroleum Engineers, Richardson, TX, 1982; paper SPE 10971.
10. Zorin, Z.M.; Churaev, N.V. *J. Adv. Colloid Interface Sci.* **1992**, *40*, 86–108.
11. Radke, C.J.; Kovscek, A.R.; Wong, H. *Proceedings of the 67th Annual Technical Conference and Exhibition of SPE*; Society of Petroleum Engineers, Richardson, TX, 1992; paper SPE 24880.
12. Anderson, W.G. *J. Pet. Technol.* **1986**, 1246–1262.
13. Amott, E. *Trans. AIME* **1959**, *216*, 156–192.
14. Boneau, D.F.; Clampitt, R.L. *J. Pet. Technol.* **1977**, 501–506.
15. Ma, S.; Morrow, N.R.; Zhou, X.; Zang, X. *Proceedings of Petroleum Society of CIM 45th Annual Technical Meeting*; 1994; paper CIM 94-47.
16. Donaldson, E.C.; Thomas, R.D.; Lorenz R.B. *Soc. Pet. Eng. J.* **1969**, 9, 13–20.
17. Spinler, E.A. *Proceedings of the Annual Technical Conference and Exhibition of SPE*; Society of Petroleum Engineers, Richardson, TX, 1997; paper SPE 38733.
18. Rapoport, L.A. *Trans. AIME* **1955**, *204*, 143.
19. Mattax, C.C.; Kyte, J.R. *Soc. Pet. Eng. J.* **1962**, 177–184.
20. Adamson, A.W. *Physical Chemistry of Surfaces*; Wiley-Interscience: New York, 1976; pp 342–343.
21. Roa, D.N.; Girard, M.G. *J. Can. Pet. Technol.* **Jan. 1996**, 31
22. Wilhelmy, L. *Ann. Phys.* **1863**, 119.
23. Andersen, M.A.; Thomas D.C.; Teeters D.C. *SPE/DOE Enhanced Oil Recovery Symposium*; Society of Petroleum Engineers, Richardson, TX, 1988; paper SPE/DOE 17368.
24. Teeters, D.; Andersen, M.A.; Thomas, D.C. *Oil Field Chemistry Enhanced Recovery and Production Stimulation*; American Chemical Society: Washington, DC, 1989; pp 560–576.
25. Mennella, A.; Morrow, N. *J. Colloid Interface Sci.* **1995**, 48–55.

26. Longeron, D.; Hammervold, W.L.; Skjaeveland, S.M. *Proceedings of the International Meeting on Petroleum Engineering*; Society of Petroleum Engineers, Richardson, TX, 1995; paper SPE 30006.
27. Hassler, G.L.; Brunner, E. *Trans. AIME* **1945**, *160*, 114–123.
28. Ruth, D.W.; Chen, A.Z. *The Log Analyst* **Sept.–Oct. 1995**, 21–32.
29. Spinler, E.A.; Baldwin, B.A. *Proceedings of 1997 International Symposium of the Society of Core Analysts*; 1997; paper SCA 9705.
30. Fassi-Fihri, O.; Robin, M.; Rosenberg, E. *Proceedings of the 66th Annual Technical Conference and Exhibition of SPE*; Society of Petroleum Engineers, Richardson, TX, 1991; paper SPE 22596.
31. Howard, J.J.; Spinler, E.A. *Proceedings of the 68th Annual Technical Conference and Exhibition of SPE*; Society of Petroleum Engineers, Richardson, TX, 1993; paper SPE 26471.
32. Medout-Marere, V.; Malandrini, H.; Zoungrana, T.; Partyka, S.; Doullard, J.M. *The 4th International Symposium on Evaluation of Reservoir Wettability and its Effect on Oil Recovery*; Montpelier, France, 1996; paper 564.
33. Robin, M.; Combes, R.; Degreve, F.; Cuiec, L. *Proceedings of the 1997 International Symposium on Oilfield Chemistry*; Society of Petroleum Engineers, Richardson, TX, 1997; paper SPE 37235.
34. Varadaraj, R.; Bock, J.; Brons, N.; Zushma, S. *J. Colloid Interface Sci.* **1994**, *167*, 207–210.
35. Holbrook, O.C.; Bernard, G.C. *Petrol. Trans. AIME* **1958**, *213*, 261–264.
36. Mitchell, A.G.; Hazell, L.B.; Webb, K.J. *Proceedings of the 65th Annual Technical Conference and Exhibition of SPE*; Society of Petroleum Engineers, Richardson, TX, 1990; paper SPE 20505.
37. Celik, M.S.; Somasundaran, P. *Proceedings of the 5th International Symposium on Oilfield Chemistry*; Society of Petroleum Engineers, Richardson, TX, 1980; paper SPE 9002.
38. Rosen, M.J. *Surfactants and Interfacial Phenomena*; 2nd ed., Wiley-Interscience: New York, 1989; pp 6–7.
39. Varadaraj, R.; Bock, J.; Zushma, S.; Brons, N.; Valint, P. Jr. *J. Phys. Chem.* **1991**, *95*, 1679–1681.
40. Schechter, R.S.; Wade, W.H. *US Dept. Energy Int. Energy Agency Enhanced Oil Recovery Workshop Proc.*; 1980, 96–104.
41. Somasundaran, P.; Fuerstenau, D.W. *J. Phys. Chem.* **1966**, *70*, 90–96.
42. Scamehorn, J.F.; Schechter, R.S.; Wade, W.H. *J. Colloid Interface Sci.* **1982**, *85*, 463–478.
43. Rosen, M.J. *Surfactants and Interfacial Phenomena*; 2nd ed., Wiley-Interscience: New York, 1989; pp 55–56.
44. Alveskog, P.L.; Holt, T.; Torsaeter, O. *Evaluation of Reservoir Wettability & Its Effect on Oil Recovery Int. Symp. Proc.*; Montpelier, France, 1996.
45. McGuiggan, P.M.; Pashley, R.M. *J. Colloid Interface Sci.* **1988**, *124*, 560–569.
46. Chesters, A.K.; Elyousfi, A.; Cazabat, A.M.; Vilette, A. *Evaluation of Reservoir Wettability & its Effect on Oil Recovery Int. Symp. Proc.*; Montpelier, France, 1996.

47. Damania, B.S.; Bose, A. *J. Colloid Interface Sci.* **1986**, *113*, 321–335.
48. Princen, H.M.; Cazabat, A.M.; Heslot, F.; Nicolet, S.; Cohen Stuart, M.A. *J. Colloid Interface Sci.* **1988**, *126*, 84–92.
49. Churaev, N.V. *Progr. Colloid Polymer Sci.* **1996**, *101*, 45–50.
50. Churaev, N.V.; Ershov, A.P.; Zorin, Z.M. *J. Colloid Interface Sci.* **1996**, *177*, 589–601.
51. Treiber, L.E.; Archer, D.L.; Owens, W.W. *Proceedings of the 46th Annual Technical Conference and Exhibition of SPE*; Society of Petroleum Engineers, Richardson, TX, 1972; paper SPE 3526.
52. Taber, J.J. In *Surface Phenomena in Enhanced Oil Recovery*; Shah, D.O., Ed.; Plenum Press: New York, 1981; p 22.
53. Morrow, N.R. *J. Can. Pet. Technol.* **1976**, *15*, 49.
54. Maini, B.; Wassmuth, R.; Schramm, L.L. In *Suspensions Fundamentals and Applications in the Petroleum Industry*, Schramm. L.L., Ed.; Advances in Chemistry Series 251, American Chemical Society: Washington DC, 1996; p 334.
55. van Olphen, H. *An Introduction to Clay Colloid Chemistry*; Wiley-Interscience: New York, 1977; pp 92–93.
56. Trantham, J.C.; Clampitt, R.L. *SPE-AIME 4th Symposium of Improved Oil Recovery*; Society of Petroleum Engineers, Richardson, TX, 1976, paper SPE 5802.
57. Buckley, J.S.; Liu, Y. *Evaluation of Reservoir Wettability & Its Effect on Oil Recovery Int. Symp. Proc.*; Montpelier, France, 1996.
58. Yan, J.; Plancher, H.; Morrow, N.R. *Proceedings of the 1997 SPE International Symposium on Oilfield Chemistry*; Society of Petroleum Engineers, Richardson, TX, 1997, paper SPE 37232.
59. Xie, X. Ph.D. Thesis, University of Wyoming, Laramie, WY, 1996.
60. Clementz, D.M. *Clays and Clay Min.* **1976**, *24*, 312–319.
61. Baldwin, B.A. *SPE Formation Evaluation* **1988**, *3*, 125–130; paper SPE 14108.
62. Burki, P.M.; Glasser, L.S.D.; Smith, D.N. *Nature* **1982**, *7*, 145–147.
63. Wellington, S.L.; Richardson, E.A. *Soc. Pet. Eng. J.* **1997**, *2*, 389–405.
64. Scamehorn, J.F.; Schechter, R.S.; Wade, W.H. *J. Colloid Interface Sci.* **1982**, *85*, 494–501.
65. Mannhardt, K.; Novosad, J.J. *J. Pet. Sci.* **1991**, *5*, 89–103.
66. Noll, L.A.; Gall, B.L.; Crocker, M.E.; Olsen, D.K. *US DOE Fossil Energy Rep.* No. NIPER-385 (DE89000745), 1989.
67. Schramm, L.L.; Mannhardt, K.; Norosad, J.J.; *Colloids and Surfaces* **1991**, *55*, 309–331.
68. Bitting, D.; Harwell, J.H. *J. Colloid Interface Sci.* **1987**, *3*, 500–511.
69. Ziegler, V.M.; Handy, L.L. *Proceedings of the 54th Annual Technical Conference and Exhibition of SPE*; Society of Petroleum Engineers, Richardson, TX, 1979, paper SPE 8264.
70. Adamson, A.W. *Physical Chemistry of Surfaces*; Wiley-Interscience: New York, 1976; p 346.
71. Morrow, N.R. *J. Can. Pet. Technol.* **Oct.–Dec. 1975**, 42–53.
72. Bayramli, E. Ph.D. Thesis, McGill University, Montreal, Quebec, Canada, 1980.

73. Lorenz, P.B.; Donaldson, E.D.; Thomas, R.D. *U.S. Bureau of Mines Rpt.* Bartlesville Energy Technology Center, Report 7873, 1974.
74. Anderson, W.G. *J. Pet. Technol.* **1987**, 1605–1622.
75. Cuiec, L.E. *Interfacial Phenomena in Petroleum Recovery*; Surfactant Science Series vol. 36; Marcel Dekker, Inc.: New York, 1991; pp 319–342.
76. Austad, T.; Milter, J. *SPE Oilfield Chemistry International Symposium Proceedings*; Society of Petroleum Engineers, Richardson, TX, 1997; paper SPE 37236.
77. Sanchez, J.M.; Hazlett, R.D. *64th Annual SPE Technical Conference Proceedings*; Society of Petroleum Engineers, Richardson, TX, 1989; paper SPE 19687.
78. Schramm, L.L.; Mannhardt, K. *J. Pet. Sci. Eng.* **1996**, *15*, 101–113.
79. Mannhardt, K.; Novosad, J.J. In *Foams, Fundamentals and Applications in the Petroleum Industry*, Schramm, L.L., Ed.; American Chemical Society: Washington, DC, 1994; pp 259–316.
80. Smith, D.H.; Comberiati, J.R. *Annual AICHE Meeting*; 1990, paper No. 243C.
81. Chisholm, J.L.; Kashikar, S.V.; Knapp, R.M.; McInerney, M.J.; Menzie, D.E. *65th Annual Technical Conference SPE*; Society of Petroleum Engineers, Richardson, TX, 1990, paper SPE 120481.
82. Bryant, R.S.; Douglas J. *SPE Res. Engr.* **1988**, 489–495.
83. Thomas, C.P.; Bala, G.A.; Duvall, M.L. *66th Annual Technical Conference SPE*; Society of Petroleum Engineers, Richardson, TX, 1991, paper SPE 22844.
84. Lin, S.C. *J. Chem. Tech.* **1996**, *66*, 109–120.
85. Kim, H.S.; Yoon, B.D.; Lee, C.H.; Suh, H.H.; Oh, H.M.; Katsuragi, T.; Tani, Y. *J. Fermentation and Bioeng.* **1997**, *84*, 41–46.
86. Marsh, T.L.; Zhang, X.; Knapp, R.M.; McInerney, M.J.; Sharma, P.K.; Jackson, B.E. *The Fifth International Conference on Microbial Enhanced Oil Recovery and Related Biotechnology for Solving Environmental Problems*; National Technical Information Service, Springfield, VA, 1995, pp 593–610.
87. Bala, G.A.; Duvall, M.L.; Barrett, K.B.; Robertson, E.P.; Pfister, R.M.; Thomas, C.P. In *Mineral Bioprocessing*; Smith, R.W.; Misra, M., Eds.; The Minerals, Metals & Materials Soc., 1991; pp 100–130.
88. Thomas, D.C. *59th Annual Technical Conference SPE*; Society of Petroleum Engineers, Richardson, TX, 1984, paper SPE 13097.
89. Sharma, M.M.; Wundelich, R.W. *60th Annual Technical Conference SPE*; Society of Petroleum Engineers, Richardson, TX, 1985, paper SPE 14302.
90. McDonald, J.A.; Buller, D.C. *J. Pet. Sci. Engr.* **1992**, *6*, 357–365.
91. Sanner, D.O.; Azar, J.J. *SPE International Symposium on Formation Damage Control*; Society of Petroleum Engineers, Richardson, TX, 1994; paper SPE 27354.
92. Christensen, K.C.; Davis, N. II; Nuzzolo, M. *World Pct. Gaz.* **1990**, *21*, 6346.
93. Downs, H.H.; Hoover, P.D. In *Oil Field Chemistry*; Borchardt, J.K.; Yen, T.F., Eds.; American Chemical Society: Washington, DC, 1989; pp 577–595.

94. Blair, C.M.; Stout, C.A. *Oil and Gas J.* **1985**, *83*, 55–59.
95. Castor, T.P.; Somerton, W.H.; Kelly, J.F. In *Surface Phenomena in Enhanced Oil Recovery*; Shah, D.O., Ed.; Plenum Publishing Corp.: New York, 1981; pp 249–306.
96. Arteaga-Cardona, M.; Pineda-Munoz, A.; Islas-Juarez, R. *SPE International Petroleum Conference of Mexico Proceedings*; Society of Petroleum Engineers, Richardson, TX, 1996; paper SPE 35340.
97. Carriay, J.; de Lautrec, J. U.S. Patent 4 233 732, September 9, 1980.
98. Penny, G.S. U.S. Patent 4 702 849, October 27, 1987.
99. Dobbs, J. B. *2nd Royal Society Chemistry Industrial Division Chemistry in the Oil Industry International Symposium Proceedings*; 1985, pp 107–119.
100. Kimball, S.L. Ph.D. Thesis, Oklahoma State University, Stillwater, OK, 1993.
101. Bowman, R.S.; Haggerty, G.M.; Huddleston, R.G.; Neel, D.; Flynn, M.M. In *Surfactant-Enhanced Subsurface Remediation Emerging Technologies*; Sabatini, D.A.; Knox, R.C.; Harwell, J.H., Eds.; American Chemical Society: Washington, DC, 1995; pp 54–64.
102. West, C.C.; Harwell, J.H. *Environ. Sci. Technol.* **1992**, *26*, 2324–2330.
103. Morrow, N.R.; McCaffery, R.G. In *Wetting, Spreading and Adhesion*; Padday, J.F., Ed.; Academic Press, New York, 1978, p 299.

RECEIVED for review November 4, 1998. ACCEPTED revised manuscript January 12, 1999.

6

Surfactant Flooding in Enhanced Oil Recovery

Tor Austad[1] and Jess Milter[2]

[1] Stavanger College, Ullandhaug, N-4004 Stavanger, Norway
[2] Statoil, N-4035 Stavanger, Norway

Improvements of surfactant flooding in enhanced oil recovery during the last 10 years are discussed. The review starts by giving a short introduction to the principles of traditional surfactant/polymer flooding in sandstone reservoirs. Progress to improve the flooding technique by simplifying the chemical formulation in a viscous displacement of oil is reviewed. Factors related to surfactant-polymer properties, interfacial tension, interaction between chemicals and rock, phase behavior, and possible phase gradients are discussed. The experience of performing chemical floods at three-phase and two-phase conditions without alcohol present is presented. The status of spontaneous imbibition of water and aqueous surfactant solution into low-permeable chalk material containing oil is brought up to date. Special focus is given to interfacial tension, wettability, and height of the chalk material. The imbibition mechanism is discussed in terms of forces related to capillary pressure, gravity, and possible gradients in surface tension.

Introduction

Scope. From a technical point of view, more so in the lab than in the field, chemical flooding of oil reservoirs is one of the most successful methods to enhance oil recovery from depleted reservoirs at low pressure. It is, however, well documented in the literature that chemical flooding is only marginally economical, or in most cases directly uneconomical. Initially, the objective of chemical flooding was to recover additional oil after a waterflood, and it is therefore described as a tertiary oil recovery process. A lot of papers and reviews, both laboratory work and field tests, have been published on this subject since the first work by Marathon Oil Company in the early 1960s [1]. Even though enormous effort by oil company, university, and government researchers during the 1970s and

1980s increased our knowledge about the chemical flooding process, it was more or less accepted or concluded by the oil companies at the end of the 1980s that the method was not economical, or the economical and technical risk was too high with the present oil prices. The research declined drastically during the 1990s, but still some research groups were active trying to improve the technique by:

- simplifying the flooding process
- improving the efficiency of surfactants
- developing new chemicals (surfactants)

Surfactants and polymers are the principal components used in chemical flooding. The surfactant lowers the interfacial tension (IFT) between the reservoir oil and the injected water, while the polymer will create favourable viscosity conditions and good mobility control for the surfactant slug. The oil is then displaced by the viscous forces acting on the oil by the flowing water. For this reason, chemical flooding is also denoted as:

- micellar/polymer flooding
- surfactant /polymer flooding
- microemulsion flooding

Secondary oil recovery by spontaneous imbibition of water into low-permeable fractured chalk is a well accepted method to improve the oil recovery from water- to mixed-wet rock material [2]. Normally, this process is driven by capillary forces, and it may seem a little strange to lower the capillary forces by adding surfactants to the injected water. In the same way as the viscous forces will mobilize capillary trapped water-flooded oil, the gravity forces may be active in displacing the oil by spontaneous imbibition at low IFT [3]. A crossover from a capillary forced spontaneous imbibition (counter-current flow) to a gravity forced imbibition (cocurrent flow) is observed by decreasing the IFT. The status and recent advances in the research in this area will be included in this presentation as well.

The objectives of the present work are to bring the topic up to date and to focus on the recent laboratory developments in chemical flooding of oil reservoirs. For the historical development of the process, interested readers are referred to noteworthy reviews by Pope and Bavière [4], Thomas and Farouq [5], Healy and Reed [6], and Ling et al. [7]. The present review will be related to new developments within:

- traditional chemical flooding of sandstone reservoirs
- imbibition of aqueous surfactant solution into low-permeable chalk

Other areas where surfactants are used in flooding of oil reservoirs, which are not included in the present review, are:

- Alkaline/surfactant/polymer process, termed ASP [8]. The ASP flooding is hoped to be a low-cost improvement over micellar/polymer flooding.
- Surfactant based mobility control, i.e. foam generation during gas injection.
- Partial or complete blocking of high-permeable regions, forcing injected displacing fluid into low-permeability areas of high oil content [9].

Traditional Surfactant/Polymer Flooding.   In general terms, Figure 1 illustrates the various regions of immiscible flow during a typical displacement of oil by a surfactant solution. Provided that a water flood was performed prior to the chemical flood, the various zones are described as:

Region 1: Waterflooded residual oil saturation, only water is flowing.
Region 2: An oil bank is formed, both oil and water are flowing.
Region 3: Surfactant slug forming the low IFT region, two- or three-phase flow of oil, brine, and microemulsion depending on the actual phase behavior.
Region 4: Polymer solution for mobility control, single phase flow of water.

The capillary number, N_c, is related to the residual oil saturation through the desaturation curve illustrated by Figure 2. N_c is defined as the ratio between the viscous and local capillary forces and can be calculated from:

$$N_c = \frac{v\mu_w}{\sigma} \qquad (1)$$

where v is the effective flow rate, μ_w is the viscosity of displacing fluid, and

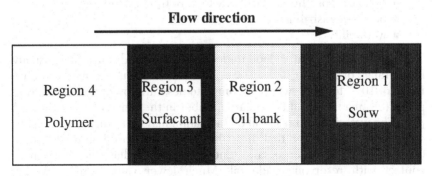

Flow direction

| Region 4 | Region 3 | Region 2 | Region 1 |
| Polymer | Surfactant | Oil bank | Sorw |

Figure 1.   Phase position in a typical chemical flood.

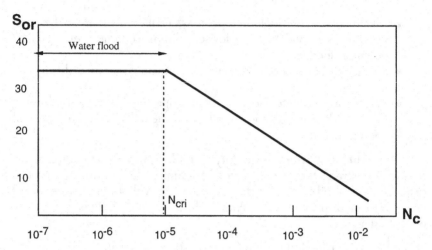

Figure 2.   Schematic capillary desaturation curve for a nonwetting phase.

σ is the IFT. If the wettability preference of the rock is taken into account, the formula for N_c becomes:

$$N_c = \frac{v\mu_w}{\sigma \cos \theta} \qquad (2)$$

where θ is the contact angle measured through the fluid with the highest density. The capillary number, corresponding to the break in the desaturation curve, is designated as the critical capillary number, N_{cri}. Thus, to improve the oil recovery relative to a water flood by using chemicals, N_c must be significantly higher than the critical capillary number, $N_c \gg N_{cri}$. The critical capillary number and the shape of the desaturation curve depend on the rock properties such as:

- aspect ratios, the ratio of body to pore throat diameter
- pore size distribution
- wettability

For an ordinary water flood under water-wet conditions, N_c is usually in the range of 10^{-7} to 10^{-5}. The critical capillary number may be in the range of 10^{-5} to 10^{-4}, whereas complete desaturation of the nonwetting phase (oil) may occur at a capillary number in the range of 10^{-2} to 10^{-1} [10]. The waterflooded residual oil saturation may be in the range of 30 to 40%. It must be noticed that these data are mainly based on model cores (Berea and other outcrop sandstone cores) which have never been in contact with reservoir crude oil. Much lower values are, however, observed under mixed-wet conditions [11]. This implies that it is about

10 times more difficult to remobilize capillary trapped discontinuous oil, compared to continuous oil. In order to be able to mobilize a significant amount of the waterflooded residual oil, it is expected that the capillary number must be increased by a factor of 10^3 to 10^4. The only practical way to do this is to reduce the IFT between the reservoir oil and the injected water by the same factor using surfactants, which normally means that the IFT should be between about 0.01 and 0.001 mN/m.

Until recently, no single surfactant or mixture of surfactants was able to decrease the IFT between water and oil to this level by adding small concentrations of surfactant (about 0.1 to 0.5 wt%) to the water injected in an ongoing water flood. Instead, complicated formulations of injection strategies were developed by forming a thermodynamically stable micro-emulsion phase between the oil and the water. At optimum conditions, i.e. equal solubilization of oil and water into this middle phase, the IFT between the microemulsion phase and the two excess phases, oil and water, is utralow and equal, fulfilling the conditions for displacing most waterflooded residual oil. A variety of chemicals, mixtures of surfactants and low molecular weight alcohols, were needed to design a gel-free chemical formulation. In order to work properly, phase gradients, usually a negative salinity gradient, must be established in the reservoir. The amount of polymer needed for maintaining mobility control must be high because the middle phase microemulsion is normally rather viscous. Good mobility control is essential to protect the integrity of the small slug of chemicals injected against dissipation because of fingering, channeling, mixing, crossflow and other reservoir mechanisms. Especially, fingering is a problem in a low tension flooding process because the capillary forces are low.

Because of the complexity of the chemical formulation, the slug is affected by mass transfer and changes in phase behavior as the fluids propagate through the resevoir. The process is sensitive to many parameters including:

- rock type
- mineral content
- interstitial brine salinity and composition
- pH
- injection rate
- slug composition
- polymer concentration and type
- oil viscosity and composition
- pressure
- temperature
- heterogeneities of the formation

The most common anionic surfactants (petroleum-, alfa-olefin-, and alkylaryl-sulfonates) have great potential for extracting adsorbed multi-valent cations, Mg^{2+}, Ca^{2+} etc., into the micellar slug, with the consequence of a drastic change in phase properties. Therefore, a preflush of water containing monovalent cations is often performed to obtain a cation exchange. In order to compensate for reservoir parameters, which disturb the phase behavior of the surfactant slug during the process, an imposed phase gradient is normally used. If the phase gradient does not behave properly, a great loss of surfactants may take place due to phase trapping, i.e. the surfactant is trapped in the oil or in the middle phase. From a reservoir engineer's point of view, it is very difficult to handle the flow of three liquid phases through an inhomogeneous porous medium. Remember that the fluids have different viscosities and mass transport is taking place between the different phases depending on the reservoir conditions. Thus, technically it would be a step forward if the chemical flooding could be performed in the two-phase mode.

The goal of the research on surfactant flooding during the 1990s was to develop surfactants that can recover additional oil in a cost-effective manner during a normal water flood using produced brine (due to environmental aspects) or seawater as injection fluid. In order to avoid many of the problems associated with complicated chemical slugs with high concentration of surfactants and cosurfactants/alcohols, the following criteria should apply:

- the only chemicals used are surfactant and polymer
- low chemical concentration (surfactant 0.1–0.5 wt%; polymer <500 ppm)
- no imposed salinity gradient or other phase gradients
- the chemicals should be insensitive to multivalent cations
- the flooding conditions should be a two-phase flood with the surfactant and polymer present in the aqueous phase, forming an oil-in-water microemulsion, termed Type II($-$)

The flooding performance is termed Low Tension Polymer Water Flood, LTPWF, or low surfactant concentration enhanced water flood, and illustrated in Figure 3.

Surfactant loss due to phase trapping is minimized if only oil and water phases are present at all times during the flood. Surfactant loss is only related to adsorption onto the mineral surface provided that the surfactant tolerates multivalent cations, i.e. no precipitation. In the following sections we will discuss recent published work which is relevant in reaching this goal.

Possible combinations of oil, water/salt, and surfactant/cosurfactant/alcohol will give different phase behavior depending on actual conditions. The liquid phase containing the surfactant is a thermodynamically stable

Flow direction

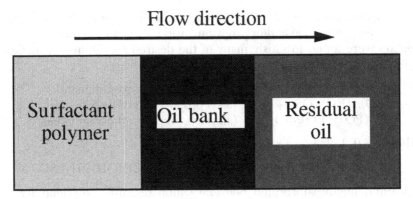

Figure 3.   Schematic illustration of a LTPWF.

phase usually termed a microemulsion. The following phase terminology will be used in this chapter:

Type II(−): An oil-in-water microemulsion in equilibrium with excess oil.

Type III: A middle phase microemulsion in equilibrium with excess oil and water.

Type II(+): A water-in-oil microemulsion in equilibrium with excess water.

In general, the phase property of the Type II(−) is described as a water-continuous phase with dissolved oil in micellar aggregates. Likewise, the Type II(+) is described as an oil-continuous phase with dissolved water in reversed micellar aggregates. The Type III is described as a bi-continuous phase, i.e. it is continuous both in oil and water with the surfactants located in the interface and, at the optimum condition, the volume of oil and water is the same.

Chemicals

Surfactant.   Based on a two-phase flood condition, Type II(−) phase behavior, the interfacial surface between oil and water must be covered by at least a monolayer of surfactants. High surface coverage is needed in order to obtain low enough IFT. Thus, the surfactant molecules must have strong lateral intermolecular association without forming liquid crystals and gels. Futhermore, the surfactant must also grade smoothly from being oil- to water-soluble over a sufficient length of the molecule. In this way it will be a smooth transition from oil- to water-like fluid along the interphase. A lot of papers have shown that it is possible to synthesize anionic surfactants with these properties that tolerate high concentrations of multivalent cations [12–15]. Anionic surfactants containing multiple units of ethylene oxide and/or propylene oxide (EO and PO) in their mid-section were found to satisfy many of the desired conditions. Examples are:

R-$(PO)_y$-$(EO)_x$-SO_3^- (alkyl propoxy-ethoxy-sulfonate)
R-$(PO)_y$-$(EO)_x$-OSO_3^- (alkyl propoxy-ethoxy-sulfate)
R-Ph-$(PO)_y$-$(EO)_x$-SO_3^- (alkylaryl propoxy-ethoxy-sulfonate)

where $y = 0, 1, 2, 3, \ldots, x = 1, 2, 3, \ldots$

R_{12-15}-O-$(CH(CH_3)CH_2O)_7$-$(CH_2CH_2O)_2$-$(CH_2CH(OH)CH_2)$-SO_3^- Na^+

Pure and highly substituted benzene sulfonates, n-C_{12}-o-xylene-SO_3^-, have similar properties regarding low IFT in the two-phase region without using alcohol [16].

Sanz and Pope [17] have screened combinations of ethoxylated sulfonates and alkylaryl sulfonates for alcohol-free chemical flooding purposes. They observed difficulties in obtaining clean or gel-free micro-emulsions at the phase transition from Type II($-$) to Type III, which means that surfactant retention and pore plugging may take place. Thus, it appears to be an advantage to stay in the two-phase region during the flood process.

Analysis.   Normally, commercial anionic surfactants of the propoxy and ethoxy type are polydisperse in the PO- and EO-groups. Furthermore, the products usually contain significant amounts of non-ionic alcohols. In the case of ethoxylated aromatic sulfonates, HPLC analysis is found to give a good quality check of the product [18]. Mixed-mode reversed-phase/ion-exchange columns of the type C18 and C4 from Alltech were found to separate the different EO-oligomers of the anionic surfactant, Figure 4. The distribution of the EO-groups in the commercial product, C_9-Ph-$(EO)_6$-SO_3Na, was in the range of 2 to 13. The figure also shows that the nonionic alcohols and possible impurities of the type R-$(EO)_x$-R can be detected and quantified.

Due to the formation of mixed micelles with a low value of critical micelle concentration, CMC, mixtures of ethoxylated sulfonates and alkyl aryl sulfonates are found to tolerate hard water and high salinities. Mixed-mode columns of the type C4 can also be used to analyse mixtures of C_9-Ph-$(EO)_6$-SO_3Na and C_{12}-Ph-SO_3Na [19].

Stability.   Ethoxylated anionic sulfonates are fairly stable regard-ing desulfonation by breakage of the C—S bond at ordinary reservoir conditions [20]. Water solvolysis, H^+ catalyzed hydrolysis, and nucleo-philic (HS^- and Cl^-) displacement reactions have, however, been observed. Each and every one of these reactions can dominate the decomposition rate under different conditions.

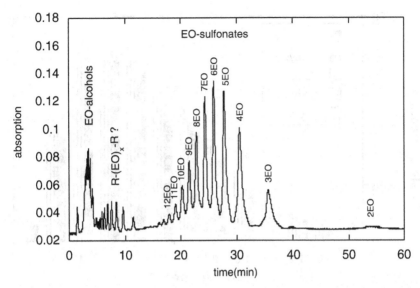

Figure 4. HPLC analysis of a commercial product of C_9-Ph-$(EO)_6$-SO_3Na. (Reproduced with permission from reference 23, copyright 1993 Elsevier.)

Even though oil reservoirs have a reducing environment, air or oxygen is usually not excluded from solutions in surfactant flood experiments, and it is therefore important to be sure that the loss of surfactant is not due to chemical decomposition during the experimental period. Under aerobic conditions, oxygen is important in the decomposition of the EO-groups. The main mechanism is believed to be cleavage of the ether bonds in the same way as in the decomposition of polyethylene oxide [21]. The ether bonds are broken by formation of hydroperoxides as intermediates. The peroxides are then decomposed by a radical mechanism, which may initiate chain scission reactions. The decomposition may also be catalyzed by metal ions [22]. It is, however, illustrated in Figure 5 than no significant decomposition of C_9-Ph-$(EO)_6$-SO_3Na takes place at 80 °C at seawater salinities under aerobic conditions during a test period of 156 days [23].

The EO-sulfates are cheaper than the corresponding EO-sulfonates, but they are hydrolysed at high temperatures and low pH. The pH of injected seawater is normally changed from about 8 to 4–6 due to solubilization of CO_2 and ion-exchange between water and the reservoir rock. At 60 °C, the halflife time for EO-sulfates is estimated to be about 7 and 30 years at pH $\approx$ 5 and pH $\approx$ 8, respectively [24]. The rate of hydrolysis increases exponentially with increasing temperature.

Polymer. The polymer must be water-soluble and of low flexibility to give high viscosity at low polymer concentration and high

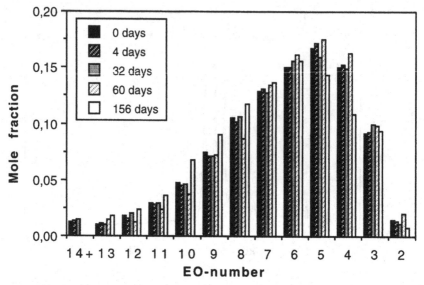

*Figure 5.   Stability test of C_9-Ph-$(EO)_6$-SO_3Na at 80 °C. (Reproduced
with permission from reference 23, copyright 1993 Elsevier.)*

salinities. The two biopolymers, xanthan and scleroglucan, are good
candidates. Xanthan acts as a negatively charged double helix in saline
solution, while scleroglucan acts as an uncharged triple helix in solution.
Polymers forming a helix usually hide their hydrophobic sections in the
interior of the helix, minimizing surfactant–polymer complex formation.
Both of the polymers tolerate high salinity and can be used in seawater.

Hydrolysed polyacrylamide, HPA, is a good alternative at low salinities, but is not recommended to be used in hard water at high salinities.
Copolymers containing sulfonate groups (acrylamide and sodium 2-
acrylamido-2-methylpropane sulfonate from Floerger) are designed to
tolerate high temperatures and seawater salinities.

Surfactant–Polymer Interaction in Solution.   In the traditional way of micellar flooding, the surfactant is present most of the time
in the microemulsion phase, i.e. the middle phase, and the polymer is
present in the excess water phase. Due to the negative salinity gradient
usually applied, a high concentration of surfactant is also present in the
aqueous phase, Type II(−) phase behavior, at the back of the surfactant
slug. A rather high concentration of polymer must initially be injected to
obtain mobility control of the surfactant slug. Thus, at the rear of the
surfactant slug, both surfactant and polymer are present in the aqueous
phase in significant concentrations (5–10 wt% surfactant and more then
1000 ppm of polymer). It is well documented in the petroleum and

chemical literature that mixtures of anionic surfactants and different water-soluble polymers tend to phase separate in saline aqueous solutions [25, 26]. The incompatibility phenomena can be chemically explained by considering the micelle–polymer system as a colloidal system according to the DLVO theory for the stability of colloids [27, 28]. Therefore, the micellar–polymer systems usually contain alcohol in order to obtain gel-free microemulsions and to improve the compatibility of the surfactant–polymer solution. High salinities require larger concentrations of alcohol to prevent phase separation.

Piculell and Lindman [29] recently discussed the phase separation of aqueous mixtures of polymer/polymer and polymer/surfactant solutions in terms of association and segregation. When one of the phases is concentrated with both of the components, the phase separation is termed associative, and when the separating phases contain components of comparable total concentrations it is called a segregative phase separation. Mixtures of nonionic polymer and ionic surfactant mainly show an associative phase separation. However, this may be due to the fact that most studies performed in the chemical literature have been specifically concerned with systems where P–S association has been important. Systematic experiments on P–S systems where both are negatively charged are reported to show a segregative phase separation.

In a LTPWF, the concentration of surfactant and polymer is much lower than in a traditional micellar slug flood (surfactant 0.1–0.5 wt% and polymer <500 ppm) and incompatibilities between the chemicals resulting in associative or segregative phase separation are normally not observed even at high salinities [14]. It is, however, very important that no association between surfactant and polymer takes place in solution. In the presence of excess polymer, the surfactant monomer concentration will then become lower than the CMC. The monomolecular packing of surfactants at the interface decreases, and the IFT will increase drastically.

It is important to note that the association between ethoxylated sulfonates [30] and sulfates [31] and nonionic polymers decreases as the ethoxylation degree increases. Practically no interaction was observed with EO-groups higher than 3–4. This means that the free energy for normal micelle formation is more favorable than for forming micellar aggregates on the polymer.

Interfacial Tension.   Another property of the polymer is the ability to bind up water. Depending on the nature of the polymer, charged or neutral, the hydrodynamic volume or the electrical double layer will be affected by both salinity and temperature. Usually it will decrease as salinity and temperature increase. Kalpakci et al. [13] suggested that a dissociative surfactant–polymer interaction would have synergistic effects on the IFT, i.e. they measured a decrease in the IFT (between the excess phases (oil and water)) and in the CMC for different surfactant systems in the presence of 1500 ppm xanthan compared to a polymer free system. Austad and Taugbøl [16] found that a solution containing 500 ppm xanthan did not affect the IFT significantly compared to a polymer free surfactant system, Figure 6. It should be noted that the authors have not found any documentation in the literature for pure components that polymer in the presence of surfactant will affect the IFT significantly. Thus, at the concentration levels for the chemicals used in LTPWF, it appears that the polymer will not change the IFT provided that it is a dissociative interaction between the surfactant and polymer.

It is reasonable to expect that the average time a surfactant molecule spends in the oil–water interface is of the same order as the average monomer life-time in a micelle, i.e. about 6 μs [32]. A stronger association between the surfactants at the interface will increase the coverage of the interface with surfactants. The electrostatic repulsion between the negatively charged head groups of the surfactants will prevent high coverage of the surface. The intermolecular repulsion will, however, decrease if the system contains a small mole fraction of cationic surfactant. Wellington and Richardson [15] observed synergistic interactions

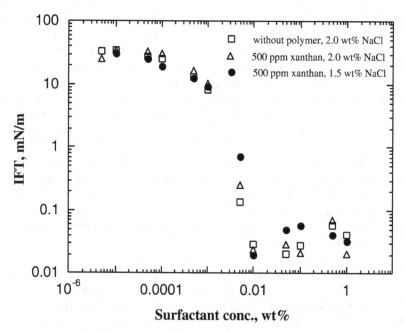

Figure 6. IFT between n-C7 and brine at 50 °C. (Reproduced with permission from reference 16, copyright 1995 Elsevier.)

Table 1.   IFT Decrease with Dilution (data from reference 15)

Anionic Surfactant (ppm)	Cationic Surfactant (ppm)	IFT (mN/m)
4000	400	0.024
1000	100	0.004
500	50	0.0002

between anionic and cationic surfactants when adding fractional amounts of PO–EO quaternary ammonium cations to PO–EO sulfonates (mole ratio of 1:10). The physicochemical properties of these surfactant mixtures are in many ways comparable to nonionic surfactants. At a given salinity and a surfactant concentration above the CMC, a phase separation temperature or a cloud point is reached by increasing the temperature of the solution. The cloud point may also be reached by dilution. In general, as the surfactant solution approaches the cloud point, the IFT will decrease. More surfactants will move to the interface because of poor dissolution in the aqueous phase. The mixed anionic–cationic surfactant system appears to behave in this way, as indicated in Table 1. This property is very useful during an enhanced waterflooding process.

The propoxy ethoxy sulfonate used by Taugbøl et al. [*14*] also showed an IFT value close to 10^{-2} mN/m towards *n*-heptane in seawater, Figure 7. Thus, the present PO–EO-surfactant systems are able to lower the IFT between water and oil by a factor of more than three magnitudes in a Type II($-$) phase behavior. The corresponding increase in the capillary number suggests that a significant amount of waterflooded residual oil will be recovered by a chemical flood performed in the two-phase region.

Another interesting feature about these surfactant systems is that the PO and EO groups need not be on the same molecule, i.e. mixtures of propoxylated and ethoxylated surfactants could be used. This is an advantage in designing and manufacturing the chemicals because the EO and PO groups may be on separate molecules. Several parameters can then be varied in order to design the wanted hydrophilic and lipophilic balance, HLB, like the relative number of PO and EO groups and the relative amount of the two surfactant types.

One characteristic of the PO/EO sulfonate-type surfactants is their remarkable performance in the presence of high concentrations of Ca^{2+} and Mg^{2+} ions. This is taken to mean that calcium and magnesium salts of these surfactants have a limited solubility in oil, which will reduce possible trapping of surfactant at low concentration in the residual oil phase. As illustrated by Figure 8, a relatively small change in IFT is observed by doubling the salinity of seawater [*14*].

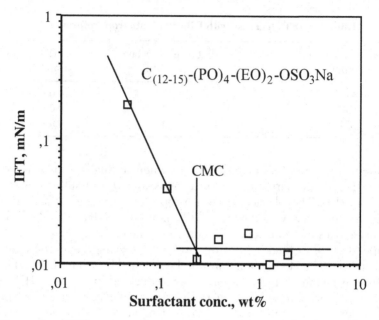

Figure 7.   IFT between n-heptane and seawater at 50 °C. (Reproduced with permission from reference 14, copyright 1995 Elsevier.)

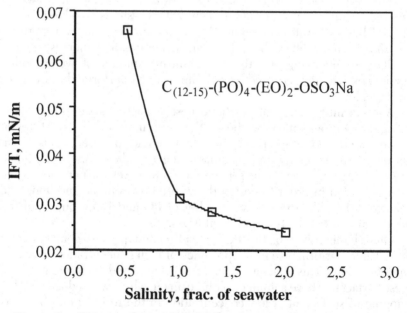

Figure 8.   IFT vs. salinity described as fraction of seawater. (Reproduced with permission from reference 14, copyright 1995 Elsevier.)

Interactions Between Chemicals and Rock

Normally, retention of surfactants, which involves adsorption, precipitation, and phase trapping, has been regarded as one of the main factors for the unfavorable economics in chemical flooding. Adsorption at the solid–liquid interface should be at minimum and be the only retention mechanism for a properly designed surfactant system for a Type II($-$) phase behavior. Commercial products of actual surfactants are polydisperse in the PO and EO groups, and mixtures of them are potential flooding chemicals.

The questions to be asked are:

1. Will there be a selective adsorption or chromatographic separation of the surfactant mixture as it propagates through the porous medium?
2. Will the polymer act as a sacrificial chemical when coinjected with the surfactant?
3. Why is adsorption of surfactant at anaerobic conditions (reservoir conditions) lower than adsorption at aerobic conditions (laboratory conditions)?

The answers to these questions will be discussed below.

Chromatographic Separation/Adsorption.   The HPLC analysis of the commercial product C_9-Ph-$(EO)_6$-SO_3Na showed that the number of oligomers can be high, Figure 4. Also, mixtures of surfactants containing different numbers of PO- and EO-groups have been used to optimize the surfactant formulation at a given salinity [12]. It has been verified experimentally that surfactant mixtures show properties that are beneficial for oil recovery compared to single components. Under a dynamic flood process, the formation may act as a chromatographic column towards the surfactant mixture. Different models for chromatographic separation of surfactant mixtures have been presented [33–35]. The models are based on idealized adsorption (Henry's Law), electrostatic interaction in the adsorbed layer combined with the pseudophase separation model, and a pseudobinary mixture of surfactant in brine (ideal mixed micelle theory). The models qualitatively describe the chromatographic movement of a surfactant mixture (usually two components) in sandstone cores. It is, however, unreasonable to believe that the models will predict the movement of a multicomponent surfactant mixture. Some experimental results concerning EO surfactants and mixtures of ethoxylated and nonethoxylated surfactant will be presented.

Surfactant flooding under a Type II($-$) phase condition will have retention mechanisms mainly related to adsorption at the water–solid interface. No trapping in the oil phase or precipitation by multivalent cations is expected at a surfactant concentration above the CMC. Austad et al. [36] circulated a solution of C_9-Ph-$(EO)_6$-SO_3Na through clay containing reservoir cores for more than two weeks to obtain adsorption equilibrium. The surfactant concentration was above the CMC all the time. The surfactant contained EO-oligomers in the range of 2–13. In the case of a precleaned core with no oil present, an increase in the adsorption of the low molecular weight EO-sulfonate oligomers relative to the high molecular weight oligomers was observed. If oil was present, no significant preferential adsorption of the various EO-sulfonate oligomers could be detected.

Static adsorption onto kaolinite at surfactant concentrations below the CMC showed a drastic change in the selective adsorption of the different oligomers. Relative to the initial distribution, the surfactant mixture became richer in the higher molecular weight oligomers, i.e. the low EO-number oligomers have a stronger adsorption to the clay. However, the main conclusion from the work is that ethoxylated sulfonates, and probably PO–EO type compounds too, have a structural similarity which will show minimal chromatographic separation during a chemical flood provided that the surfactant concentration is above the CMC. Similar conclusions are made by others [12, 15].

In order to reduce the cost of the chemical formulation, mixtures of ethoxylated and alkylaryl sulfonates have been tested as potential chemicals both in the field and in the laboratory [37, 38]. Analysis of produced fluids from a pilot test suggested that no significant chromatographic separation had occurred between the components of the surfactant slug that had been injected. Without any experimental documentation, Miller et al. [38] suggested that the interaction between surfactants in such mixtures tends to equalize the adsorption of the components onto the reservoir rock.

The low CMC of the mixed micelles is the reason for the stability of these surfactant mixtures towards multivalent cations, and the surfactant system can be applied even in seawater. The chromatographic separation of the dual surfactant system C_9-Ph-$(EO)_6$-SO_3Na (6EOS) and sodium dodecyl-benzene-sulfonate (SDBS) in the mole ratio of 1:1, was tested by Fjelde et al. [39]. Both of the chemicals were polydisperse, but the system was studied as a pseudobinary surfactant mixture. This dual surfactant system showed a strong negative deviation from ideality, which means that a minimum in the CMC versus surfactant mole fraction is observed. Long term dynamic circulation experiments in a reservoir core, and dynamic slug injection in a Berea core at waterflooded residual oil showed that a fast selective adsorption took place until the surfactant composition of the most stable mixed micelles was reached. Further decrease in the individual concentrations of the two groups of surfactants appears to be governed by a simple linear relationship between concentrations of the chemicals in solution, provided that the concentration is above the CMC. Thus, in order to avoid drastic changes in the relative composition of the two surfactants, the surfactant flooding should be performed with a concentration ratio of the two surfactants corresponding to the minimum in the CMC. Below the CMC, the adsorption of the nonethoxylated surfactant is significantly higher than the ethoxylated surfactant.

Effect of Polymer. In recent years, as described previously, much attention, both experimental and theoretical, has been focused on surfactant–polymer interaction in solution. Less experimental work, however, is done on the interaction between polyelectrolyte and surfactant of similar charge at the solid–liquid interface. Static adsorption experiments from the chemical literature indicate that the polymer does not affect the adsorption of the surfactant onto solid material as long as the surfactant concentration is above the CMC, apparently owing to the availability of sufficient surface sites for adsorption of the surfactant molecules [40, 41].

The adsorption process of surfactant and polymer under a two-phase dynamic flow condition in a porous medium, being heterogeneous in the mineral composition, showing flow constrictions due to small pore throats, and having stagnant volumes, i.e. zones bypassed by the injected fluid, is quite different from what is observed in a stationary adsorption process. The mechanism of a dynamic surfactant adsorption process onto clay-containing reservoir sandstone cores has been studied by Austad et al. as a function of time by circulating the solution through the core and determining the decrease in the surfactant concentration. Several adsorption regimes were observed due to the complexity of the mineral composition [42], residual oil saturation [43], and the presence of polymer [44]. Due to diffusion of surfactant monomers into the micropores of clay minerals, the time to reach adsorption equilibrium was in many cases rather long, about 50 days.

In the pre-equilibrium stage, the interaction between the surfactant and the polymer at the solid–liquid interface during the flooding process is also affected by chromatographic separation between the chemicals due to size exclusion phenomena, relative diffusion rate between surfactant monomers and polymers and between polymers of different molecular weight, and the access to the surface area. The polymer will move ahead of the surfactant when the chemicals are coinjected into a porous medium. Will an anionic polymer act as a sacrificial adsorbate towards PO–EO anionic surfactants? Experimentally, this was tested by circulating the injected solution containing xanthan and a PO–EO sulfate through reservoir cores of rather high clay content, approximately 20 wt%, [45].

Different adsorption regimes of the surfactant were observed with and without polymer present.

In Figures 9(a) and (b), the S-core only contains surfactant, while the SP-core contains surfactant and polymer. The different adsorption regimes responsible for the decrease in the surfactant concentration are described as:

1. Dilution with brine from the core/tubing/pump and fast adsorption.
2. Adsorption where there is a surfactant concentration gradient between the mixing bottle and the solution inside the core.
3. A temporary delay in the adsorption process.
4. Diffusion controlled adsorption where the surfactant concentration in the mixing bottle is nearly equal to the surfactant concentration in the solution inside the core.
5. A constant surfactant concentration, indicating that final adsorption equilibrium is obtained.

The temporary delay in the adsorption process may be explained by the time it takes for the surfactant to "clean up" sufficient openings of the micropores where adsorption of surfactants still can take place.

In the presence of xanthan, however, a minimum in the surfactant concentration is observed prior to the regime 3. This is interpreted as a fast surfactant adsorption onto easily accessible surface area, followed by a slow displacement by the polymer. Final adsorption equilibrium for the polymer was also obtained close to 2000 PV. The polymer will act as a sacrificial adsorbate towards the surfactant, causing a 20% decrease in equilibrium adsorption of the surfactant. In model cores of low clay content, small slug experiments, i.e. at surfactant concentrations well below the CMC in parts of the slug, the surfactant adsorption also decreases in the presence of xanthan, probably due to competitive adsorption between the surfactant and the polymer. In the large slug experiments, i.e. at surfactant concentrations well above the CMC, the surfactant adsorption is not significantly affected by the presence of xanthan or AN 125 (anionic copolymer containing sulfonate groups) [45].

Thus, to sum up, the polymer may decrease the adsorption of surfactant in reservoir rock of high clay content, approximately 20 wt%. In cores of low clay content, approximately 5 wt%, the polymer will probably have negligible effect on the surfactant adsorption onto the rock.

Adsorption Anaerobic/Aerobic. A very interesting paper, suggesting that surfactant adsorption onto reservoir rock is related to the reduction/oxidation potential of the system, was presented by Wang [46] using the PO–EO sulfates from in the Loudon field test. Conventional laboratory core floods consistently resulted in higher surfactant adsorp-

(a)

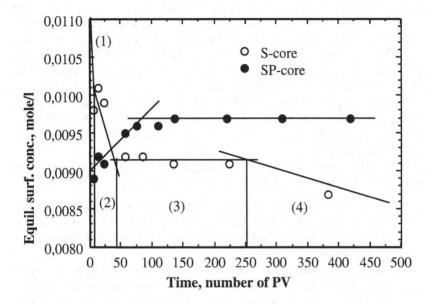

(b)

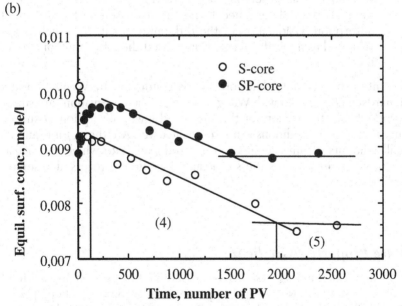

Figure 9.   Surfactant concentration vs. time (PV). (a) Early flooding stage, (b) late flooding stage. (Reproduced with permission from reference 45, copyright 1997 Elsevier.)

tion or retention levels than observed in the field tests. There are, however, great uncertainties in determining the adsorption levels from a field test. Laboratory core floods under anaerobic reduced conditions using reservoir brine and sodium dithionite, $Na_2S_2O_4$, as oxygen trapper, appeared to give similar results. Static adsorption experiments on different clay minerals showed that the surfactant adsorption level was unaffected by using either synthetic brine or Loudon field brine. In the presence of dithionite, the adsorption of surfactant was reduced, and the reduction was partly reversible when the clays were reexposed to oxygen. Wang also pointed out that the adsorption of the PO–EO sulfates was primarily due to the presence of clays, and the contribution from quartz and silica was minimal.

The adsorption results from Wang's core flood experiments are interesting, but some critical comments are relevant:

1. Dithionite may act as a sacrificial agent towards the anionic surfactants and reduce the adsorption in the same way as carbonate, phosphate, and silicate do. This effect is not discussed in Wang's paper.
2. Why is only dithionite found to be active in reducing the surfactant adsorption onto clay minerals? No effect was observed by washing with anaerobic reservoir brine. Remember also that the clay minerals were determined to be the main source for surfactant adsorption. Oxidatation of the dithionite to sulfate, which is not regarded as a sacrificial agent, increased the adsorption of surfactant.

It is interesting to note that core floods conducted by Wellington and Richardson [15] agreed with Wang's findings. The observations are surely important, but the reason for the suggested lower adsorption of surfactants at reservoir conditions is not quite clear yet. Are there, for example, small amounts of anionic species in a reduced state which can act as sacrificial agents towards the surfactants in a flood process at reservoir conditions?

Phase Properties/Gradients

The chemical flooding systems discussed in this section are without alcohol present as cosurfactant. As pointed out by Sanz and Pope [17], a major difficulty was to preclude gels, liquid crystals, macroemulsions, and precipitates along the compositional path during a chemical flood if cosurfactants/alcohols are not part of the chemical formulation. Phase trapping and blockage of the porous medium must be avoided.

Parameters Affecting Phase Behavior. In general, the phase behavior at reservoir conditions for oil–water–surfactant systems without alcohol present is very sensitive to the oil composition. Polar components in the crude oil may act as cosurfactants. On the other hand, model oils like n-alkanes do not contain this type of material. Different papers using the Exxon surfactant product termed RL-3011 (dodecyl-o-xylene sulfonate) illustrate the importance of oil composition on the phase behavior regarding changes in reservoir parameters like temperature, pressure, and salinity [47–50]. Conclusions from these papers may be summarized in the following way:

Effects of temperature:

- For model oils, the multiphase system moves towards the II($-$) state as the temperature is increased.
- For crude oil systems, the mutiphase behavior is complicated. At moderate temperatures $<90\,°C$ the system appears to move towards a II($-$) state as the temperature is increased. At higher temperatures, the system will move towards the II($+$) state. For a given system, the phase transition temperatures are dependent on pressure.
- At constant salinity and pressure, the phase behavior on changing the temperature may lead to two optimum conditions (equal solubilization of oil and water into the middle phase).
- Two middle phases are frequently observed in certain temperature ranges.
- The optimal solubilization of oil and water in the middle phase decreases as the temperature increases.

Effects of pressure:

- For all types of oils, the multiphase system moves towards the II($-$) state as the pressure is increased.
- At constant salinity and temperature, two middle phases and two optimum pressures are frequently observed.
- The optimum solubilization in the middle phase increases as the pressure increases.

Salinity:

- Nonclassical phase behavior in a salinity scan is often found for crude oil systems. In this case, no optimal salinity exists. As the salinity increases in the three-phase region approaching optimal salinity, suddenly the system turns into the II($+$) state.

Examples of the multiphase behavior using a reservoir crude oil are shown in Figures 10 and 11. Thus, in both cases (temperature and

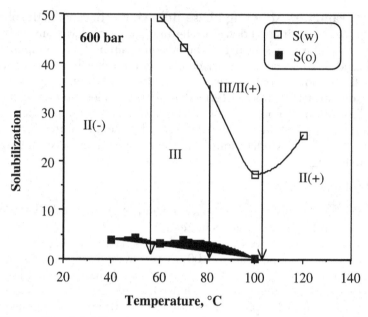

Figure 10. Solubilization parameters vs. temperature for a crude oil system. (Reproduced with permission from reference 50, copyright 1996 Elsevier.)

pressure scans) a nonclassical phase behavior is observed, i.e. no optimum (defined as $S(w) = S(o)$) in the three-phase region is observed.

Although EO and PO–EO sulfonates and sulfates have been used in field tests, no systematic phase studies of these systems are presented in the literature, especially at reservoir conditions. Maerker and Gale [12] observed that the PO–EO-sulfates designed for the Loudon field showed a normal phase behavior using diesel oil, but a nonclassical phase behavior using crude oil in a salinity scan. It appears to be a general trend that these surfactants have low ability to form middle phases at low surfactant concentration.

The PO–EO sulfonate formulation composed by Wellington and Richardson [15] contained small amounts of a cationic surfactant with EO groups. Contrary to traditional anionc surfactants, an aqueous solution of this formulation passes through a cloud point by increasing the temperature, i.e. a phase separation takes place. In the presence of oil, the system will move towards the II(+) state as the temperature is increased. This is the same behavior as for nonionic surfactants.

Phase studies at reservoir conditions were conducted using a branched ethoxylated sulfonate termed AS-142 which was used in a single well test in the Gullfaks field [51]. The phase properties were studied as a function

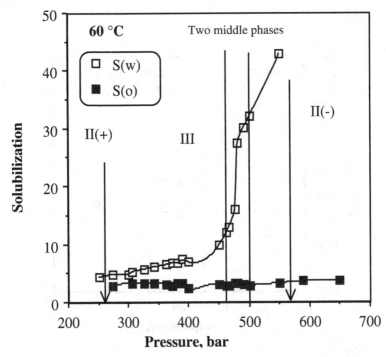

Figure 11.   Solubilization parameters vs. pressure for a crude oil system. (Reproduced with permission from reference 50, copyright 1996 Elsevier.)

of pressure, temperature, and gas–oil ratio (GOR) at seawater salinity using reservoir crude from Statfjord [52]. The general conclusions were:

- The system has a preference for forming either II($-$) or II($+$) phase behavior.
- In the two-phase area, the phase behavior is hardly affected by changes in pressure.
- The system moves towards the II($+$) state as temperature is increased.
- The system moves towards the II($+$) state as the GOR decreases.

The variation in the volume fraction of the aqueous phase versus temperature is shown in Figures 12(a) and (b) at 400 and 300 bar. A phase transition from a II($-$) state to a II($+$) state takes place at 400 bar by increasing the temperature. At 300 bar, the system stayed in the II($+$) state in the temperature range studied.

Osterloh and Jante [53] reported that classical phase transitions from II($-$) → III → II($+$) were observed using PO–EO sulfates in combination with a blend of stock tank oil and a light fraction from the crude oil, probably in a salinity scan.

(a)

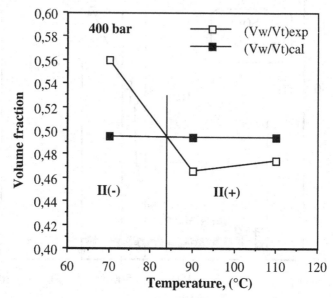

(b)

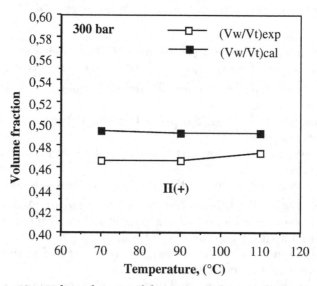

Figure 12.   Volume fraction of the aqueous phase vs. temperature. (Vw/
Vt)exp is experimental data, while (Vw/Vt)cal is calculated data based on
no phase solubilization. (The data are from reference 52, copyright 1996,
with permission from NPD.)

Phase Gradients. A flooding process under multiphase conditions requires a correct phase gradient for optimal displacement of oil. It is relevant to discriminate between natural phase gradients which are caused by changes in reservoir parameters like temperature, pressure, and salinities and imposed phase gradients due to changes in the concentration or composition of the injected fluids. Under traditional chemical flooding under multiphase conditions, the negative salinity gradient concept is regarded as the most effective flood process. A II($-$) phase state is then created at the rear of the microemulsion in order to minimize the surfactant loss due to phase trapping.

Natural phase gradients. *Temperature.* The temperature will increase from the injector to the producer which will have an influence on the multiphase behavior for a chemical formulation. Anionic surfactant systems without PO and EO groups will, under moderate temperatures, move towards the II($-$) state as temperature is increased. At temperature gradients above about 100 °C, a similar system may move towards the II($+$) state. Anionic surfactants containing PO and/or EO groups will move towards the II($+$) state with increasing temperature. Thus, it is very important to take into account possible temperature gradients when designing the HLB of the surfactant system in order to prevent unfavorable phase conditions.

Pressure. The reservoir pressure will decrease from the injector to the producer. Most anionic surfactant multiphase systems will move towards the II($+$) state as the pressure is decreased. The pressure gradient in most of the reservoir is small, and it will normally not influence the phase behavior during the flood process.

Salinity. A possible salinity gradient is related to the salinity of the injected water, salinity of reservoir brine, and composition and amount of adsorbed cations onto the formation. It is verified that the phase behavior of anionic surfactants of the sulfonate type without PO and/or EO is much more affected by changes in the ionic strength than the PO–EO sulfonates.

Imposed phase gradients. *Salinity.* The chemical slug is injected with a salinity corresponding to the optimal salinity in the three-phase state. The salinity of the injected water is then decreased to obtain a II($-$) state at the rear of the microemulsion zone [54].

Alcohol. Based on numerical simulation data, Baker [55] suggested that self-sharpening behavior, and robustness to variations in reservoir conditions, could also be achieved by inclusion of an alcohol in the chemical slug, provided that the alcohol is chosen to have an appreciable effect on the phase behavior. This will, however, increase the chemical cost and complicate the flood performance.

Multiple micellar slugs.   Thomas and Farouq [56] suggested the use of multiple chemical slugs with graded miscibility characteristics to improve the flood efficiency. The slugs should be injected in the order of increasing water content, with the compositions falling along an appropriate line on the phase diagram. They used, however, surfactant systems containing various amounts of alcohols in their experiments.

Polymer.   Austad and Taugbøl [57] discussed the possibility of using a polymer gradient at constant salinity in a three-phase low tension polymer flood process. Polymer has effects on the $II(-) \rightarrow III$ phase transition for a single surfactant system as indicated by Figure 13. The $II(-) \rightarrow III$ phase transition moves to a lower salinity in the presence of 500 ppm xanthan. At a salinity of 2.1 wt% NaCl, the system will move into a three-phase state if the concentration of xanthan exceeds approximately 150 ppm, Figure 14. Thus, by decreasing the polymer concentration, a type III system will move into a type $II(-)$ state at the rear of a three-phase surfactant slug.

Viscous Displacement of Oil

Some recent laboratory experiments using mostly low concentration surfactant systems without alcohol will be presented in this section. The experiments were conducted either as a three-phase flood with a phase gradient or simply as a two-phase flood without any need for phase gradient. There has been a discussion in the literature whether the polymer and surfactant should be coinjected or the polymer should be injected after the surfactant solution. References will also be given to actual field tests/pilots where the new generation of surfactants was used.

Three-Phase Displacement.   Provided that the alcohol free surfactant system shows a classic phase behavior, $II(-) \rightarrow III \rightarrow II(+)$ by increasing the salinity without forming gels or stable macroemulsions during the phase transitions, the chemical flood can be performed as a three-phase flood using a negative salt gradient.

Kalpakci et al. [13] introduced the concept "Low Tension Polymer Flood", LTPF. The flood was conducted by coinjection of surfactant and polymer and, due to chromatographic effects, the polymer moved ahead of the surfactant. In such an application, mobility of the water at the front will be reduced and the activity of the surfactant will be enhanced. The LTPF system should, according to Kalpakci, exhibit type $III/II(-)$ (slightly under optimum) phase behavior at the injection. Ethoxylated

(a)

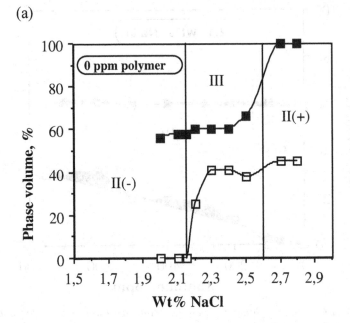

(b)

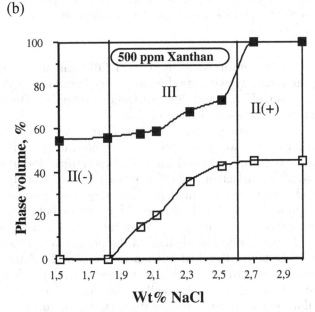

Figure 13. Phase volume vs. NaCl concentration. (a) Without polymer, (b) 500 ppm xanthan. The lower and upper curves represent the interface towards the excess water and oil phase, respectively. (Reproduced with permission from reference 57, copyright 1997 Elsevier.)

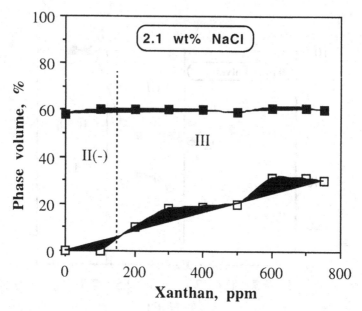

Figure 14.   Phase volume vs. polymer concentration. The lower and upper curves represent the interface towards the excess water and oil phase, respectively. (Reproduced with permission from reference 57, copyright 1995 Elsevier.)

sulfonate and scleroglucan were used as LTPF chemicals at high temperature and salinity with good recovery characteristics (66% of waterflooded residual oil) and low surfactant retention (less than 0.18 mg/g of rock) in core lengths up to 2.9 m.

One of the great advantages of coinjection of surfactant and polymer is probably the improved sweep efficiency. Some interesting studies performed by Amoco in the middle of the seventies confirm the idea that polymer ahead of the surfactant slug will improve the surfactant efficiency [58, 59]. Flood tests in physical models of heterogeneous porous media showed that preinjection of polymers could result in better flooding efficiency because of increased volumetric sweep. It was also found that the presence of waterflooded residual oil in the porous media increased the water-flow resistance. Residual resistance factors of 2–3.5 times the value when the rock was free of residual oil were observed. Furthermore, polymer preinjection had no effects on oil displacement characteristics of the micellar fluid and appeared to reduce the surfactant adsorption on the rock for the polymer–micellar system studied. Although these experiments were conducted to improve the macroscopic surfactant flooding efficiency, similar effects may be seen at a lower scale in a homogeneous medium like a Berea core.

Austad et al. [60] discussed the physicochemical principles of LTPF in relation to surfactant–polymer interaction, chromatographic separation of surfactant and polymer, effects of polymer on the II(−)/III phase transition and interfacial tension.

Reservoir simulators in combination with an economics model are useful to optimize the design of a chemical flood using surfactant and polymer. Project profitability is found to vary significantly at different surfactant concentrations. Recently, Wu et al. [61] found that the best results were obtained for the case where low concentrations of both surfactant and polymer were simultaneously injected, i.e. under a LTPF condition. Sensitivity analysis on the optimum design showed that the most important economic variables were oil price, discount rate, operating cost and chemical prices.

Single surfactant (n-C$_{12}$-o-xylene sulfonate) core flood experiments using Berea cores and model oil have been performed in order to compare the oil recovery from floods using polymer and salt gradients [57]. The surfactant and the polymer were injected simultaneously. About 70% of waterflooded residual oil was recovered, and the recovery from the salinity gradient was slightly higher compared to the polymer gradient flood. Normally, in order to lower the chemical costs, the concentration of polymer in the injection fluid after the surfactant/polymer slug is reduced in a LTPF and a II(−) phase state is then created in an optimized system. Due to the lower mobility of the less concentrated polymer solution, the polymer gradient zone may be rather large. In that case, the phase transition is not as sharp as for a salinity gradient. This may have an effect on the self-sharpening behavior of the surfactant slug using a polymer gradient in this way. However, the polymer gradient concept does not involve chemicals other than surfactant and polymer and should be studied further because of low operational costs.

Alcohol-free chemical floods using an equimolar blend of an olefin sulfonate and a petroleum sulfonate were reported to give a final oil recovery of 94% with a 13% of PV slug size using 3 vol.% surfactant concentration. When the slug size was reduced to 3% of PV, the oil recovery was still 80% [17]. The mobility was controlled by adding polymer so the minimum slug viscosity, μ_s, was at least equal to the reciprocal value of the water mobility at residual oil saturation, S_{or}; μ_w is viscosity of water and k_{rw} is relative permeability of water, i.e.:

$$\mu_s > \mu_w/k_{rw}$$

Dual surfactant systems, ethoxylated sulfonate/sulfate and alkane/aromatic sulfonate, as potential flooding chemicals have been studied in the laboratory by Miller et al. [38] and in the field by Holley and Caylas [37]. In a laboratory experiment at low surfactant concentration, the residual oil saturation decreased to about 5% with increasing surfactant concentration. In the field experiment, some alcohol was added to the chemical formulation. The incremental oil recovery was only 25% of the remaining oil after waterflooding in the entire pilot area. It was concluded, however, that better sweep efficiency in the pilot area by the surfactant would have increased the oil recovery. It is interesting to note that during the pilot test, no significant chromatographic separation of the surfactant slug was detected.

In some cases, the ethoxy groups may play an active role in surfactant adsorption towards certain minerals. A natural question to be asked is then, "Will polyethylene glycol, PEG, containing many EO-groups, act as a sacrificial adsorbate in the presence of actual PO–EO sulfonates/sulfates and in this way improve the oil displacement performance?" Osterloh and Jante [53] have studied the effect of adding 0.5 wt% PEG-1000 to a blend of PO–EO sulfates at a brine salinity of 190 g/L and a temperature of 47 °C. Static adsorption of the surfactants onto kaolinite was lowered to undetectable levels, while dynamic adsorption of the surfactants onto Berea sandstone was lowered by a factor of four when PEG-1000 was added to the microemulsion. The oil recovery factor was also increased. Independently, Austad et al. [62] did similar studies using 0.4 wt% PEG-4000 at seawater salinities and 80 °C in combination with an ethoxylated sulfonate containing 12 mole% nonionic unconverted material. Short-term static adsorption studies confirmed that PEG-4000 had sacrificial adsorbate effects towards the sulfonate for both kaolinite and quartz. pH-variations between 7.5 and 3.5 had small effects on the behavior of PEG-4000. Long-term dynamic adsorption studies were performed by circulating the surfactant solution through Berea and reservoir cores. A temporary decrease in the surfactant adsorption corresponding to values between 7 and 35% was observed by adding PEG-4000, illustrated by Figure 15. The decrease in adsorption was lowest for the oil-containing reservoir core. Furthermore, a pressure build-up was observed for the oil-free Berea cores. The observations were related to formation of PEG-microgels which are verified from light scattering experiments. Increase in temperature caused the weight fraction of PEG-microgel particles to increase. Thus, long-term experiments are needed in order to confirm the sacrificial adsorbate effects of PEG towards potential PO–EO surfactants.

A favorable phase gradient for oil displacement can also be obtained by using multiple chemical slugs in a sequence from oil-rich to water-rich [56]. The injected slugs were selected from equilibrium phase formed in pseudo-ternary oil–water–surfactant systems and followed by a mobility buffer. The chemical systems used contained alcohol and appeared to be rather sensitive to multivalent cations, which requires a preflush if used in reservoir field situations. Compared to single slugs of either oil-rich or water-rich microemulsions, multiple slugs of similar chemical size were observed to improve the recovery of waterflooded residual oil. With

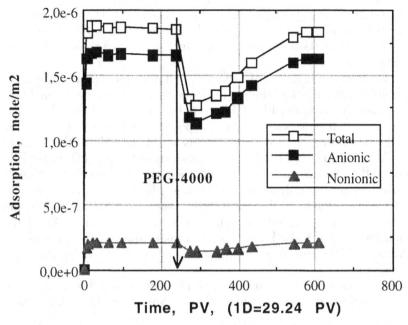

Figure 15. Dynamic adsorption of ethoxylated surfactant mixture onto Berea vs. time. At 250 PV 0.4 wt% PEG-4000 was added (T = 80 °C; brine = synthetic seawater; pH = 6.9–7.1). (Reproduced with permission from reference 62, copyright 1992 Elsevier.)

process optimization, very high process efficiency was noticed, i.e. a graded 2% PV slug yielded 15 times as much oil. Similar experiments should be performed using PO–EO surfactants free from alcohols.

Chemical floods are usually performed by phase gradients in the III/II(−) area for two reasons. (1) To obtain low enough IFT-value. (2) To obtain a self-sharpening of the surfactant slug. At large well distances, it is difficult to control the self-sharpening behavior of the chemical slug by proper phase gradients. Technically, a low tension polymer waterflood is more easy to perform as discussed in the next section.

Low Tension Polymer Water Flood. In oil reservoirs, where the critical capillary number is relatively low, a significant amount of waterflooded residual oil can be displaced by surfactants of high efficiency even at two-phase flood conditions. This was demonstrated by the successful second Ripley surfactant flood pilot test in the Loudon field where approximately 68% of waterflooded residual oil was recovered by injecting a 0.3 PV microemulsion bank [63]. The microemulsion bank was followed by 1.0 PV of higher viscosity polymer drive. The chemical formulation consisted of a blend of two PO–EO sulfates, i.e. i-$C_{13}H_{27}O(PO)_4(EO)_2SO_3Na$ and i-$C_{13}H_{27}O(PO)_3(EO)_4SO_3Na$. The retention of surfactant was confirmed to be less than 0.08 mg/g of rock, and more than 93% of the injected surfactant was recovered in the producing wells.

Simultaneous injection of surfactant and polymer with a hydrophilic–lipophilic balance, HLB, close to the three-phase region was termed Low Tension Polymer Water Flood, LTPWF, by Austad and Taugbøl [14, 16]. Due to chromatographic effects, the polymer will move ahead of the surfactant during the flooding process as illustrated by the effluent profiles in Figure 16. The chemical system consisted of a PO–EO-sulfate and xanthan in seawater [16]. The oil recovery is, as expected, strongly dependent on the polymer concentration, Figure 17. Close to 60% of waterflooded residual oil in Berea core was recovered using 500 ppm xanthan in the surfactant slug. In Bentheim cores, nearly 80% of the waterflooded residual oil was recovered under the same conditions. No pressure build-up or flow restrictions were observed during the flood experiments.

Similar experiments were performed by using dodecyl-*o*-xylene sulfonate as surfactant together with xanthan in NaCl-brine [14]. In this case,

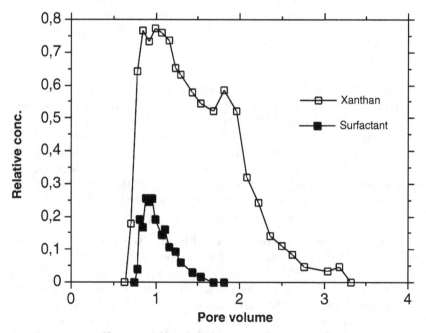

Figure 16.   Effluent profile of surfactant and polymer vs. pore volume from a 60 cm Berea core. (Reproduced with permission from reference 14, copyright 1995 Elsevier.)

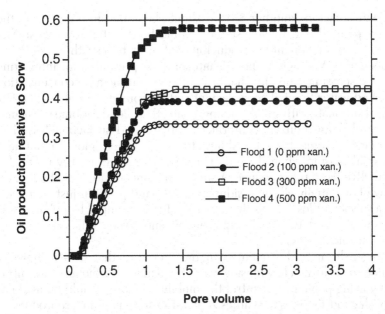

Figure 17. Oil recovery vs. polymer concentration. Flood conditions: 0.5 PV of 0.5 wt% surfactant in 60 cm 500 mD Berea cores. (Reproduced with permission from reference 14, copyright 1995 Elsevier.)

the presence of xanthan caused the oil recovery to decrease, and it was verified from the effluent profiles that the polymer had a negative effect on the flow performance of the surfactant in the porous medium. NMR diffusion studies and small angle neutron analysis showed that the micellar aggregates of this surfactant system are very large at conditions close to the II($-$)/III phase transition, and this may be the reason for the special behavior. Remember that the surfactant and the polymer are both present in the aqueous phase during the flood process.

Enhanced waterflooding design with dilute surfactant concentrations for North Sea conditions was evaluated by Shell [64]. It was concluded that alkyl–PO–EO glyceryl sulfonate surfactants could be used in a dilute (0.1 wt%) surfactant flood at North Sea reservoir temperatures ($<120\,°C$) and seawater salinities without polymer drive but with a sacrificial agent. The oil viscosity should be less than 3 mPa s. Increase in the temperature of the surfactant/water/oil system will give a cloud point or a three-phase system. The IFT will decrease, and at $95\,°C$ it was 5×10^{-3} mN/m, which is in the range required for tertiary oil remobilization. The estimated technical cost for application in the North Sea oil fields ranged from \$81 to \$94/incremental m^3 of oil. Taking into account uncertainty factors, another $31/m^3 should be added to the cost. As the authors pointed out, further studies should be directed towards surfactant–rock interaction and reproducibility of surfactant synthesis.

Recently, Wellington and Richardson [15] presented an interesting paper discussing the mechanism of low surfactant concentration enhanced water flood. The surfactant system consisted of alkyl–PO–EO glyceryl sulfonate with small amounts of an ethoxylated cationic surfactant to control phase behavior, interfacial activity, and surfactant loss. The surfactant systems had the ability to reduce their cloud point and interfacial tension when diluted, which was regarded as very useful for an effective flood performance. A surfactant concentration of about 0.4% removed essentially all the residual oil from sand packs in just over 1 PV with a surfactant loss of less than 0.1 PV. Mobility control by polymer was strongly required for good displacement and sweep efficiency and to reduce surfactant loss.

Obviously, it is documented in the literature that it is technically possible to perform a low tension water flood at low surfactant concentration by using polymer to control the mobility. Efforts should be made to establish good routines to synthesize PO–EO sulfonates in a reproducible and cost effective way. Mobility control is very important, and routines for preparing low cost polymers are important as well.

Recently, Taber and co-workers [65] have published screening criteria for all enhanced oil recovery (EOR) methods and their applications and impact of oil prices. About 3% of the worldwide production now comes from EOR. There are relatively few chemical flooding projects in the world, and these projects contribute very little to worldwide EOR production when compared to steamflooding and gas injection. A LTPWF, as a secondary flood method, may drain the reservoir to a residual oil saturation in the range of 15–20% rather fast and this may have an impact on the economics of the process. Future research on chemical flooding should move in this direction.

Displacement of Oil by Spontaneous Imbibition of Aqueous Surfactant Solution

Imbibition of water is a physical process caused by adsorption of water to hydrophilic ion-groups forming a hydrophilic surface. A porous reservoir medium, consisting of a hydrophilic surface, may contain lipophilic liquid such as oil. When such an oil filled reservoir rock is exposed to water, the water may spontaneously be sucked into the pores and displace the oil. This physical process is a result of forces acting in the individual pores.

The pores are often imagined as capillaries and the forces acting in the pores called capillary forces. However, pores in reservoir media are far from capillaries and very irregular [66] but in the absence of a better term, the term "capillary forces" will be used in this chapter. Capillary forces in a porous medium are related to the wettability of the minerals making up the porous rock [67–70], fluid/fluid [71, 72] and fluid/rock chemistry [73, 74], and saturation history of the reservoir rock [75, 76].

Cuiec et al. [77, 78] investigated the role of capillary forces, influence of length, boundary conditions and variable interfacial tension, IFT, using high porosity and low permeability chalk, model oils and brine. They found that imbibition rate decreased when IFT decreased, and final recovery was higher for low-IFT systems.

The interfacial tension between the hydrophilic and the lipophilic fluids can be modified by surfactants. In addition, adsorption of surfactants onto a solid surface affects wettability. The wettability is affected by salting out, hydrophobic bonding, solubility, Point of Zero Charge (PZC), mono-layer adsorption, and electrostatic forces [79–82].

According to the Laplace equation, the capillary pressure is given as:

$$P_c = \frac{2\sigma_{o/w}\cos\Theta}{R} \tag{3}$$

where R is the radius of the pore, Θ is the contact angle, and $\sigma_{o/w}$ is the interfacial tension between oil and water. From this equation, the capillary pressure is proportional to the interfacial tension. Different methods can be used to obtain the P_c-values if $P_c > 0$ [83–85]. A non-linearity between P_c and IFT could be caused by use of chemical additives acting on the solid surface, i.e. wettability alteration [86, 87].

Mattax and Kyte [88] discussed the imbibition driven by only capillary forces. This was done in relation to the dimensionless time scale for flow given by:

$$t_d = t\sqrt{\frac{k}{\phi}}\,\frac{\sigma_{o/w}}{\mu_w L^2} \tag{4}$$

Here, t_d is dimensionless time, t is time, ϕ is porosity, k is permeability, $\sigma_{o/w}$ is interfacial tension, IFT, μ_w is viscosity of water, and L is block dimension (length). They assumed that gravity effects are negligible, and that the shape of the matrix blocks, wettability, initial fluid distributions, relative permeabilities, and capillary pressures are the same. From equation 4 it is seen that the imbibition rate decreases if interfacial tension $\sigma_{o/w}$ decreases.

In experiments where an oil saturated rock is surrounded by a water phase, both capillary and gravity pressure gradients may be active in the displacement process. A capillary to gravity ratio expression, Ω, was derived by Iffly et al. [89]:

$$\Omega = \frac{\sigma_{o/w} \cos\Theta \sqrt{\phi/k}}{\Delta\rho g H} \tag{5}$$

High Ω-values mean that the flow is dominated by capillary forces and low Ω-values mean that the flow is governed by gravity segregation.

A transition from capillary-dominated flow to gravity-dominated flow occurs as IFT is reduced. From a balance between capillary and gravity forces, examined by du Prey [90], Schechter et al. [3, 91] derived an inverse Bond number, N_B^{-1}:

$$N_B^{-1} = 0.4 \frac{\sigma_{o/w} \sqrt{\phi/k}}{\Delta\rho g H} \tag{6}$$

where $\sigma_{o/w}$ is the IFT (mN/m), ϕ is the porosity, k is the permeability (cm^2), $\Delta\rho$ is the density difference between the two immiscible phases (g/cm^3), g is acceleration due to gravity (cm/s^2), H is the core length (cm), and C is a constant related to the pore geometry ($C = 0.4$ for cylindrical capillaries). Schechter et al. found that at low values of N_B^{-1} ($\ll 1$), gravity segregation dominates the flow and at high values of N_B^{-1} (> 5) a counter-current flow based on capillary forces dominates the flow. Snap-off is partially suppressed for $N_B^{-1} < 1.0$.

Recovery factors from oil reservoirs with use of surfactants and water injection with surfactants can be affected strongly by the rate and level of spontaneous imbibition. Improved oil recovery from low permeability rock may consequently be possible by decreasing the capillary to gravity force ratio, i.e. decrease N_B^{-1} and Ω. This could be done by decreasing IFT between oil and water if the displacement rate does not end up too slow for commercial use. In the following sections, displacement of oil by spontaneous imbibition at high and low IFT are considered for each wettability state; water-wet, mixed-wet and oil-wet.

Water-Wet Systems.   At high IFT, a fast imbibition process starts immediately after completely oil saturated water-wet rock is surrounded by brine [92]. Oil is displaced by a counter-current process. This can be observed visually because rather large oil drops grow from all sides of the porous rock before they release from the core surface [93]. Very high values of the inverse Bond numbers, according to equation 6, signify that capillary forces dominate the flow. Besides a high imbibition rate, the final oil production may be high, depending on the rock dimensions [93–95] and boundary conditions [77, 78]. A high recovery can, besides favourable boundary conditions, be related to the pore geometry characterized by a rather small aspect ratio, decreasing the amount of oil being trapped by a snap-off mechanism. After a production plateau is reached, it is difficult to recover any significant additional oil by decreasing the IFT. Morrow and Songkran [11] have estimated that mobilization of trapped oil blobs is about five times more difficult to achieve than prevention of trapping. Trapped oil due to snap-off and bypass may not be mobilized even by decreasing the IFT by a factor of about 10^3 [93].

At low IFT, the displacement process is different from the displacement process at high IFT. At the beginning of the displacement process, a counter-current imbibition takes place, but later a cocurrent flow takes over [93]. The cocurrent displacement process is slow, and final oil production equilibrium may be gained after a very long time. The oil production curve will show a break, which indicates a change in the oil expulsion mechanism, see Figure 18. The longer the core, the sooner this crossover from capillary-forced imbibition to a gravity-dominated flow will happen. In the slow part of the displacement process, N_B^{-1}-values less than 1 indicate that gravity forces are active in the cocurrent displacement mechanism. Visual observations of the released oil drops from the core

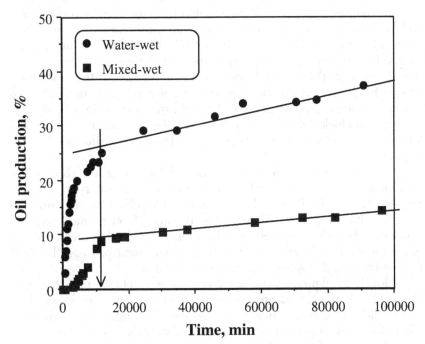

Figure 18. Imbibition curves for the long water-wet and mixed-wet core experiments at low IFT. (Reproduced with permission from reference 97, copyright 1996 Elsevier.)

top surface confirm that gravity forces are active in the displacement process [93].

If the imbibition process at low IFT is governed from the beginning by capillary forces only, the oil recovery should be related to the factor $(\sigma/\mu)t$, as described by equation 4. The oil production in the presence of surfactant is much too high in the early stage to be scaled as a capillary-forced imbibition only [86, 93]. Due to adsorption of surfactant onto the rock surface, adsorption at the oil–water interface, and lateral displacement at the liquid–liquid interface, a gradient in the surfactant concentration is established as the water invades the pore system. At the water front, the surfactant concentration is below the CMC, and a relatively higher IFT-value will result in a relative increase in capillary pressure and higher imbibition rate.

Cuiec et al. [77] also discussed this effect as a gradient in the capillary pressure over the oil blobs. The oil will then move in the direction of lower capillary pressure until equilibrium is established. The effect is a higher oil displacement rate than accounted for by equation 4 because equation 4 does not include gradients in the IFT.

Mixed-Wet Systems.   At high IFT conditions, oil from oil-filled rock surrounded by water will be expelled from the vertical rock surface and from the top and bottom surfaces in line with a counter-current flow mechanism governed by capillary forces [97]. The imbibition rate and the oil production rate are much faster from a water-wet rock than from a mixed-wet rock. The size of the core is very important for the production profile. For bigger blocks, the oil production plateau will be higher for the mixed-wet case compared to the water-wet case [97], contrary for smaller blocks. The block size in the reservoir may therefore be important, and lab experiments on small sized rock may lead to a too pessimistic recovery estimate [98].

A drastic change in the oil expulsion will happen at low IFT. In the beginning, the oil will be expelled from all sides in a counter-current flow regime. After a short time, the displacement mechanism will turn to a cocurrent flow regime based on gravity. Because of low permeability and small density difference between the fluids, this displacement process may be extremely slow. During such a displacement process, the fluid distributions inside the rock look as given in Figure 19.

At low IFT, oil production from a small mixed-wet core may stop even after just a small percentage recovery [97]. On the contrary, oil production would continue in a small water-wet core. In long core experiments, where the gravity forces are about 10 times larger than in small cores, the imbibition curves for mixed-wet and water-wet cores may be quite similar in shape. The production profiles have a break after a certain time, see Figure 18. This break is related to change in the imbibition mechanism,

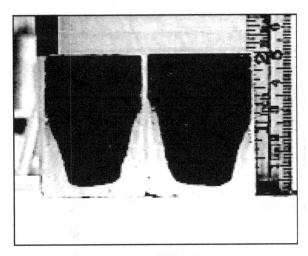

Figure 19. Cleaved mixed-wet small chalk core during a low IFT oil displacement experiment. (Reproduced with permission from reference 91, copyright 1991 SPE.)

from a counter-current flow, governed by capillary forces and surface tension gradients, to a cocurrent flow based on gravity forces.

The lower imbibition rate in the gravity-dominated regime for a mixed-wet system compared to a water-wet system may be explained by adsorption of surfactant onto the chalk surface. The equilibrium adsorption of surfactant at concentrations above the CMC onto water-wet patches, in between patches with adsorbed organic matter from the oil, may create oil lenses. If oil lenses are formed and the oil pins to the surface, then brine imbibition would be drastically reduced.

Oil-Wet Systems. Imbibition of water into an oil-wet material is not possible by definition without a change in wettability towards a more water-wet system. A drainage process governed by gravity forces may take place, but this process is very slow. A strategy to initiate an imbibition of water into an initially oil-wet rock is to use a surfactant system that creates water-wettability in the porous medium and gives a moderate decrease in IFT. In this way, the flow would turn out to be counter-current based on forces related to capillary and surface tension gradients. Recent results indicate that this vision may be possible to realize [92] by adding a surfactant to the aqueous phase. The IFT between the surfactant solution and the oil decreases and some of the surfactant dissolves in the oil phase. A possible mechanism was suggested which involves that some surfactant cross the oil–water interface and form unstable reversed micelles, where the water may act as a powerful nucleophile towards the rock surface. By

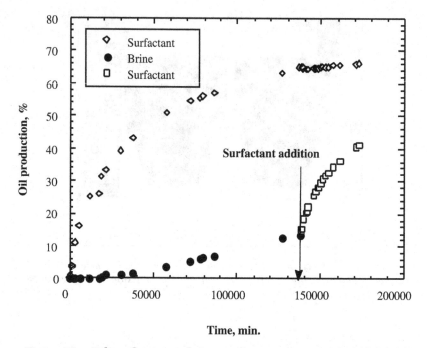

Figure 20.   Oil production as function of time in spontaneous imbibition experiments with and without the cationic surfactant present. (Reproduced with permission from reference 92, copyright 1997 SPE.)

using this technique a drastic increase in oil expulsion rate can be achieved, as illustrated by Figure 20.

Previously in this chapter, the imbibition at low IFT in a water-wet chalk system was described as a two-step process. Visual observations of the oil expulsion from the almost oil-wet system using surfactant indicated that the oil expulsion was governed by capillary forces and gradients in surface tension. At high IFT, the counter-current imbibition rate was very small, and the rate increased in the presence of surfactant. According to equation 5, one way to increase the counter-current flow rate is to make the rock surface more water-wet. The increase in the capillary forces (equation 3) by making the rock water-wet must overcome the decrease in the capillary forces due to the decrease in the IFT. The distribution of the oil saturation in the core at a water saturation of 41% was visually observed by cutting the core vertically, Figure 21. No segregation in the oil density in the vertical direction was observed, confirming that the displacement took place in a counter-current flow process. Thus, a reasonable explanation for the improved imbibition in the presence of surfactant is that the surfactant makes the rock more water-wet. This

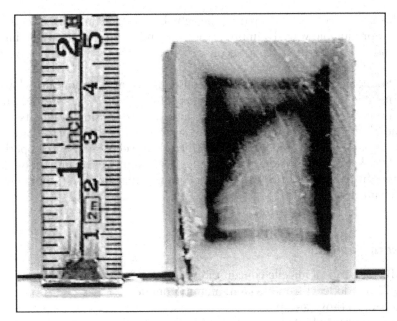

Figure 21. Picture of the cleaved core from the spontaneous imbibition experiment with brine followed by C_{12}TAB surfactant solution. The water saturation is approximately 41%. (Reproduced with permission from reference 92, copyright 1997 SPE.)

means that IOR can be possible from an almost oil-wet rock by use of surfactants.

Concluding Remarks

It is documented both in laboratory and field tests that more than 50% of waterflooded residual oil can be produced in a simple low tension polymer water flood, i.e. without using complicated phase gradients in the three-phase region. Anionic surfactants of the propoxy–ethoxy type will lower the interfacial tension between oil and water by a factor of more than 1000 in the two-phase area. The surfactants are stable at most reservoir conditions, and negligible chromatographic separation of the different PO–EO oligomers takes place during the flooding process. Polymers are needed for mobility control of the surfactant slug, and coinjection of surfactant and polymer appears to be the most favorable flooding process. The flood can be conducted at low chemical concentrations (0.1–0.5 wt% of surfactant and about 500 ppm of polymer for a low-viscous oil). For field applications, future work should be focused on industrial processes for preparing the actual surfactants in a cost effective and reproducible way. To improve the economics, the chemical flooding should be performed as a secondary flood process if the reservoir description is well known.

Spontaneous imbibition of aqueous surfactant solution into low-permeable oil saturated chalk material is complex due to the presence of different forces, i.e. capillary, gravity, and surface tension gradients. In general, it is not recommended to add surfactants to the injection water for a water-wet system. For mixed-wet to oil-wet systems, a properly designed surfactant system may in some cases improve the imbibition of water. In this case, more work is needed to understand the imbibition mechanism.

List of Symbols

CMC	critical micelle concentration
$C_{12}TAB$	dodecyl-trimethylammonium bromide
EO	ethoxy-group
GOR	gas–oil ratio
HLB	hydrophilic–lipophilic balance
HPLC	high pressure liquid chromatography
HPA	hydrolyzed polyacrylamide
IFT	interfacial tension
LTPWF	low tension polymer water flood
PO	propoxy-group
S(o)	solubilization of oil
S(w)	solubilization of water
$II(-)$	oil-in-water microemulsion
$II(+)$	water-in-oil microemulsion
III	middle phase microemulsion
C	constant related to pore geometry
g	acceleration due to gravity
k	permeability
k_{rw}	relative permeability of water at S_{or}
H	height
L	length of block
N_B	Bond number
N_c	capillary number
N_{cri}	critical capillary number
P_c	capillary pressure
R	radius
S_{or}	residual oil saturation
t	time

t_d dimensionless time
v effective flow rate

Greek

μ_w viscosity of water
μ_g viscosity of gas
ϕ porosity
$\Delta\rho$ density difference
σ interfacial tension
θ contact angle
Ω capillary to gravity force ratio

References

1. Gogarty, W.B. *J. Pet. Techn.* **1976**, Dec., 1475–83.
2. Thomas, L.K.; Dixon, T.K.; Evans, C.E.; Vienot, M.E. *J. Pet. Tech., Trans., AIME* **1987**, *283*, 221–32.
3. Schechter, D.S.; Zhou, D.; Orr Jr., F.M. *J. Pet. Sci. Eng.* **1994**, *11*, 283–300.
4. Pope, G.A.; Bavière, M. In *Basic Concepts in Enhanced Oil Recovery Processes*; Bavière, M., Ed.; Elsevier Applied Science, 1991; pp 89–122.
5. Thomas, S.; Farouq, S.M. *J. Can. Pet. Techn.* **1992**, *31*, 53–60.
6. Healy, R.N.; Reed, R.L. *SPEJ* **1974**, Oct., 491–501.
7. Ling, T.F.; Lee, H.K.; Shah, D.O. In *Industrial Applications of Surfactants*; Karsa, D.R., Ed.; The Royal Society of Chemistry, Special publication No. 59., 1986, pp 126–78.
8. Bavière, M.; Glénat, P.; Plazanet, V.; Labrid, J. *SPE Res. Eng.* **1995**, *10*, 187–93.
9. Hankins, N.P.; Harwell, J.H. *J. Pet. Sci. Eng.* **1997**, *17*, 41–62.
10. Lake, L.W. *Enhanced Oil Recovery*; Prentice Hall, Inc., New Jersey, 1989; p 70.
11. Morrow, N.R.; Songkran, B. In *Surface Phenomena in Enhanced Oil Recovery*; Shah, D.O. Ed.; Plenum Press, New York, 1981, pp 387–411.
12. Maerker, J.M.; Gale, W.W. *SPE Res. Eng.* **1992**, *7*, 36–44.
13. Kalpakci, B.; Arf, T.G.; Barker, J.W.; Krupa, A.S.; Morgan, J.C.; Neira, R.D. *Proceedings of the 7th Symposium on Enhanced Oil Recovery of SPE*; Society of Petroleum Engineers: Tulsa, OK, 1990, paper SPE 20220.
14. Taugbøl, K.; Van Ly, T.; Austad, T. *Colloids Surfaces A: Physicochem. Eng. Aspects* **1995**, *103*, 83–90.
15. Wellington, S.L.; Richardson, E.A. *Proceedings of the Annual Technical Conference of SPE*; Society of Petroleum Engineers: Dallas, TX, 1995, paper SPE 30748.
16. Austad, T.; Taugbøl, K. *Colloids Surfaces A: Physicochem. Eng. Aspects* **1995**, *103*, 73–81.
17. Sanz, C.A.; Pope, G.A. *Proceedings of The SPE International Symposium on Oilfield Chemistry*; Society of Petroleum Engineers: San Antonio, TX, 1995, paper SPE 28956.

18. Austad, T.; Fjelde, I. *Analytical Letters* **1992**, *25*, 957–71.
19. Fjelde, I.; Austad, T. *Colloids Surfaces A: Physicochem. Eng. Aspects* **1994**, *82*, 85–90.
20. Tally, L.D. *Proceedings from the International Symposium on Oilfield Chemistry of SPE*; Society of Petroleum Engineers: Houston, TX, 1989, paper SPE 18492.
21. Baviére, M.; Bazin, B.; Labrid, J. *In Situ* **1989**, *13*, 101–20.
22. McGary, C.W. *J. Polymer Sci.* **1960**, *XLVI*, 51–7.
23. Austad, T.; Fjelde, I. *Colloids Surfaces A: Physicochem. Eng. Aspects* **1993**, *81*, 263–67.
24. Tally, L.D. *SPE Res. Eng.* **1988**, *3*, 235–42.
25. Pope, G.; Tsaur, K.; Schechter, S.; Wang, B. *Proceedings from the First Symposium on Enhanced Oil Recovery of SPE*; Society of Petroleum Engineers, Tulsa, OK, 1980, paper SPE 8826.
26. Lindman, B.; Thalberg, K. In *Interactions of Surfactants with Polymers and Proteins*; Goddard, E.D.; Antanthapadmanabhan, K.P., Eds., CRC Press, Boca Raton, 1993, pp 203–76.
27. Vervey, E.J.W.; Overbeek, J.Th. *Theory of the Stability of Lyophobic Colloids*; Elsevier, Amsterdam, 1948.
28. Yang, C.Z. *Proceedings from the 5th Symposium on Enhanced Oil Recovery of SPE*; Society of Petroleum Engineers: Tulsa, OK, 1986, paper SPE 14931.
29. Piculell, L.; Lindman, B. *Advances in Colloid and Interface Sci.* **1992**, *41*, 149–78.
30. Veggeland, K.; Nilsson, S. *Langmuir* **1995**, *11*, 1885–92.
31. Saito, S.J. *Colloid Interface Sci.* **1960**, *15*, 283–6.
32. Lindman, B. In *Surfactants;* Tadros, Th.F., Ed., Academic Press, Inc, London, 1984, pp 83–109.
33. Trogus, F.J.; Schechter, R.S.; Pope, G.A.; Wade, W.H. *J. Pet. Technol.* **1979**, *31*, 769–78.
34. Harwell, J.H.; Hoskins, J.C.; Schechter, R.S.; Wade, W.H. *Langmuir* **1985**, *1*, 251–62.
35. Mannhardt, K.; Novosad, J.J. *J. Pet. Sci. Eng.* **1991**, *5*, 89–103.
36. Austad, T.; Fjelde, I.; Rolfsvåg, T.A. *J. Pet. Sci. Eng.* **1992**, *6*, 277–87.
37. Holley, S.M.; Caylas, J.L. *SPE Res. Eng.* **1992**, Feb., 9–14.
38. Miller, D.J.; von Halasz, S.P.; Schmidt, M.; Holst, A.; Pusch, G. *J. Pet. Sci. Eng.* **1991**, *6*, 63–72.
39. Fjelde, I.; Austad, T.; Milter, J. *J. Pet. Sci. Eng.* **1995**, *13*, 193–201.
40. Moudgil, B.M.; Somasundaran, P. *Colloids Surfaces* **1985**, *13*, 87–95.
41. Esumi, K.; Masuda, A.; Otsuka, H. *Lamgmuir* **1993**, *9*, 284–95.
42. Austad, T.; Bjørkum, P.A.; Rolfsvåg, T.A. *J. Pet. Sci. Eng.* **1991**, *6*, 125–35.
43. Austad, T.; Bjørkum, P.A.; Rolfsvåg, T.A.; Øysæd, K.B. *J. Pet. Sci. Eng.* **1991**, *6*, 137–48.
44. Austad, T.; Fjelde, I.; Veggeland, K. *J. Pet. Sci. Eng.* **1994**, *12*, 1–8.
45. Austad, T.; Ekrann, S.; Fjelde, I.; Taugbøl, K. *Colloids Surfaces A: Physicochem. Eng. Aspects* **1997**, *127*, 69–82.
46. Wang, F.H.L. *Proceedings from the 66th Annual Technical Conference of SPE*; Society of Petroleum Engineers: Dallas, TX, 1991, paper SPE 22648.

47. Austad, T.; Hodne, H.; Staurland, G. *Progr. Colloid Polym. Sci.* **1990**, *82*, 296–310.
48. Austad, T.; Staurland, G. *In Situ* **1990**, *14*, 429–54.
49. Austad, T.; Strand, S. *Colloids Surfaces A: Physicochem. Eng. Aspects* **1996**, *108*, 243–52.
50. Austad, T.; Hodne, H.; Strand, S.; Veggeland, K. *Colloids Surfaces A: Physicochem. Eng. Aspects* **1996**, *108*, 253–62.
51. Nordbotten, A.; Maldal, T.; Gilje, E.; Svinddal, S.; Kristensen, R. *Proceedings from the 8th European IOR Symposium*, Vienna, 1995, pp 86–95.
52. Austad, T.; Hodne, H.; Starand, S.; Veggeland, K. In *RUTH. A Norwegian research program on improved oil recovery. Program summary*; Skjæveland, S.M.; Skauge, A.; Hinderaker, L.; Sisk, C.D., Eds.; Norwegian Petr. Directorate, Stavanger 1996; pp 387–98.
53. Osterloh, W.T.; Jante, M.J. *Proceeding from the 8th Symposium on Enhanced Oil Recovery of SPE*; Society of Petroleum Engineers: Tulsa, OK, 1992, paper SPE 24151.
54. Hirasaki, G.J.; van Domeslaar, H.R.; Nelson, R.C. *Soc. Pet. Eng. J.* **1983**, *23*, 486–500.
55. Baker, J.W. *Proceedings from the 6th European IOR Symposium*, Stavanger, Norway, 1991, pp 777–87.
56. Thomas, S.; Farouq Ali, S.M. *J. Can. Pet. Techn.* **1990**, *29*, 22–8.
57. Austad, T.; Taugbøl, K. *Colloids Surfaces A: Physicochem. Eng. Aspects* **1995**, *101*, 87–97.
58. Dabbous. M.K.; Elkins, L.E. *Proceedings from the Improved Oil Recovery Symposium of SPE*; Society of Petroleum Engineers: Tulsa, OK, 1976, paper SPE 5836.
59. Dabbous, M.K. *Soc. Pet. Eng. J.* **1977**, Oct., 358–68.
60. Austad, T.; Fjelde, I.; Veggeland, K.; Taugbøl, K. *J. Pet. Sci. Eng.* **1994**, *10*, 255–69.
61. Wu, W.; Vaskas, A.; Delshad, M.; Pope, G.A.; Sepehrnoori, K. *Proceedings from the 10th Symposium on Improved Oil Recovery of SPE*; Society of Petroleum Engineers: Tulsa, OK, 1996, paper SPE 35355.
62. Austad, T.; Rørvik, O.; Rolfsvåg, T.A.; Oysæd, K.B. *J. Pet. Sci. Eng.* **1992**, *6*, 265–76.
63. Reppert, T.R.; Bragg, J.R.; Wilkinson, J.R.; Snow, T.M.; Maer Jr., N.K.; Gale, W.W. *Proceedings from the 7th Symposium on Enhanced oil Recovery of SPE*; Society of Petroleum Engineers: Tulsa, OK, 1990, paper SPE 20219.
64. Michels, A.M.; Djojosoeparto, R.S.; Haas, H.; Mattern, R.B.; van der Weg, P.B.; Schulte, W.M. *SPE Res. Eng.* **1996**, *11*, 189–95.
65. Taber, J.J.; Martin, F.D.; Seright, R.S. *SPE Res. Eng.* **1997**, *12*, 189–98 and 199–205.
66. McCaffery, F.G.; Sigmund, P.M.; Fosti, J.E. *Canadian Well Logging Society Procedings Formation Evaluation Symposium*, Calgary, 1977.
67. Cuiec, L.E. In *Evaluation of Reservoir Wettability and Its Effect on Oil Recovery. Interfacial Phenomena in Petroleum Recovery*, N.R. Morrow Ed.; 1991, Marcel Dekker Inc.: New York, 1991, pp 319–73.
68. Anderson, W.G. *J. Pet. Technol.* **1987**, Dec., 1605–19.

69. Ma, S. *The Petroleum Society of CIM 45th Annual Technical Meeting and Aostra 1994 Annual Technical Conference*, Calgary, June 12–15, 1994.

70. Graue, A.; Tonheim, E.; Baldwin, B. *The 3rd International Symposium on Evaluation of Reservoir Wettability and Its Effects on Oil Recovery*, Laramie, WY, Sept. 21–23, 1994.

71. Ghedan, S.G.; Poettmann, F.H. *Seventh Symposium on Enhanced Oil Recovery of SPE*; Society of Petrolum Engineers: Tulsa, OK, 1990, April 22–25, paper SPE 20244.

72. Perez, J.M.; Poston, S.W.; Sharif, Q.S. *Eighth Symposium on Enhanced Oil Recovery of SPE*; Society of Petroleum Engineers: Tulsa, OK, 1992, April 22–24, paper SPE 24164.

73. Jadhunandan, P.P.; Morrow, N.R. *In Situ* **1991**, *15*, 319–45.

74. Morrow, N.R.; McCaffery, F.G. In *Displacement Studies in Uniformly Wetted Porous Media*, Paddy, G.F., Ed. Academic Press: New York, 1978; pp 289–319.

75. Milter, J.; Øxnevad, I.E.I. *Petroleum Geoscience* **1996**, *2*, 231–40.

76. Kovscek, A.R.; Wong, H.; Radke, C.J. *AIChE Journal* **1993**, *39*, 1072–85.

77. Cuiec, L.; Bourbiaux, B.; Kalaydjian, F. *SPE Formation Evaluation* **1994**, *9*, 200–8.

78. Cuiec, L.E.; Bourbiaux, B.; Kalaydjian, F. *Seventh Symposium on Enhanced Oil Recovery of SPE*; Society of Petrolum Engineers: Tulsa, OK, 1990, paper SPE 20259.

79. Lahann, R.W.; Cambell, R.C. *Geochimica et Cosmica Acta* **1980**, *44*, 629–34.

80. Zullig, J.J.; Morse, J.W. *Geochimica et Cosmica Acta* **1988**, *52*, 1667–78.

81. Mannhardt, K.; Schramm, L.L.; Novosad, J.J. *Colloids and Surfaces* **1992**, *68*, 37–53.

82. Mannhardt, K.; Novosad, J.J. *IEA Collaborative Project on Enhanced Oil Recovery Workshop and Symposium*, Banff, Alberta, Canada, Sep. 27–30, 1992.

83. Torsæter, O. *The 3rd International Symposium on Evaluation of Reservoir Wettability and Its Effect on Oil Recovery*, Laramie, WY, Sep. 21–23, 1994.

84. Hammervold, W.L. Ph.D. Thesis, Stavanger College, Stavanger, Norway, 1994.

85. Anderson, W.G. *J. Pet. Techn.* **1987**, Oct., 1283–99.

86. Thiebot, B.; Barroux, C.; Bouvier, L.; Heugas, O.; Plazanet, V. *The 3rd North Sea Chalk Symposium*, Copenhagen, Denmark, 1990.

87. Christoffersen, K.R.; Whitson, C.H. *SPE Formation Evaluation* **1995**, Sep., 153–9.

88. Mattax, C.C.; Kyte, J.R. *Soc. Pet. Eng. J.* **1962**, *12*, 177–84.

89. Iffly, R.; Rousselet, D.C.; Vermeulen, J.L. *The 47th Annual Technical Conference of SPE*; Society of Petroleum Engineers: San Antonio, TX, 1972, paper SPE 4102.

90. du Prey, L.E. *Soc. Pet. Eng. J.* **1978**, *18*, 927–35.

91. Schechter, D.S.; Zhou, D.; Orr, F.M. *The 66th Annual Technical Conference and Exhibition of SPE*; Society of Petroleum Engineers: Dallas, TX, 1991, paper SPE 22594.

92. Austad, T.; Milter, J. *International Symposium on Oilfield Chemistry of*

SPE; Society of Petroleum Engineers: Houston, TX, 1997, paper SPE 37236.

93. Milter, J.; Austad, T. *Colloids Surfaces A: Physicochem. Eng. Aspects* **1996**, *113*, 260–78.

94. Torsaeter, O.; Silseth, J.K. *North Sea Chalk Symposium*, Stavanger, Norway, 1985.

95. Torsæter, O. *An experimental study of water imbibition in North Sea Chalk*, Ph.D. Thesis, NTH, Trondheim, Norway, 1993.

96. Keijzer, P.P.M.; de Vries, A.S. *Seventh Symposium on Enhanced Oil Recovery of SPE*; Society of Petroleum Engineers: Tulsa, OK, 1990, paper SPE 20222.

97. Milter, J.; Austad, T. *Colloids Surfaces A: Physicochem. Eng. Aspects* **1996**, *117*, 109–15.

98. Sylte, J.E.; Hallenbeck, L.E.; Thomas, L.K. *Technical Conference and Exhibition of SPE*; Society of Petroleum Engineers: Houston, TX, 1988, paper SPE 18276.

RECEIVED for review June 1, 1998. ACCEPTED revised manuscript October 16, 1998.

Scale-Up Evaluations and Simulations of Mobility Control Foams for Improved Oil Recovery

Fred Wassmuth, Laurier L. Schramm, Karin Mannhardt, and Laurie Hodgins

Petroleum Recovery Institute, 100, 3512 – 33rd Street N.W., Calgary, AB, T2L 2A6, Canada

Foam experiments were duplicated in both short (20 cm) and long (2 m) Berea cores to ascertain how to scale-up foam performance. Gas mobility reduction factors were measured at pseudo-steady state as a function of foam quality, and foam velocity in oil free cores and at residual oil saturation, at room temperature and at 7000 kPa system pressure.

The experimental results indicate that different water fractional flows, for particular frontal advance rates, are needed to generate strong foams. This effect is much more pronounced in the presence of oil, i.e. higher fractional flow of water was needed to establish significant mobility reduction factors when residual oil was present. Foams generated in the presence of residual oil produced consistently lower mobility reduction factors than foams generated in cores without oil.

When no oil was present, the scale-up work experiments show good correspondence between the short and long core lengths. The increased pressures experienced in the upstream section of the long core, during foam flow, do however affect the mobility reduction capacity of the foam and need to be taken into consideration. Injecting foam steadily into a short core at water-flood residual oil lowered the oil saturation significantly and subsequently allowed for strong foams to be established. Repeating this flooding sequence on the long core caused a blocking emulsion (gas/surfactant solution/oil) to be formed in-situ, which completely blocked the long core.

Three prevalent, steady state foam models, all based on modification of the gas phase relative permeability, are reviewed. Two supplemental correlations were derived in order to account for the effect of ambient pressure on foam performance. Thus the reduced mobility of the gas phase, when foam is present, can be effectively modeled for long and short core experiments, with and without oil. Subsequently, the results from this work can be extended to model a field application.

Introduction

A large percentage (80%) of Canadian enhanced oil production comes from hydrocarbon miscible flooding [1]. The low density and viscosity of the injected fluids cause hydrocarbon miscible EOR processes to suffer from poor sweep efficiency, due to viscous fingering and gravity override. Mobility control foams provide a means for improving the sweep efficiency and could significantly increase oil production from Canadian reservoirs.

Many steam-foam field tests [2–4] and three hydrocarbon solvent-foam field tests [5–7] have shown that foams can be used successfully in the field. Most of the surfactants used in foam studies are unsuitable for western Canadian reservoirs. Through previous efforts, surfactants were identified that are soluble and form strong foams in the high salinity brines encountered in many western Canadian reservoirs currently subjected to hydrocarbon miscible flooding [8]. Extensive studies were conducted with respect to aspects important to the application of foams in these reservoirs: gas mobility reduction in porous media [8], foam/oil interactions [9], surfactant loss through adsorption [10], and the effects of hydrocarbon solvents and wettability on foam performance [11]. Our continuing research is aimed at improving foam flood design through a combination of experimental and computer modelling approaches. The next logical step, modelling and experimental scale-up work, has been addressed in this project with the goal to transfer this technology to the field. The experiments were designed such that the results of the foam floods were used as input to calibrate the numerical model. A robust foam model was developed since the foam behaviour was studied and simulated under a wide variety of conditions.

In the first phase of this work, foam experiments were performed in short and long cores in the absence of oil. In short cores, three different foam qualities were investigated at four different advance rates. The experiments proceeded until a steady state pressure drop was measured across the core at a given quality and flow rate. Historically most foam work in our laboratories has been performed under such conditions, so that these experiments can be compared to cases existing in the literature. To understand the scaling effects on foam behaviour the same corefloods that were performed in short cores, 0.2 m in length, were repeated in long cores, 1.8 m in length. Due to the longer duration of long core experi-

Table 1. Fluid Properties at 6900 kPa and 23 °C

Judy Creek Oil	
dead oil density (g/cm^3)	0.8293
live oil density (g/cm^3)	0.8258
dead oil viscosity (mPa·s)	2.866
live oil viscosity (mPa·s)	2.484
Surfactant Solution	
Chevron Chaser GR-1080	0.5 wt%
Injection Brine (2.1%) TDS	
density (g/cm^3)	1.0162
viscosity (mPa·s)	0.85
TDS	2.1%
Nitrogen	
viscosity (mPa·s)	0.0175

ments, the pressure drops generated by only one frontal advance rate, at three different foam qualities, were investigated.

Oil can usually destabilize foam lamellae and foam collapse is much more prevalent in this instance. In the second phase of this work, the set of foam flood experiments (oil free) was repeated in short and long cores in the presence of oil, to achieve more realistic reservoir conditions.

The foam experiments were tailored to generate data suitable for simulation studies. Three foam models are compared in this work, each of which relies on modifying the gas relative permeability in the presence of foam.

Experimental

In order to differentiate between foam effects, the effects of surfactant transport, and multiphase flow, a number of peripheral experiments were conducted. Through additional corefloods the surfactant adsorption level was measured and the relative permeabilities between the different phases gas/oil/water were determined, as outlined in the Appendix.

Materials. Several fluid properties, listed in Table 1, were determined at a temperature of 23 °C and a pressure of 1000 psig (dead oil indicates no gas saturation while live oil indicates nitrogen gas saturated). A compromise was chosen between extreme Canadian reservoir conditions and the limiting operating conditions of the laboratory equipment. All experiments were conducted at a pressure of 6900 kPa (1000 psig) and at room temperature (23 °C). Berea rock was used as core material.

The surfactant chosen was Chevron Chaser GR-1080, a proprietary commercial mixture of surfactants proven to form effective mobility control foams in high salinity and hardness conditions, and in the presence of crude oil [11]. The surfactant concentration was kept at 0.5 wt% throughout the experiments.

Apparatus. Essentially the coreflood equipment used for the long and short core experiments was very similar. For the short core experiments, only a single pressure drop across the whole core was measured. In the long core apparatus five separate pressure taps were mounted along the 1.8 m long core (see Figure 1).

Foam Flooding in Oil Free Cores. *Short Core Experiments.* The bulk of the short core experiments consist of measurements of pressure drops and mobility reduction factors (MRFs) generated by foams in porous media. The MRF is determined by comparing the pressure drop across a core during simultaneous injection of surfactant solution and gas with that during injection of brine (without surfactant) and gas at the same experimental conditions. The MRF is defined as follows:

$$\text{MRF} = \frac{\Delta P_f}{\Delta P_n} \tag{1}$$

Baseline pressure drops, ΔP_n, were measured during co-injection of brine and gas into a previously brine saturated core (no foam present in

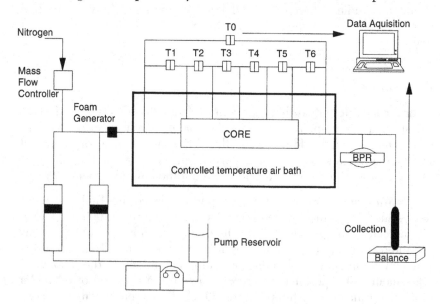

Figure 1. Long coreflood apparatus.

Table 2.   Effect of Flow Rate on Foam ΔP

Foam ΔP (kPa)	FAR at 95% Quality (m/day)							
	0.5	1	3	3	4	5	6	7
Chaser GR-1080 (This work)	18	24	33		2595			
Dow XSS-84321.05 (Reference [12])			86	193		377		848
Flourad FC-751 (Reference [12])			520	627	755	758	707	

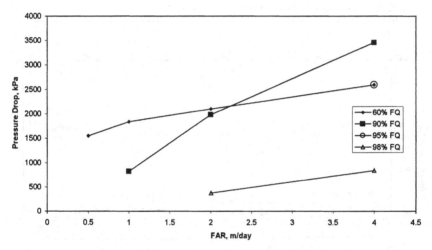

Figure 2.   Foam pressure drops in an oil free short core.

the core). The co-injection of surfactant solution and gas followed at varying foam qualities and frontal advance rates (FAR). Figure 2 shows the results for ΔP_f, measured in the presence of surfactant. The trend towards increasing pressure drop due to foam with increasing foam flow velocity agrees with previous results obtained for Flourad FC-751 and Dow XSS-84321.05 foams flowing in Berea sandstone cores. Table 2 shows such a comparison for 95% quality foams [12]. Figure 3 shows the variations in MRF with changing foam flow rate and quality. Note that in the low foam quality region (fg = 0.6), the lower frontal velocities demonstrate higher MRFs than the higher frontal velocities. In the high foam quality region, the MRFs are of similar magnitude for all frontal advance rates.

Long Core Experiments.   Experiments designed to match those performed in the short core were conducted using a 1.8 m long Berea

Table 3.   Flood History for the Long Core (MCF4), Oil Free Experiments

Injection History:
  saturated core with brine
Base Line Experiment: total injection rate 40 ml/h of gas and brine
  injected 19.8 PV of gas and brine with tracer at 95% quality, ΔP = 24.6 kPa
  injected 6 PV of gas and brine at 90% quality, ΔP = 29.6 kPa
  injected 5 PV of gas and brine at 85% quality, ΔP = 47.2 kPa
  injected 5 PV of gas and brine at 80% quality, ΔP = 61.8 kPa
  injected 5 PV of gas and brine at 98% quality, ΔP = 19.6 kPa
Foam Experiments: total injection rate 40 ml/h of gas and brine with surfactant
  injected 10 PV of gas, brine + surfactant + tracer at 95% quality, system overpressured
  injected 40 PV of gas, brine + surfactant at 98% quality
  injected 13 PV of gas, brine + surfactant at 60% quality

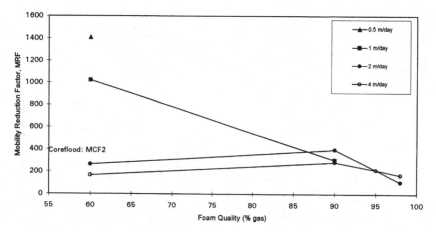

Figure 3.   Variation in MRF with foam flow rate and quality (oil free, short core).

sandstone core. In the long core experiments, following the progress of a sharp pressure-front also monitors the advance of foam generation and foam transport.

The flood history is summarized in Table 3. First baseline pressure drops were determined across the 1.8 m core during simultaneous injection of brine (without surfactant) and gas at a fixed frontal advance rate and varying gas fractional flows. A frontal advance rate of 4.0 m/day was selected to ensure a flow rate higher than the critical rate for effective foam formation and propagation. A non-adsorbing tracer, tritiated water, was added to the brine so that the breakthrough of the tracer could be compared with the breakthrough of the gas.

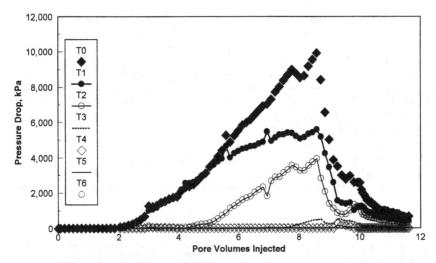

Figure 4. Foam generation at 95% quality in a long, oil free core.

In Figure 4 the pressure build-up during foam generation, for each 30 cm section of core, is followed as a function of total pore volumes injected. Transducer T0 measures the overall pressure drop across the whole core. After an initial delay, approximately 2.5 PV total injection, foam is generated in the front part of the core as registered by T1 (first 30 cm). The foam propagates into the second core section. After 5 PV of total injection foam progress is registered in the second section by transducer T2, and after 7.75 PV pressure starts to build-up in the third section, as registered by T3. After 8.5 PV total injection the absolute pressure in the core started to exceed the limiting pressure of the core holder and the experiment had to be cut short. Steady state foam flow across the core was not achieved. Subsequently, the foam was washed out with brine and the 98% and 60% foam quality floods were conducted. The latter experiments approached steady state conditions across the whole core. The pressure drops obtained during the foam experiments are presented in Figure 5.

Overall, a trend toward increasing pressure drop due to foam with increasing system pressure was observed for all foam qualities (see Figure 5). Exceptions to this trend are core sections T4 to T6 for the 95% quality; foam was not established in these sections since the experiment was aborted at an early stage. Section T6 for the 60% quality case proved to be an exception. This pressure trend agrees with previous results [12] obtained for Flourad FC-751 and Dow XSS-84321.05 foams flowing in Berea sandstone cores. Table 4 shows some comparisons for 90–95% quality nitrogen foams flowing at constant flow rate (4 m/day) at various

Table 4.   Effect of System Pressure on 95%, 4 m/day, Foam ΔP

Flowing Foam ΔP (kPa)	Approximate System Pressure (kPa)					
	100	800	3500	7000	7400	11,300
Chaser GR-1080 (This work)				469	3937	5586
Dow XSS-84321.05 (Reference [11])	69	99	322	206		
Flourad FC-751 (Reference [11])	170	470	760	1120		

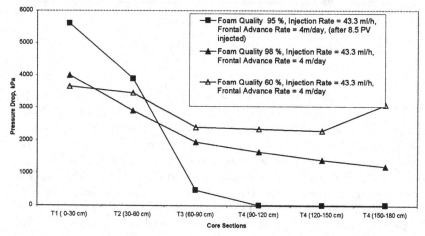

Figure 5.   Foam pressure drops from long core experiment (no oil).

system pressures. The results for the present work represent the pressure drops across sections T3, T2, and T1 and are referenced to the total of the system backpressure plus the accumulated pressure drops in downstream sections of the core.

Calculated mobility reduction factors are also shown in Figure 6. For the 60% foam quality case an experimental baseline pressure drop was not available, so we used the results of the modelling work described in a later section to estimate the pressure drop expected for gas/brine flow at the appropriate fractional flow. Since the 95% quality foam flood did not reach steady state foam flow conditions, and since the pressure drops in individual sections of the long core were influenced by the pressure drops due to foam flowing in downstream sections of the core, we cannot make exact comparison between the MRFs generated in the long core with

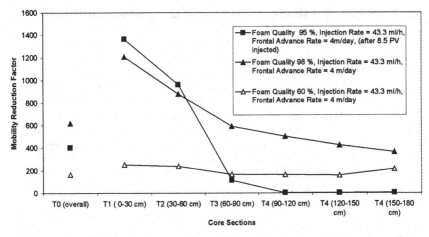

Figure 6. Mobility reduction factors from long core experiment (no oil).

those generated in the short cores studied in the next section. For the two long core experiments, where steady state foam flow was achieved, the MRF values for the final section (T6) of the long core were somewhat higher in comparison to the short core experiments (215 versus 165 at 60% quality, 364 versus 169 at 98% quality). Qualitative agreement between the long and short core MRFs was thus achieved.

Foam Flooding at Residual Oil Saturation. *Short Core Experiments.* The core was waterflooded to residual oil saturation before the foam experiments were conducted. We have already gained experience with the behaviour of the Judy Creek oil from previous research [9, 13]. The residual oil saturation after waterflooding was 24%. Baseline pressure drops, ΔP_n, were measured, at residual oil saturation, during co-injection of water and gas, with no foam present in the core (see Table 5).

First, the 98% foam quality experiments were executed at velocities of 1, 2, and 4 m/day. No significant foaming behaviour was observed. Next, the foam quality was decreased to 60%; no significant foaming behaviour was observed for frontal velocities of 1 and 2 m/day. However, at 4 m/day strong foaming tendencies were observed through a sharp increase in pressure drop. Along with the formation of foam, additional oil was produced. The oil saturation was lowered from $S_{orw} = 24\%$ to $S_{orf} = 13\%$. At the residual oil saturation to foam, S_{orf}, the foam experiments were repeated using qualities from 98% down to 10% with three different frontal advance rates. Table 5 shows the results for ΔP_f, measured in the presence of surfactant. One must differentiate between the foam pressure drops, ΔP_f, evaluated at the two different oil saturations, S_{orw} and S_{orf}. Since the oil saturation decreased after strong foam formation, the

Table 5.   Summary of Short Core Foam Experiments, at Residual Oil Saturation

Foam Quality (% gas)	Injection Rate = 43.3 ml/h, Frontal Advance Rate = 4 m/day		Injection Rate = 21.6 ml/h, Frontal Advance Rate = 2 m/day		Injection Rate = 10.8 ml/h, Frontal Advance Rate = 1 m/day	
	Base ΔP_n (kPa)	Foam ΔP_f (kPa)	Base ΔP_n (kPa)	Foam ΔP_f (kPa)	Base ΔP_n (kPa)	Foam ΔP_f (kPa)
0.98	13.4	45.3	9.2	25.0	6.9	12.7
0.98		32.0		17.3		9.9
0.98		8.8		5.1		3.3
0.95	21.6		11.1		5.4	
0.95		79.2		53.7		36.8
0.8		1168.2		163.5		52.6
0.6	81.6	1728	59.9	113	37.8	55.7
0.6		1446.8		879.6		175.4
0.4		1488.1		1113.2		251.8
0.2		1236.9		1065.1		749.0
0.1		916.6		676.2		451.5

Values in the shaded areas were obtained at $S_{orf} \approx 0.13$

baseline pressure drops evaluated at S_{orw} can not be compared to the foam pressure drops evaluated at S_{orf}. Thus, it is difficult to establish MRFs after the oil saturation has been lowered.

Figure 7 shows the variations in MRF with changing foam flow rate and quality at residual oil saturation. At S_{orw} the MRFs ranged from 1.5 to 3.5; once the oil saturation was lowered to S_{orf} the MRFs increased by an order of magnitude. This MRF data further strengthens the observation that foams are ineffective at higher oil saturations. The effect of residual oil saturation on the effectiveness of Chaser GR-1080 foams can also be compared with previous experience, as long as one also considers the magnitude of the residual oil saturation. Table 6 shows a comparison for 95% quality foams flowing at 4 m/day. Other work (see Schramm [14]) suggests that oil sensitive foams can be relatively effective in porous media as long as the residual oil saturation is below some critical value, which appears to generally lie in the range 10 to 20%. This is consistent with the results [11, 15] brought together in Table 6.

Long Core Experiments.   Experiments designed to match those in the short core, at residual oil, were also conducted in the long core apparatus using a 1.8 m long Berea core. The flood history and baseline pressure drops are summarized in Table 7. During the 60% baseline flood the residual oil saturation was reduced slightly, from 35% to 33%.

Table 6.   Effect of Residual Oil Saturation on Foam MRF

GR-1080 Flowing Foam (95% quality at 4 m/day)	Residual Oil Saturation (%)	MRF Oil Free	MRF at S_{orf}
Judy Creek & Berea Core (This work)	13–14	215	≈ 5
Oseberg & Berea Core15	13–14	671	21
Keg River & Carbonate Core11	$\ll 13$	8–18	27–31

Table 7.   Core Data and Flood History for the Long Core, Residual Oil Experiments

Injection History:
    saturated core with brine
    flooded with oil to S_{wc}
    flooded with brine to S_{orw}
Base Line Experiment: total injection rate 42.6 ml/h of gas and brine
    injected 5 PV of gas and brine at 98% quality, $\Delta P = 220$ kPa
    injected 3 PV of gas and brine at 60% quality, $\Delta P = 2407$ kPa
    injected 2 PV of gas and brine at 95% quality, $\Delta P = 593$ kPa
Foam Experiments: total injection rate 42.6 ml/h of gas and brine with surfactant
    injected 27 PV of gas, brine + surfactant at 98% quality
    injected 3 PV of gas, brine + surfactant at 60% quality, system overpressured
    injected 1 PV of gas, brine + surfactant at 60% quality, reduced rate to 2 m/day
    injected 3 PV of gas, brine + surfactant at 60% quality, reduced rate to 1 m/day
    injected 56 PV of gas, brine + surfactant at 95% quality, restored rate to 4 m/day

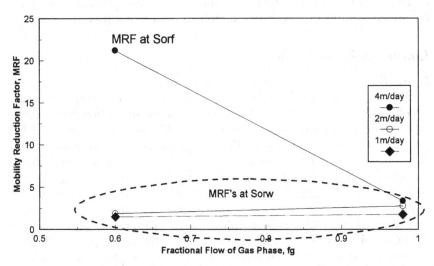

Figure 7.   Variation in MRF with foam flow rate and quality at S_{or}.

The long core foam experiments were started at a 98% foam quality and yielded very low pressure drops across the core. The full core pressure drop was 249 kPa, only 30 kPa higher than the baseline pressure drop measured under similar flow conditions. This difference is considered insignificant (yielding an MRF of 1.1) and it was judged that no effective foam was formed at a quality of 98%. Similarly, in the short core the foam effectiveness in the presence of an oil saturation of 24% was considered negligible; as indicated by the low mobility reduction factor of 3. The residual oil saturation in the long core was 33% resulting in the lower MRF for the long core.

The next foam experiments were conducted at 60% quality. In this case, unstable pressure drops were observed over the first two segments of the core (several MPa per section), a very large pressure drop was noted in the third segment of the core (over 7 MPa), and diminishing and unstable pressure drops were found in the final sections of the core. By reducing the flow rate, flooding was able to continue, some oil having a waxy appearance was produced, and the residual oil saturation was reduced to about 32%. It appears that oil was mobilized in the front part of the core and that in some fashion it was involved in the increased resistance to flow experienced in the middle segment of the core. Since reducing the flow rates did not solve the problem, the foam was washed out with brine.

The final foam experiments were conducted at 95% quality. In this case reasonably stable pressure drops were achieved across the first two segments of the core (T1: 234 kPa and T2: 91 kPa) and increasing pressure drops were measured across the remaining segments of the core (ca. 4.8 MPa). Under these conditions, foam was apparently formed, but steady state foam flow across the full core was not achieved.

Empirical Foam Modelling

Due to the extensive research that has been conducted in the area of foam application in enhanced oil recovery, simulation of foam behaviour has become more feasible. Several methods of foam simulation have been developed: population balance models [16, 17], fractional flow models [18, 19], and models that alter the gas phase permeabilities [20, 21]. Although the population balance models treat the foam generation mechanisms in a detailed fashion, they may be impractical to apply on large field scale simulations. Both the fractional flow model and the models that alter the gas phase permeabilities rely on history matching experimental data. The fractional flow model provides insight into one-dimensional foam flow, but it may be more difficult to apply in three-dimensional situations. In the following section, the application of relative permeability alterations to model foam flow is investigated.

The empirical approach to modelling foam behaviour in porous media is based primarily on laboratory observations. In addition, it makes several simplifying assumptions. The first of these is that the time involved in both the generation and destruction of lamellae is negligible compared to the time scale employed in the simulation. Secondly, it is assumed that whenever gas and aqueous surfactant coexist at a given point in the medium, foam exists, provided all other conditions are suitable for the formation of foam. It is generally assumed that foam alters only the relative permeability of the gas phase while the relative permeability of the aqueous phase remains unaltered. In addition, the amount of water that is needed to generate the foam is considered negligible. Foam effects are therefore modeled using an interpolation scheme designed to extend gas phase mobility data in case of foam formation.

It is difficult to distinguish between foam affecting the relative permeability or the viscosity of the gas phase. Thus, the preference of the individual dictates whether foam effects are attributed to the relative permeability or the viscosity. In this chapter, the effects due to foam are attributed to the relative permeability.

Fractional Flow Model at "Limiting Capillary Pressure".
Rossen et al. [*18, 19, 22*] applied fractional flow theory to foam processes. At the center of their application lies the concept of "limiting capillary pressure", P_c^*. Experimental evidence shows that the higher the capillary pressure the more unstable the foam. If P_c^* is surpassed rapid bubble coalescence destroys the foam as capillary suction withdraws the water out of the foam lamellae. In two phase flow (water and gas), the capillary pressure is related directly to the water saturation. For water wet porous media, as the water saturation increases the capillary pressure decreases. Therefore, P_c^* can be related to a limiting water saturation, S_w^*. Once the water saturation decreases below the limiting water saturation, existing foam should rapidly destabilize.

It is assumed that the foam remains at the limiting capillary pressure independently of pressure gradient and gas and liquid flow rates. This implies that the water saturation remains at S_w^* over a range of water fractional flows (approximately $0 < \mathrm{fw} < 0.2$). The equations of multiphase flow can be manipulated to yield expressions for the pressure gradient and gas mobility when the water saturation equals S_w^*.

$$\nabla P = \frac{u_w}{\lambda_w^*} \quad \text{where} \quad \lambda_w^* = \frac{K k_{rwg}(S_w^*)}{\mu_w} \tag{2}$$

$$\lambda_g = \lambda_w^* \frac{\mathrm{fg}}{\mathrm{fw}} = \lambda_w^* \frac{(1 - \mathrm{fw})}{\mathrm{fw}} \quad \text{when} \quad S_w = S_w^* \tag{3}$$

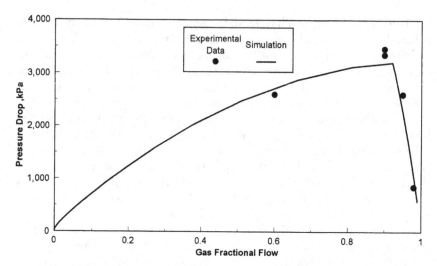

Figure 8. Foam pressure drop using the limiting capillary pressure concept (FAR = 4 m/day).

The remainder of the fractional flow curve, $S_w > S_w^*$ is constructed to match the experimental data.

In Figure 8, the experimental results from the (4 m/day frontal advance rate, oil free) short core flood are compared to the simulated pressure drops which were based on the limiting capillary pressure principle. In this particular case S_w^* was chosen at 0.35 over a range of water fractional flows from 0.01 to 0.15 to closely match the experimental data. For $S_w > S_w^*$, a fractional flow curve was chosen which matched the experimental data closely by appropriately adjusting the gas phase relative permeability curve. The water relative permeability curve remains the same as defined in the Appendix under gas/water relative permeabilities. The composite foam fractional flow curve can be seen in Figure 9. Notice the vertical section in the curve for the foam flow case lies at $S_w^* = 0.35$.

The close match between experimental and simulated data does not continue when the same fractional flow curve is used to simulate the experimental pressure drop results at a slower frontal advance rate (2 m/day, oil free). A new fractional flow curve had to be constructed to give a closer match. In Figure 10 the experimental pressure drops are compared to the simulated curves and in Figure 9 the contrast between the new and old fractional flow curves is made clear. Due to the shear thinning nature of the foam, at slower frontal advance rates a steeper fractional flow curve is required at the same critical water saturation, $S_w^* = 0.35$.

STARS Foam Interpolation. STARS [21] is a commercially available reservoir simulator created by the Computer Modelling Group,

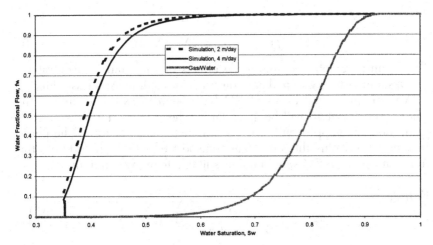

Figure 9. Shear thinning effect on water fractional flow under foaming conditions.

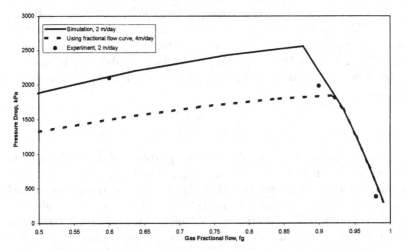

Figure 10. Foam pressure drop using the limiting capillary pressure concept (FAR = 2 m/day).

CMG, in Calgary, Canada. Since STARS was already equipped with a foam interpolation parameter, its functional form was explored through history matching short and long core experiments.

A dimensionless foam parameter, FM, can be formulated to adjust the gas phase relative permeability to foaming conditions.

$$k_{rgf} = k_{rgw} \, FM \tag{4}$$

$$FM = f(C_s, S_o, S_g, S_w, N_{cw}, N_{cg}, P) \tag{5}$$

This foam parameter combines the effects of surfactant concentration, phase saturations, water and gas flow rates into a dimensionless value.

Various functional forms of FM have been investigated during this study; explicit details will be given later. The primary advantage of this foam modelling approach lies in its simplicity; use is made of the phase flow equations plus the surfactant transport equation. In addition, the functional form of FM should mimic suitable foaming conditions.

$$FM_{CMG} = \cfrac{1}{1 + FF_{max} \left(\dfrac{C_s}{C_{smax}}\right)^{es} \left(\dfrac{S_{omax} - S_o}{S_{omax}}\right)^{eo} \left(\dfrac{N_{cref}}{N_{cp}}\right)^{ev} \left(\dfrac{N_{cp} - G_{cref}}{G_{cref}}\right)^{en}} \tag{6}$$

where: C_s = surfactant concentration, S_o = oil saturation, N_{cp} = pressure based capillary number, N_{cref} = reference capillary number for shear thinning, G_{cref} = critical foam formation capillary number.

The foam factor FF_{max} is a scaling factor that weighs the overall foam effects. FF_{max} is related to the foam mobility reduction factor "MRF", but there exists no straightforward correlation between the two factors. In the STARS formulation, the single capillary number N_{cp} is based on the local pressure drop, and the length over which the pressure drop is effective:

$$N_{cp} = \frac{K \, \Delta P}{\sigma \, \Delta x} \tag{7}$$

The reference capillary numbers for the purposes of determining shear thinning and generation effects are N_{cref} and G_{cref} respectively. The various exponents are used to weigh the relative contributions of each mechanism. The parameter FM predicts the behaviour of foam after the various experimentally observable coefficients in equation 6 have been appropriately weighted. To ascertain true foam performance, it is important to possess an understanding of flow behaviour in the porous medium in the absence of foam, most notably three phase relative permeability behaviour and the transport properties of the surfactant (adsorption).

As illustrated by equation 6 the value of the dimensionless interpolation factor, FM, depends on several dimensionless ratios. The first of these is the surfactant concentration term. Experimental observations suggest that foam often becomes more effective with increasing surfactant concentration. Clearly, as the concentration of surfactant in the system approaches C_{smax}, the interpolation factor will more closely approximate the inverse maximum mobility reduction factor. If the exponent es is assigned a positive value, decreasing the surfactant concentration has a detrimental effect on foam behaviour as evidenced by an increase in FM. A negative exponent will cause an increase in foam effectiveness as the surfactant concentration is decreased. Switching the sign of the exponent could prove useful in the situation where the perceived strength of a foam passes through a maximum at a given surfactant concentration and foam quality.

The next term of interest is the oil saturation term. The presence of oil is known to destabilize some foams. This is exemplified by the oil saturation term, which becomes smaller as the saturation of the oil phase approaches a given maximum value, S_{omax}. Should the maximum saturation be equaled or exceeded by the actual saturation of the oil phase, foam will be unable to form. The exponent eo will always be positive, reflecting the fact that greater oil saturations impede the formation of foam.

The capillary number of the system at experimental conditions affects the performance of foam in two ways. A shear thinning effect is modeled by the third dimensionless term in equation 6. As the capillary number is increased relative to the reference capillary number of the experiment, it can lead to increased shear thinning and degradation of foam. The exponent ev is therefore positive in most cases. A negative exponent would indicate shear-thickening behaviour. Secondly, a critical capillary number needs to be exceeded, below which foam will not form. Foam will first appear when this capillary number of generation is surpassed, and as the capillary number continues to increase beyond the critical value, the foam will become stronger. The fourth dimensionless term in equation 6 models this behaviour. The exponent en is generally positive, though a negative exponent can indicate an increasing instability in the foam as the capillary number increases. In most instances, where both ev and en are positive, increasing the velocity of the experiment will act both to stabilize the foam, as the generation capillary number is further exceeded, and reduce its effectiveness, as shear thinning effects come into play.

Comparing the experimental pressure drop at steady state (4 m/day frontal advance rate, oil free, short core) to the simulated pressure drop indicates the weakness of STARS foam correlation (see Figure 11). At high gas fractional flow (fg > 95%) STARS predicts very high pressure drops. Yet the experimental results show a weakening in the foam, i.e. lower pressure drops at high gas fractional flow. The simulated results in Figure 11 were obtained by setting FF_{max} to 500 and setting the exponents (es, eo, en) equal to zero (Table 8). This effectively shuts down any impact of the variables C_s, S_o. During this particular short core experiment the surfactant concentration was not varied, so the effectiveness of the surfactant concentration was not probed; also, no oil was present during this experiment. Only the capillary number N_{cp} varied

Table 8.   Correlation Parameters used in the STARS Foam Simulation

Relative Permeability Curves		Foam Correlation Parameters	
S_{wrg}	0.3	FF_{max}	500
S_{grw}	0.07	N_{cref}	0.05
k_{rwg}^{o}	0.5	ev	0.4
k_{rgw}^{o}	0.2	es	0
z_{wg}	3.2	eo	0
z_{gw}	2.2	eg	0

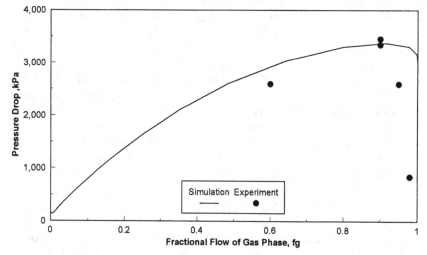

Figure 11.   Comparison of experimental to simulated pressure drop using STARS.

during the course of the oil free experiments. Shear thinning behaviour was modeled by setting the parameter ev equal to 0.52.

A limiting capillary pressure concept, i.e. critical water saturation (introduced in the previous section), can also be applied in the STARS simulator, through the input of a composite foam/no-foam relative permeability curve. If the water saturation is below S_w^*, then no foam exists, and the gas/water relative permeability curves are followed. On the other hand, if S_w is greater than S_w^*, then foam is formed and the foam correlation takes effect. A composite gas relative permeability curve is shown in Figure 12; the water relative permeability curve remains unchanged. In this composite gas permeability curve, S_w^* was set equal to 0.35. Furthermore, the foam section of the relative permeability curve equaled the gas relative permeability curve divided by 2.0×10^4. Figure 13 shows the corresponding foam simulation in comparison to the

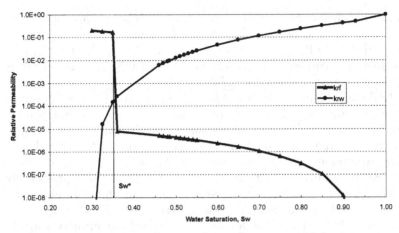

Figure 12. Composite foam/no-foam relative permeability curves.

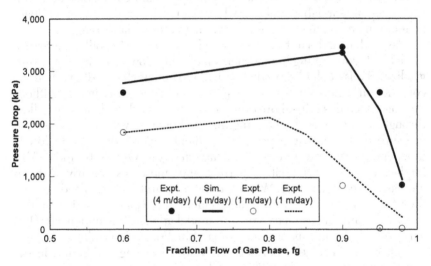

Figure 13. STARS simulation with limiting capillary pressure concept.

experimental data. Two separate experiments are presented, one with a frontal advance rate of 4 m/day, and one with an advance rate of 1 m/day. An excellent match was achieved between simulated and experimental results for the 4 m/day case. Applying the critical water saturation tends to correct the problem of overpredicting the pressure drop in the high gas fractional flow area (compare Figure 11 and Figure 13). For the 1 m/day case, the simulated pressure drops are greater in comparison to the experimental values, in the high gas fractional flow region. Shifting S_w^* to a larger value on the gas relative permeability curve may lead to a better history match for the 1 m/day experiment. The foam correlation parameters for the last two simulations are the same as shown in Table 8, except that the FF_{max} factor was set equal to 1000, in the 1 m/day case. The foam experiment with an advance rate of 0.5 m/day could not be modeled because of simulator stability problems.

In the experiments conducted here, the pressure drop was always the dependent variable, dependent on foam formation. The independent variables were the flow rates of gas and water; they are easier to control than running the experiments at fixed pressure drops. It seems that for these experiments the capillary number N_{cp} is a rather poor choice of variable to predict foam generation. Conversely, in field projects the injection is often controlled by pressure limitations, thus for field studies the N_{cp} variable may be more suitable. de Vries and Wit [23] concluded in their modelling efforts that for foam modelling the primary variables should be changed from pressure and saturation to gas and water velocities. It should be possible to develop a foam correlation based on the independent lab variables, which can predict the onset of foam generation and foam effectiveness over a greater variable range.

Vassenden and Holt [24] extended the limiting capillary pressure model by incorporating the mechanism of foam flow at a critical pressure gradient. Falls et al. [25] predicted that a foam lamella, positioned near a constriction, could support a pressure drop (up to a maximum) without flowing. Once this critical pressure drop is surpassed the lamella will flow through the pores, and increasing the gas flow rate will not increase the pressure drop. Thus, the pressure gradient generated by foam flow is independent of flow rate and not affected by water saturation. This implies that the gas relative permeability increases linearly with an increase in foam flow rate. When the water saturation is reduced near S_w^* then the lamellae may rupture due to increased capillary suction. In this region, the limiting capillary pressure model applies. Vassenden and Holt constructed a convenient exponential function, which allows for the reduction in gas relative permeability under foaming conditions, incorporating the notions of limiting capillary pressure and critical pressure gradient.

$$
FM = \begin{cases} e^{(S_w^* - S_w)s_1} + \left(\dfrac{u_g}{u_{go}}\right)^{ev} F_o e^{(S_w^* - S_w)s_2} & \text{for } S_w > S_w^* \\ 1 & \text{for } S_w < S_w^* \end{cases} \tag{8}
$$

Only when the water saturation exceeds the critical water saturation will foam be allowed to form. In the critical capillary pressure regime the gas relative permeability is reduced exponentially with the slope s_1; in the critical pressure gradient region, the gas relative permeability is reduced by a factor of $(u_g/u_{go})^{ev}F_o$. This simple model was applied successfully to simulate the foam experiments in the short core without oil present. In Figure 14, pressure drops are simulated using the Vassenden and Holt model. An excellent match between simulated and experimental pressure drops is obtained for all of the frontal advance rates (FAR = 0.5, 1, 2, 4 m/day). The corresponding fractional flow curves in Figure 15 demonstrate the shear thinning behaviour of the foam in the critical pressure gradient region. The parameters for the foam model are presented in

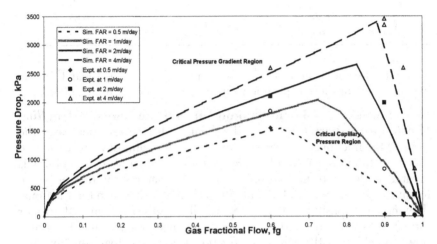

Figure 14.   Pressure drops simulated with critical pressure gradient and capillary pressure models.

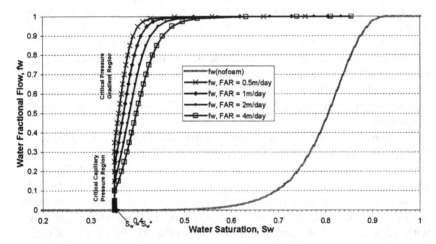

Figure 15.   Fractional flows simulated with critical pressure gradient and capillary pressure models.

Table 9.   Correlation Parameters Used in the Critical Pressure Gradient and Capillary Pressure Model (no oil present)

Relative Permeability Curves		Foam Correlation Parameters	
S_{orf}	0.13	Fo	2.1×10^{-4}
S_{wrg}	0.3	u_{go}, m/day	4
S_{grw}	0.02	S_w^*	0.35
k_{rwg}^o	0.5	s_1	50,000
k_{rgw}^o	0.2	s_2	-10
z_w	3.3	ev	0.65
z_g	2.3		

Table 9. Vassenden and Holt used a value of 1 for the shear thinning exponent ev; for these experiments ev equal to 0.65 was more suitable.

Modelling of Foam Flooding in Oil Free Cores.   *Modelling Short Core Experiments.*   As the frontal advance rate decreases the slope of the fractional flow curves approaches 90° (almost vertical); due to the shear thinning nature of the foam, the reduction of foam mobility is greater at slower frontal advance rates. The water saturation at which foaming occurs was set at 35%, identical to the S_w^* chosen in the fractional flow model. Typically, in the critical capillary pressure regime the water saturation is virtually constant over a large range of foam qualities. Foam flow, coinciding with high water fractional flow (low foam quality), has received limited attention in the literature. Additional experimental foam data should be gathered in this region to fill out the picture.

Experimental evidence [35] suggests that, in the presence of foam, the irreducible phase saturations of water and oil (S_{wrg}, S_{org}) can be lowered significantly. Thus, the irreducible water saturation, S_{wrg}, can actually be lower in the presence of foam than the measured value of 46%, determined from the gas/water relative permeability experiment (Appendix). For all of the foam simulations, an irreducible water saturation of 30% was chosen. Since the in-situ water saturation was not measured, the critical water saturation value S_w^* (set to 0.35) could not be determined accurately.

In Figure 16 the MRF data was constructed from the simulations. In the high foam quality region (high gas fractional flow), the MRF for foams propagated at various frontal advance rates is nearly identical. Thus, the shear thinning nature of foams seems to become effective only when the foam quality decreases below 95%, in the critical pressure gradient region. The shear thinning behaviour dictates that the MRFs for foams with a low FAR are higher in comparison to the MRFs for foams with a high FAR. Qualitatively, the MRFs agree with the experimental data (compare Figure 3 and Figure 16).

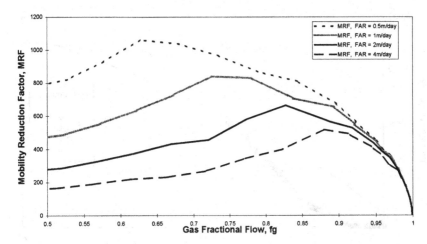

Figure 16.   Constructing MRF data using the critical pressure gradient and capillary pressure models.

Modelling Long Core Experiments.   The long core, initially saturated with brine, was foam flooded at a 95% foam quality. As discussed in the experimental section, the 95% foam flood could not continue until a steady state was reached, since the generated pressure drop exceeded the safety margin of the core apparatus.

The simulations for this long core experiment, at 95% quality, focused on testing the validity of instantaneous foam generation assumed in STARS, provided sufficient surfactant and gas co-exist. Therefore, the foam front in the STARS model advances as the surfactant front advances, taking into account surfactant adsorption.

The same input parameters as for the short core simulations were used for the long core simulations. In Figure 17, the experimental overall pressure drop is compared to the simulated pressure drop. The simulated pressure drop increases with the advancing of the surfactant front. During the simulation, as soon as surfactant is injected into the core, and adsorption has been satisfied, foam is generated at the injection front. After 20 PV of total injection, i.e. 1 PV of surfactant injection, the pressure drop levels out because surfactant has been transported through the length of the core and in the simulation foam formed everywhere. In the experiment, the actual foam generation lags behind by 2.5 PV of total injection. At approximately 8.5 PV injection the simulated pressure drop and the experimental pressure drop seem to match. This match is achieved erroneously, because the simulation underestimates the pressure gradient generated by the foam.

After the foam was partially broken, the foam quality was changed to 98%. At this quality, foam could be set up for the whole length of the core

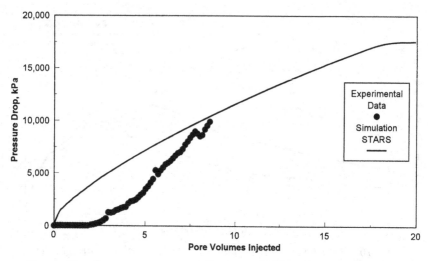

Figure 17.   Foam generation in a long core (oil free, foam quality 95%).

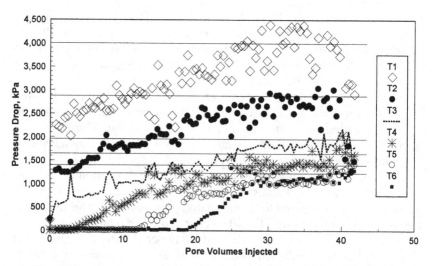

Figure 18.   Foam transport in a long core (oil free, foam quality 98%).

without overpressuring the system. Steady state foam flow was obtained after injecting 35 PV. Figure 18 shows the build-up of foam over the whole length of the core, where the horizontal lines indicate the steady state pressure drops for the different sections. To simulate the steady state pressure drops across the 30 cm long sections (T1 to T6), the Vassenden and Holt model was used with the same foam parameters as obtained for the short core. The compressibility of the gas and its effect on the foam quality was taken into account for the long core simulation. Section T6 was modeled first since the backpressure of the system was known and the specified experimental foam quality was set at backpressure conditions. Using the model with the specified parameters, the foam pressure drop for section T6 was calculated. Knowing the foam pressure drop for section T6 specifies the end condition of section T5: ambient pressure, foam quality, frontal advance rate. As the ambient section pressure increases the foam quality is reduced (due to the gas compressibility) and so is the frontal advance rate. A new foam correlation was established for each section. Sequentially, the pressure drop is then calculated for all the sections (from T6 to T1) and compared to the experimental data (see Figure 19). Excellent agreement is obtained between experimental and simulated pressure drops for the last section of core; larger deviations are apparent near the front of the core

Sanchez and Schechter [26] demonstrated that the local pressure and surface tension affect the rate of lamellae generation. The higher the ambient pressure and the lower the surface tension the faster the lamellae snap off process proceeds, thus generating a denser foam texture. Holt et al. [27] presented experimental results, which showed that the surface tension of a C_{16} AOS surfactant is significantly lowered when the ambient pressure is increased. Thus, the combined effect of pressure and lower surface tension generates a finer foam texture resulting in increased pressure gradients. A competing process is also effective; displacing newly formed lamellae out of the constriction at extremely rapid rates will cause the lamellae to collapse, reducing density of the foam texture. An

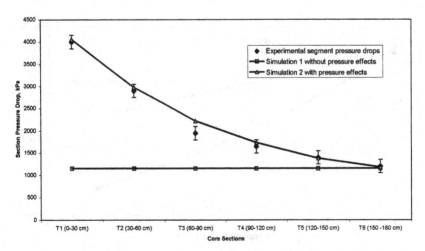

Figure 19.   Core section pressure simulations with and without ambient pressure effect at high foam quality, 98%.

optimum ambient pressure should exist at which the foam generates a maximum resistance to flow. For the long core experiment discussed here this optimum pressure was not surpassed, the mobility reduction factor kept increasing with increasing pressure. Vassenden and Holt [24] explained the effect of system pressure in a different way. They assumed that the critical capillary pressure would remain independent of system pressure, however, the critical water saturation S_w^* would decrease as the surface tension was lowered under increasing system pressure. For our simulation purposes we assumed that both mechanisms were in effect:

(i) the critical water saturation is reduced with ambient pressure logarithmically:

$$S_w^* = 0.3471 - 0.0213 \ln\left(\frac{P}{P_o}\right) \tag{9}$$

(ii) the foam quality changes with ambient pressure, such that F_o decreases linearly for the range of test conditions investigated here:

$$F_o(P) = -6.000E - 05(P/P_o) + 2.751E - 04 \tag{10}$$

A lower value in F_o results in lower foam mobility, i.e. increased pressure gradients during foam flow. Here P_o serves as the reference pressure for F_o, set at 6894 kPa. The overall change in the critical water saturation and the foam mobility factor are presented in Figure 20. Since the ambient pressure in each core section changes, so does the critical water saturation S_w^*, the foam mobility multiplier F_o, the FAR, and the foam quality. A family of pressure drop vs. foam quality curves was established, for each

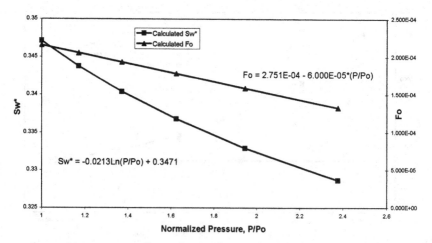

Figure 20.   Pressure dependence of critical water saturation and foam mobility multiplier.

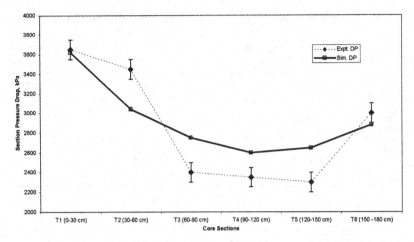

Figure 21. Long core section pressure simulations at low foam quality, 60%.

core section. With the implementation of the ambient pressure effects, it is possible to obtain an extremely close match between experimental and simulated values (see Figure 19). Due to the gas compressibility, the FAR drops from 4 m/day at the end of the core to 1.7 m/day at the front of the core. This corresponds to a change in foam quality from 98% (at the end of the long core) to 95% (at the front of the long core). The pressure correction term may need additional refinement since the pressure for maximum mobility reduction was not reached. The numerical data was calibrated in the region of high foam quality, which also corresponds to the critical capillary pressure regime of foam flow.

The final long core foam injection experiment (without oil) was carried out at a foam quality of 60%. The experimental pressure drops measured across each segment are presented in Figure 21 and compared to the simulated pressure drops. The previously determined functions for S_w^* and F_o, equations 9 and 10, were also used for this simulation. Working backwards from sections T6, T5, and T4, the segmental pressure drop decreases because the frontal advance rate and foam quality decrease, due to increased compression of the gas (ambient pressure increases from back towards front of the core). As the ambient pressure increases, the foam texture becomes finer, shifting the foam mobility multiplier to lower values. In sections T3, T2, and T1 we assume that the change in foam texture is responsible for the observed increase in segmental pressure drop.

Modelling of Foam Flooding at Residual Oil. *Modelling Short Core Experiments.* Further simulations were carried out to

Table 10.   Correlation Parameters Used in the Critical Pressure Gradient and Capillary Pressure Model (oil present)

Relative Permeability Curves		Foam Correlation Parameters	
S_{orf}	0.13	Fo	5.0×10^{-4}
S_{wrg}	0.3	u_{go}, m/day	4
S_{grw}	0.02	S_w^*	0.355
k_{rwg}^o	0.5	s_1	350
k_{rgw}^o	0.2	s_2	0
z_w	3.3	ev	0.65
z_g	2.3		

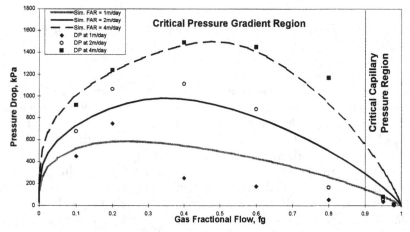

Figure 22.   Matching foam pressure drop data at S_{orf}.

match the experimental foam flood results in the presence of residual oil using the Vassenden and Holt model. In Figure 22, the simulated pressure drop generated for a range of foam qualities is superimposed onto the experimental results; each curve represents a different frontal advance rate. A much closer match between experimental pressure drop and simulated pressure drop can be obtained for a singular advance rate. However, the foam correlation parameters were chosen such that an overall agreeable fit could be obtained. The following simulations are evaluated at a constant oil saturation of 13% with the optimized foam correlation parameters noted in Table 10. The most notable difference in the model parameters between the foam case with oil and the foam case without oil is the value of the exponent s_1. As described earlier, the value s_1 dictates the gas permeability reduction in the critical capillary pressure gradient region. If s_1 is large, the gas relative permeability is reduced dramatically by foam over a small saturation change, near the critical water saturation, S_w^*. With residual oil present, s_1 is relatively small (<1000) and the critical capillary pressure gradient region is defined by a much larger saturation change. This is evident in the corresponding fractional flows curves presented in Figure 23. The critical capillary pressure region ranges from a water saturation of 0.355 to 0.373 with respective foam qualities above 90%. Again, the slower frontal velocities demonstrate a steeper fractional flow curve in the critical pressure gradient region due to the shear thinning nature of the foam. The mobility reduction factors are easily constructed using the simulated foam and base case pressure drops, see Figure 24. In the critical capillary pressure region, the simulated MRF values are equal, while in the critical pressure gradient region the difference in MRF arises due to the shear thinning nature of the foam.

Discussion

Most laboratory investigations of improved oil recovery processes involve coreflood experiments using short core lengths, usually on the order of 9 to 20 cm. There are very few accounts of work involving significantly longer core lengths that also include comparisons with the results from conventional short core lengths. In earlier work [28], we studied the effect of increasing the core length on a chemical flooding process. Increasing the scale by a factor of 4.5 (from 9 cm to 41 cm cores) decreased the amount of chemical dispersion (by increasing the Péclet number) but had

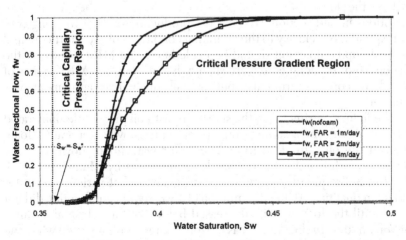

Figure 23. Water fractional flow under foaming conditions at S_{or}.

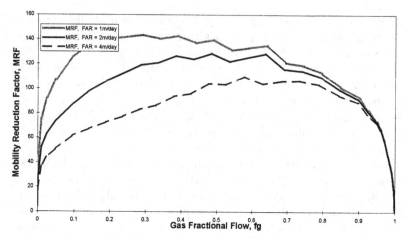

Figure 24.   Simulated MRF data using mechanistic approach (at S_{or}).

no significant effect on tertiary oil recovery. Chung and Hudgins [29], however, found that in foam flooding there seemed to be an increased randomness, or poor reproducibility, for foam flow with increasing distance along a 10 m packed-bed slim tube. In their work foam effectiveness and reproducibility were good near the foam injection end of the slim tube, but were poor and non-reproducible near the outlet end.

In this project, the experimental scale-up work for the oil free case showed a good reproducibility between the short and long core experiments. However, it is important to compare those sections of the long core to the short core, which are operating under similar conditions. Under foaming conditions, the increased pressures experienced in the upstream part of the long core definitely influence the foam behaviour. The conditions in end section T6 (near the production end) of the long core correspond more closely to the operating conditions of the short core and any scale-up comparison should be made here. The MRF values for the section T6 of the long core were slightly higher than the MRF values of the short core experiments, but overall qualitative agreement was achieved.

The foam experiments in the short core with oil present demonstrated that no significant foaming took place until slowly generated foam reduced the oil saturation from 24% (after waterflood) down to 13%. Similar experimental evidence has been recorded by Mannhardt et al. [11, 15]. Subsequently, strong foam was generated at various flow rates and foam qualities. At an oil saturation of 13%, significant foaming did not occur until the foam quality decreased below 98% at a frontal advance rate of 4 m/day. In the oil free case, strong foams were observed when the foam quality was equal to or below 98%. The maximum foam pressure drops generated in the presence of oil (short core) were consistently lower than the foam pressure drops measured without oil present.

During the waterflooding stage of the long core experiments, the oil saturation was only reduced to 33%. Upon co-injection of surfactant solution and gas, some of the oil was mobilized, similarly as in the short core. Whereas the foam/oil emulsion was produced without significant problem in the short core experiment, during the long core experiment this emulsion generated a blockage inside the long core. This blockage could not be displaced under the pressure limitations of the long core apparatus. Although the long core foam tests at residual oil had to be abandoned, the obtained results are still significant. A strong blocking emulsion can be formed when co-injecting gas and surfactant solution into a porous medium with a limited amount of mobile oil. If this emulsion needs to be transported a significant distance (in our case 2 m), then the medium can be completely blocked. This effect may be desirable or undesirable but it should not be ignored. Furthermore, this effect was not observed during the short core experiment.

Three foam models were investigated in the course of this project. All three models relied on modifying the gas relative permeability in the presence of foam. The foam model by Vassenden and Holt [24] was the most versatile platform to match steady state foam results at various frontal advance rates and foam qualities. With this steady state foam model, it was possible to history match the foaming behaviour investigated on the long and short cores.

As other investigations had proven previously, the critical capillary pressure concept is effective in modelling foam transport at high foam qualities, greater than 90% for the system investigated here. Without oil present, the critical capillary pressure region was defined by a singular saturation. However, with residual oil present, the same model suggested that a band of critical saturations was active, $0.355 < S_w^* < 0.372$. Additional foam experiments in the presence of residual oil need to be conducted to ascertain if this band of critical water saturations is a modelling artifact or reality.

In the critical pressure gradient regime (foam qualities <90%) two types of pressure responses during foam propagation were observed. The pressure drop developed by foam transport can be monotonically increasing with an increase in foam quality, as was the case in the oil free system (see Figure 14). With residual oil present, the pressure drop was first increasing, reached a maximum, and then decreased with an increase in foam quality (see Figure 22).

The experimental evidence from the long core floods indicates that ambient pressure effects on foam behaviour need to be taken into account. Two additional pressure correlations were established (no oil present) specific to the foam system investigated:

1. the critical water saturation, S_w^*, is lowered with increasing ambient pressure
2. the foam mobility multiplier, F_o, is decreased with increasing ambient pressure

Thus, it was possible to history match the pressure responses obtained from the long core foam floods, which were operated in the high quality (critical capillary pressure region) and in the low quality (critical pressure gradient region) foam regions. Overall, the foam correlation used in this project achieved good agreement between experimental and simulated results. Furthermore, the model seemed to be applicable for both cases with and without oil present. A population balance model to predict foam behaviour may be more fundamentally correct [16], especially when investigating transient foam effects, however it also requires a tremendous amount of additional computation time. Modelling and history matching steady state foam results with relative permeability modifications of the gas phase, provides a convenient and efficient tool to investigate foam transport.

Recommendations

Future work should focus on foam floods in the presence of oil, monitoring in-situ saturations of all three phases: gas, water, and oil. Currently we are imaging foam floods using nuclear magnetic resonance techniques in order to elucidate these points.

Acknowledgments

The authors are grateful to Monty Hans for his assistance with the experimental work. Furthermore, the authors thank Roy Woo for his work in creating the data acquisition system on the long core flood apparatus and the real-time hardware/software used in the scintillation counter data collection.

Appendix

Core Data.   A detailed list of core properties is presented in Table 11.

Surfactant Adsorption.   Surfactant adsorption was measured by flooding one core under the same conditions as used in the MRF measurements.

Measured amounts of Chaser GR-1080 surfactant (5.006 g/l) and tritiated water in a solution of 2.1% TDS brine were injected into a

Table 11. Core Data

Core ID	MCF1	MCF2	MCF4	MCF6	MCF5
Rock Type	Berea	Berea	Berea	Berea	Berea
Dry Weight (g)	450.0	451.0	4426.4	449.3	4398.4
Length (cm)	20.0	19.8	183.0	19.9	182.90
Diameter (cm)	3.76	3.77	3.87	3.76	3.849
Area (cm^2)	11.1	11.1	11.8	11.1	11.63
Bulk Volume (cm^3)	222	220	2153	221	2128
Porosity (%)	23.1	23.3	20.5	23.7	22.0
Pore Volume (cm^3)	51.1	51.3	441.3	52.2	468.0
Abs. Permeability: Air (mD)	1006	998	1158	916	1226
Abs. Permeability: Brine (mD)	646	259	297	251	241
Oil Perm. at S_{wc} (mD)				169	138
Brine Perm. at S_{or} (mD)				34	84
S_{or} (%)				24	35
S_{orf} (%)				13	

brine-saturated core at a temperature of 23 °C and a pressure of 6.89 MPa (1000 psig), and effluent concentrations were determined. Figure 25 shows the effluent profiles. A material balance from the effluent profiles yielded the amount of surfactant adsorption at 0.23 mg/g (based on the leading edge only) and the tracer balance was closed within 1.6%. The critical micelle concentration, needed for the adsorption simulation, of the surfactant in 2.1% TDS brine was determined by surface tension titration. The interpolated cmc was 0.084 g/l at 23 °oC under atmospheric pressure. The core flood results have been simulated using a PRI adsorption model and methods illustrated in reference [10], to determine the adsorption isotherm.

The simulation resulted in the following adsorption model parameters:

S	14,000
m_1 (mg/g)	1.9
m_1/m_2	8.6
k_a (hr^{-1})	0.15
k_d (hr^{-1})	0.001
D/v(tracer) (cm)	0.15
D/v(surfactant) (cm)	0.15

The calculated adsorption isotherm is shown in Figure 26. The plateau adsorption determined from the calculated adsorption isotherm was 0.23 mg/g (mass adsorbed per unit mass of rock) or 0.83 μmol/m^2 (moles adsorbed per unit rock surface area).

Water/Oil Relative Permeabilities. For simple two phase flow, the relative permeability is a function of saturation only. The relative

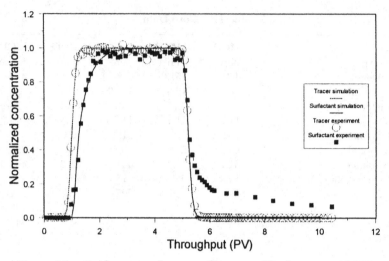

Figure 25.   Surfactant and tracer effluent profiles from core MCF1.

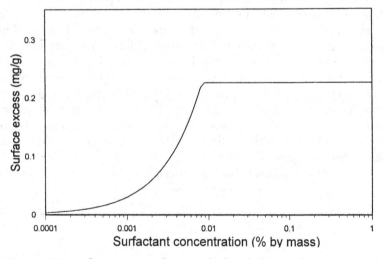

Figure 26.   Adsorption isotherm calculated from adsorption model parameters.

permeability–saturation relationship can be approximated using the end-point relative permeabilities (k_{wo}, k_{ow}), the residual saturations (S_{wc}, S_{orw}) and empirical exponents (z_w, z_o).

$$k_{rw} = k_{rwo}^{o} \left(\frac{S_w - S_{wc}}{1 - S_{orw} - S_{wo}} \right)^{z_w} \qquad k_{ro} = k_{row}^{o} \left(\frac{S_o - S_{orw}}{1 - S_{orw} - S_{wc}} \right)^{z_o} \qquad (11)$$

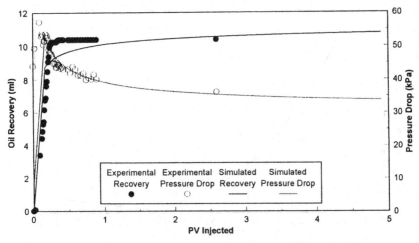

Figure 27.   Experimental and simulated waterflood results.

The experimental pressure and effluent curves were history matched using an analytic simulation package, "PRIsm" [*30, 31*], which is based on the Buckley–Leverett theory.

An initially brine saturated core was flooded with oil to irreducible water saturation (S_{wc}), and subsequently waterflooded. During the water-flooding stage, the pressure drop and oil productions were continuously monitored until the residual oil saturation was reached. In Figure 27 the comparison between experimental and simulated values of the waterflood recovery and concurrent pressure drop (across the core) is made. The history matched relative permeability curves are shown in Figure 28; the corresponding parameters are listed here:

$$K = 459.4 \text{ mD} \qquad k_{wo} = 0.018 \qquad k_{ow} = 0.22$$
$$S_{wc} = 0.25 \qquad S_{orw} = 0.43 \qquad z_w = 3.4 \qquad z_o = 3.1$$

Gas/Oil Relative Permeabilities at Connate Water Satura-tion.   The same core that was used for water/oil relative permeability evaluations, core MCF3, was re-saturated with oil after the waterflood. Next, nitrogen was injected at a flow rate of 6 ml/hr. The oil recovery was determined at consistent time intervals. Problems were experienced with the backpressure regulator, such that the pressure readings were too erratic to be useful in the relative permeability curve evaluation. Only the residual oil saturation to gas and the gas endpoint relative permeability were determined in the presence of connate water:

$$K = 459.4 \text{ mD} \qquad S_{wc} = 0.25 \qquad S_{org} = 0.5 \qquad k^o_{rgo} = 0.05$$

Gas/Water Relative Permeabilities.   The gas/water relative permeabilities were not measured directly. However, a suitable set was

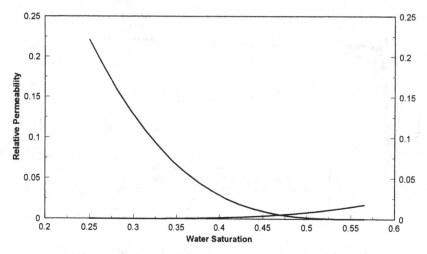

Figure 28.   Oil/water relative permeability curves.

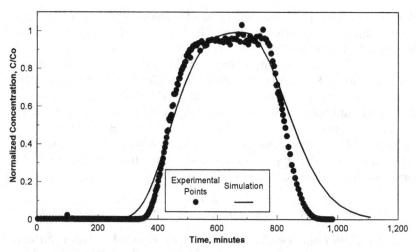

Figure 29.   Comparison of simulated and experimental tracer effluent profiles.

deduced from history matching the displacement of tracer in the 100% brine saturated long core at a gas fractional flow of 95% (see Figure 29). The STARS simulator was used for the history matches. Predicted pressure drops from the relative permeability curves were also compared to the baseline pressure drops generated at various foam fractional flows (see Figure 30). The relative permeability parameters, deduced from simulations, are summarized as follows:

$$k_{wg} = 0.5 \quad K_{gw} = 0.22 \quad S_{wrg} = 0.46 \quad S_{grw} = 0.07 \quad z_{wg} = 3.2 \quad z_{gw} = 2.2$$

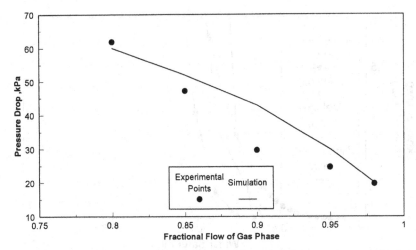

Figure 30. Long core baseline pressure drops (oil free, FAR = 4 m/day).

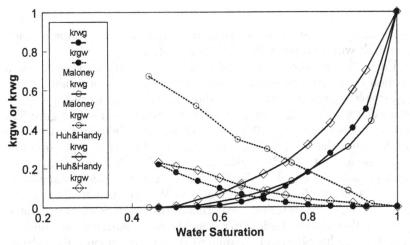

Figure 31. Comparison of gas/water relative permeabilities.

A number of investigations have reported three phase relative permeability data for Berea sandstone [32, 33]. In Figure 31, the gas/water relative permeabilities deduced from the simulations are compared to similar relative permeability data presented by the authors Maloney and Brinkmeyer [34] and Huh and Handy [35]. The water relative permeabilities (k_{rwg}) seem to group tightly together. Larger discrepancies are observed for the gas relative permeabilities (k_{rgw}). The gas relative permeabilities obtained from our calculations seem to group more closely with the data by Huh and Handy. In any case, the gas/water relative

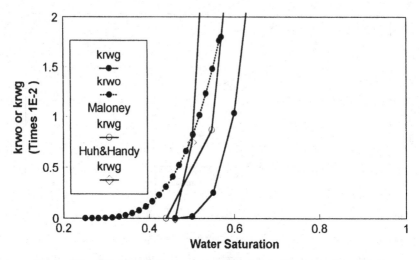

Figure 32.   Water relative permeabilities as a function of saturation.

permeability curves deduced by our history match compare quite favourably with those established by other authors.

Under the assumption that the water relative permeability is only a function of the water saturation, the water relative permeability curve, k_{rwg}, from the gas/water experiment should overlap with the water relative permeability curve, k_{rwo}, from the oil/water experiment (see Figure 32). In the region where $S_w > 0.5$ the relative permeabilty curve seems to match within the limits of experimental variation. However, large discrepancies are noted in the region of irreducible water saturation. In the oil/water case, S_{wc} equals 0.25, while the irreducible water saturation in the gas/water experiment, S_{wrg}, equals 0.46. We speculate that during the gas/water relative permeability experiment the pressure drop generated by the gas flow was insufficient to reduce the water saturation any further. Since the irreducible water saturations do not correspond, the relative permeability curves do not match in the low water saturation region.

List of Symbols

A	cross sectional area of core
C_s	surfactant concentration
C_{smax}	maximum surfactant concentration
D	dispersion coefficient
eg, ev, ew, ep	exponents
es, eo, en	exponents
FAR	frontal advance rate, $q/A\phi$

FF_{max}	foam scaling factor
FM	foam mobility factor
fg	gas fractional flow
F_o	foam mobility multiplier
fw	water fractional flow
G_{cref}	critical foam generation capillary number
K	absolute permeability
k_a	adsorption rate constant
k_d	desorption rate constant
k_1, k_{-1}	lamella generation and decay rate constants
k_1^*	net foam generation rate constant
k_{rgf}^o	gas endpoint relative permeability at S_{wrf} under foam conditions
k_{rgo}^o	gas endpoint relative permeability at S_{org}
k_{rgw}^o	gas endpoint relative permeability at S_{wrg}
k_{rwg}^o	water endpoint permeability at S_{grw}
k_{rwo}^o	water endpoint permeability at S_{orw}
k_{row}^o	oil endpoint relative permeability at S_{wc}
L	length of core
MRF	$\Delta P_f/\Delta P_n$, mobility reduction factor
m_1	monolayer coverage of surfactant
m_2	monolayer coverage of solvent
N_{cp}	$K\,\Delta P/\sigma L$, capillary number based on pressure
N_{cref}	reference capillary number for shear thinning
N_{cw}	$\mu_w v_w/\sigma$, water capillary number
N_{cg}	$\mu_g v_g/\sigma$, gas capillary number
n_f	lamellae number per unit volume
P	pressure
P_c	capillary pressure
P_c^*	limiting capillary pressure
ΔP_f	pressure drop during foam (surfactant solution plus gas) injection
ΔP_n	pressure drop during brine plus gas (no surfactant) injection
P_o	reference pressure
q	total injection rate
S_g	gas saturation
S_{grf}	residual gas saturation under foam conditions
S_{grw}	residual gas saturation after water flooding
S_o	oil saturation
S_{orf}	residual oil saturation after foam flooding
S_{omax}	maximum oil saturation after gas flooding for foam to be effective
S_{org}	residual oil saturation after gas flooding

S_{orw}	residual oil saturation after waterflooding
S_w	water saturation
S_{wc}	connate water saturation
S_{wrf}	residual water saturation under foam conditions
S_{wrg}	residual water saturation after gas flooding
s_1, s_2	exponents
u	Darcy velocity
u_g, u_w	Darcy velocity of gas and water phase
u_{go}	reference velocity of gas phase
u_{gc}, u_{wc}	critical velocities of gas and water phases
v_{total}	$q/A\phi$, total frontal advance rate
v_g	$q_g/A\phi$, gas phase advance rate
v_w	$q_w/A\phi$, water phase advance rate
z_f	exponents for gas relative permeability curve under foam conditions
z_{gw}, z_{wg}	exponents for gas/water relative permeability curves
z_w, z_o	exponents for water/oil relative permeability curves

Greek

λ_g	gas phase mobility
λ_w	water phase mobility
μ_g	gas phase viscosity
μ_w	water phase viscosity
σ	water/gas surface tension
ϕ	porosity

References

1. Moritis, G. *Oil & Gas Journal* **1992**, *90(16)*, 51.
2. Hirasaki, G.J. *J. Petrol. Technol.* **1989**, *41(5)*, 449.
3. Castanier, L.M.; Brigham, W.E. *SPE Res. Eng.* **1991**, *6(1)*, 62.
4. Eson, R.L.; Cooke, R.W. *Proceedings of the California Regional Meeting of SPE*; Society of Petroleum Engineers: Richardson, TX, 1989, paper SPE 18785.
5. Chad, J.; Matsalla, P.; Novosad, J.J. *Proceedings of the 39th Annual Technical Meeting of the Petroleum Society of CIM*; Canadian Institute of Mining, Metallurgical, and Petroleum Engineers: Calgary, AB, 1988, paper CIM 88-39-40.
6. Liu, P.C.; Besserer, G.J. *Proceedings of the 63rd Annual Technical Conference of SPE*; Society of Petroleum Engineers: Richardson, TX, 1988, paper SPE 18080.
7. Arra, M.G.; Skauge, A.; Sognesand, S.; Stenhaug, M. *Petrol. Geoscience*, **1996**, *2*, 125–132.
8. Novosad, J.J.; Ionescu, E.F. *Proceedings of the 38th Annual Technical Meeting of the Petroleum Society of CIM*; Canadian Institute of Mining,

Metallurgical, and Petroleum Engineers: Calgary, AB, 1987, paper CIM 87-38-80.

9. Schramm, L.L.; Novosad, J.J. *J. Petrol. Sci. Eng.* **1992**, 7, 77–90.
10. Mannhardt, K.; Schramm, L.L.; Novosad, J.J. *Colloids and Surfaces* **1992**, 68, 37–53.
11. Mannhardt, K.; Hans, M.J.; Masata, V.; Randall, L. *Core flood evaluation of solvent compositional and wettability effects on hydrocarbon solvent foam performance*, PRI Report 1995/96-03, Petroleum Recovery Institute: Calgary, AB, 1995.
12. Mannhardt, K.; Hodgins, L.A.; Schramm, L.L.; Novosad, J.J. *J. Petrol. Sci. Eng.* **1996**, 14, 183–195.
13. Schramm, L.L.; Turta, A.; Novosad, J.J. *SPE Res. Eng.* **1993**, 8(3), 201–206.
14. Schramm, L.L. In *Foams: Fundamentals and Applications in the Petroleum Industry*, Schramm, L.L., Ed.; American Chemical Society: Washington, DC, 1994, pp 165–197.
15. Mannhardt, K.; Novosad, J.J.; Schramm, L.L. *Proceedings of the SPE/DOE Improved Oil Recovery Symposium*; Society of Petroleum Engineers: Richardson, TX, 1998, paper SPE/DOE 39681.
16. Kovscek, A.R.; Patzek, T.W.; Radke, C.J. *Proceedings of the SPE/DOE Improved Oil Recovery Symposium*; Society of Petroleum Engineers: Richardson, TX, 1994, paper SPE/DOE 27789.
17. Falls, A.H.; Hirasaki, G.J.; Patzek, T.W.; Gauglitz, D.A.; Miller, D.D.; Ratulowski, T. *SPE Res. Eng.* **1988**, *August*, 884–892.
18. Zhou, Z.; Rossen W.R. *Proceedings of the SPE/DOE Improved Oil Recovery Symposium*; Society of Petroleum Engineers: Richardson, TX, 1992, paper SPE/DOE 24180.
19. Rossen, W.R.; Zhou, Z.H.; Mamun, C.K. *Proceedings of the 66th Annual Technical Conference of SPE*; Society of Petroleum Engineers: Richardson, TX, 1991, paper SPE 22627.
20. Fisher, A.W.; Foulser, R.W.S.; Goodyear, S.G. *Proceedings of the SPE/DOE Improved Oil Recovery Symposium*; Society of Petroleum Engineers: Richardson, TX, 1990, paper SPE/DOE 20195.
21. Mohammadi, S.S.; Coombe, D.A.; Stevenson, V.M. *J. Can. Petrol. Technol.* **1993**, 32, 49–54.
22. Rossen, W.R.; Zhou, Z.H. *Proceedings of the Symposium on Field Applications of Foams for Oil Production*; U.S. Dept. of Energy: Bartlesville, OK, 1993, pp 173–184.
23. de Vries A.S.; Wit K. *SPE Res. Eng.* **1990**, *May*, 185–192.
24. Vassenden, F.; Holt, T. *Proceedings of the SPE/DOE Improved Oil Recovery Symposium*; Society of Petroleum Engineers: Richardson, TX, 1998, paper SPE 39660.
25. Falls, A.H.; Musters, J.J.; Tatulowski, J. *SPE Res. Eng.* **1989**, *May*, 155–164.
26. Sanchez M.J.; Schechter, R.S. *J. Petrol. Sci. Eng.* **1989**, 3, 185–199.
27. Holt, T.; Vassenden, F.; Svorstol, I. *Proceedings of the SPE/DOE Improved Oil Recovery Symposium*; Society of Petroleum Engineers: Richardson, TX, 1996, paper SPE 35398.

28. Nasr-El-Din, H.A.; Green, K.A.; Schramm, L.L. *Rev. L'Inst. Français Pétrole.* **1994**, *49(4)*, 359–377.
29. Chung, F.T.H.; Hudgins, D.A. *Development of an Engineering Methodology for Applying Foam Technology*, Report NIPER-374, DE89000740, U.S. Dept. of Energy: Bartlesville, OK, May, 1989, 52 pp.
30. Rao, D.N.; Girard, M.; Sayegh, S.G. *SPE Res. Eng.* **1992**, *May*, 204.
31. Claridge, E.L.; Bondor, P.L. *SPE J.* **1974**, *December*, 609.
32. Saraf, D.N.; Batycky, J.P.; Jackson, C.; Fisher, D.B. *Proceedings of the California Regional Meeting of SPE*; Society of Petroleum Engineers: Richardson, TX, 1982, paper SPE 10761.
33. Oak, M.J.; Baker, L.E.; Thomas, D.C. *J. Petrol. Technol.* **1990**, *August*, 1054–1061.
34. Maloney, D.; Brinkmeyer, A. *Three Phase Permeabilities and Other Characteristics of 260-mD Fired Berea*, NIPER-581, U.S. Dept. of Energy: Bartlesville, OK, 1992.
35. Huh, D.G.; Handy, L.L. *Proceedings of the California Regional Meeting of SPE*; Society of Petroleum Engineers: Richardson, TX, 1986, paper SPE 15078.

RECEIVED for review August 20, 1998. ACCEPTED revised manuscript January 11, 1999.

OILWELL, NEAR-WELL, AND SURFACE OPERATIONS

8

The Use of Surfactants in Lightweight Drilling Fluids

Todd R. Thomas and Ted M. Wilkes

Clearwater Inc., Pittsburgh, Pennsylvania, USA

An overview of lightweight drilling fluids is presented with a discussion of the application of foam as a drilling fluid. The primary classes of surfactants used in such fluids, of varying quality, are described. An overview of specialty surfactant conditioning additives is presented. A novel foam control system is offered and several lightweight fluid case studies and field examples are showcased.

Introduction

Figure 1 depicts four fluids that are commonly used in "lightweight" or "underbalanced" drilling applications. Shown in each depiction is the drilling fluid, as it may appear down-hole with cuttings being carried from the borehole. These fluids include:

- dry gas, typically: air, pure nitrogen, oxygen depleted air (membrane nitrogen), carbon dioxide or methane
- mist, fog or high quality foam (HQF), low liquid volume fraction (LVF), <4%
- gasified or aerated fluid, generally less than 55% gas volume fraction (GVF)
- foam, generally 55–96% GVF

Each of these fluids has distinct surfactant requirements. In dry gas applications, there is typically no surfactant used. In mist, fog, and in all foams, several classes of surfactants are used. Gasified (aerated, nitrified) fluid typically employs standard polymer and clay mud systems lightened with a gas to lower hydrostatic pressure on the formation.

The focus of this chapter will be:

1. To present an overview of lightweight drilling fluids.
2. To discuss the application of foam as a drilling fluid.

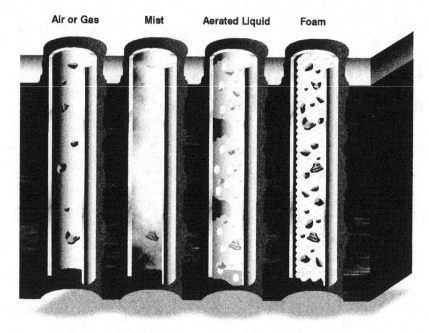

Figure 1.   Depiction of lightweight drilling fluids.

3. To describe the primary classes of surfactants used in lightweight drilling fluids from qualities of 55% through 99.9%.
4. To present an overview of specialty surfactant conditioning additives.
5. To offer a novel foam control system.
6. To showcase several lightweight fluid case studies and field examples.

Overview of Lightweight Fluids

The application of high quality fluids (>96% GVF) is commonly referred to as mist drilling. In traditional mist drilling, the water present is treated as a cutting and for calculation purposes, the density and mass flow rate of the water vapor replace those for the cuttings. However, field experience has indicated that in most instances, when foaming surfactants are employed, the "mist" actually becomes high quality foam. When the quality increases beyond a critical level, the foam does not collapse and become droplets of surfactant laden water. Rather, the foam becomes discontinuous. The point at which this happens is determined by the elasticity and surface tension of the foam structure itself. Only when non-foaming surfactants are used, will the water appear as a continuous phase as in the appearance of a "fog," or a fine dispersion of micro water droplets in the gas. Govier and Aziz [1] depict the physical appearance of this fog. In mist drilling with a "fog," calculations are made with the assumption that air–water is a continuum. Introduction of a non-foaming surfactant to the water that is being pumped downhole can create this "fog." The surfactant reduces the water's surface tension. The agitation realized in the expanding air–water–surfactant mixture across the bit nozzles, or in fog generators at surface, permits the formation of the "fog" or "mist."

Foam, as a drilling fluid, has been in use since the 1960s and has been the subject of several journal articles [2, 3]. Foam, as typically used in drilling, is a mixture of water, a surfactant and air or nitrogen. Field terms that are generally applied to foam are foam quality and foam texture. Foam quality is the ratio of gas volume to the total volume (GVF) and foam texture relates the size and distribution of the gas bubbles.

Foam Utilization as a Drilling Fluid

If gas and a liquid are mixed together in a container, and then shaken, examination will reveal that the gas phase has become a collection of bubbles that are dispersed in liquid: a foam has been formed, as in Figure 2. Foams serve many important functions in the petroleum production process from their use as a drilling fluid, through reservoir stimulation, to secondary and tertiary recovery. Foam has gained widespread acceptance as a viable drilling fluid alternative. Foam, as a

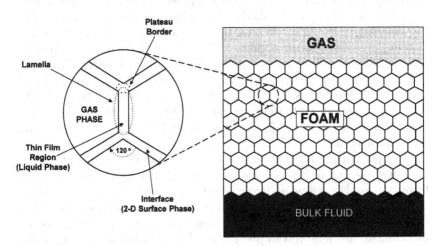

Figure 2.   Overview of a foam bubble. (Reproduced from reference 4. Copyright 1994 American Chemical Society.)

drilling fluid, has endured a rather unrefined gestation period. Only in the very recent past has the industry recognized the importance of foam drilling science and engineering. Presently, with the evolution of under-balanced drilling technology, foam has become a more frequently considered option, as a lightweight drilling fluid. Today, wells are being drilled with foam, that, until recently, were not considered foam drilling candidates. Although limitations still exist, many of the obstacles facing foam drilling have been conquered through chemical and mechanical innovation.

Overview of the Foam Bubble [4]

Foams consist of polyhedral gas bubbles separated by thin liquid films or lamellae as shown in Figure 2. A single bubble, like a balloon, has a very thin skin, or film, surrounding a volume of gas. When an individual bubble is formed it takes the shape of a sphere. The surface tension of the film maintains a form, which minimizes surface area and requires the least amount of energy. Of all possible shapes, a sphere has the largest volume in comparison to its surface area. Thus, "a soap bubble is a thin film that embraces a fixed volume of gas ... through its need to minimize stored energy, it assumes the shape of a sphere" [5].

Aside from being entertaining, a single soap bubble has little practical application in the drilling industry, whereas persistent, or stable, foams have unlimited possibilities. In a physical system, when three or more soap bubbles come together they immediately transform into foam cells, polyhedra, separated by almost flat liquid films. The polyhedra are almost, but not quite, regular dodecahedra. These two-sided thin films are called the lamellae of the foam. Where three or more gas bubbles meet, the lamellae are curved, at an angle of 120°, concave to the gas cells, forming what is called the Plateau border or Gibbs triangles. Figure 2 illustrates this principle. This honeycomb shape is stabilized by the presence of surface-active materials at the liquid/gas interface, which can retard the loss or drainage of the liquid from the area between the bubbles. Thus, created is a somewhat rigid, mechanically strong bi-layer that maintains the foam structure.

The previous definitions aid in the understanding of the chemistry and engineering of foams and are critical when dealing with the formation and stability of foam structure. Persistent or stable foams consist of a network of thin liquid films, which exhibit complex hydro-dynamics. For a drilling foam to remain persistent, or stable, several mechanisms are required to prevent the loss of liquid and gas from the foam and to prevent premature collapse of the foam, when subjected to environmental stresses.

Benefits of Foam as a Drilling Fluid

A properly prepared foam fluid has several features which would make it an excellent drilling fluid. The following is a brief summary of the benefits of foam.

Lightweight Drilling Fluid.   Specific gravities, as low as 0.15 (1.25 lb/gal, equivalent circulating density, ECD), can be effectively achieved with foam. Low density has two profound advantages. The first being increased penetration rate, and the second being the ability to achieve bottom hole pressures below that of the productive reservoir. This condition is called "underbalance." With recent emphasis focused on drilling "underbalanced," the effective density of the drilling fluid should be below the pore pressure of the reservoir. In depleted, or low-pressure zones, the choice of fluids, which function effectively at these densities, is rather limited. Conversely, relatively high fluid densities, up to a specific gravity of 0.78 (6.5 lb/gal, ECD) can also be achieved with foam. Foam in a dynamic system produces significant friction factors. When calculating bottom hole pressure, in dynamic foam flow, the sum of the hydrostatic weight and the friction pressure provides the total bottom hole pressure (BHP) and equivalent circulating density.

To eliminate unwanted influxes, many wells have been intentionally drilled overbalanced with foam. Proper fluid chemistry is essential in these situations. If the foam "breaks-down," or specifically, if the viscosity of the foam is altered by the influxes, the friction factor decreases and, in turn, the bottom hole pressure is reduced. A general understanding of fluid chemistry, and the selection of the proper foaming agent, is essential in assuring a successful project. Many projects have failed in the past due to inadequately applied chemistry.

Cuttings Transport.   The application of foam has shown it to possess an exceptional ability to remove cuttings from the wellbore. When created with the proper surfactant, viscous foam creates a motive fluid, with the ability to transport up to ten times the amount of cuttings, when compared to a single-phase fluid. This translates to increased penetration rates, by reason of efficient hole cleaning. Along with the ability to carry cuttings, foam has the ability to displace large volumes of downhole fluid influxes. Recent foam drilling jobs have continued to operate effectively in the presence of 25 bbl/min of water influx. Certain "oil foaming" surfactants have performed remarkably well on horizontal drilling operations, with as much as 50 bbl/hr oil influx [6].

Annular Velocity.   Recently, attention has been focused upon the detrimental effects of wellbore erosion, due to high annular velocities. It is a general drillers' guideline that, for a given formation, there is a critical velocity which should not be exceeded. Fluid velocities beyond a certain level will adversely affect borehole stability. Annular velocities as low as 100 feet/min have provided sufficient hole cleaning in many foam drilling operations. This gentle flow medium is quite beneficial in velocity sensitive shales and semi-unconsolidated formations.

Down-hole pressure fluctuations, due to the termination of flow in the annulus, are unquestionably of great concern. When circulation is stopped, during a connection or survey, for example, the foam can "break-back" to soapy water and its gas phase. The gas can then expand toward the surface. In order to re-establish circulation of foam, the "broken-back" water in the bottom of the well must be displaced by newly introduced foam through the drill pipe. The BHP increases until the fluid begins to move up the wellbore. As the fluid is lifted out of the well, the BHP decreases, subjecting the formation to a dramatic pressure drop. In pressure sensitive formations this may have an unfavorable effect on borehole stability. In these cases, special additives can be used to create exceptionally stable foams. These foams are many times referred to as "stiff," or "stable," foams.

Lost Circulation.   Unlike the flow of foam within the drill string and casing, circumstances are quite different within a porous medium, in which immobile containing boundaries are near to every avenue of flow. The motion of the two-dimensional array of foam cells limits the entry into the porous zone. This is shown in Figure 3. When drilling with a single-phase fluid, lost circulation is both an economic and operational consideration. Drilling with foam can eliminate these concerns. Elimination of lost circulation greatly reduces the likelihood of differential sticking of the drill string.

Flexibility of Foam Drilling.   Foam is the most versatile of all the low density drilling fluids. The water (LVF) and percentage of gas (GVF) can be easily controlled at the surface to achieve bottom hole pressures sufficient to drill the well. Various surfactant chemicals can be added to the water phase to address such problems as influxes, shale

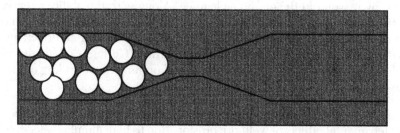

Figure 3.   Immobile boundaries containing lost circulation of foam.

control, corrosion, and borehole stability. Thereby, the foam drilling fluid can be customized to meet the requirements of each foam drilling application.

Surfactant Testing

Surfactant testing and reagent screening can be performed in a variety of ways. The simplest and most common technique used in the petroleum industry is the standard "blender test" [7]. Other test methods [4] include Ross-Miles foam height test, flow loop testing, capillary foam testers, contact angle measurements, tensiometers and actual field evaluations. Due to its low cost, easy duplication and ability to be used in the field, blender testing has become the norm in the industry for foam based fluids.

A standard blender test consists of the following:

1. To a 1000 ml Waring Commercial Blender add 100 ml of the test water.
2. With a pipette add the desired amount of surfactant to be tested. The standard concentration is 0.5% v/v.
3. Operate blender on "high" for 30 seconds. Start timing the half-life immediately upon completion of the agitation.
4. Immediately pour the foam from the blender into a 1000 ml graduated cylinder.
5. Measure the foam height in milliliters, when all of the foam is transferred to the graduated cylinder.
6. Record the foam half-life as the time required for drainage of 50 ml of water in the bottom of the cylinder, as measured at the point when the meniscus rises above the 50 ml graduate.

Interpretation of Blender Tests

Interpretation of blender test results is critical to product selection and formulation.

For example:

A given foamer shows test results indicating a foam height of 500 ml and a half-life of four minutes and thirty seconds in fresh water at 0.5% v/v surfactant concentration.

What do these numbers mean?

Generally, foam height in a blender test describes the ease of foam formation, due to surface tension lowering and surface elasticity, the ability to create very thin lamellae and the speed with which foam can be created or recreated. Half-life gives an indication of foam stability. Both measurements are important in drilling operations. Stability is required during static periods, such as "air off" connections, surveys and in low flow areas, and surface tension lowering and elasticity are required to create foam viscosity.

Foam viscosity and elasticity are required for cuttings transport and hole cleaning, particularly at high qualities. Returning to our example, a foam height of 500 ml, after starting with 100 ml of water, gives a foam quality of only 80% at atmospheric pressure. Actual bottom hole conditions show foam qualities of above 98% are common. This would mean that the foam height created from 100 ml of fluid would be 5000 ml at 98% quality.

How does this happen?

In flow loop testing, continuous foams at qualities of up to 99.6% have been observed. The energy created by the pumping pressure, shear and friction with the drill pipe, allows the lamellae to thin to a point where the water is dispersed over a greater volume. In the case of our example, a 10 ml aliquot of water would create a foam height of 500 ml.

As quality is increased beyond a certain point, the rate of bubble bursting exceeds the rate of bubble creation and stabilization, caused by the dynamic flow conditions. The foam then becomes discontinuous. Discontinuous foams do not carry cuttings as well as a solid foam matrix. Therefore, hole cleaning is compromised.

Correlating blender test results with other test mechanisms and actual field observations is critical for proper foamer application.

Surfactant Selection

Five broad surfactant groups are used in lightweight drilling fluids.

- primary foaming surfactants
- conditioning additives
- corrosion inhibitors
- fog drilling surfactants
- defoamers

Extensive field experience, controlled flow loop testing and foam circulation in test well projects [8] has clearly shown proper surfactant selection to be the dominant factor controlling drilling foam performance. Many different surfactants are used in oilfield applications; anionic, cationic nonionic and amphoteric. Anionic surfactants have a polar group, which is negatively charged. Due to their cost performance in drilling operations, anionic surfactants are the most broadly used. Cationic surfactants conversely have a polar group, which is positively charged, and are used much less often. Nonionic surfactants carry no charge. Surfactant species that can be either cationic or anionic, depending upon the pH of the solution, are called amphoteric, or zwitterionic.

Several criteria must be considered when evaluating surfactants.

- foaming capability – foam drilling
- surface tension reduction – fog drilling
- ionic charge – compatibility with other additives
- salt tolerance – compatibility with brine water influxes
- hydrocarbon tolerance – compatibility with hydrocarbon influxes
- temperature limitations – effectiveness in high temperature wells
- environmental concerns – biodegradability, toxicity
- cost

Primary Foaming Surfactants

Primary foaming surfactants are typically selected from any of the groups shown in Tables 1–4. They can be used alone or in combination with other primary surfactants or conditioners. They are selected for their foaming or surface tension reduction properties.

Conditioning Additives

Conditioning additives are used to enhance specific properties of a foaming agent solution. Foam stabilizers extend the drainage time of the foam. Stabilizers can be polymeric additives, that add base fluid viscosity, or other surfactants, which work synergistically with the primary surfactant. Some conditioning additives can be used alone as base surfactants, but due to lack of versatility, or their cost, are mostly used as secondary additives in the oilfield.

Many times, conditioning additives are used to impart particular properties to a foam system. The use of primary surfactants blended with conditioning additives, in demanding applications, can be significantly more economical than the use of primary surfactants alone. Desirable properties and examples of their application include:

- *High Temperature Stability*: Foam was recently used to drill a well in Texas with a bottom hole temperature (BHT) of 370 °F.
- *Resistance to Degradation by Oil Contamination*: Many wells, both vertical and horizontal, have been drilled using stable foam, with up to 40% by volume oil influx. Foam stability and BHP were maintained even despite this influx.
- *Salt Contamination (Chlorides)*: Surfactants are available to effectively maintain stable foam in saturated aqueous solutions. An example application is using foamed saturated brine to drill through troublesome salt sections.

Table 1. **Primary Foaming Surfactants – Anionic**

	Chemistry	Application	Comment	Relative Cost Index
Sulfates of linear alcohol ethoxylates	C6 through C12, with between 2.0 and 4.0 moles of ethylene oxide per mole of alcohol	Wide application	Most versatile and most widely used	Moderate
Sulfates of linear alcohols	C8 through C16	Mostly fresh water	Highly effective, but not as versatile	Moderate
Sulfates of nonyl phenol ethoxylates	C9 alkyl group plus phenol ring	Broad effectiveness	Not environmentally preferred	Low–Moderate
Sulfonates of alpha olefins	C10–C16	Mostly fresh water	Commonly used as a foam booster	Moderate
Sulfonates of linear alcohol ethoxylates	Sulfonate vs. sulfate	Better acid tolerance, will not degrade as quickly as sulfates	Not widely used	High
Sulfonated alkyl benzene	Soft or hard acid, DDBSA, typically Na, NH_3 or MEA salts	Limited in drilling, due to water hardness intolerance	Limited brine (NaCl) tolerance, low hardness (Ca + Mg) tolerance	Low
Fatty (C12–16) acid salts	Soap	Limited in drilling, due to hardness and low pH intolerance	Limited brine tolerance, low hardness tolerance	Low

Table 2. Primary Foaming Surfactants – Cationic

	Chemistry	Application	Comment	Relative Cost Index
Fatty (C10–C18) quaternaries	Basic cationic foamer	Some fresh water use	Cationic systems can have improved compatibility with certain corrosion inhibitors and shale stabilizer systems	High
Fatty quaternary ethoxylates	Better performance than straight quaternaries	Ethylene oxide gives better hard and salt-water performance	Works synergistically with other cationic surfactants	High
Fatty amine ethoxylates and propoxylates	Amount and type of alkoxylation affects solubility and performance	Not commonly used alone	Usually used as boosters in cationic systems	High
Fatty amine oxides	Less cationic functionality gives better compatibility	Good fresh water enhancement	Used as foam boosters in all systems	Moderate

Table 3. Primary Foaming Nonionic Surfactants

	Chemistry	Application	Comment	Relative Cost Index
Alcohol ethoxylates	i.e. C9–11 + 8 moles EO	Some usage, when anionics or cationics are not used	Not as efficient as the anionics, but not bad for not having a charge	Low–Moderate
Nonyl or octyl phenol ethoxylates	Similar HLBs to the above	Limited usage	Environmental questions	Low–Moderate
Ethoxylated esters	i.e. PEG (20) mono-sorbitan oleate	Good emulsifiers, can add oil tolerance to anionic systems	Niche products in drilling foams	High
PEGs, polyethylene glycols	Highly water soluble polymers	Can add base fluid viscosity and stability to other systems	Good boosters or stabilizers, also used for shale stabilization	Low–Moderate

Table 4. Primary Foaming Amphoteric Surfactants

	Chemistry	Application	Comment	Relative Cost Index
Betaines	Coco-amido-propyl betaine	Not usually used alone	Shows synergy with sultaines	Moderate–High
Sulfaines	Coco-amido sulfo propyl betaine	Good versatility can be used alone	Shows synergy with anionic systems	Moderate–High
Proprionates	Imidazoline derivatives	Not common	Can enhance foam heights – elasticity	High
Glycinates	Imidazoline derivatives	Not common	Can enhance foam heights – elasticity	High

Table 5.   Foam Boosting Properties of AOS

Weight % AOS	Weight % AES	Foam Performance in FW
25	0	500/4:08
20	5	500/4:01
15	10	510/4:13
10	15	520/4:18
5	20	520/4:23
0	25	500/3:55

Synergistic effect of mixing 40% active alpha olefin sulfonate solutions (AOS) with 80% active C6–C10 alcohol ether sulfate solutions (AES), balance is water

- *Depth Limitations*: Foam has been used as the drilling fluid to depths over 21,000 feet.

Examples of specialty surfactant conditioners include:

- *Amides*: Commonly used as foam stabilizers. Coconut or lauryl diethanolamide shows excellent performance for fresh water systems.
- *Sodium diphenyl oxide disulfonate*: Dual anionic functionality can provide superior hydrocarbon tolerance.
- *Sulfosuccinates*: Typically, dioctyl sulfosuccinate. Commonly referred to as DOSS – can exhibit good wetting and foam stabilizing abilities.
- *Alpha Olefin Sulfonates, AOS*: Frequently used conditioning additive, standard C-14/16 product gives excellent fresh water foam boosting properties. Lower molecular weight AOS can provide hard and/or salt-water performance, as well as oil tolerance. Table 5 demonstrates the synergy of AOS with an alcohol ether sulfate (AES) solution. In this example, 5 parts AOS together with 20 parts AES clearly shows synergy.

Hydrotropes

Hydrotropes are frequently used to improve the solubility, and therefore the effectiveness, of marginally soluble surfactants. This is particularly beneficial in high electrolyte (salty) waters. Commonly used hydrotropes include: aryl sulfonates (SXS, STS, SCS), glycol ethers, and phosphate esters.

Widely used because of their relatively low cost, glycol ethers, such as ethylene glycol mono butyl ether, can have beneficial effects as demonstrated below, in Table 6. In this table, the effect of various glycol ethers on the performance of a "typical" commercial foam drilling product is

Table 6. Effect of Glycol Ethers in Brine Water Foaming

Base Product A Formulation

Material	Weight percent
C6–C10 linear alcohol with 3–4 moles ethylene oxide, sulfated	45%
Sodium xylene sulfonate, 45% solution	8%
Alpha olefin sulfonate, C14–16, 40% solution	10%
Isopropanol	12%
Water	Balance

To Base Product A, a series of glycol ethers were added to measure foam performance in a standard blender test. The test utilized synthetic brine made by dissolving excess Allberger #3 salt into distilled water, at 90 °C. This brine is then allowed to cool to 25 °C, whereupon salt crystals precipitate out of solution. The supernatant water is then used for the test.

Control – No glycol ether added, Base Product A foam height 370 ml. Half-life 3:25

Glycol ethers added – ethylene glycol mono butyl ether (EGMBE), diethylene glycol mono butyl ether (DEGMBE), and triethylene glycol mono butyl ether (TEGMBE).

Concentration EGMBE, w/w	Foam Height, ml/Half-Life, min:sec
0%	360/3:25
2%	370/3:31
4%	370/3:36
6%	380/3:43
8%	380/3:51
10%	390/3:59
12%	410/4:11
14%	400/4:14
16%	400/4:12
18%	400/4:11
20%	400/4:03

Concentration DEGMBE, w/w	Foam Height, ml/Half-Life, min:sec
0%	360/3:25
2%	360/3:31
4%	360/3:33
6%	370/3:37
8%	370/3:41
10%	380/3:45
12%	380/3:47
14%	370/3:44
16%	370/3:45
18%	370/3:42
20%	370/3:43

Continued

Table 6.   (*Cont.*)

Concentration TEGMBE, w/w	Foam Height, ml/Half-Life, min:sec
0%	360/3:25
2%	360/3:26
4%	360/3:26
6%	370/3:22
8%	370/3:24
10%	370/3:26
12%	370/3:31
14%	370/3:32
16%	360/3:32
18%	370/3:35
20%	370/3:32

Table 7.   Effect of Coupling Agents on Fresh Water Foam

Base Composition	
C10–12 linear alcohol ether sulfate	20%
C14–16 alpha olefin sulfonate	5%
IPA	5%
Water	Balance
Foam height/half-life in Neville Island Tap Water (NITW)	330 ml/2:09
Addition to formula with corresponding reduction in water:	
8% by weight of sodium xylene sulfonate, 100% basis	310 ml/2:16
8% by weight of sodium toluene sulfonate, 100% basis	420 ml/2:48
8% by weight of sodium cumene sulfonate, 100% basis	370 ml/2:32

observed. The product in this test is being evaluated in synthetic brine. Significant performance improvements are observed with the addition of glycol ethers.

This effect is not only seen in high electrolyte systems. Certain beneficial hydrotropic effects can be observed in fresh water systems as well. This is shown in Table 7, where sodium sulfonates of xylene, cumene and toluene are added to a fresh water foaming system.

Foam Stability and Foam Destruction Mechanisms

The understanding of foam persistence, or stability, and bubble coalescence is an analytic dilemma. Unquestionably, this is a dynamic phenomenon of stunning complexity. Study in a controlled laboratory is, as we all know, quite different than the environment within a wellbore. Several of the major factors affecting foam persistence are:

- thin film elasticity
- drainage of liquid in the lamellae
- bulk liquid and surface viscosity
- surfactant selection

These aspects are discussed in more detail in Chapter 1 of this book.

Thin Film Elasticity.   As imagined, the liquid thin film layer is basically stretched around the gas to form the foam cell or bubble. The presence of aqueous solution of surface-active agents facilitates the stretching of the film. As the stretching, or thinning, occurs, equal and opposing forces must immediately occur to counteract the thinning of the membrane to avoid rupture of the foam cell. As thinning occurs in one isolated location, liquid flows, due to surface tension increase, and equalizes the thin film in the thinning area. The change in the surface tension with the change in the concentration of surface-active agent is known as the Gibbs effect (Gibbs, 1878). The change in surface tension with time is known as the Marangoni effect (Marangoni, 1872).

Drainage of Liquid in the Lamellae.   The lamella contains the liquid film separating the gas bubbles. In relatively thick lamellae, or wet foams, gravitational effects dominate drainage of this liquid. In the absence of, or with insufficient quantities of surface-active agents the fluid drains from the lamellae. This causes the film to thin, reaching a critical thickness, at which time the film ruptures or gas is diffused through the film and coalescence occurs. High viscosity of the lamella fluid, surface rheological effects, and electrical interactions can all have varying consequences on drainage. The selection of the appropriate surfactant can beneficially alter all of these attributes and contribute to foam stability.

Purushottam reports that drainage half-life is directly proportional to base fluid viscosity and roughly inversely proportional to the square of the bubble size [9].

In drilling fluid applications, several types of viscosity modifiers can be used to enhance foam stablity. These include: partially hydrolyzed polyacrylamide (PHPA); clays, such as bentonite; xanthan gum; guar gums; starches; cellulosic polymers, such as hydroxy ethyl cellulose (HEC) and sodium carboxy methyl cellulose (SCMC). In Table 8, base fluid viscosities and drainage times for one-half and four-fifths of the base fluids are recorded in a foam system consisting of an anionic foamer, sodium lauryl sulfate (SLS) and a nonionic foamer, TX100 – ethoxylated octyl phenol. It can be seen that in these systems base fluid viscosity alone

Table 8.   Effect of Base Fluid Viscosity on Foam Stability [10]

Solutions	Viscosity, mPa.s	$T\frac{1}{2}$, seconds	T 4/5, seconds
SLS, 0.23%	1.00	110	370
SLS, 0.23% SCMC, 0.75%	36.50	990	1930
SLS, 9.2%	1.32	110	350
SLS, 9.2% SCMC, 0.75%	37.00	170	420
TX-100, 0.02%	1	65	225
TX-100. 0.02% SCMC, 0.1%	28	245	850
TX-100, 0.8%	1.05	95	260
TX-100, 0.8% SCMC, 0.1%	30.00	360	1260

does not determine foam stability, rather, it is a synergistic combination of the foaming agent and the viscosifying agent. This is especially true in the case of the anionic (SLS) surfactant. The nonionic (TX-100) is more prone to base fluid viscosity dominant stability profiles.

It is common belief that the addition of viscosifying agents to the bulk fluid greatly increases foam persistence or stability. In general, slowing down the drainage rate can indeed promote foam stabilization, but this is a complex issue. Adding viscosity to the base fluid doesn't necessarily add mechanical strength to the thin film. Particularly viscous thin films do not produce especially persistent foams. By increasing viscosity, elasticity is often sacrificed and the net result is detrimental, relative to persistence. High viscosity of the lamellae fluid may retard the flow toward the thinning area and minimize the equalization effect. At the elevated pressures consistent with drilling conditions, the bulk fluid is extremely viscous. As a basic rule of thumb, with the correct surfactant selection and proper concentrations, the addition of foam stabilizing agents such as polymers, gels, etc. is not necessary. In only exceptional instances are they required to enhance the performance of the foam as a drilling fluid.

Film Rupture.   This effect is essentially the coalescence of bubbles, i.e. lots of small bubbles joining to form larger bubbles. The large bubbles, being less electrostatically stable, create higher drainage rates and eventual loss of foam performance. Film rupture is the method by which defoamers work. This can also be caused by the presence of cuttings, and other substances unfriendly to the foam, such as oil, condensate, salt, sulfur or water hardness.

Gas Diffusion.   Gas diffusion is the process whereby, in static conditions, small bubbles being of higher pressure than large bubbles, lose their gas to larger bubbles. Smaller bubbles have greater interfacial curvature and by Laplace's law, higher internal pressure. Therefore, small bubbles shrink and big ones grow with time. Diffusion is ultimately driven by surface tension and the movement towards equilibrium. Generally, over time the average bubble gets bigger [11–13]. At some point this effect can create discontinuous flow or slugging in drilling operations.

Rheology.   Drilling foam rheology is extremely complex [14]. Drilling fluid foams are many times described as a Bingham plastic, however, they share the rheological properties of a gas, a liquid and a solid. Foams have a finite shear modulus, as does a solid, however, they respond elastically to a small shear stress. When the applied stress is increased beyond the yield point, the foam flows in a laminar fashion like a viscous liquid. Finally, to totally complicate the matter, foams are compressible like a gas and are greatly affected by changes in pressure.

Corrosion Inhibitors

Lightweight drilling fluids can employ several types of surfactants used for corrosion abatement in drilling operations. Typical targets of corrosive attack include the drill pipe, casing and bottom hole assembly. Surface equipment is usually much less affected as corrosion rates are greatly increased by the increases in temperature experienced down hole. In particular, oxygen generated or catalyzed corrosion is the prime concern, especially when compressed air is being used. Membrane generated nitrogen systems (oxygen depleted air) reduce the oxygen content of the gas being used to 5% or less. This has some limited beneficial effect, but does not completely control corrosion. The presence of acid gases, H_2S and CO_2, greatly enhances the corrosivity of any system.

Corrosion inhibiting surfactants used in lightweight fluids for general and oxygen-related corrosion include:

- Phosphate esters – typically used with anionic systems for compatibility reasons. Generally, effective below 200 °F.
- Phosphonates – used as passivating agents at higher temperatures.
- Surface-active filming amines – can be incompatible with anionic base foaming agents, but can be effective at temperatures above 200 °F and in the presence of acid gases.
- Neutralizing amines – used as scavengers or neutralizers for acid gases.

Fog Drilling Surfactants – Surface Tension Reducers

The goal in fog drilling applications is to create an extremely fine dispersion of liquid into the gas phase, essentially, a true mist. This is best accomplished by reduction of the interfacial tension to below 25 mN/m (see also reference [15]). In aqueous systems this can be accomplished by the use of various types of surfactants. Preferably, non-foaming materials are used. These can be chosen from a list including:

- fluorocarbons, cationic, anionic, and nonionic
- acetyl surfactants and alcohols
- phosphate esters
- low HLB ethoxylates of alcohols and alkyl phenols

The primary surfactant is typically diluted with isopropanol or methanol for field use products.

Defoamers and Foam Control Systems

Historically, antifoaming agents or, more commonly, defoamers were single component liquid systems or homogeneous solutions derived from vegetable oils and fatty acids. Other useful materials included mineral oils, aliphatic and aromatic hydrocarbons [16, 17].

Defoamers in use in the petroleum industry today are formulated to meet more diverse and demanding applications in the field. Four broad classifications can be distinguished.

1. Liquid single component or homogeneous solutions – hydrocarbon oils.
2. Dispersions of hydrophobic solids in a carrier oil – hydrophobic silica dispersions.
3. Aqueous or water containing suspensions or emulsions – silicon oil emulsions.
4. Solid materials – stearates.

A brief overview of some typical oilfield defoamers follows.

1. Carrier Oils.   These are typically water insoluble paraffinic and naphthenic mineral oils, preferred because of their low cost and versatility. Alternatively, vegetable oils, such as tall oil, castor oil, soybean oil, or peanut oil are used. These oils themselves have foam control capabilities as well as the ability to work synergistically with the ingredients they carry.

2. Silicone Oils.   Widely used and highly effective because of their low surface tension, chemical inertness and total water insolubility, these compounds have become predominate in the industry. The most common silicone oil used is dimethylpolysiloxane, with 2 to 2000 repeat-ing siloxane units. These oils are commonly blended with small mesh size hydrophobic solid particles, with which synergy is developed. The oils provide a carrier fluid and the spreading mechanism for the foam killing solid particles. Field strength products are typically sold as 5–50% emulsions.

3. Hydrophobic Silica. Amorphous, precipitated silica (particle size 1–2 μm) is commonly added to silicon oils, fats, waxes and aqueous mixtures for preparation of cost effective foam control products [18, 19].

4. Hydrophobic Fat Derivatives and Waxes. This category includes a broad variety of compounds, waxes and derivates.

   a. Fatty acid esters
   b. Fatty acid amides and sulfoamides
   c. Paraffinic hydrocarbon waxes
   d. Phosphoric acid mono-, di-, and tri-esters of short- and long-chain fatty alcohols (tri-*n*-butyl phosphate)
   e. Short- and long-chain natural or synthetic fatty alcohols
   f. Water insoluble soaps of long-chain fatty acid, such as aluminum stearate, magnesium stearate and calcium behenate
   g. Perfluorinated fatty alcohols

5. Water Insoluble Polymers. There are a number of poly-meric substances reported to have functionality as defoamers. These include fatty acid modified alkyloid resins, copolymers of vinyl acetate and long-chain maleic and fumaric acid diesters, as well as polypropylene and butylene oxide polymers and addition products [19–21].

6. Amphiphilic Components. This category includes a variety of substances that are activated by either temperature increases that create insolubility near the cloud point (low HLB nonionic ethoxylates, fatty acids, fatty amines) or in situ formation of insoluble calcium salts [22, 23].

Defoamers, such as those listed above, are typically introduced into the drilling fluid system upon return of the fluid to the surface. This can be accomplished by direct injection in the blooie (return) line, surface application in mud pits or by direct injection into closed loop separators.

Alternative Foam Control Systems [24, 25]

Trans-Foam[®] is a novel foaming agent system developed by Clearwater Inc., which is greatly affected, and thereby, controlled by changing the pH of the water being treated. Trans-Foam consists of components that by themselves are effective at broad pH ranges (components #25 and #12, for example). However, when mixed together these materials interfere with each other at low pH and complement each other at high pH, see Figure 4.

As can be seen in Figure 4, at pH levels of 11 or above, the formula #7, when agitated in a standard blender test creates a viscous stable foam.

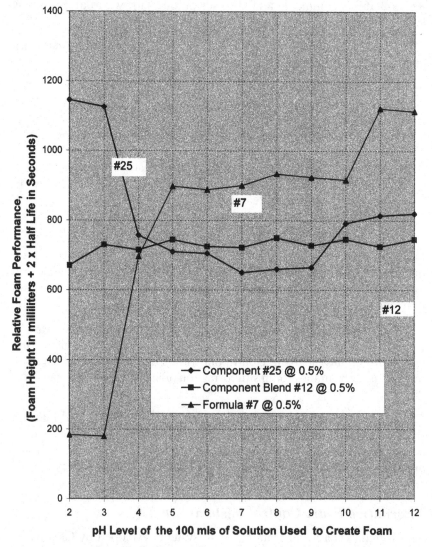

Figure 4.   Foam performance of two components (#25 and #12), individually and blended together.

When the pH is dropped to below 3.5, the foaming potential of the water is greatly reduced. It should also be noted that this formula is essentially a mixture of two surfactants, Components #25 and #12, which are themselves largely unaffected by the pH of the system. Component #12 actually shows slightly increased foam performance at reduced pH levels. It is the interaction, or incompatibility, of the two components at low pH that cause the reduced foam performance.

The foam is essentially killed by the addition of an acidic solution. Of commercial importance is the fact that this fluid can then be recycled, as this interference is reversible. Once the pH is elevated the foaming action returns and the fluid can be reused. This foam, defoam, refoam cycle can continue on for many cycles. However, makeup foaming agent must be added to replace the foaming agent lost due to surface adsorption of the drilled solids.

Benefits of Using a Recyclable Foam System.   *1. Containment.* One problem with using foam in the past has been the necessity to excavate large earthen pits to capture the volumes of foam returning out of the blooie line. On large diameter wells the equivalent of up to 600 gal/min of foam is needed to adequately clean the well bore. In one hour this equates to 36,000 gallons of foam. Some of the foam will break down over this period of time. However, a desirable characteristic of a foaming agent is its ability to maintain foam volume over time. Therefore, the pits fill up. Recyclable foam systems are designed to be used in the confines of an above ground standard mud system, thus eliminating the effort and expense of excavating large pits. In many situations installing earthen pits is not an option. This is especially true in offshore environments and in environmentally sensitive areas.

2. Water Consumption. The overall consumption of water is greatly reduced, since the water is being continuously recycled. On many drilling operations, the supply of fresh water is of foremost concern. The economics of water handling vary greatly depending on location. When using recyclable foam systems, the benefits are twofold, the water is continually recycled and the system operates in the absence of conventional defoamers, such as silicone oils. Disposal of water containing such materials can be difficult and expensive. With pH controlled foaming systems, the continual adding of defoamer (acid) and activator (alkali) forms an inert, non-toxic material that is highly water soluble (salt); therefore no significant environmental concerns are added to the system. At the conclusion of the drilling project, the drilling water can be disposed of in the normal fashion or recycled for the next well. With proper fluid cleaning procedures, the fluid may be used repeatedly, thereby positively affecting the bottom line economics of the project.

Table 9.   Specific Surface Areas of Some Reservoir Rocks and Clays

Solid	Specific Surface Area (m^2/g)
Clays	
Kaolinite	9–23
Smectite	35–110
Illite	40–110
Chlorite	14–42
Reservoir Rocks	
Berea sandstone	0.8–1.2
Indiana limestone	0.37–0.54
Baker dolomite	0.22–0.29
Quartz sand	0.16

3. Chemical Consumption.   In a closed loop recyclable system, a large majority of the foaming agent is recovered and available to be used again. Experience shows a 50–90% surfactant recovery rate depending on well conditions. Due to surface adsorption on the drilled solids and the wellbore surface some surfactant is continually lost. The amount of overall adsorption is determined by parameters such as the following [26–28]:

- solid: rock type (sandstone or carbonate), mineral composition (clays, other minor or trace minerals), wettability, surface charge, and specific surface area
- surfactant type, surfactant composition, and surfactant solubility
- temperature

Basically, the absolute amount of surfactant adsorbed per unit mass depends on the drilled rock's specific surface area. The larger the specific surface area the greater the surfactant loss as in Table 9.

4. Adaptability.   The single most outstanding feature of a recyclable system is its simplistic operation. By being able to kill the foam rapidly, the system will operate within the confines of nearly all conventional drilling mud or aerated fluid systems.

5. Well Site Safety.   Containment of a foam drilling fluid into a closed loop system provides a measure of safety that open systems do not allow. Hydrocarbons, acid gases and oxygen depleted air are kept safely away from personnel. Additionally, open "foam pits" can create a hazard, as they are typically plastic lined pits, that could allow the unknowing person to slip in and possibly suffocate under a blanket of foam.

An essential piece of equipment in recyclable systems is a well-designed mud–gas separator. The function of the separator is to disengage the gas and liquid phases. When the foam control additive is introduced prior to entering the separator, the foam is destroyed leaving fluid, cuttings, and gas. There exist many new designs for mud–gas separators, ranging from the simple to the complex. However, the most important feature of the mud–gas separator, in respect to a recyclable foam system, is the ability to maintain a liquid level in the bottom of the unit. The function of this "liquid leg" is to increase the efficiency of the liquid–gas separation process. As the returned fluid/gas mixture enters the unit, the velocity of the multi-phase fluid decreases rapidly, the gas phase exits out the top of the unit, the liquid and cuttings exit the bottom as in Figure 5.

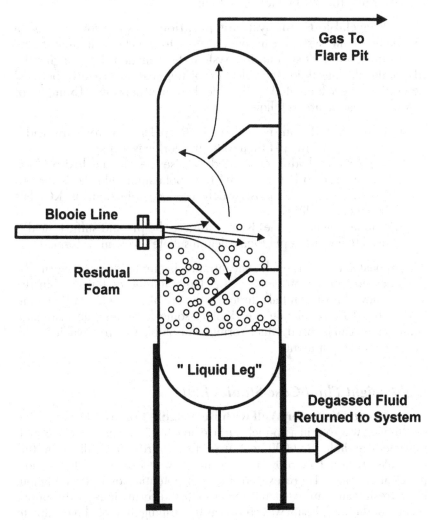

Figure 5.   Mud–gas separator schematic, showing foam "break-out".

By maintaining a "liquid leg" in the unit, the gas more efficiently follows the path of least resistance, which is out the top. This "liquid leg" allows for near complete destruction of the foam, by disallowing any remaining foam to flow out of the bottom of the unit onto the shaker screens. This is extremely important when drilling in producing oil reservoirs. Often, the reservoir pressure is slightly above the bubble point and, as the oil is brought to the surface by the foam, the pressure is reduced to below the bubble point. By utilizing an efficient mud–gas separator, the operator will ensure that the gas contained in the oil "breaks out" within the unit, thus eliminating the safety and handling concerns of excess natural gas carryover into the mud handling system.

Recyclable Foam System Operation. As previously stated, a recyclable foam system can be adapted to function in almost any conventional mud pit system. In a system that functions by changing the pH of the drilling fluid, the selection of the acid and base to be used determines the salt created by the acid–base neutralization. Examples of this salt formation are as follows:

1. If sodium hydroxide is used to raise the pH and hydrochloric acid is used to lower the pH then sodium chloride is formed.
2. If potassium hydroxide is used to raise the pH and hydrochloric acid is used to lower the pH then potassium chloride is formed (KCl). This is the most widely used combination, as KCl has beneficial qualities.
3. If lime (calcium oxide) is used to raise the pH and sulfuric acid is used to lower the pH then calcium sulfate (gypsum) is formed.

After repeated cycles, the salt concentration begins to increase. Normally, this does not pose a problem, as the large volume of fluid within the system allows for substantial dilution. If the use of salts in the drilling fluid is prohibited by regulation (i.e. KCl in some locations), an acid–base combination can be used, which forms an insoluble salt, removable by the mud cleaning equipment.

Lightweight Fluid Case Studies [29]

Large Diameter Well with Recyclable Foam. The objective on this well was to drill a 26-inch surface hole to 1500 feet and a 17.5-inch intermediate hole to 6000 feet with foam. Previous drilling in this particular region had proven to be excessively expensive due to low penetration rates. There existed a known potential for lost circulation, large fresh water influxes, and unconsolidated formations in the upper section of the well. Foam was chosen as the drilling fluid of choice, due to its ability to function effectively in the presence of high water influxes and its capability to drill effectively in formations with lost circulation potential. The location was in an environmentally sensitive area. Excavation of large earthen pits was not considered an option. Also, availability of fresh water to generate foam was limited. Transporting water to this remote location was quite expensive. The decision was made by to use a recyclable closed loop foam system. Existing surface equipment was utilized and few modifications were required to retrofit the mud pits to effectively operate the foam system.

The mud system capacity totaled 1200 bbl and consisted of four three-hundred-barrel standard surface mud tanks. The fluid in the first pit, below the shaker, was maintained at pH 3 by the addition of hydrochloric acid. A centrifugal pump was installed to pump the low pH fluid from the shaker pit into the blooie line, downstream of the choke manifold. This low pH fluid intermixed with the foamed fluid returning from the well, destroying the foam. The fluid was then run through a mud–gas separator. The gas phase was split off and sent to a flare pit. The fluid and drilled solids were sent over two parallel double deck shakers. Fluid level in the shaker pit was controlled by the use of a floating suction line and maintained to approximately 75% of working capacity. The fluid, a low pH foamer solution, was allowed to flow over into the second pit at the same quantity that was being pumped down the well. Simply stated, a recirculation loop was established, pumping from the shaker pit, into the blooie line, through the mud gas separator, across the shaker and back into the shaker pit. Once the fluid flowed out of the shaker pit, the pH was increased to 10 by the addition of potassium hydroxide (caustic potash). At this point the fluid was pumped through a desilter, followed by a centrifuge to eliminate the drilled solids. As with any drilling fluid, solids control is of utmost importance in a foam system.

Once the returned fluid had been cleaned, blender tests were run to determine the makeup concentrations of foamer and polymer. Due to natural adsorption of surfactant on the cuttings surface, maintenance concentrations of foamer and polymer were added to maintain the desired performance. The percentage of makeup depended on many different factors, such as penetration rate, temperature, rock type, etc. On this particular well makeup percentages were in the 0.1–0.2% range out of the original 0.8% loading. The foam system was considered to be very controllable, economical and successful by all those involved. Penetration rates of 5–10 times those achieved with conventional mud systems were realized.

Coiled Tubing Horizontal Re-entry with Recyclable Foam. This project was one of the first coiled tubing horizontal re-entries into a depleted oil reservoir with foam. The productive zone is a blanket sandstone reservoir varying from 10 to 100 feet in thickness. Over 700 wells were drilled in the field on 20 acre spacing. The wells vary from 2200 to 3500 feet in depth. The reservoir pressure is nearly constant throughout the field at 500 psi, due to uniform depletion. The objective of the project was to utilize a coiled tubing drilling unit to carry out the entire operation.

Foam was selected as the drilling fluid for the horizontal section. The necessity to maintain a low bottom hole pressure in order to remain "underbalanced," relative to formation pore pressure, along with the imminent danger of lost circulation, made this project an excellent candidate for foam. Minimizing formation damage was of prime concern.

Prior to beginning the project, a major issue to be addressed was the need for a recyclable surfactant, which would perform satisfactorily in the presence of a high hydrocarbon influx. A formulation was developed to withstand up to 50% oil influx by weight of the injected foam solution. This was accomplished through the use of oil tolerant amphoteric surfactants. Figure 6 depicts the surface handling facility and mud pit configuration used on this project.

The horizontal section was drilled with a $4\frac{3}{4}$-inch tri-cone bit to 4100 feet measured depth, MD. The bottom hole assembly (BHA) consisted of a bit, $3\frac{3}{8}$-inch positive displacement mud motor, adjustable bent housing, non magnetic dual float sub, non-magnetic collars (as required for non-magnetic spacing), coiled tubing quick connect, bi-directional orienting tool, steering tool, and coiled tubing connector.

Nitrogen and foamer solution (1.5% surfactant, 0.4% shale control polymer) were pumped at a rate of 500 scfm and 25 gpm respectively. The rate of penetration averaged 50 feet/hour. The horizontal section was drilled in 14 hours. Over the course of drilling, 300 bbl of foam solution were pumped and over 250 bbl of 35° oil were produced. This was viewed as a technological success, as the well was effectively drilled with foam in the presence of nearly a 50% oil influx.

A surface foam handling unit was specially built for this project (Figure 6). The unit is an efficient, compact, trailer mounted fluid system, designed to be used with recyclable foams. The unit is basically a condensed version of a standard mud handling system, complete with all of the necessary fluid conditioning capabilities. Due to the unique simplicity of the recyclable foam system, the scale of surface handling equipment is greatly reduced. The unit is comprised of one main square vessel compartmentalized into three separate pits. The unit contains a double deck shale shaker, desilter, centrifuge, and integrated 30-inch diameter, mud–gas separator. Two fully automated electronic pH controllers maintain fluid pH within the unit. The unit functioned as designed without operational problems.

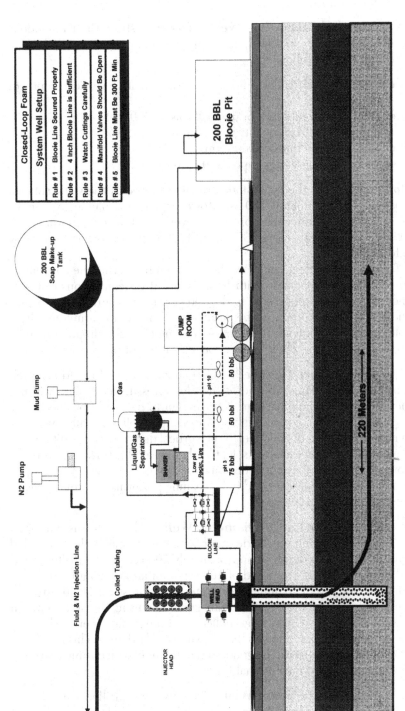

Figure 6.   Coiled tubing drilling, closed loop recirculating foam system surface layout.

Extended Reach Gas Well.   To demonstrate the tremendous friction that can be created by foam in small diameter, extended reach holes, consider the following:

- Re-entry well drilled into a gas reservoir at a vertical depth of 12,000 feet
- Total measured depth of the well was 14,500 feet
- Bottom hole foam quality was maintained at 98%
- Nitrogen flow rates of 900 scfm
- Foam solution was pumped at 9 gpm

With this geometry, the hydrostatic pressure exerted by the static column of foam was approximately 300 psi. However, under dynamic conditions, the observed bottom hole pressures were in the neighborhood of 2100 psi. This indicates that 1800 psi of friction pressure was created by the foam.

This frictional effect could also be observed by varying the concentration of foaming surfactant. With foaming agent concentrations of 0.4%, standpipe pressures registered in the 500 psi range. A surfactant concentration of 0.6% was found to be sufficient to create a continuous phase stable foam. Without adjusting other parameters, standpipe pressures increased to 1500 psi when foamer concentration was increased.

Low Pressure Gas Well.   Certain gas bearing formations, such as shales or coal seams, can be of such low pressure that foam will not generate a bottom hole pressure low enough to be "underbalanced". In these cases the fluid of choice is fog. An example well was drilled with fog to a measured depth of 3500 feet. True vertical depth was only 900 feet. A fog generating surfactant was selected that created a surface tension of below 25 dynes/cm at a concentration of 0.5% v/v. By increasing the GVF, observed cuttings transport and hole cleaning performance were deemed to be similar to that of foam.

Exploratory Large Diameter Well.   This well was equipped with 28-inch ID casing set at a depth of 30 feet. A 26-inch hole was drilled from this point to a depth of approximately 2000 feet. The drilling fluid used in this well was composed of 1000 scfm of air and 25 gpm of water containing 1% v/v of foaming surfactant. Back-pressure of 10 psig was maintained at the blooie line. The foam generated using this system was in excess of 96% GVF. The cuttings collected at the surface were as large as $\frac{1}{2}''$ in diameter compared to less than $\frac{1}{4}''$ with standard mud. This illustrates that high quality foams can be generated in the field with cuttings carrying capacity superior to conventional fluids.

High Quality Foam Well.   This well was drilled with dry air until a water-bearing zone was encountered, and the hole became wet. At this point, the circulating medium was changed from the dry air to a high quality foam by adding to the air 16 gpm of water containing 1.78% foaming agent and 0.34% shale control polymer. As a result of this change in the circulating medium, three effects were immediately noted. The diameter of the cutting size increased from $\frac{1}{8}''$ to $\frac{3}{8}''$; the standpipe pressure increased from 210 psig to 245 psig; and the penetration rate dropped from 40 feet/hour to 15 feet/hour. It should be noted that a portion of the reduction in penetration rate was attributable to a change in the formation penetrated.

Summary

Lightweight drilling fluids are experiencing ever-broadening exposure in the oil and gas industry. This is driven by the expansion of "under-balanced" type drilling operations, which become increasingly important in the face of mature reservoirs, reduced pressures and depleted fields. The design, selection and application of surfactant systems for these fluids will play a critical role in determining the extent and success of this expansion.

Abbreviations

AOS	alpha olefin sulfonate
bbl	barrel, 42 US gallons
BHA	bottom hole assembly
BHP	bottom hole pressure
BHT	bottom hole temperature
DDBSA	dodecyl benzene sulfonic acid
ECD	equivalent circulating density
EO	ethylene oxide
GVF	gas volume fraction, volume percent gas in a compressible fluid
HEC	hydroxy ethyl cellulose
HLB	hydrophile–lipophile balance
HQF	high quality foam
LVF	liquid volume fraction, volume percent liquid in a compressible fluid
MD	measured depth
MEA	mono-ethanolamine
PEG	polyethylene glycol
PHPA	partially hydrolyzed polyacrylamide
PO	propylene oxide
ROP	rate of penetration

SCMC sodium carboxy methyl cellulose
SCS   sodium cumene sulfonate
SLS   sodium lauryl sulfate
STS   sodium toluene sulfonate
SXS   sodium xylene sulfonate

References

1. Govier, G.W.; Aziz, K. *The Flow of Complex Mixtures in Pipes*; Van Nostrand Reinhold: New York, 1972.
2. Beyer, A.H.; Millhone, R.S.; Foote, R.W. *Proceedings of the 47th Annual Technical Conference of SPE*; Society of Petroleum Engineers: Richardson, TX, 1972, paper SPE 3986.
3. Okpobiri, G.A.; Ikoku, C.U. *SPE Drilling Engineering* **1986**, Feb., 71–88.
4. Schramm, L.L.; Wassmuth, F. In *Foams: Fundamentals and Applications in the Petroleum Industry*; Schramm, L.L. Ed.; American Chemical Society: Washington, DC, 1994, pp 3–45.
5. Boys, C.V. *Soap Bubbles: Their Colours and the Forces Which Mold Them*, Dover Publications: New York, 1959.
6. *Field Testing Report Trans-Foam O*, Clearwater Inc., Pittsburgh, 1995.
7. *Foam Engineers Manual*, Clearwater Inc., Pittsburgh, p 17.
8. Clearwater Inc., Unpublished research documents, Clearwater Inc., Pittsburgh, 1997.
9. Purushottam, Y., *Network Modeling of Foam Drainage*, B. Tech. Thesis, Indian Institute of Technology, Bombay, India, 1995.
10. Laheja, A.P.; Basak, S.; Patil, R.M.; Khilar, K.C. *Langmuir* **1998**, *14*, 560–564.
11. Markworth, A.J. *J. Coll. Interface Sci.* **1984**, *107*, 569.
12. Mullins, W.W. *J. Appl. Phys.* **1986**, *59*, 1341.
13. Glazier, J.A.; Gross, S.P.; Stavans, J. *Phys. Rev. A* **1987**, *36*, 306.
14. Graham, R.L. *Report of Foam Drilling for Petrobras Well Sabalo X-1*, ECD Northwest Inc., Pittsburgh, Dec. 26, 1998.
15. Schramm, L.L., Ed. *Foams: Fundamentals and Applications in the Petroleum Industry*, American Chemical Society: Washington, DC, 1994.
16. Philipp, C. *Allg. Oel Fett Ztg.* **1942**, *39*, (5) 167–170; (6) 203–207; (7) 235–239.
17. Ohl, F. *Seifen Oele Fette Wachse* **1953**, (5) 114–115; (6) 141–143.
18. Iler, R.K. U.S. Patent 2 801 185, 1952.
19. Simpson, E.A.; Doyle, C.F. U.S. Patent 3 720 532, 1972.
20. Fock, J. German Patent DE 3 201 478, 1982.
21. Fock, J.; Fink, H.F. German Patent DE 3 201 479, 1982.
22. Hofer, R. "Foams and Foam Control", In *Ullmann's Encyclopedia of Industrial Chemistry*, Gerhartz, W., Ed.; Wiley: New York, 1988, Vol. A, 11.
23. Friberg, S.; Liang, P. In *Encyclopedia of Chemical Processing and Design*, McKetta, J.J., Ed.; Marcel Dekker: New York, 1985, Vol. 23, 312–333.

24. Thomas, T.R. U.S. Patent 5 385 206, Jan. 31, 1995.
25. Thomas, T.R. U.S. Patent 5 591 701, Jan. 7, 1997.
26. Mannhardt, K.; Schramm, L.L.; Novosad, J.J. *Colloids and Surfaces* **1992**, *68*, 37–53.
27. Mannhardt, K.; Schramm, L.L.; Novosad, J.J., *SPE Adv. Technol. Ser.* **1993**, *1(1)*, 212–218.
28. Mannhardt, K.; Novosad, J. In *Foams: Fundamentals and Applications in the Petroleum Industry*, Schramm, L.L., Ed.; American Chemical Society: Washington, DC, 1994, 259–316.
29. Wilkes, T.M.; Watson, R.W.; Graham R.L. *Proceedings of the IADC/SPE Drilling Conference*; Society of Petroleum Engineers: Richardson, TX, 1972, paper IADC/SPE 39301, 1998.

RECEIVED for review December 2, 1998. ACCEPTED revised manuscript February 3, 1999.

Surfactant Use in Acid Stimulation

Hisham A. Nasr-El-Din

Lab Research and Development Center, Engineering Services, Saudi
Aramco, PO Box 62, Dhahran 31311, Saudi Arabia

*Acids are extensively used to enhance the performance of various
types of wells. Several chemicals are added to the acid to
minimize adverse effects of the acid and enhance the overall
efficiency of the acidizing treatment. Surfactants are commonly
included in the acid formulation to perform one or more
important tasks. Surfactants encounter various chemical species
that can affect their performance.*

*This chapter discusses several applications where surfactants
can be used as anti-sludge agents, acid retarders, acid diverters
and surface tension reducers. The chapter also highlights some
chemical interactions, which may result in phase separation of
nonionic surfactants, which are widely used in acid stimulation.*

Introduction

Acidizing treatments are used to remove wellbore damage, enhance
matrix permeability or both. The acid reacts with the rock matrix and, as
a result, the permeability of the formation will increase. Reagents
commonly used to stimulate carbonate reservoirs are hydrochloric (15 or
28 wt%), formic (9 wt%), and acetic (10 wt%) acids [1]. Recently, Fredd
and Fogler [2, 3] introduced ethylenediamine tetraacetic acid (EDTA) as
an alternative fluid to acidize carbonate formations.

Acids commonly used in sandstone formations include hydrofluoric-
based acids. Full strength mud acid (12 wt% HCl + 3 wt% HF) has
been used to stimulate sandstone reservoirs for several decades [1].
Because of the fast reaction of hydrofluoric acid with clay minerals,
various retarded mud acids were introduced. In one system, aluminum
chloride was added to mud acid systems to ensure deep acid penetration
[4]. In a second system, boric acid was used to achieve the same goal [5].
Recently, Lullo and Rae [6] introduced a new acid system, consisting of
phosphonic acid and hydrofluoric acid. For high temperature applications, mixtures of organic acids (formic or acetic) and hydrofluoric acids
are used [7].

Surfactants are used in almost all acidizing treatments to do one or more of the following functions: water-wet the formation, break emulsions or sludges, reduce surface or interfacial tension, and help to remove fine particles. Other functions of surfactants include forming foams for acid diversion and preparing emulsified acids for deep acid penetration.

Surfactants are chemicals containing an oil-soluble group (lipophilic) and a water-soluble group (hydrophilic). These chemicals have the ability to accumulate (adsorb) at various interfaces and alter their properties. Surfactants are classified according to their ionic nature into four categories: anionic, cationic, nonionic and amphoteric. Various surfactants from these groups are used in well acidizing. Some specific applications of these surfactants in well stimulation are discussed in this chapter.

The type and concentration of the surfactant used during acid stimulation depends on the characteristics of the treated well (temperature, pressure, lithology, salinity) and the function of the surfactant. For oil producing wells, surfactants are added to prevent the formation of acid–oil sludge [8–10]. They are also used to water-wet the formation after the acid job [1]. For gas wells in low permeability reservoirs, the spent acid can be trapped around the wellbore by the effect of the capillary forces. Accumulation of water around the wellbore area (known as water blockage) will reduce the relative permeability to gas and hence gas production [11]. To overcome this problem, surfactants are added to the acid to reduce the surface tension of the spent acid, allowing the spent acid to flow easily from tight formations. For water disposal wells, surfactants are used to lower interfacial tension between the native oil and treating fluids and, as a result, the trapped oil will be mobilized and the injectivity of the well will be increased [12].

Besides the above important functions, surfactants are used to enhance acid penetration into the formation and improve sweep efficiency during well stimulation. It is well known that acidizing tight carbonate formations will result in surface wash-out only [13]. One way to enhance acid propagation is to use emulsified acids [14, 15]. In this case, a suitable surfactant is used to emulsify the acid (typically 15 wt% HCl) and a light hydrocarbon phase (crude oil or diesel). Injection of the acid in this form (acid-in-diesel emulsion) will minimize acid–rock contact, which will result in the formation of deep wormholes (i.e. channels with high permeability). Emulsifying the acid will also increase its viscosity and improve acid distribution in heterogeneous reservoirs [16].

Stimulation of horizontal and multi-lateral wells presents a very challenging problem. With a very long target zone(s), acid diversion plays a very important role in determining the efficiency of the acidizing job. Without a proper means for acid diversion, most of the injected acid will only flow into high permeability zones. This uneven flow will result in improving the permeability of the already permeable zone and leave the tight zones nearly unaffected. One way to improve sweep efficiency during well stimulation is to use foams [*17*]. In this case, a surfactant is used to form viscous foam. The foam will flow into the high permeability zones, causing the acid to flow into the tight zones.

As outlined above, surfactants are added to acids to perform one or more of several needed functions. However, other chemicals are also added to the acid. These additives include corrosion inhibitors [*18*], iron control agents [*19, 20*], hydrogen sulfide scavengers [*21*], scale inhibitor [*22*] and clay stabilizers [*23*]. It is very important to perform compatibility tests of the selected surfactant with the acid formula, especially in this complex environment. Also, some of the surfactants are used in high temperature and high salinity applications. Therefore, it is necessary to ensure thermal stability of these surfactants under these harsh conditions.

As indicated above, surfactants are added to acids during well stimulation to perform several tasks. The overall efficiency of stimulation jobs depends to a great extent on the ability of these surfactants to perform their functions effectively. This chapter will discuss some specific applications of surfactants in well stimulation and examine the effect of acids and stimulation additives on the properties of some classes of surfactants that are commonly used during well acidizing.

Surfactants as Anti-Sludge Agents

Sludge Formation. One of the main functions of acid stimulation is to remove formation damage, hence, to enhance the productivity of the oil producing zones. Some of the major problems encountered during stimulation of oil producing wells are the formation of acid-in-oil emulsions and precipitation of asphaltene particles, which induce sludge formation. These viscous emulsions and sludges can plug the formation and cause further damage to the well [*24–26*]. They also cause operational problems in the surface facilities following acid stimulation [*26–29*].

Both oil-in-acid and acid-in-oil emulsions form during well acid stimulation [*10*]. The latter type of emulsions, however, can cause serious problems because of its high viscosity. These viscous emulsions are slow to return into the wellbore and result in loss of production, especially in low-pressure reservoirs.

It is important to understand the causes of acid sludge formation. One mechanism that is frequently cited is precipitation of asphaltene particles and heavy hydrocarbons when oil contacts strong acids. Asphaltenes are colloidal particles dispersed in crude oils that are composed of condensed aromatic ring structures containing a significant number of heteroatoms such as oxygen, sulfur and nitrogen [*30*]. They exist in a micelle form with natural resins (maltenes) adsorbed on the surface of the asphaltic particle (Figure 1). The adsorbed resins stabilize asphaltene micelles [31].

Because asphaltene particles are negatively charged [24], the highly charged positive acid protons (H^+) can have a significant effect on the surface charge of asphaltene particles. Neutralization of these particles with H^+ allows larger aggregates of asphaltenes to form [32]. These aggregates can form very rigid emulsions during acidization and can also deposit at the pore throats, reducing well productivity (Figure 2). The emulsion droplets are stabilized by the asphaltene or resin fraction of crude oils, and these can reduce the interfacial tension [31]. Moreover, the problem becomes more complicated if other particulate solids are present. Sand, silt, and metal oxides can also accumulate at the oil–acid interface and stabilize acid-in-oil emulsions [24, 27].

Crude oils have different emulsion stability, viscosity and density [33]. Houchin et al. [34] studied the effect of acids on 231 crude oils from 27 formations. Depending on oil characteristics, acids can form sludges or

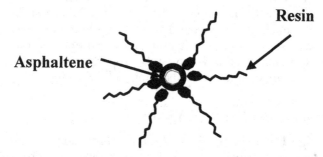

Figure 1.   Asphaltene micelle.

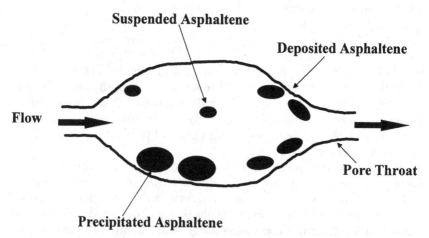

Figure 2.   Formation damage due to asphaltene precipitation.

rigid film emulsions. Acid-induced sludge occurs primarily in crudes with °API ⩾ 27 and asphaltene content ⩽ 3 wt%. Strassner [35] found that rigid film emulsions occurred when crude containing high asphaltenes contacts an aqueous phase of pH < 6. Houchin et al. [34] found that rigid film emulsions form when crude oils with °API ⩽ 22 and asphaltene content ⩾ 4 wt% come into contact with strong acids.

There are several factors which can enhance acid sludging [10]. Increasing HCl strength will aggravate sludging of asphaltic crudes, especially at high temperatures. Therefore, asphaltic crudes should not be exposed to 28 wt% HCl [34]. Iron contamination, especially ferric iron, also plays a very important role in sludge formation [36]. The influence of iron on sludge formation can be explained by flocculation of asphaltene particles, which follows charge neutralization. The latter occurs due to coordination of iron with porphyrin, pyrroles, pyridines or possibly phenolic hydroxyl groups present in the crude oil. Iron can also catalyze polymerization processes of heavy hydrocarbons at high temperatures [32]. Iron in acid can originate by dissolution of corrosion products and some iron-rich minerals (chlorite, ankerite, and magnetite) [19].

Full strength mud acid (12 wt% HCl + 3 wt% HF) appears to cause more severe sludging than HCl at the same acid concentration. Other factors that control sludge formation include reservoir temperature, viscosity of the native oil [10] and acid additives (e.g., corrosion inhibitors) [31, 37].

Based on the above discussion, every effort should be made to avoid the formation of acid–oil sludge. Anti-sludge agents, surfactants and mutual solvents, are usually added to the injected acid to minimize the impact of acid–oil sludge [26]. These agents disperse the asphaltene particles and prevent their precipitation.

It is worth mentioning that mutual solvents are multi-functional, nonionic agents soluble in oil, water and acids. They contain strong ether and alcohol groups, which provide a wide range of solvent properties. Ethylene glycol monobutyl ether (EGMBE) is one of the mutual solvents that is commonly used in the oil industry. It is used to achieve several tasks, including reducing surface tension and enhancing water-wetting characteristics of formation rocks after acidizing.

Field Application.   Sludge formation during acid stimulation of oil wells was noted in the United States and central Alberta, Canada, and its formation was detected in the field by one or more of the following symptoms [38]:

1. Acid cannot be squeezed deep into the formation without applying high pressures
2. The presence of viscous material in the produced fluids
3. Stimulation results below expectations

4. Formation does not respond to acidizing
5. Communication behind casing

The American Petroleum Institute recommends conducting acid sludge test API RP 42 before performing acidizing jobs in the field [39]. These tests are used to determine the type and concentration of the surfactant (nonionic or anionic) needed to break acid-in-oil emulsions in a reasonable period of time. It is very important to perform the tests using live and spent acids at reservoir temperature. Spent acid should be prepared in the lab using formation rock. The live and spent acids should include acid additives as per field formula. The crude oil sample should be fresh, and free from water and oilfield chemicals (e.g., scale inhibitors, demulsifiers, etc.).

Because of the strong effect of iron ions on sludge formation, especially Fe^{3+}, several researchers [9, 36, 40] modified the API test procedure by adding up to 10,000 mg/L of iron (Fe^{2+} and Fe^{3+}) to the acid. This amount of iron is equivalent to the iron that would be dissolved by the acid during pumping.

It should be mentioned that the standard method API RP 42 has drawbacks [9]. For example, it does not take into account loss of surfactant due to adsorption on the rock surface [8, 9]. Once the concentration of the surfactant in the injected acid decreases due to adsorption or phase separation (precipitation), acid–oil sludge will form [32]. Rietjens [41] recommended performing acid sludge tests using flow systems (coreflood experiments) to overcome some of the problems encountered with the API RP 42 procedure.

Besides adding a suitable anti-sludge agent to the injected acid, the concentration of ferric ion in the injected acid should be reduced. This can be achieved by using properly cleaned mixing tanks, and pickling the coiled tubing used to place the acid. Moreover, iron reducing agents should be added to the acid when sweat wells are acidized [19].

Another method to reduce sludge formation is to minimize acid contact with the native crude oil. One way to achieve this goal is to use solvent preflush (xylene or toluene) that can act as a barrier to separate sensitive crudes and the acid [26, 34]. A second way is to use acid emulsified in solvent [42]. The solvent in this emulsion will minimize acid–oil contact.

Surfactants as Acid Retarders

Hydrochloric acid is commonly used to stimulate carbonate (calcite and dolomite) reservoirs [1]. The reaction rate of HCl acid with calcite is very fast [43, 44]. In conventional stimulation when 15 wt% HCl is used at low flow rates, the acid reacts with carbonate rocks and causes a face dissolution or surface wash-out [13]. This means that the acid is spent on the formation surface and cannot penetrate the damaged zone to enhance the formation permeability. Face dissolution leads to consumption of large volumes of acid [2]. One way to overcome this problem is to use acid-in-diesel emulsions [14, 15, 45, 46]. This emulsion consists of diesel (continuous phase), HCl acid (dispersed phase), and an emulsifier. Diesel acts as a diffusion barrier between the acid and the rock [47–49], slowing the reaction rate of the acid with carbonate rock. This gives the acid the ability to penetrate deeper into the formation by creating wormholes and enhancing well productivity [13, 50–52].

Method of Preparation in the Lab. The emulsified acid consists of three main components: HCl acid, a hydrocarbon phase (diesel or light crude oil) and an emulsifier (a cationic surfactant). Typically the acid to diesel volume ratio is 70:30 [14]; however, other volume ratios can be used. First, the emulsifier is mixed with diesel in a Waring blender at a medium speed. The corrosion inhibitor and other acid additives are mixed with 15 wt% HCl acid in a separate beaker. The acid is gently added to diesel in the blender. Then, the two phases are mixed in the blender for 10 minutes at a high speed. The concentration of the emulsifier and mixing time should be adjusted until the electrical conductivity of the emulsified acid is zero. High conductivity readings mean that the acid remains the external phase, and more mixing or additional emulsifier is needed. The same procedure can be applied in the field using proper mixing equipment.

Characteristics of Emulsified Acid. Al-Anazi et al. [14] measured the apparent viscosity of the acid-in-diesel emulsion as a function of shear rate at various temperatures by using a Brookfield viscometer Model DV-II. Figure 3 shows that the apparent viscosity decreased as the shear rate was increased. This result indicates that the acid-in-diesel emulsion is a non-Newtonian fluid (shear-thinning behavior). Crowe and Miller [45] and Krawietz and Rael [53] reported a similar behavior. The apparent viscosity (η) can be predicted over the shear rate ($\dot{\gamma}$) examined using the power-law model given by the following equation:

$$\eta = k(\dot{\gamma})^{n-1} \tag{1}$$

This model yields good predictions (straight lines shown in Figure 3) for the apparent viscosity. The model parameters are needed to calculate the pressure required for pumping the emulsified acid. The power-law index (n) slightly increases (from 0.62 to 0.68) as temperature is increased to 45 °C [14]. As a result, the emulsified acid approaches Newtonian behavior at higher temperatures. Field data [14, 15] indicate that the

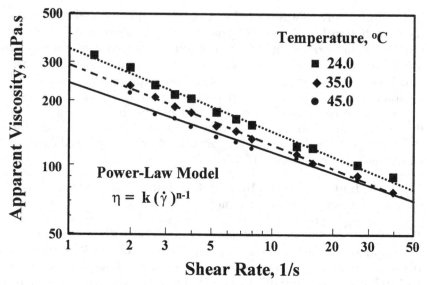

Figure 3.   Apparent viscosity of emulsified acid as a function of shear rate. (Reproduced with permission from reference 14. Copyright 1998.)

emulsified acid can be pumped using a 1.25 or 1.5 inch coiled tubing. The emulsified acid can also be placed into the target zone by bullheading [15].

Stability of Acid-in-Diesel Emulsions.   Al-Anazi et al. [14] showed that acid-in-diesel emulsion is stable for more than three days at room temperature. However, at high temperatures it breaks down and an aqueous (acidic) phase was noted at the bottom of the test tube. Figure 4 depicts the volume of the separated aqueous phase as a function of time at $96\,^{\circ}C$. The aqueous phase first appears after 85 minutes. The volume of separated acid gradually increases until complete phase separation occurs after nearly 220 minutes. In the presence of reservoir rock, the aqueous phase appears after approximately 20 minutes, and complete phase separation takes place after an hour. These results indicate lower emulsion stability in the presence of calcite. The acid reaction with the carbonate rock produces water (which causes the pH to rise) and calcium chloride. It appears from these results that the surfactant moves away from the acid–diesel interface as the pH or ionic strength increases, which causes the emulsion to break.

The droplet size of the dispersed phase plays a key role in the effectiveness of the acidizing job. Too fine or coarse droplets will adversely affect the efficiency of the stimulation job. The acid-in-diesel emulsion was examined under the microscope and the drop size distribution of the dispersed phase determined using a phase contrast technique.

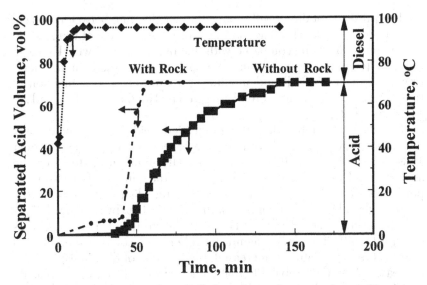

Figure 4.   Separation of emulsified acid as a function of time. (Reproduced with permission from reference 14. Copyright 1998.)

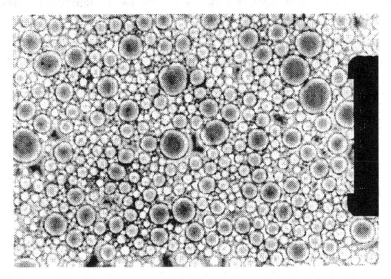

Figure 5.   A photomicrograph (×250) of the emulsified acid. (Reproduced with permission from reference 14. Copyright 1998.)

Figure 5 is a photomicrograph (250 ×) of an acid-in-diesel emulsion. The average droplet size for this acid-in-diesel emulsion is nearly 77 μm [*14*]. Excellent field results were claimed when this acid was used to stimulate carbonate formations with permeability less than 100 mD [*14, 15*].

The presence of diesel will reduce the reaction rate between the acid and carbonate rock. The reaction rate (dissolving power) of the acid-in-diesel emulsion can be compared with that of a regular 15 wt% HCl acid at room temperature by placing the rock (calcite) slices in the acids (regular and emulsified) and monitoring their weight as a function of time. More details on this method are given by Hoefner et al. [13] and Al-Anazi et al. [14].

Figure 6 illustrates that the emulsified acid has a much lower dissolution rate compared with regular 15 wt% HCl [14]. The regular 15 wt% HCl dissolved 90 wt% of the rock after 6 minutes. However, the acid-in-diesel emulsion dissolved 2 wt% of the sample after the same period of time. This result indicates that the dissolution rate of the emulsified acid is slower than that of the 15 wt% HCl by a factor of 45.

The reaction rate of regular HCl with calcite is fast. However, mass transfer by diffusion plays a significant role in the case of the emulsified acid. As the temperature is increased, the viscosity of the emulsified acid decreases. Also, the probability that an acid drop will contact the calcite rock increases. Both factors enhance mass transfer of the acid to the rock surface. This point was examined by Al-Anazi et al. [14] who showed that the dissolution rate of the emulsified acid increases at high temperatures (Figure 7).

Coreflood Tests. The effectiveness of the emulsified acid is also evaluated by performing coreflood experiments [14, 15]. In each

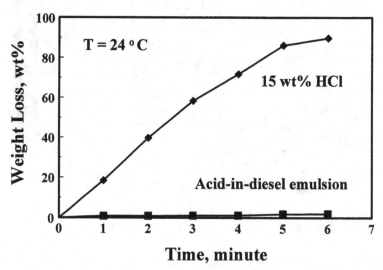

Figure 6. Weight loss of carbonate rock in the presence of regular and emulsified HCl. (Reproduced with permission from reference 14. Copyright 1998.)

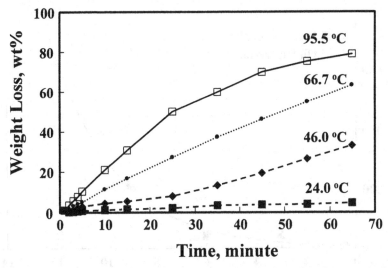

Figure 7.   Effect of temperature on the dissolution of carbonate rock. (Reproduced with permission from reference 14. Copyright 1998.)

experiment, the core (carbonate) is first saturated with the brine under vacuum. Either brine or oil is injected until the pressure drop across the core becomes constant. Then, the acid (either regular or emulsified) is injected and followed by the formation brine or oil as a postflush. A new core plug sample is used in each experiment. The effect of injection flow rate on acid propagation in the core is studied at flow rates from 0.5 to 12.0 cm^3/min.

Core permeability is determined before (K_o) and after (K) acid injection by using Darcy's for linear flow. Permeability ratio, K_r, is also calculated as a function of the cumulative core effluent using the following equation:

$$K_r = \frac{K}{K_o} \tag{2}$$

Figure 8, from reference [14], shows the permeability ratio as a function of the acid injection rate. For brine saturated cores, the final core permeability exponentially increases with the acid injection rate. The same trend is noted in the case of oil saturated cores.

Field Application of Emulsified Acids.   Surfactants are used to form emulsified acid, where acid is the dispersed or internal phase. Pumping the acid in this form reduces acid contact with the rock. This enables the acid to form deep wormholes. It also reduces acid contact with the native crude. This will reduce sludge precipitation and associated problems [42].

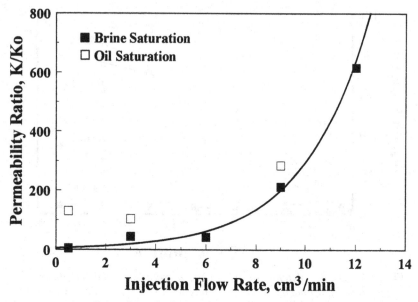

Figure 8.   Effect of emulsified acid on the permeability ratio. (Reproduced with permission from reference 14. Copyright 1998.)

The apparent viscosity of this acid is relatively high, which will enhance sweep efficiency during acidizing [16]. The acid can be pumped using a coiled tubing or bullheaded. The fact that the acid does not come in contact with the well tubulars minimizes corrosion, which will reduce iron concentration in the acid. Other advantages of this acid are that only small amounts of corrosion inhibitors and iron control agents are usually needed.

There are specific properties required for using emulsified acids [14]. The emulsified acid should be stable for a long period of time at room temperature, in case of operational problems, and should be stable until it reaches the target zone. Obviously, a longer period of time will be needed when acidizing horizontal wells. Once inside the formation, the acid should be soaked for a long period of time until it completely reacts with the formation. It should be also mentioned that the soaking time should be longer when stimulating injectors or disposal wells, because of their lower temperature [15].

Because the acid is injected in an emulsified form, extreme care should be taken in the design of the preflush and postflushes [14]. The presence of demulsifiers or mutual solvent may break the emulsified acid.

Emulsified acid was successfully applied in oil producing wells [14], water injectors and water disposal wells [15]. Its use is recommended when the permeability of the target zone is less than 100 mD. Emulsified acid should be used with care when stimulating weak carbonate formations where there is a potential for producing fine particles.

Surfactants to Reduce Surface Tension During Well Acidizing

The surface tension of the aqueous phase is an important consideration during acid stimulation of gas wells. Low surface tension is required to reduce capillary forces that trap the aqueous phase in the pores of formation rock. Accumulation of the aqueous phase near the wellbore area, known as water blockage, leads to significant reduction in gas production [11].

In tight gas reservoirs, the spent acid is usually trapped at the pore throats by the capillary forces (Figure 9). The pressure drop (Δp) required to mobilize a drop of trapped spent acid can be calculated using the Laplace equation, as follows:

$$\Delta p = 2\sigma \left[\frac{1}{r_1} - \frac{1}{r_2} \right] \tag{3}$$

where σ is the surface tension, r_1 and r_2 are the radii of the water droplet. According to equation 3, the surface tension should be reduced so that the spent acid droplet can be mobilized and produced from the formation. Lower surface tension values of the spent acid can be achieved by adding special surfactants to the injected acids. Nonionic surfactants and mutual solvents are commonly added to the acid to lower its surface tension [54–57].

Surface tension of stimulating fluids is a complex function of acid type, concentration, ionic strength, and additive type and concentration [58, 59]. Figure 10 shows a schematic diagram of the variation of surface

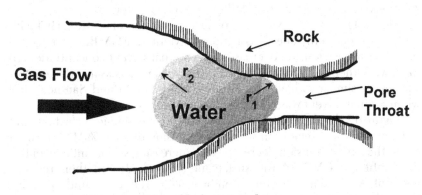

Figure 9.   *Water blockage in tight gas reservoirs.*

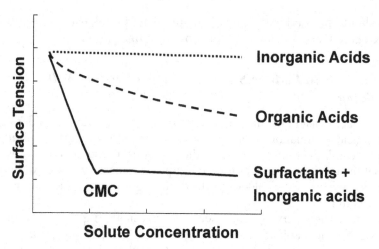

Figure 10.   *Surface tension of various stimulating fluids.*

tension for inorganic acids, organic acids and inorganic acids + surfactants. The addition of inorganic acid (e.g., HCl) to water slightly decreases its surface tension [59], as a function of acid concentration (Figure 10). Organic acids commonly used in well stimulation (acetic and formic) significantly reduce surface tension. The relationship between surface tension and acid concentration is not linear [58–62]. Addition of surfactants to a strong acid reduces its surface tension until the critical micelle concentration, CMC, after which increasing surfactant concentration does not affect surface tension.

Nasr-El-Din and coworkers [57, 58] conducted detailed studies on the effect of acids and stimulation additives on surface tension, using an automated Krüss (Model K-12) Tensiometer. The measurements were carried out at 25 °C using the Wilhelmy plate method. All aqueous solutions were prepared using Millipore Milli-Q water.

Figure 11 shows the variation of surface tension of 15 wt% HCl with mutual solvent (ethylene glycol monobutylether, EGMBE) concentration. The surface tension decreases with mutual solvent concentration up to 10 wt%, then remains constant. Mutual solvent acts as a surface-active species. A similar behavior was noted by D'Angelo and Santucci [63] when mutual solvent was added to distilled water.

Figure 12 depicts the influence of TX-100, a nonionic surfactant, and SDS, an ionic surfactant, on the surface tension of 15 wt% HCl. In both cases, the surface tension decreases with increasing surfactant concentration until the CMC of the surfactant is reached, and then remains constant. At the same concentration by weight, SDS was found to reduce the surface tension more than TX-100 [58]. However, SDS can interact

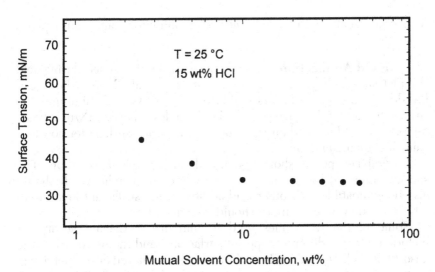

Figure 11.   Effect of mutual solvent on the surface tension of HCl solutions.

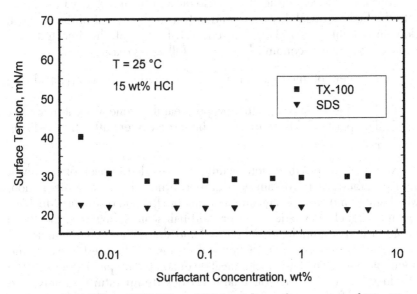

Figure 12.   Effect of TX-100 and SDS on the surface tension of 15 wt% HCl.

with other acid additives, e.g., corrosion inhibitors, and cause phase separation.

Field Application.   Low surface tension is needed during acid stimulation of tight reservoirs. Surfactants, especially fluorocarbons, can be added to the injected acid to achieve this goal. As a rule of thumb, the surface tension of the injected acid should be less than 30 mN/m at room temperature. Fluorocarbon surfactants can reduce surface tension to a value less than 20 mN/m.

Several key points should be noted during field application. The surfactant should be thermally stable at reservoir conditions. It should also be compatible with other acid additives, and surfactant loss due to adsorption on the rock surface should be minimal.

Some of the techniques used to minimize surfactant adsorption include mixing of different types of surfactants and the use of sacrificial adsorbent [64]. At the low pH environment encountered during acidizing, mutual solvent was found to minimize the loss of surfactant due to adsorption in sandstone reservoirs [65].

Surfactants to Divert Acid Using Foams

Most oil reservoirs (sandstone and carbonate) are heterogeneous with permeability varying from one layer to another. Acidizing such heterogeneous media is very difficult, because most of the injected acid will flow into the high permeability zones and only a small amount of the acid will flow into tight zones. This uneven distribution of the injected acid represents a major economic loss for the following reasons:

1. A larger volume of acid is needed to conduct the matrix-acidizing job
2. The flow from or into the low permeability zone will not increase
3. The permeability contrast will be more evident after the acidizing job

An effective acid diversion technique is needed to overcome uneven acid distribution and obtain good sweep efficiency during stimulation. Mechanical and chemical means are available for acid diversion. Mechanical means include straddle packers and ball sealers, however, they have limited use in openhole, gravel packed and slotted liner completions and are normally expensive [66]. Chemical means can be used in cased and openhole wells. The type of chemical diversion technique depends on the lithology and other reservoir characteristics (temperature, salinity, and hydrogen sulfide content). In carbonate reservoirs, emulsified acids [16] and viscosity controlled acids [67] have been used to improve sweep efficiency during well stimulation. In sandstone reservoirs, oil-soluble resins or benzoic acid particles were used in some cases.

Another effective method of acid diversion is the use of foams. Foam is a mixture of a liquid phase and a gas phase. A suitable surfactant is added to the liquid phase to reduce surface tension and stabilize foam lamellae. Nitrogen and carbon dioxide are exclusively used in the field to form foams [17]. The volume percent of the gas phase in the foam is known as foam quality (α), and is calculated using the following equation:

$$\alpha = \frac{V_{g_{T,P}}}{V_{g_{T,P}} + V_l} \tag{4}$$

where V_g and V_l are the volumes of the gas and liquid phases at bottom hole pressure and temperature.

Acid diversion using foams was introduced over ten years ago [66]. Foams can be used in sandstone and carbonate reservoirs for both cased and openhole completions [68–82]. The goal of such diversion processes is to reduce acid flow into the high permeability layers where less acid is needed, and thereby divert the acid into the low permeability layers where more acid is needed for stimulation.

Foam Mobility. The objective of foam is to control acid mobility in high permeability zones. The mobility of a given fluid (M) in a reservoir zone or layer is simply the ratio of the permeability of this zone to the effective fluid viscosity, which is given by Darcy's law as:

$$M = k/\eta = QL/(\Delta PA) \tag{5}$$

where:

M = mobility (D/mPa·s)
k = absolute permeability (D)
η = viscosity (mPa·s)
Q = flow rate (cm^3/s)
L = length (cm)
ΔP = differential pressure (atm)
A = cross-sectional area (cm^2)

Once the foam flows into a specific zone, the mobility of the fluid in this zone will decrease. This is referred to as the mobility reduction factor, MRF.

$$\text{MRF} = M_{\text{brine}}/M_{\text{foam}} \tag{6}$$

Foam Diversion in Parallel Cores. To assess effectiveness of foams in reducing fluid mobility, coreflood tests are usually conducted in two parallel cores (dual-core) of different permeability. The pressure drop is the same across the two cores. Darcy's law can be applied to each core to determine the flow rate as follows:

For the low permeability core:

$$Q_l = k_l A_l \, \Delta P / (\eta_l L_l) \tag{7}$$

For the high permeability core:

$$Q_h = k_h A_h \, \Delta P / (\eta_h L_h) \tag{8}$$

where the subscripts l and h refer to the low and high permeability cores, respectively. If the two cores have the same length $(L_l = L_h)$ and cross-sectional area $(A_l = A_h)$, then the ratio of flow rates into the two cores can be determined from the following equation:

$$Q_l / Q_h = (k_l / \eta_l) / (k_h / \eta_h) \tag{9}$$

$$= M_l / M_h \tag{10}$$

Coreflood experiments.   Acid diversion using foams was thoroughly investigated using limestone cores [68, 74] and Berea sandstone [80–82]. Figure 13 illustrates the apparatus used by Khamees et al. [68] to study foam diversion during acidization experiments. The apparatus has the capability of injecting both gases and liquids into two cores connected in parallel. Brine, acid and surfactant solutions were injected by an Eldex^{TM} pump. Nitrogen, the foaming gas, was stored in floating-piston cells, which were driven by hydraulic oil pumped through a Quizix^{TM}

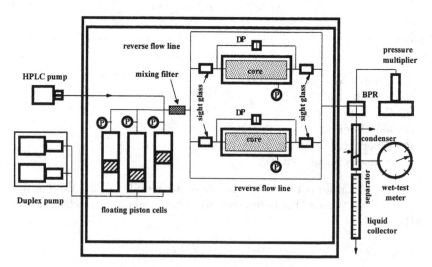

Figure 13.   Dual-core apparatus to study foam diversion during acid stimulation.

computer-controlled positive displacement pump. The apparatus is equipped with two high-pressure visual cells before and after the cores. These cells enabled the observation of the foam injected into and produced from the cores.

Two core plugs from a carbonate reservoir (calcite) were used [68]. One core holder held a high permeability core (initial brine permeability = 535 mD) and the second core holder had a low permeability core (initial brine permeability = 3.1 mD). The cores and floating-piston cells were placed inside a temperature-controlled oven. Foam quality was varied by adjusting the flow rates of nitrogen and the surfactant solution while keeping the total flow rate at 5 cm^3/min.

Coreflood Procedure. The cores were flushed with several pore volumes of trichloroethane and toluene to remove residual oil. The cores were then evacuated and saturated with brine. Nitrogen was injected into the high permeability core at 4 cm^3/min together with a 1 wt% solution of a nonionic surfactant at 1 cm^3/min. This generated a foam of 80% quality. Foam injection was continued until a constant differential pressure across the cores was obtained. Acid injection at 1 cm^3/min was simultaneously started, and nitrogen injection was stopped. The acid was 15 wt% HCl with 0.2 wt% corrosion inhibitor. Each core was then individually flushed with brine to ensure that the acid, surfactant, and gas were displaced out of the cores. A preflush of surfactant solution was carried out to equilibrate the rock surface with the nonionic surfactant. More details on the experimental procedure are given by Khamees et al. [68].

Brine Permeability After Acidization. Brine permeability was measured on each core after the foam/acidization process. Foam diversion was successful in stimulating the low permeability core where its permeability increased from 3.1 mD before acidization to 1133 mD after treatment. At the same time, the permeability of the high permeability core did not significantly change.

CT Scanning of Cores Before and After Acidization. Figures 14 and 15 show the results of the CT scans of the two cores, before and after acidization. Each figure shows two columns of cross-sections of a core, roughly equally spaced, with the inlet at the bottom and the outlet at the top. The column on the left is for the pre-treatment images, and the column on the right is for the post-treatment sections.

The scales show the color intensity and its equivalent bulk density. The acid did not cause significant changes to the high permeability core as shown in Figure 14. It appears that the foam was successful in preventing the acid from invading this core. On the other hand, Figure 15 shows that the low permeability core did change substantially after acidization, which

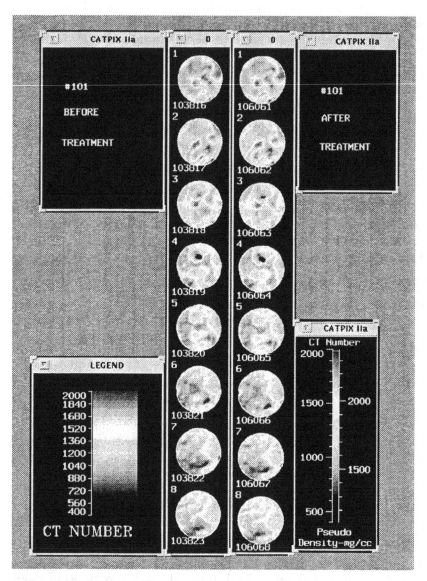

Figure 14.　CT scans for a high permeability carbonate core before and after acid contact.

is a further indication that the acid successfully invaded this core. The porosity at the center of the low permeability core increased (darker color at the center, Figure 15). Also, a dark small circle appeared at the top of all the cross-sections. This indicates that the acid formed a "wormhole" that extended throughout the length of the core.

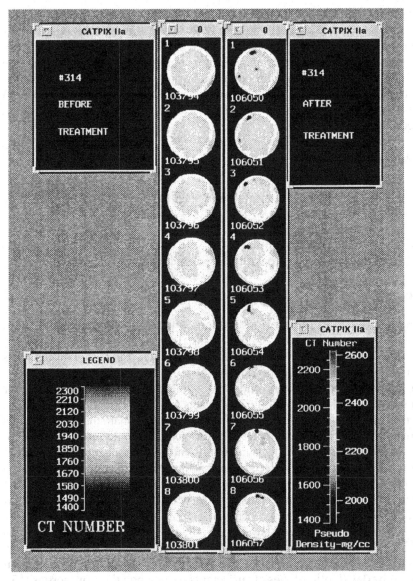

Figure 15.   CT scans for a low permeability carbonate core before and after acid contact.

Field Application.   Foams have been used to divert acids both in carbonate and sandstone reservoirs. However, there are several factors that should be carefully considered before field application. Compatibility of the selected surfactant (foaming agent) with reservoir oil must be examined. It is known that oil can adversely affect foam propagation in the reservoir [83]. Surfactant loss due to adsorption should be carefully considered [64]. If the surfactant concentration decreases due to adsorption, then foam will not propagate far from the wellbore area. Selection of a suitable surfactant is a difficult task. Surfactants used in this application should be stable at high salinity, low pH values, and high temperatures. A very limited number of surfactants can perform under these harsh conditions.

There are several factors to consider during field application [66, 69], some of these are:

1. Permeability contrast
2. Injection mode (continuous or staged)
3. Injection rate
4. Foam quality
5. Preflush and postflush
6. Pumping mode (coiled tubing or bullheading)

Surfactant Separation During Well Acidizing

As mentioned in the previous sections, surfactants are included in acid formulations to perform specific tasks. In acid stimulation treatments, surfactants encounter various chemical species. First, the surfactant is mixed with the acid and its additives. Some of these additives are cationic, e.g., corrosion inhibitors and clay stabilizers. Others are anionic or nonionic species. Second, the acid reacts with the formation and releases several cations. Hydrochloric acid reacts with carbonate minerals and, as a result, the spent acid contains calcium, magnesium and iron. Hydrofluoric-based acids react with clay minerals and release silicon and aluminum in addition to those dissolved by hydrochloric acid. The presence of these chemicals together with surfactants can cause phase separation of the surfactants. As a result, surfactants will not perform their task as anticipated.

Nonionic surfactants have been added to various acid formulations [54, 56, 84]. These surfactants are chosen because they maintain low interfacial tension between the acid and oil. They are inexpensive, and can be mixed with other types of surfactants to enhance their properties [85]. However, in common with other types of surfactants, they should be carefully tested before field applications. Nonionic surfactants can separate out of solution under certain conditions, as will be explained in the next section.

Solubility of Nonionic Surfactants.   Nonionic surfactants owe their solubility in water to hydration of the polyethylene oxide chains. The solubility of these surfactants in water increases as the length of the hydrophilic part of the surfactant increases. As the temperature of a nonionic surfactant solution is increased, hydrogen bonds break [86] and the solubility of the surfactant in water diminishes. At a certain temperature, known as the cloud point, the surfactant molecules separate out of solution, causing it to become cloudy. Ultimately, the surfactant solution separates into two immiscible phases: a surfactant-rich phase and a surfactant-lean phase. In oilfield operations, the cloud point of nonionic surfactants is an important parameter that determines the efficiency of such operations. For example, in stimulation, or during drilling operations, separation of surfactant from solution in the injected fluid can plug the formation, hence the productivity or injectivity of the well will diminish [87].

The cloud point of nonionic surfactants is a function of ionic strength [87–89]. Several studies have indicated that simple inorganic salts, e.g., NaCl and $CaCl_2$, lower the cloud point of nonionic surfactants, with sodium chloride causing more depression than calcium chloride [90, 91]. A summary of these studies is given by Hinze and Pramauro [86] and Sadaghiania and Khan [92].

The effect of inorganic acids on the cloud point of nonionic surfactants was considered by several investigators [54, 57, 84]. Travalloni-Louvisse and Gonzalez [57] found that hydrochloric acid raised the cloud point of TX-100. Nasr-El-Din and Al-Ghamdi [54, 84] examined the effect of acids and other stimulation additives on the cloud point of nonionic surfactants over a wide range of parameters, and this is discussed in the next section.

Experimental Studies.   Nasr-El-Din and Al-Ghamdi [54, 84] measured the cloud point of ethoxylated octyl phenyl alcohols known as the Triton-X series. These surfactants have the general chemical formula of $R'-C_6H_4-(OC_2H_4)_n-OH$ where R' is a branched octyl group and n is the number of ethylene oxide groups. Four surfactants having various numbers of ethylene oxide groups were examined.

The cloud point of various surfactant solutions was determined visually by noting the temperature at which the continuously heated solution suddenly became cloudy. In most cases, the surfactant solution under investigation was heated starting from room temperature. The repeatability of the cloud point measurements using this method was $+0.5\,^{\circ}C$.

The cloud point varies with the number of ethylene oxide groups of the surfactant. As the number of ethylene oxide groups increases, the solubility of the surfactant increases, hence the cloud point becomes higher. The cloud point of nonionic surfactants is also a function of surfactant concentration [92, 93]. Therefore, the surfactant concentration in all experiments conducted by Nasr-El-Din and Al-Ghamdi [54] was kept constant at 20,000 ppm.

Effect of Acids.   To examine the effect of acids on the cloud point, various acids were added to several nonionic surfactants at acid concentrations from 0 to 15 wt% [54]. The cloud point of TX-45 (5 ethylene oxide groups) solutions was less than 0 °C, and addition of HCl up to 15 wt% did not raise its cloud point above room temperature. On the other hand, the cloud point of TX-405 (number of ethylene oxide groups = 40) solutions was higher than 100 °C at all acid concentrations examined. Only TX-114 and TX-100 (number of ethylene oxide groups = 7.5 and 10, respectively) have cloud points that can be measured with the method employed by Nasr-El-Din and Al-Ghamdi [54].

Figure 16 shows the effect of hydrochloric, acetic, citric and formic acids on the cloud point of TX-100 [54]. The cloud point of neutral TX-100 solutions at a surfactant concentration of 2 wt% was 64.5 °C, which agrees with literature values [94]. At HCl concentrations greater than 1 wt%, the cloud point monotonically increased with HCl concentration, and exceeded 100 °C at an acid concentration of nearly 10 wt%. The effect of HCl on the cloud point of TX-100 was similar to that observed by Travalloni-Louvisse and Gonzalez [57].

At acid concentrations less than 1 wt%, the effect of acid type on the cloud point was not significant, as shown in Figure 16. However, at higher acid concentrations, the cloud point obtained with acetic or citric acid was less than that observed with HCl and the difference increased at higher acid concentrations. Formic acid, a weak organic acid, did not raise the

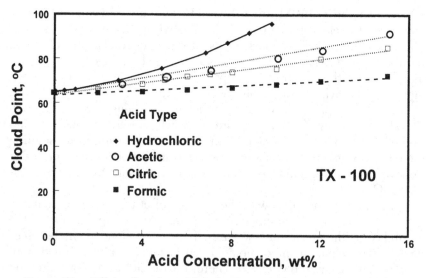

Figure 16.   Effect of acids on the cloud point of TX-100. (Reproduced with permission from reference 54. Copyright 1997.)

cloud point significantly. The relationship of the cloud point and acid concentration was linear for the acids tested, except HCl.

The effect of HCl acid on the cloud points of TX-100 can be explained as follows: the hydrogen ion causes salting-in of nonionic surfactants [95], whereas the chloride ion induces a salting-out effect [57]. It appears from the results shown in Figure 16 that the salting-in effect dominates, and becomes important at higher acid concentrations. Obviously, the concentration of hydrogen ions in the case of HCl is much higher than that obtained with the weaker organic acids. Consequently, the cloud point is higher with hydrochloric acid.

Combined Effect of Acids and Salts. The effect of simple inorganic salts, which are commonly encountered in acid stimulation processes, was examined [54]. Sodium and calcium chlorides are present in almost all formation brines. Potassium and ammonium chlorides are used in sandstone reservoirs as temporary clay stabilizers [96, 97]. Aluminum chloride is used in some mud acid formulations to retard hydrofluoric acid for deep acid penetration [4].

Figure 17 depicts the effect of these salts on the cloud point of TX-100 solutions which contained 5 wt% HCl [54]. The influence of salt type on the cloud point was not significant at salt concentrations less than 1 wt%. At higher salt concentrations, the five salts depressed the cloud point of

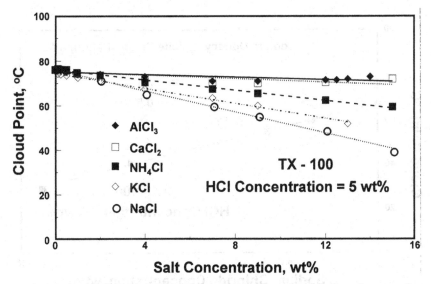

Figure 17.   Effect of simple salts on the cloud point of acidic solutions of TX-100. (Reproduced with permission from reference 54. Copyright 1997.)

TX-100. Mono-valent cations depressed the cloud point more than di- or tri-valent cations. On a weight percent basis, the cloud point was nearly the same in the presence of calcium or aluminum chloride. Sodium ion caused the steepest drop in the cloud point among the three mono-valent cations examined. The relationship between the cloud point and salt concentration was linear for the five salts. It is interesting to note that the effect of salt type on the cloud point was similar to that observed with nonionic surfactants at neutral pH conditions [95].

Effect of Anionic Surfactants.   Mixing two surfactants of different hydrophilic groups is commonly used to enhance the properties of the surfactants. Several researchers [84, 85, 92, 98–101] have indicated that the addition of a small amount of an ionic surfactant to a nonionic surfactant increases the cloud point of the latter. The rise in the cloud point of nonionic surfactants, due to the addition of an ionic surfactant, depends on the ratio of the ionic and the nonionic surfactants and salt concentration. Figure 18 illustrates the effect of NaCl on the cloud point of TX-100 solutions containing 5 wt% HCl at SDS concentrations of 0, 0.1 and 0.5 wt%. Unlike the effect of SDS on the cloud point of nonionic surfactants at neutral pH conditions [93, 99–101], SDS depressed the cloud point of acidic TX-100 solutions at all sodium chloride concentrations examined. The depression in the cloud point continued as the concentration of SDS was increased to 0.5 wt%.

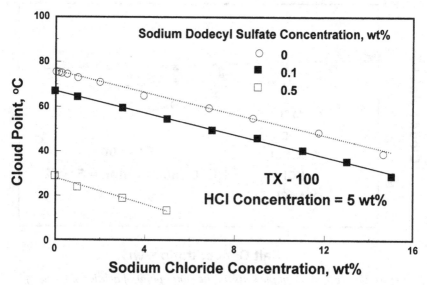

Figure 18.   Effect of SDS on the cloud point of acidic solutions of TX-100. (Reproduced with permission from reference 54. Copyright 1997.)

The effect of SDS on the cloud point of TX-100 solutions at neutral pH conditions can be explained as follows. SDS monomers carry negative charges and form mixed micelles when they are added to TX-100 solutions. The mixed micelles will also have negative charges, which will generate electrostatic repulsion between various micelles. As a result, the cloud point of solutions containing the two surfactants will be higher. The addition of sodium chloride will lower the cloud point of TX-100/SDS solutions, because the sodium ion will shield the negative charges of the mixed micelles. As a result, the electrostatic repulsion diminishes. The size of the micelles increases, hence the cloud point decreases. At low pH values, nonionic surfactants carry positive charges. These charges attract SDS monomers, which carry negative charges. As a result, the size of the mixed micelles increases, and the cloud point of the nonionic surfactant decreases. The presence of sodium ion will reduce the cloud point by the salting-out effect explained earlier.

Effect of Mutual Solvents. Mutual solvents are commonly included in acid formulations [102, 103]. They are used as water-wetting agents, demulsifiers, and surface/interfacial tension reducers. The mutual solvent examined by Nasr-El-Din and Al-Ghamdi [54] was ethylene glycol monobutyl ether (EGMBE) which is commonly added to acid formulations at 5 to 15 vol%. The mutual solvent was added to HCl formulations, which contained 2 wt% surfactant. The concentration of mutual solvent was varied from 0 to 40 wt%. The variation of the cloud point as a function of mutual solvent concentration depended on the number of ethylene oxide groups of the surfactant, and the concentrations of mutual solvent and acid, as discussed below.

Figure 19 illustrates the variation of the cloud point of TX-45 with HCl concentration at mutual solvent concentrations between 5 and 20 wt%. At a mutual solvent concentration of 0 wt%, the cloud point was less than room temperature. At a mutual solvent concentration of 5 wt% the cloud point was greater than room temperature at HCl concentrations greater than 11 wt%. At 10 wt% mutual solvent, the cloud point was greater than room temperature at HCl concentrations greater than 1 wt%. The cloud point further increased at higher mutual solvent concentrations of 15 and 20 wt%. At a given HCl concentration, the effect of mutual solvent on the cloud point of TX-45 diminished at higher mutual solvent concentrations. The relationship of the cloud point and HCl concentration was linear at all mutual solvent concentrations examined.

The variation of the cloud point of TX-45 with the concentrations of the acid and mutual solvent should be considered when designing an acid formulation. At a mutual solvent concentration of 10 wt%, acidic solutions of TX-45 will be clear in live acid (15 wt% HCl), however, they will become cloudy once the acid is spent. Therefore, it is important to

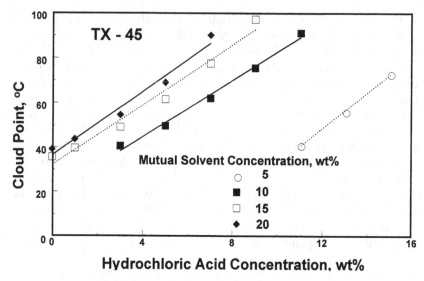

Figure 19.   Effect of mutual solvent on the cloud point of acidic solutions of TX-45. (Reproduced with permission from reference 54. Copyright 1997.)

measure the cloud point of nonionic surfactants both in live and spent acids before field application.

Unlike the results obtained with TX-45, the cloud points of TX-405 decreased with the addition of the mutual solvent [54]. The effect of the mutual solvent on the cloud point of TX-405 was significant, as illustrated in Figure 20. At 0 wt% HCl, the cloud point sharply decreased with the addition of 5 wt% mutual solvent. This drop continued as the concentration of the mutual solvent was increased up to 20 wt%. It should be mentioned that the cloud point at 5 wt% mutual solvent and HCl concentrations greater than 0 wt% was higher than 100 °C. Therefore, the cloud point was measured at neutral pH conditions only. Similar to the results obtained with TX-100, the cloud point of TX-405 was depressed with increasing the concentration of mutual solvent at acid concentrations less than 5 wt%. At mutual solvent concentrations greater than 7.5 wt% and acid concentrations greater than 5 wt%, the cloud point was higher than 100 °C. Therefore, it was not possible to examine the effect of mutual solvent on the cloud point of these surfactant solutions.

The results shown in Figures 19 and 20 indicate that the cloud point depends on acid concentration, mutual solvent concentration, and the number of ethylene oxide groups of the nonionic surfactant. There are strong interactions between the mutual solvent and nonionic surfactants in acidic solutions. Mutual solvent contains an alcohol group, however, its

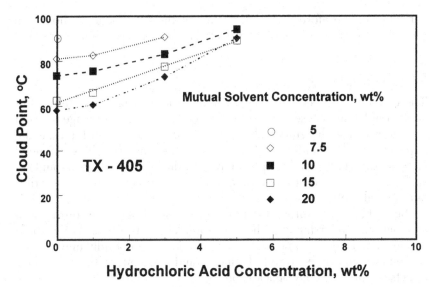

Figure 20.   *Effect of mutual solvent on the cloud point of acidic solutions of TX-405. (Reproduced with permission from reference 54. Copyright 1997.)*

effect on the cloud point is more complicated than that observed with short-chain alcohols. Short-chain alcohols raise the cloud point of nonionic surfactants. However, in the case of mutual solvent, the cloud point of the nonionic surfactant may increase or decrease, depending on the number of ethylene oxide groups of the surfactant. The solubility of mutual solvent in water is higher than nonionic surfactants with a small number of ethylene oxide groups. Therefore, the mutual solvent raised the cloud points of TX-45 and TX-114. The solubility of mutual solvent is less than nonionic surfactants with a high number of ethylene oxide groups, therefore, mutual solvent depressed the cloud points of TX-100 and TX-405. At higher mutual solvent concentrations, the properties of the solvent are different, and the effect of the number of ethylene groups diminishes. The interactions between the mutual solvent and nonionic surfactants are more complicated than described above, and more work is needed to explain these interactions.

Field Application.   Nonionic surfactants have several advantages over other classes of surfactants [54, 56, 84]. This explains their extensive use in the area of acid stimulation. However, these surfactants can separate out of solution at temperatures greater than their cloud point. Because phase separation can cause formation damage [87], these surfactants should be used in the field at temperatures below their cloud point. Several additives may depress the cloud point of nonionic surfactants. Therefore, the cloud point of these surfactants should be measured in the presence of all expected additives as per field formula.

Concluding Remarks

There is a wide spectrum of applications where surfactants are used in acid stimulation. These applications range from anti-sludge agents, acid retarders, to acid diverters. Selection of a suitable surfactant for a specific application is a difficult task. This is due to the many variables that need consideration before field application. In addition, using large amounts of surfactants can lead to emulsion formation, precipitation and other operational problems.

Surfactants encounter large numbers of chemical species during well stimulation. Understanding chemical interactions is a must before field application. To date, some of these interactions are not fully understood. More research is needed to fully understand these interactions and their effects on the efficiency of acidizing.

List of Symbols

A	cross-sectional area (cm^2)
EDTA	ethylenediamine tetraacetic acid
EGMBE	ethylene glycol monobutyl ether
L	length (cm)
ΔP	differential pressure (atm)
k	consistency factor, $mPa \cdot s^n$
K	permeability, md
K_o	initial permeability, mD
K_r	permeability ratio, dimensionless
M	mobility ($D/mPa \cdot s$)
MRF	mobility reduction factor
n	power-law index, dimensionless
p	pressure, psi
Q	flow rate (cm^3/s)
r	radius, m
SDS	sodium dodecyl sulfate
V_g	gas volume, m^3
V_l	liquid volume, m^3

Greek

α	foam quality
$\dot{\gamma}$	shear rate, s^{-1}
η	apparent viscosity, $mPa \cdot s$
σ	surface tension, mN/m

Subscripts

l   low
h   high

Acknowledgments

The author wishes to acknowledge the Saudi Arabian Ministry of Petroleum and Mineral Resources and the Saudi Arabian Oil Company (Saudi Aramco) for granting permission to publish this paper. Thanks are also due to the professional and technical staff of the LR&DC of Saudi Aramco without whose work this publication would not have been possible.

References

1. McLeod, H.O., Jr. *J. Pet. Technol.*, Dec. **1984**, 2055–2069.
2. Fredd, C.N.; Fogler, H.S. *Proceedings of the SPE International Symposium on Oilfield Chemistry of SPE*; Society of Petroleum Engineers: Richardson, TX, 1997, paper SPE 37212.
3. Fredd, C.N.; Fogler, H.S. *Proceedings of Formation Damage Control Symposium of SPE*; Society of Petroleum Engineers: Richardson, TX, 1996, paper SPE 31074.
4. Gdanski, R.D. *Oil & Gas J.*, Oct. 28, **1985**, 111–115.
5. Kunze, K.R.; Shaughnessy, C.M. *Soc. Pet. Eng. J.*, Feb. **1983**, 65.
6. Lullo, G.D.; Rae, P. *Proceedings of the 6th International Asia Pacific Oil & Gas Conference of SPE*; Society of Petroleum Engineers: Richardson, TX, 1996, paper SPE 37015.
7. Shuchart, C.E.; Gdanski, R.D. *Proceedings of the European Petroleum Conference of SPE*; Society of Petroleum Engineers: Richardson, TX, 1996, paper SPE 36907.
8. Houchin, L.R.; Foxenburg, W.E.; Usie, M.J.; Zhao, J. *Proceedings of the International Symposium on Formation Damage Control of SPE*; Society of Petroleum Engineers: Richardson, TX, 1992, paper SPE 23817.
9. Ali, S.A.; Durham, D.K.; Elphingstone, E.A. *Oil & Gas J.*, Mar. 28, **1994**, 47–51.
10. Nasr-El-Din, H.A.; Al-Rammah, A. *Proceedings of the Second International Conference on Chemistry in Industry*, ACS, Manama, Bahrain, October 24–26, 1994.
11. Bennion, D.B.; Thomas, F.B.; Bietz, R.F. *Proceedings of the Gas Technology Conference of SPE*; Society of Petroleum Engineers: Richardson, TX, 1996, paper SPE 35577.
12. Nasr-El-Din, H.A.; Al-Taq, A.A. *Proceedings of the International Symposium on Formation Damage Control of SPE*; Society of Petroleum Engineers: Richardson, TX, 1998, paper SPE 39487.
13. Hoefner, M.L.; Fogler, H.S., Stenius, P.; Sjöblom, J. *J. Pet. Technol.*, Feb. **1987**, 203–208.

14. Al-Anazi, H.A.; Nasr-El-Din, H.A.; Safwat, M.K. *Proceedings of the International Symposium on Formation Damage Control of SPE*; Society of Petroleum Engineers: Richardson, TX, 1998, paper SPE 39418.
15. Nasr-El-Din, H.A.; Al-Anazi, H.A.; Safwat, M.K. *Proceedings of the Oilfield Chemistry of SPE*; Society of Petroleum Engineers: Richardson, TX, 1999, paper SPE 50739.
16. Buijse, M.A.; van Domelen, M.S. *Proceedings of the International Symposium on Formation Damage Control of SPE*; Society of Petroleum Engineers: Richardson, TX, 1998, paper SPE 39583.
17. Chambers, D.J. In *Foams: Fundamentals and Applications in the Petroleum Industry*, Schramm, L.L., Ed.; American Chemical Society: Washington, DC, 1994, p 355.
18. McLeod, H.O., Jr.; Ledlow, L.B.; Till, M.V. *Proceedings of the 58th Annual Technical Conference of SPE*; Society of Petroleum Engineers: Richardson, TX, 1983, paper SPE 11931.
19. Taylor, K.C.; Nasr-El-Din, H.A.; Al-Alawi, M. *Proceedings of the International Symposium on Formation Damage Control of SPE*; Society of Petroleum Engineers: Richardson, TX, 1998, paper SPE 39419.
20. Taylor, K.C.; Nasr-El-Din, H.A. *Proceedings of the International Oilfield Chemistry of SPE*; Society of Petroleum Engineers: Richardson, TX, 1999, paper SPE 50772.
21. Al-Humaidan, A.Y.; Nasr-El-Din, H.A. *Proceedings of the International Oilfield Chemistry of SPE*; Society of Petroleum Engineers: Richardson, TX, 1999, paper SPE 50765.
22. Nasr-El-Din, H.A.; Al-Anazi, H.A.; Hopkins, J.A. *Proceedings of the International Symposium on Oilfield Chemistry of SPE*; Society of Petroleum Engineers: Richardson, TX, 1997, paper SPE 37215.
23. Nasr-El-Din, H.A.; Al-Mulhem, A.; Lynn, J.D. *Proceedings of the International Symposium on Formation Damage Control of SPE*; Society of Petroleum Engineers: Richardson, TX, 1998, paper SPE 39584.
24. Moore, E.W.; Crowe, C.W.; Hendrickson, A.R. *J. Pet. Technol.*, Sept. **1965**, 1023–1028.
25. Picou, R.A.; Ricketts, K.; Luquette, M.; Hudson, L.M. *Proceedings of the International Symposium on Formation Damage of SPE*; Society of Petroleum Engineers: Richardson, TX, 1992, paper SPE 23818.
26. Dunlap, D.D.; Houchin, L.R. *Proceedings of the Formation Damage Control Symposium of SPE*; Society of Petroleum Engineers: Richardson, TX, 1990, paper SPE 19425.
27. Copple, C.P. *J. Pet. Technol.*, Sept. **1975**, 1060–1066.
28. Bansal, K.M. *Proceedings of the International Symposium on Oilfield Chemistry of SPE*; Society of Petroleum Engineers: Richardson, TX, 1994, paper SPE 25199.
29. Hebert, P.B.; Khatib, Z.I.; Norman, W.D.; Acock, A.; Johnson, M. *Proceedings of the Annual Technical Conference and Exhibition of SPE*; Society of Petroleum Engineers: Richardson, TX, 1996, paper SPE 36601.
30. Houchin, L.R.; Hudson, L.M. *Proceedings of the International Sym-*

posium on Formation Damage of SPE; Society of Petroleum Engineers: Richardson, TX, 1986, paper SPE 14818.

31. Lichaa, P.M.; Herrra, L. *Proceedings of the International Symposium on Oilfield Chemistry of SPE*; Society of Petroleum Engineers: Richardson, TX, paper SPE 5304.

32. Jacobs, I.C.; Thorne, M.A. *Proceedings of the 7th Symposium on Formation Damage Control of SPE*; Society of Petroleum Engineers: Richardson, TX, 1986, paper SPE 14823.

33. Schramm, L.L. Petroleum emulsions: basic principles, in *Emulsions: Fundamentals and Applications in the Petroleum Industry*, Schramm, L.L., Ed.; Advances in Chemistry Series 231, American Chemical Society, Washington, DC, 1992.

34. Houchin, L.R.; Dunlap, D.D.; Arnold, B.D.; Domke, K.M. *Proceedings of the Formation Damage Control Symposium of SPE*; Society of Petroleum Engineers: Richardson, TX, 1990, paper SPE 19410.

35. Strassner, J.E. *J. Pet. Technol.*, Mar. **1968**, 303–312.

36. Vinson, E.F. *Proceedings of Formation Damage Control Symposium of SPE*; Society of Petroleum Engineers: Richardson, TX, 1996, paper SPE 31127.

37. Jacobs, I.C. *Proceedings of the International Symposium on Oilfield Chemistry of SPE*; Society of Petroleum Engineers: Richardson, TX, 1989, paper SPE 18475.

38. Suzuki, F. *Proceedings of the Western Regional Meeting of SPE*; Society of Petroleum Engineers: Richardson, TX, 1993, paper SPE 26036.

39. API RP 42: API recommended practices for laboratory testing of surface-active agents for well stimulation, American Petroleum Institute, 2nd Edition, 1977, 6–8.

40. Johnson, D.E. *Proceedings of the SPE Oil and Gas Recovery Conference of SPE*; Society of Petroleum Engineers: Richardson, TX, 1994, paper SPE 27695.

41. Rietjens, M. *Proceedings of the European Formation Damage Control of SPE*; Society of Petroleum Engineers: Richardson, TX, 1997, paper SPE 38163.

42. Figueroa-Ortiz, V.; Cazares-Robles, F.; Fragachan, F.E. *Proceedings of Formation Damage Control Symposium of SPE*; Society of Petroleum Engineers: Richardson, TX, 1996, paper SPE 31124.

43. Nierode, D.E.; Williams, B.B. *Soc. Pet. Eng. J.*, Dec. **1971**, 407.

44. Lund, K.; Fogler, H.S.; McCune, C.C.; Ault, J.W. *Chem. Eng. Sci.* **1975**, *30*, 825.

45. Crowe, C.W.; Miller, B.D. *Proceedings of the Rocky Mountain Regional Meeting of SPE*; Society of Petroleum Engineers: Richardson, TX, 1974, paper SPE 4937.

46. Bergstrom, J.M.; Miller, B.D. *Proceedings of the 50th Annual Fall Meeting of SPE*; Society of Petroleum Engineers: Richardson, TX, 1975, paper SPE 5648.

47. Peters, F.W.; Saxon, A. *Proceedings of the Asia-Pacific Conference of SPE*; Society of Petroleum Engineers: Richardson, TX, 1989, paper SPE 19496.

48. Daccord, G.; Touboul, E.; Lenorm, R. *Proceedings of the 62nd Annual Technical Conference and Exhibition of SPE*; Society of Petroleum Engineers: Richardson, TX, 1987, paper SPE 16887.
49. Williams, B.B.; Nierode, D.E. *J. Pet. Technol.*, Jul. **1972**, 849.
50. Hoefner, M.L.; Fogler, H.S.; *Chem. Eng. Prog.*, Apr. **1985**, 40.
51. Guidry, G.S.; Ruiz, G.A.; Saxon, A. *Proceedings of the Middle East Oil Technical Conference and Exhibition of SPE*; Society of Petroleum Engineers: Richardson, TX, 1989, paper SPE 17951.
52. Buijse, M.A. *Proceedings of the European Formation Damage Conference of SPE;* Society of Petroleum Engineers: Richardson, TX, 1997, paper SPE 38166.
53. Krawietz, T.E.; Rael, E.L. *SPE Production Facilities*, Nov. **1996**, 238–243.
54. Nasr-El-Din, H.A.; Al-Ghamdi, A. Accepted for publication, *SPE Production and Facilities*, **1997**.
55. Paktinat, J. *Proceedings of the International Symposium Oilfield Chemistry of SPE*; Society of Petroleum Engineers: Richardson, TX, 1991, paper SPE 21011.
56. Gdanski, R.D. *Proceedings of the International Symposium Oilfield Chemistry of SPE*; Society of Petroleum Engineers: Richardson, TX, 1995, paper SPE 28971.
57. Travalloni-Louvisse, A.M.; Gonzalez, G. In *Surfactant-Based Mobility Control*, Smith, D.H., Ed.; ACS Symp. Series 373, American Chemical Society, Washington, DC, 1988, Chapter 11.
58. Dabbousi, B.O., Nasr-El-Din, H.A., Al-Muhaish, A.S. *Proceedings of the Oilfield Chemistry of SPE*; Society of Petroleum Engineers: Richardson, TX, 1999, paper SPE 50732.
59. Dabbousi, B.O., Al-Muhaish, A.S.; Nasr-El-Din, H.A. submitted to *J. Physical Chemistry*, 1998.
60. Paluch, M.; Rybska, *J. Colloid Interface Science*, **1991**, *1455*, 219–223.
61. Alvarez, E.; Sanchez-Vilas, M.; Sanjoujo, B.; Navazn, J.M. *J. Chem. Eng. Data* **1997**, *42*, 957–960.
62. Lord, D.L.; Hayes, K.F.; Demon, A.H.; Salehzadeh, A. *Environ. Sci. Technol.* **1997**, *31*, 2045–2051.
63. D'Angelo, M.; Onori, G.; Santucci, A. *Chem. Phys. Lett.* **1994**, *220*, 59–63.
64. Mannhardt, K.; Novosad, J.J. In *Foams: Fundamentals and Applications in the Petroleum Industry*, Schramm, L.L., Ed.; American Chemical Society, Washington, DC, 1994, p 259.
65. Hall, B.E. *J. Pet.Technol.*, Dec. **1975**, 1439–1442.
66. Thomas, R.L.; Ali, S.A.; Robert, J.A.; Acock, A.M. *Proceedings of the International Symposium on Formation Damage Control of SPE*; Society of Petroleum Engineers: Richardson, TX, 1998, paper SPE 39422.
67. Yeager, V.; Shuchart, C. *Oil & Gas J.*, Jan. 20, **1997**, 70–72.
68. Khamees, A.A.; Sayegh, S.G.; Nasr-El-Din, H.A. Submitted for publication in the *Saudi Aramco J. Technol.*, June **1998**.
69. Morphy, P.H.; Greenwald, K.G.; Herries, P.E. *Proceedings of the International Symposium on Formation Damage Control of SPE*; Society of Petroleum Engineers: Richardson, TX, 1998, paper SPE 39423.

70. Smith, C.L.; Anderson, J.L.; Roberts, P.G. *Proceedings of the Annual Meeting of SPE*; Society of Petroleum Engineers: Richardson, TX, 1969, paper SPE 2751.
71. Burman, J.W.; Hall, B.E. *Proceedings of the Annual Technical Conference and Exhibition of SPE*; Society of Petroleum Engineers: Richardson, TX, 1986, paper SPE 15575.
72. Kennedy, D.K.; Kitziger, F.W.; Hall, B.E. *SPE Production Engineering*, May **1992**, 203–211.
73. Bernadiner, M.G.; Thompson, K.E.; Fogler, H.S. *SPE Production Engineering*, Nov. **1992**, 350–356.
74. Thompson, K.; Gdanski, R.D. *SPE Production Engineering*, Nov. **1993**, 285.
75. Gdanski, R.D. *Oil & Gas J.*, Sept. 6, **1993**, 85.
76. Zerhboub, M.; Ben Naceur, K; Touboul, E.; Thomas, R.L. *SPE Production Facilities*, May **1994**, 121–126.
77. Kibodeaux, K.R., Zeilinger, S.C.; Rossen, W.R. *Proceedings of the 69th Annual Technical Conference and Exhibition of SPE*; Society of Petroleum Engineers: Richardson, TX, 1994, paper SPE 28550.
78. Zhou, Z.H.; Rossen, W.R. *SPE Production Facilities*, Feb. **1994**, 29–35.
79. Persoff, P.; Pruess, K.; Benson, S.M.; Wu, Y.S.; Radke, C.J.; Witherspoon, P.A. *Energy Sources*, **1990**, *12*, 479–497.
80. Behenna, F.R. *Proceedings of the European Formation Damage Conference of SPE*; Society of Petroleum Engineers: Richardson, TX, 1995, paper SPE 30121.
81. Parlar, M.; Parris, M.; Jasiski, R. *Proceedings of the Western Regional Meeting of SPE*; Society of Petroleum Engineers: Richardson, TX, 1995, paper SPE 29678.
82. Robert, J.A.; Mack, M.G. *Proceedings of the Western Regional Meeting of SPE*; Society of Petroleum Engineers: Richardson, TX, 1995, paper SPE 29676.
83. Schramm, L.L.; Turta, A.T.; Novasad, J.J. *SPE Reservoir Engineering*, Aug. **1993**.
84. Al-Ghamdi, A.; Nasr-El-Din, H.A. *J. Colloids and Surfaces A: Physicochemical and Engineering Aspects*, **1997**, *125*, 5–18.
85. Denoyel, R.; Giordano, F.; Rouquerol, J. *J. Colloids and Surfaces A: Physicochemical and Engineering Aspects*, **1993**, 76, 141.
86. Hinze, W.L.; Pramauro, E. *Critical Reviews Analytical Chemistry*, **1993**, 24, 133.
87. Jachnik, R.P.; Green, P. *Proceedings of the International Symposium Oilfield Chemistry of SPE*; Society of Petroleum Engineers: Richardson, TX, 1995, paper SPE 28963.
88. Schott, H. *J. Colloid Inter. Sci.* **1973**, 43, 150.
89. Gu, T., Qin, S.; Ma, C. *J. Colloid Inter. Sci.* **1989**, *127*, 586.
90. Deguchi, K.; Meguro, K. *J. Colloid Inter. Sci.* **1975**, 50, 223.
91. Doscher, T.M.; Myers, G.E.; Atkins, D.C., Jr. *J. Colloid Inter. Sci.* **1951**, 6, 223.
92. Sadaghiania, A.S.; Khan, A. *J. Colloid Inter. Sci.* **1991**, *144*, 191.
93. Corti, M.; Minero, C.; Degiorgio, V. *J. Phys. Chem.* **1984**, *88*, 309.

94. Han, S.K.; Lee, S.M.; Kim, M.; Schott, H.J. *J. Colloid Interface Sci.* **1989**, *132*, 444.
95. Schott, H.; Royce, A.E.; Han, S.K. *J. Colloid Inter. Sci.* **1984**, *98*, 196.
96. Azari, M.; Leimkuhler, J. *Proceedings of the Formation Damage Control Symposium of SPE*; Society of Petroleum Engineers: Richardson, TX,1990, paper SPE 19431.
97. Khilar, K.C.; Fogler, H.S. *J. Colloid Inter. Sci.* **1984**, *101*, 214.
98. Ivanova, N.I.; Shchukin, E.D. *J. Colloids and Surfaces A: Physicochemical and Engineering Aspects*, **1993**, 76, 109.
99. Marszall, L. *J. Colloids and Surfaces*, **1987**, 25, 279.
100. Valaulikar, B.S.; Manohar, C. *J. Colloid Inter. Sci.* **1985**, *108*, 403.
101. Manohar, C.; Kelkar, V. *J. Colloid Inter. Sci.* **1990**, *137*, 604.
102. King, G.E.; Lee, R.M. *SPE Production Engineering*, May **1988**, 205–209.
103. Hall, B.E. *J. Pet.Technol.*, Dec. **1975**, 1439–1442.

RECEIVED for review August 13, 1998. ACCEPTED revised manuscript January 4, 1999.

10

Surfactants in Athabasca Oil Sands Slurry Conditioning, Flotation Recovery, and Tailings Processes

Laurier L. Schramm[1], Elaine N. Stasiuk[1], and Mike MacKinnon[2]

[1] Petroleum Recovery Institute and Chemistry Dept., University of Calgary, Calgary, AB, Canada T2L 2A6
[2] Syncrude Canada Ltd., Research Dept., Edmonton, AB, Canada

In the surface processing of oil sands, surface and interfacial phenomena involving surfactants are involved in the occurrence and properties of suspensions, emulsions, and foams of several kinds. The actions of natural surfactants originating in the bitumen, and underlying the physical chemical basis for the separation process, are reviewed in the context of individual process steps. Issues arising from the occurrence of these surfactants in the process tailings basins are also discussed.

Introduction

Slurry conditioning of oil sand and the subsequent flotation recovery of separated bitumen comprise what is known as the hot water flotation process for Canada's Athabasca oil sands, a large-scale commercial application of mined oil sands technology. As will be seen, the hot water flotation process is composed of numerous inter-linked elementary process steps many of which are rich in surfactant chemistry. We will review aspects of the surfactant science underlying this process. But first, a few words on oil sands and their early exploitation.

Oil sands are unconsolidated sandstone deposits containing a very heavy crude oil termed bitumen. Bitumen is chemically similar to conventional crude oil but has a greater density (a lower API gravity) and a much greater viscosity. Deposits of oil sand are present in many locations around the world and they appear to be similar in many respects [1–3], occurring along the rim of major sedimentary basins, mainly in either fluviatile or deltaic environments containing sands of high porosity and permeability. Reviews are available for most locations worldwide [1, 3–10]. The amount of world oil sands rivals the world's total discovered medium and light gravity oils in place [2]. Most of the bitumen (about 91%) is contained in the Canadian and Venezuelan deposits. Of the Canadian deposits, the Athabasca deposit forms the world's largest self-contained accumulation of hydrocarbons totalling 600 billion barrels. This is at least four times the size of the largest conventional oil field (Ghawar in Saudi Arabia) [2].

Accounts of early exploration and examination of the Athabasca deposit can be found elsewhere [3, 11–13] as can accounts of some of the early process development efforts [14–16]. Commercial plants now mine oil sands and then extract bitumen using the hot water conditioning and flotation process (at production levels of over 300,000 bbl/d). The extracted bitumen is subsequently upgraded by refinery type processes to produce light, sweet crude oil.

In oil sand processing the general principles of mineral flotation apply but oil sand composition and structure, and their variations, have a great impact on the way the flotation must be operated. General descriptions of the geology of the Canadian deposits can be found in several books [12, 17–20]. Considerable effort has gone into describing the geological aspects of oil sand deposits including subdivisions into depositional environments based on the principle that, for example, rivers deposit different sands in a different geometry than do lakes or oceans [18, 19, 21–26]. The Canadian oil sand deposits occur in sandstones of early Cretaceous (ca. 110 million years) [18]. Because sediments were brought in to the deposit area from different sources and at different times, the oil sands occur as a mixture of sediment types, overlain by varying thicknesses of non-oil bearing formations. In the case of the Athabasca deposit, the largest of the Canadian oil sand deposits [18, 27], the bitumen is contained mostly in the McMurray formation which lies over limestone and under marine shale. The McMurray formation is a drainage basin that filled in with sediments and at different times the sea alternately flooded and then receded so that a number of distinct depositions can be discerned [19, 25, 27, 28]. The bulk of the sediments appear to be the result of estuarine phases, with increasing marine invasion at later dates. Such sequences, each layering and disrupting with fluvial/marine movements, lead finally to a system of sediments having great diversity. Accordingly, the oil bearing sands in this deposit have great variability in their compositions and properties.

The oil sands resemble conventional oil deposits but there are some important differences [3, 29–31]. Athabasca oil sand consists mainly of quartz sand, with smaller amounts of feldspar grains, mica flakes and clays [28–30, 32]. The clays in this deposit are predominantly kaolinite and illite with some chlorite. Some tables of mineral and bitumen compositions of Athabasca oil sands are given by Camp [31]. In general the oil bearing sands are very-fine to fine grained (62.5 to 250 μm diameter). These oil sands are unconsolidated and have fairly high porosities (30–35%) due to a low occurrence of mineral cements [29].

It is generally accepted that the Athabasca oil sand grains are predominantly water-wet (see the discussion in reference [55]). Most of the literature results are consistent with the view [33] that connate water exists as pendular rings around sand grain contact points and as roughly 10 nm thick films on sand grain surfaces. The oil is thought to have migrated into this water/mineral environment and then degraded in-place due to some combination of evaporation, diffusion, oxidation, and/or bacterial degradation of components [3, 19, 20, 34]. Any combination of these factors would have resulted in a residuum of the heavier components, i.e., bitumen. The viscosity of Athabasca bitumen in-place is sufficiently high, about 1,000,000 mPa·s at reservoir temperature, that oil sand has enough material strength to be mineable. A number of studies have been published on its density [31, 35] and rheological properties [35–40].

It follows that the depositional environments, porosities, permeabilities, and bitumen saturations are related. Where sediment transport and deposition were originally slow relatively large amounts of silt and clay deposited. The strong influence of clay content on oil saturation has been emphasized by Carrigy [41] who has surmised that the ability of clays to absorb large amounts of water reduced the permeability to oil so that when oil migrated into such areas of low porosity and permeability, little was retained. For these lean oil sands water forms the continuous phase. In regions where there were originally strong currents, primarily larger grains were deposited and little fine grained material. When oil migrated into these environments of high porosity and permeability relatively large amounts of bitumen were trapped. Accordingly [22, 29], the best ore bodies are those located along deep river or estuarine channels. For these richer oil sands bitumen forms the continuous phase.[3] Carrigy [41] has related grain size distribution for a number of Athabasca oil sands to the variation in oil content as shown in Figure 1; oil sands containing progressively more clay-size (<2 μm) materials have lower oil contents.

In summary, there is a general consensus that, for the most part, the mineral grains in Athabasca oil sand are water-wet and that most of the bitumen is not in direct contact with the mineral phase, but instead separated by at least a thin film of water. There remains some reason to believe that a fraction of the solids are, however, oil-wetted. The

---

[3] Although there are various conventions for describing saturations (e.g. [27]) for oil sands amenable to mining and hot water flotation an appropriate set of definitions are as follows: rich oil sand containing 12–14% bitumen, average grade 10–11%, and lean grade 6–9%. Lower than 6% is usually not considered to be of "ore-grade" quality.

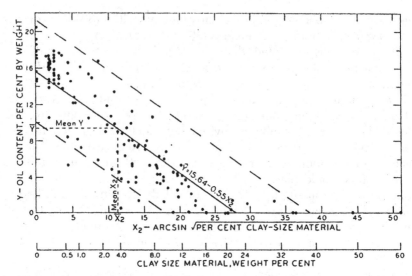

Figure 1. Oil contents of Athabasca oil sands as a function of the percentage of clay-size particles (<2 μm), according to Carrigy [41]. (Copyright 1962.)

separation of most of the oil from solids by a water film is widely held to be the characteristic difference between Athabasca oil sand and oil sand from other oil sand deposits in the world (e.g., California, New Mexico or Utah) that are thought to consist of oil-wet solids. These "oil-wet" oil sands are considered to be more difficult to beneficiate using hot water flotation because of the difficulty in dislodging bitumen from an oil-wet surface during the slurry conditioning stage [42, 43].

The Commercial Hot Water Flotation Process

We will briefly review the initial steps in the operation of an integrated commercial oil sands-synthetic crude oil production process (see Figure 2). Additional details are available in the technical and patent literature [3, 10, 31, 44–50].

Before oil sand is mined, some 30 m of overburden material must be removed. The mining of the oil sand (ore) body, which is about 60 m thick, is accomplished either by large draglines or by mobile power shovels that dig the oil sand from an open pit. Typically the mining operation must remove $\frac{1}{2}$ tonne of overburden and mine 2 tonne of oil sand of about 10% bitumen content to yield 1 barrel of oil after extraction. Obviously as the grade of oil sand decreases, additional tonnes must be mined and processed to yield the same amount of oil. Therefore a

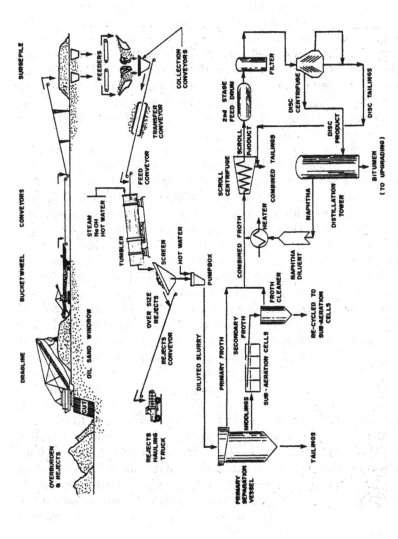

Figure 2. Diagram of a commercial oil sands mining and hot water flotation process. From Schramm and Kwak [40]. (Copyright 1988.)

commercial operation has an economic grade limit dictated by the trade-off between the mining and processing costs versus the value of the oil.

In some operations, mined oil sand is free-casted onto windrows from which bucketwheel reclaimers load the oil sand onto conveyors that carry it to surge pile/dump pockets. In other operations the oil sand is mined directly by power shovels and moved by trucks to surge bins. From the surge piles or bins and dump pockets a complex arrangement of feeding devices and conveyors is used to deliver oil sand to tumblers at a uniform feed rate. In some future operations it is probable that the conveyors and tumblers will be replaced by pipeline hydrotransport and conditioning. Each of the mining methods blends some of the oil sand and clay (lens) bands to various degrees. A certain degree of mixing also occurs during the subsequent handling and transferring operations, including transferring via conveyors and dumping into surge piles. Although modern truck and shovel operations permit more selective mining than was previously possible with draglines and bucketwheels, there is still mixing when feeding trucks and surge bins. Despite the mixing, delivered oil sand is not homogeneous. Since oil sands having different natures and compositions are associated with different conditions for optimal separation and flotation, bitumen process control strategies are very important.

Mined oil sand is first conditioned by slurrying with water in rotating horizontal tumblers (although again, in some future operations it is probable that the conditioning will be accomplished during pipeline hydrotransport instead). Here heat and shear are employed to overcome the forces holding oil sand lumps together. In this ablation process, successive layers of each lump are warmed and sheared off until everything is fairly well dispersed. Besides stirring to maintain a state of suspension, a number of other things must happen in the conditioning step. The bitumen has to be separated from the solids (which make up about 70% w/w of the slurry) and prepared for separation from the aqueous phase. Steam is added to raise the tumbler (exit) temperature to 80 °C. Air is not directly sparged in, but becomes worked in to aerate the bitumen by inclusion of about 30% v/v gas [31]. Sodium hydroxide is added to raise the solution pH. The amounts of the reagents added are typically in the proportion: oil sand/water/NaOH = 1/0.4/0.0012 by mass. An appreciable time is required to achieve a good distribution of the bitumen, minerals and reagents and to allow chemical and surface reactions to occur. Within 5 minutes or so a quasi-steady state is reached, although probably not full thermodynamic equilibrium.

The slurry is discharged from the tumblers onto vibrating screens and washed with hot spray water to remove oversized solids and undigested oil sand lumps. This process may also provide additional air entrainment and hence further aeration of bitumen. Additional hot water is added to the slurry which is then pumped to the primary separation (flotation) vessels.

The rejected solids (about 5% of the original oil sand) are conveyed out of the plant for disposal.

The diluted (flooded) slurry contains about 7% aerated bitumen droplets, 43% water and 50% suspended solids. The aerated bitumen droplets have the lowest density and rise (float) to the surface of the primary separation vessel, a large vessel with a cylindrical upper section and a conical lower section. The vessel is maintained in a quiescent condition to facilitate this flotation, as well as the settling of coarse solids to the bottom. The slurry is retained here for about 45 minutes. Since the process is continuous, the presence of fine minerals (e.g., clays) makes this vessel susceptible to solids build-up which can increase the viscosity [31, 51, 52]. To maximize the flotation and sedimentation processes the middlings region viscosity and density are kept low by adjusting the flood water addition and middlings removal rates. Mechanical rakes at the bottom of the vessel keep the coarse, rapidly settling solids moving toward the bottom from which they are withdrawn as a concentrated suspension (primary tailings).

The smaller suspended solids which do not settle rapidly, and the smaller and poorly aerated bitumen droplets which do not float rapidly, are all drawn off in a slurry from the middle of the vessel (middlings). The bitumen droplets in middlings have either too little air content or have too small diameters for rapid enough flotation. The middlings stream and primary tailings stream contain enough bitumen that they are combined and pumped to a special tailings oil recovery (TOR) flotation circuit [53]. The middlings from this TOR vessel are then pumped to a scavenging (secondary) flotation circuit for additional bitumen flotation. Here conventional flotation cells, employing vigorous agitation and air sparging, are used to cause further bitumen aeration and flotation. Meanwhile, the TOR froth is recycled into the flooded slurry that is fed into the primary separation vessels. The TOR tailings are combined with the tailings from the scavenging circuits. Other variations of this process are also practised.

Processibility and Process Control

Many sub-processes are required in order to carry out conditioning and flotation steps efficiently. Figure 3 shows some of the elementary processes. Although the real phenomena may not be entirely subdividable in this way, or take place in exactly the order assigned, it can be seen that the tumblers and primary flotation vessels combine quite a few simultaneous or nearly simultaneous elementary process steps. This makes the interaction of process variables difficult to predict. Consequently, much hot water flotation process optimization research involves test processing in a laboratory or pilot-scale apparatus. The small-scale observations are used to describe the flotation behaviour of the oil sand and infer what will

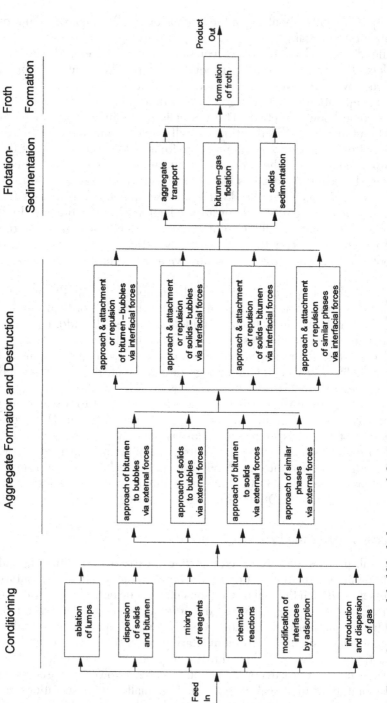

Figure 3. Simplified block diagram of elementary process steps in the hot water flotation process for Athabasca oil sands. From Shaw et al. [56]. (Copyright 1996, American Chemical Society.)

happen at the full-scale plant level. Many different laboratory- and pilot-scale investigations into oil sand processing have been conducted over the past 60 years or so.

A practical standard hot water flotation process batch extraction unit (BEU) and test procedure has evolved in which small (0.5 kg) samples of homogenized oil sand are processed. A detailed description is given elsewhere [54]. Figures 4 and 5 illustrate some of the steps and variables involved. This test is reproducible and sensitive enough to be useful for evaluating new process aids (chemicals), process variables, and determining the processibility of different oil sand samples [45, 101, 102, 104]. An example of a continuous pilot-scale experimental extraction circuit (EEC) has been described in the literature [46, 54–57]. In this particular unit larger amounts, 2000–3000 kg/h, of oil sand are processed continuously. It is a scaled-down version of the continuous commercial process although the addition of sophisticated measuring sensors and computer control allow more careful monitoring and mass-balancing than is possible in the full-scale commercial process. The smaller circuit is thus better suited to research studies. Sanford [45] has shown that results from the batch and pilot processes described above can be correlated. As shown in Figure 6 the batch test results establish trends which translate directly to the pilot scale. Absolute process recoveries are translated only with difficulty due to unavoidable differences incorporated into the processing in batch mode at such a small scale. The larger pilot process, being continuous

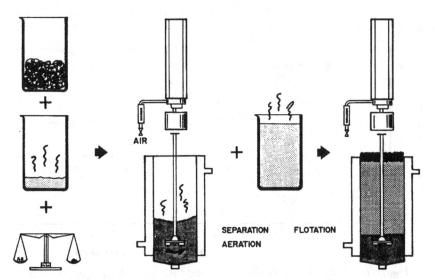

Figure 4. Illustration of the steps involved in conducting a batch hot water flotation process test, after Schramm and Smith [101]. (Copyright 1989, Alberta Oil Sands Technology and Research Authority.)

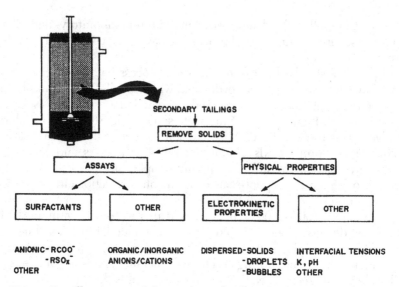

Figure 5.   Illustration of the steps involved in determining the surface, interfacial, and other properties of dispersed bitumen drops, solid particles, and gas bubbles in aqueous solutions from batch extraction tests.

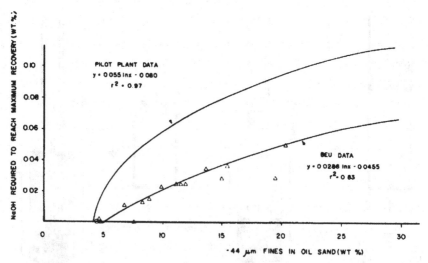

Figure 6.   Relationship between the <44 μm particle size fraction in Athabasca oil sands and the amount of NaOH addition required to optimize the hot water flotation process, as determined by Sanford [45]. (Copyright 1983.)

and closely modelled after the commercial process, yields results that compare quite well with the commercial process.

Although there are many variables, process efficiency is more sensitive to some variables than to others [47, 58–60]. Early studies led to the identification of base (NaOH) addition level as the preferred process variable and since then much work has been aimed at determining how much base is needed. It was at first thought that the process must be operated at generally "alkaline pH" [15, 61, 120]. Further research involved study of an increased number of oil sands, which led to the discovery that the process could be controlled to achieve good bitumen separation and flotation efficiency by maintaining a constant pH. This was specified at different values, for example, Bowman [62] recommended the middlings layer pH be kept in the region 7 to 8.5 while Innes and Fear [47] and Floyd et al. [63] recommended the pH range 8.0 to 8.5.

It was eventually shown by Sanford [64] that pH was not the important parameter as such but rather NaOH addition level, and that it should be regulated in response to fines level in the feed (Figure 6). Figure 7 shows processibility curves for four oil sands of differing composition. The term processibility refers to the primary bitumen (oil) recovery versus process aid (NaOH addition) relationship for a given oil sand and means, in essence, the NaOH addition level required to achieve maximum primary

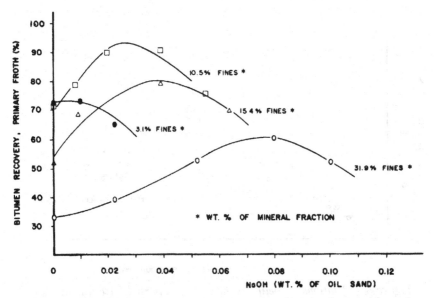

Figure 7. Processibility curves (laboratory batch extraction) for four oil sands of different composition. (From Sanford [45]. Copyright 1983.)

oil recovery. In addition to its process control origin, the concept of processibility forms a partial means for categorizing oil sands.

The commercial extraction/flotation plants are controlled using empirical relations involving oil sand grade and fine solids content information [45, 47]. Sanford [45] found several important correlations, first between the <44 μm fine solids size fraction and the <5 μm fine solids size fraction as shown in Figure 8. This correlates with the bitumen content in oil sand as shown in Figure 1, and also correlates with the amount of process aid (sodium hydroxide) addition required for optimal hot water flotation process efficiency as shown in Figure 6. Taking Figures 1, 6, and 8 together leads to the main method of commercial process control: the bitumen content of oil sand feed entering the plant is determined on-line by infrared reflectance and used to estimate the level of fine solids in the feed, thereby indicating the level of process aid addition required. Despite optimizing for each quality of feed, Figure 9 shows how oil recoveries become progressively poorer with decreasing grade of oil sand [65]. For grades of below 10% bitumen content, recoveries of less than 90% and lower energy efficiencies in the process are obtained. Improved empirical correlations are continually being discovered for these and other, anomalous oil sands [66]. Further mechanistic information could be used to develop improved process aids, process controls, and even alternate processes.

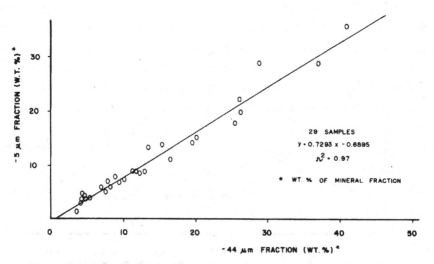

Figure 8. Relationship between the <44 μm particle size fraction and the <5 μm particle size fraction in Athabasca oil sands determined by Sanford [45]. (Copyright 1983.)

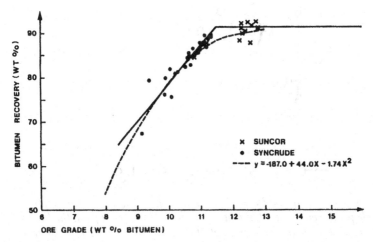

Figure 9. Bitumen recovery versus oil sand grade based on mean commercial operating plant data to 1980, according to Houlihan [65]. (Copyright 1982.)

The Role of Surfactants and Interfacial Properties

The hot water flotation process for oil sands is a separation process in which the objective is to separate bitumen from mineral particles by exploiting the differences in their surface properties. The slurry conditioning process involves many process elements as illustrated in Figure 3. Given that ablation and mixing, mass and heat transfer, and chemical reactions are accommodated, the conditioning step involves separating bitumen from the sand and mineral particles.

Disengagement of bitumen from solids will be favoured if their respective surfaces can be made more hydrophilic since a lowering of surface free energy will accompany the separation. The phase separation is enhanced by the effects of mechanical shear and disjoining pressure. Adopting the water-wet model for Athabasca oil sand, one has that a thin aqueous film already separates the bitumen from the sand. So this pre-existing separation needs only to be enhanced.

The need for alkaline conditions in the oil sand slurry has already been emphasized. The main role of the base (e.g., NaOH) is to produce (saponify) natural surfactants from the bitumen [45, 54]. In the 1960s, Bowman and co-workers used foam fractionation and spectroscopic characterization of the isolated waxy material obtained to establish that the surfactants produced in the process are primarily carboxylic salts of naphthenic acids [67–69] with the possibility of sulfonic salts as well. Figure 10 shows an example of an early identification of the possible surfactant structure and illustrates the reaction of base with the acid form

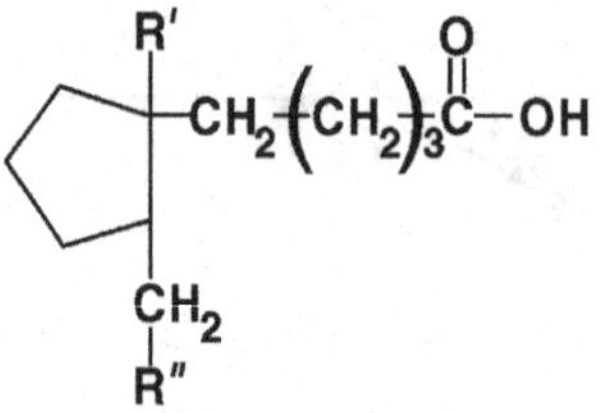

$$\sim\sim OH + NaOH \rightleftharpoons \sim\sim O^- + Na^+ + H_2O$$

Figure 10.   Simplified representation of the structure of the naphthenic acids produced during hot water processing, compiled from the literature of the 1960s. (From Schramm et al. [102]. Copyright 1984, Alberta Oil Sands Technology and Research Authority.)

of the surfactant (in the bitumen) to produce the salt form of the surfactant (in the aqueous phase).

For rich grade oil sands, the addition of a base such as NaOH is usually not necessary and simply slurrying with hot water is all that is needed to release sufficient quantities of natural surfactants into the aqueous phase. For lower grades of oil sand, NaOH process aid addition is needed to optimize bitumen recovery. In this case, only a small fraction of the NaOH added in processing reacts to produce the natural surfactants; while the major portion (ca. 90%) reacts with minerals (to produce mostly bicarbonate) [70, 102–104].

By 1987, Schramm, Smith, and Axelson [71] isolated natural surfactants from a large sample of tumbler slurry from Syncrude's continuous pilot plant. The slurry was allowed to settle for several days and the supernatant clarified by centrifuging and then filtering through 0.8 μm (nominal pore size) filters and then 1000 molar mass (nominal) ultra-filters. The clarified process solution was then foam fractionated, using the apparatus depicted in Figure 11, in six stages of collecting foam, diluting, and re-foaming. Using the final fractionate, the isolated surfactants were characterized using proton and carbon-13 NMR and were found to predominantly consist of aliphatic carboxylates having hydrocarbon chains of at least five carbons (typically C_{15} to C_{17}) and aliphatic sulfonates having hydrocarbon chains of at least five carbons. Also found were traces of species having methoxyl, aromatic and humic character. Figure 12 shows formulae for the predominant structural types. Surfactants such as sodium myristate (C_{14}) and sodium palmitate (C_{16}) have been used as model surfactants in process research. Misra, Aguilar, and

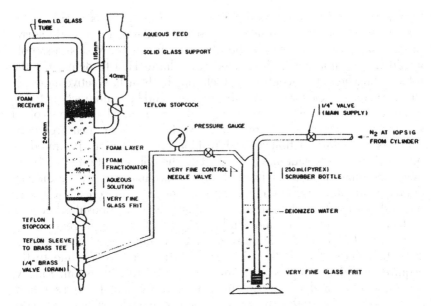

Figure 11.   Foam fraction apparatus for isolating natural surfactants.

$$CH_3 (CH_2)_{x>4} COONa$$

$$CH_3 (CH_2)_{x>4} SO_3Na$$

Figure 12.   Simplified formulae for the major structural types of naphthenic acids produced during hot water processing, as determined more recently. The total carbon numbers are likely to be about 14 to 16.

Miller [72] have made a similar identification of paraffinic carboxylate surfactants as the principal surfactant type released in the processing of Utah tar sands.

Analytical Methods for the Process Natural Surfactants. In order to experimentally verify linkages between surfactant action and process performance one needs appropriate analytical techniques. A number of books and reviews [73–83] discuss methods for the determination of anionic surfactants. Most are applicable only to the determination of sulfate- and sulfonate-functional surfactants. Difficulties associated with their application to carboxylate-functional surfactants include lack of stoichiometry in the formation of anionic–cationic surfactant pairs, interferences due to inorganic species, and the need to control the solution pH so that the surfactant remains in the salt form.

The first surfactant assays reported for oil sand process samples were conducted by Schramm, Smith, and Stone [102]. In these early assays process samples were first clarified by centrifuging. Next, a sub-sample would be potentiometrically titrated with dilute HCl to determine the total carboxylic salt concentration, including both surface active and non-surface active species. Another sub-sample would be foam fractionated to exhaustion, after which the residue would be essentially stripped of all surface active species, as verified by surface tension measurements. Aliquots of foam fractionate and residue were then acid titrated as before. From such data one can calculate the concentration of carboxylate-functional surfactants present in the original sample. Additional details including the needed equations are given in reference [102]. This method was used to establish the first quantitative relationships between surfactant concentrations and conditioning and flotation process performance [48, 102, 103, 121], but was restricted to the carboxylate-functional surfactant class. The method is also very time-consuming due to the requirement that samples be foam fractionated to exhaustion. Finally, foam fractionation does not completely remove surfactants from solution, and it was later found that the presence of unremoved surfactant from the residue contributes a consistent 10% error (underestimate) to this method [106].

In order to determine both carboxylate- and sulfonate-functional surfactants, probably the most common general method for anionic surfactants is Epton's two-phase titration method [84, 85] or one of its variations [86–88]. However, when assaying samples from the conditioning and flotation processes it is found that such methods exhibit endpoints that are very difficult to determine due to the pH at which the titration must be employed, and also due to the presence of inorganic and organic salts, finely emulsified oil, and dispersed fine solids. An improved method was developed by Schramm and Smith [106], still based on the formation of a compound between the anionic surfactant to be determined and a cationic surfactant added as a titrant, but in which a single-phase aqueous titration is carried out and monitored by means of surface tension measurements using the maximum bubble pressure technique [106]. We have since found that the titration can be equally well conducted using surface tension monitoring by the Wilhelmy plate method, which can also be automated. In either case, the method is dependent on obtaining clarified solutions with little ultra-fine solids. An added benefit is the simultaneous determination of a sample's dynamic surface tension.

A promising method for quantitation of anionic process surfactants is by cationic surfactant (e.g., Hyamine) titration monitored by a surfactant-sensitive electrode. The basic approach is described in references [76, 77, 89–92]. This technique has found application in the analysis of formulated products in the cosmetic [91] and pharmaceutical [90] industries and may show potential in the analysis of oil sand secondary tailings. This technique has the advantages that it is useful for all ionic surfactants, it is reproducible, there is no organic waste, no experience is needed on the part of the operator and low concentration electrolytes do not affect the endpoint detection. The endpoint is observed as the inflection point of a potential jump. Carboxylic moieties, having weak anions, are more difficult to assess, but assays are possible in an alkaline environment. We are currently pursuing the development of this promising method.

Several methods have been developed specifically for naphthenic acids, a class which includes the surface active carboxylate surfactants. Naphthenic acids are present as a complex mixture of a number of homologues with only a small range in molar mass (166–450 mol/g), little change in solubility character, and have been difficult to assay using conventional analytical methods. Methods such as negative ion-mode mass spectrometry using fast atom bombardment (FABMS), have been successfully applied to the analysis of naphthenic acid mixtures [93, 94]. Other promising techniques include fluoride ion chemical ionization mass spectrometry (FI-MS) [95], and electrospray ionization mass spectrometry (ESIMS), which may allow for the quantification of the various naphthenic acid fractions [96].

An FTIR method has also been developed for the determination of the naphthenic acids in oil sand process tailings [97]. In this method, a tailings sample is clarified by filtration (0.45 μm nominal pore size) then by ultra-filtration. Acidification to pH 2.5 with sulfuric acid ensures the acid form of all carboxylate functionalities and thus complete dissolution. The sample is then extracted with methylene chloride and evaporated to dryness. The naphthenic acid residue is dissolved in methylene chloride, the carbonyl stretching frequencies at 1708 and 1748 cm^{-1} are observed and the corresponding absorbance values determined by FTIR. The method determines total organic carboxylates and therefore is sensitive to a broader range of structures than the carboxylate surfactants alone, but is sometimes used as an indicator of relative carboxylate surfactant concentrations, especially in studies of oil sand tailings pond samples.

For example, in an adaptation of the standard methods [98, 99], the FTIR analytical technique just described was used to determine the octanol/water partition coefficient of naphthenic acids as a function of the aqueous phase pH. In this case, the octanol/water partition coefficient (K_{ow}) is the ratio of the equilibrium concentration of naphthenic acids in octanol to that in the aqueous phase. K_{ow} was determined by equilibrating a known total amount of naphthenic acids in a 1:1 volume mixture of octanol-saturated water (buffered to the selected pH) and water-saturated octanol, then determining the aqueous phase concentration. Figure 13 shows the variation in log(K_{ow}) with solution pH. It can be seen that K_{ow} decreases from quite high values, near 1000, to about 1 as the

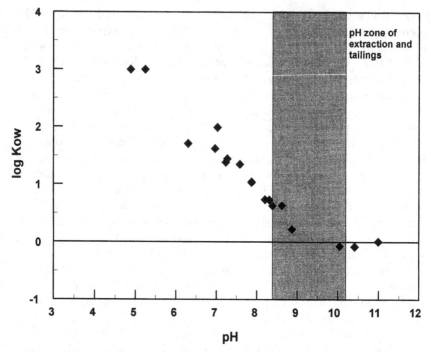

Figure 13.   Relationship between K_{ow} of naphthenic acids and solution pH, based on n-octanol:water partitioning.

aqueous phase pH increases from 5 to 9. Above pH 9, K_{ow} remains low. Thus, under the alkaline conditions typical of the water-based oil sands flotation process, in the slurry conditioning and flotation stages the naphthenates (including the carboxylate surfactants) will be partitioned about equally between the oil and aqueous phases. This represents a much lower preference for the aqueous phase than is found with most synthetic anionic surfactants, which are typically very hydrophilic. These results are, however, consistent with the long hydrocarbon chains (ca. C_{15} to C_{17}) of the process's natural carboxylate surfactants described in the preceding section.

Natural Surfactants and Interfacial Properties.   The action of the natural process surfactants has been studied in some detail [100–104]. The impact arises due to their adsorption at surfaces and interfaces, by which they alter surface electric charges and interfacial tensions. Figure 5 shows an example of the steps involved in determining the surface, interfacial, and other properties of dispersed bitumen drops, solid particles, and gas bubbles in aqueous solution. The samples analysed would be based on batch extraction tests involving different oil sand types and different process conditions.

In the conditioning process, under suitable alkaline conditions, both ionization of functional groups at the bitumen surface [33, 105] and adsorption of the natural anionic surfactant molecules at the bitumen/ aqueous interface [100, 101, 104] occur. Descriptions of the experimental techniques, including microelectrophoresis, employed to study the effects are given elsewhere [100, 102, 104, 106]. Figure 14 shows how addition of NaOH in the process increases the concentrations of surfactant in the aqueous phase, which in turn increases the extents of surfactant adsorption at all of the aqueous phase interfaces present in the system: gas/ aqueous, bitumen/aqueous, and solid/aqueous. The adsorption increases until monolayer coverage is achieved and thereafter either levels off or continues into multilayer adsorption.

The adsorption of anionic surfactant molecules directly affects the electrophoretic mobilities of dispersed bitumen droplets, gas bubbles, and fine solid particles. These electrophoretic mobilities are directly linked to the Zeta potentials at the surfaces and therefore to the surface electric charges on the drops, particles or bubbles. Reference [100] shows how to convert the mobilities into Zeta potentials or surface charges. Although, as will be seen later, the shapes of the processibility curves can vary considerably, the various surface charges are always quite negative

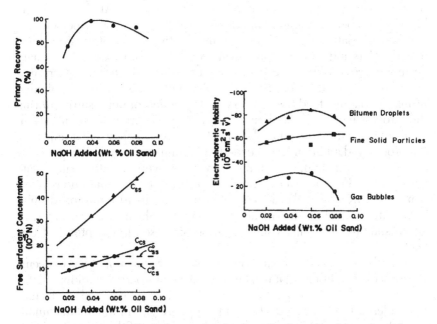

Figure 14. Example of the result of sodium hydroxide additions to the process on surfactant concentrations and, in turn, the electric charges on different surfaces.

under reasonable processing conditions. Whereas the surface charge on the solid particles reaches a plateau with increasing surfactant concentration, the surface charges on bitumen drops and gas bubbles reaches a maximum and thereafter decrease. Figure 15 shows some additional examples. Essentially the same trends have been independently confirmed by Hupka and Miller [107] and Drelich et al. [108]. The ionization of surface groups and adsorption of charged surfactants cause increased electrostatic repulsion which increases the disjoining pressure in the aqueous film separating the bitumen and solids.

In the optimized slurry conditioning step the natural surfactants have adsorbed just enough on the bitumen to impart a maximum charge, and just enough on the mineral solids to yield nearly a maximum charge there as well. Since initially the bitumen and solids are separated by a very small distance (ca. 10 nm), this charging causes a large repulsive force that results in the bitumen being pushed off the solids. Figure 16 illustrates the effects of increasing (NaOH addition) versus decreasing ($CaCl_2$ addition) this disjoining pressure in the film. (Additional details on electrostatic stabilization, the DLVO theory of colloid stability, and its contribution to disjoining pressure can be found in references [109, 110] and in many colloid chemistry textbooks.) Increased disjoining pressure together with the applied mechanical and thermal energy cause the separation of bitumen from solids, see Figure 17(a). At this stage the bitumen has been separated, but the fine solids have been dispersed.

The correlation between maximum negative Zeta potential on bitumen droplets and optimum processing conditions for maximizing primary bitumen recovery [100] has been shown to scale-up. On-line determination of the emulsified bitumen droplet Zeta potentials has been applied to primary separator feed (i.e., screened and flooded tumbler slurry) during the processing of oil sand under a variety of process conditions in Syncrude's continuous pilot-plant [111]. Figure 18 shows that adjusting the process aid addition level needed to maximize bitumen Zeta potential agreed well with both maximum primary recovery and the batch-scale correlations. For the data point corresponding to the continuous pilot-plant measurements, the actual maximum (negative) emulsion droplet Zeta potential achieved was about -35 mV, which is consistent with "good stability" guidelines such as those discussed in Chapter 1 of this book.

After bitumen–solid separation, bitumen–air attachment has to occur. The process conditions that most favour bitumen–solids separation, that is a high degree of electrostatic repulsion due to charged surfactant molecules at the interfaces, also tend to oppose gas–bitumen attachment since the gas bubbles also acquire a surface charge of the same sign [112] (see Figure 14). In comparison, mineral flotation involves gas–solid attachment without filming and such electrostatic repulsion is not as

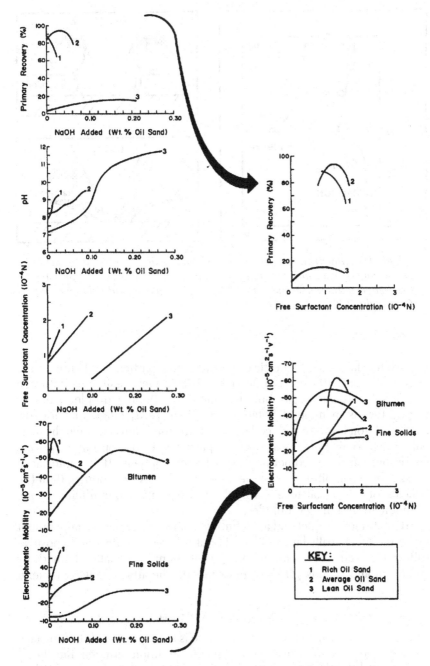

Figure 15. Examples of the connection between hot water flotation process efficiency and measured chemical and physical properties. (From Schramm and Smith [104]. Copyright 1987.)

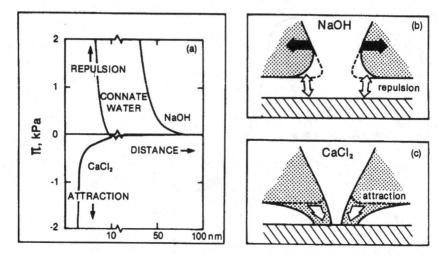

Figure 16.   Illustration of the effects of increasing (NaOH addition) or decreasing (CaCl₂ addition) on the thin aqueous film disjoining pressure (π) between bitumen and sand. (From Takamura and Chow [158]. Copyright 1983.)

important a factor as are inertia effects when the particles and bubbles are larger than, say, 10 to 40 μm diameter. It is in fact possible for bitumen droplets to attach to gas bubbles and form bubble droplet pairs or aggregates, as in mineral flotation. Houlihan [113] found that for low alkali addition levels or reduced temperature conditions bitumen droplets will attach to air bubbles as discrete particles. Under optimum process conditions however, something even better happens. If the interfacial tension between bitumen and the aqueous phase is low enough, then the balance of interfacial tensions in the system will favour filming of the bitumen around the gas bubbles.

If a bitumen droplet and a gas bubble collide, their mutual attachment is thermodynamically favourable if the "attachment coefficient" is positive (often referred to as an entering coefficient; originally defined by Robinson and Woods [114] for defoamers). The attachment coefficient, A, is given as:

$$A = \gamma^\circ_{Aq} + \gamma_{Bit/Aq} - \gamma^\circ_{Bit}$$

where the first term on the right-hand side is the aqueous solution surface tension, the second term is the aqueous solution/bitumen interfacial tension, and the third term is the surface tension of the bitumen. When bitumen attaches to a gas bubble a certain amount of gas/bitumen interface is created while some gas/aqueous and aqueous/bitumen inter-

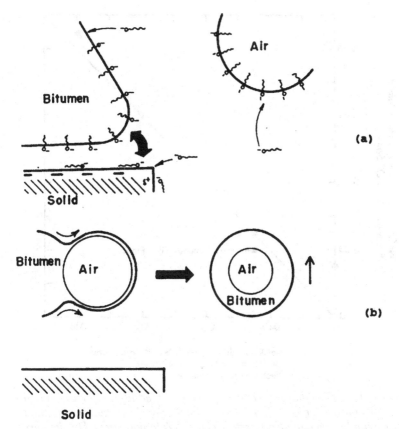

Figure 17. Illustration of two of the steps in the hot water flotation process: (a) the separation of bitumen from solids showing the adsorption of naturally produced surfactants, and (b) the attachment and filming of bitumen around gas bubbles. (From Shaw et al. [55]. Copyright 1994, American Chemical Society.)

facial areas are eliminated. The attachment is predicted to be favoured and spontaneous when $A > 0$. If $A < 0$, the bitumen should not attach.

If bitumen attaches to a gas bubble then, from thermodynamics, bitumen would be predicted to spread spontaneously over a gas bubble if its "spreading coefficient", S, is positive (Harkins [115]). The spreading coefficient is given by:

$$S = \gamma^{\circ}_{Aq} - \gamma_{Bit/Aq} - \gamma^{\circ}_{Bit} \tag{1}$$

where the symbols are defined as above. If bitumen spreads out over the gas/aqueous interface a certain amount of both gas/bitumen and aqueous/bitumen interface is created while some gas/aqueous interface is elimi-

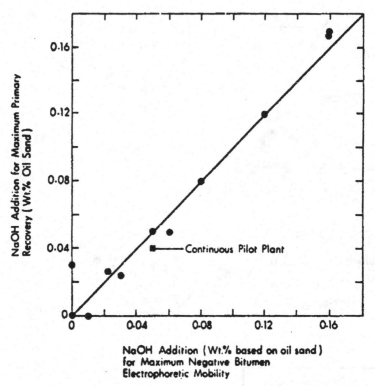

Figure 18. Correlation between process aid additions for maximum primary bitumen recovery and process aid additions required to attain maximum bitumen/aqueous surface charge (Zeta potential). The data are for continuous pilot plant operation (■) and laboratory batch extractions (●). (From Schramm and Smith [111].)

nated. The spreading is predicted to be favoured, and spontaneous, when $S > 0$. That S and A are interrelated is shown through a combination of the above equations:

$$A = S + 2\gamma_{Bit/Aq}$$

This shows that A is always greater than or equal to S, but that three combinations of effects can occur.

$A < 0$. If A is negative then S must be negative. Here the bitumen would neither attach nor spread at the aqueous solution/gas interface. Flotation is not expected in this case.

$A > 0, S < 0$. In this case the bitumen would attach but would not be expected to spread at the interfaces. This condition could cause flotation of the bitumen, depending on whether the flotation medium is sufficiently quiescent that the bitumen is not sheared away from the bubble.

$A > 0$, $S > 0$. Here the bitumen would attach and then spread over the gas bubble. Once the bitumen encapsulates a gas bubble only very high mechanical shear would cause it to be stripped away. This is the best configuration for bitumen flotation in a primary separation (flotation) vessel.

Table 1 shows some ranges of values that have been measured, at 80 °C, for froths and clarified secondary tailings solutions from batch extraction tests. Based on this data one would predict that under reasonable processing conditions bitumen will spontaneously attach to and then spread over the gas bubbles, encapsulating them. The surfactant properties that most promote this behaviour are a major lowering of bitumen/aqueous interfacial tension with minor lowering of the aqueous phase surface tension. This behaviour is consistent with the action of ionic surfactants having long hydrocarbon tails so that they will tend to partition mostly into the bitumen and only slightly into the aqueous phase. Based on actual process surfactant isolation and characterization research, the surfactant formulae shown in Figure 12, with carbon numbers in the range C_{15-17}, match these criteria very well.

Both laboratory studies [112] and Houlihan's continuous extraction circuit studies [113] indicate that under normal (good) processing conditions the bitumen does indeed preferentially encapsulate air bubbles. Figure 17(b) and the photographs in Figure 19 show the spontaneous filming of bitumen around a gas bubble brought into contact with the solution/bitumen interface. Similar observations have been made independently by Miller et al. [43, 72, 108]. These aerated bitumen globules are the species that float upwards in the flotation vessels to form froth.

Gas-bubble/mineral-particle attachment also occurs for the fraction of mineral particles that are not hydrophilic, so that some bubble-particle aggregates also form, float upwards, and become incorporated in the froth. An important goal of the flotation process is to produce bituminous froth without entraining large amounts of solids. Since the entrained

Table 1.   **Approximate Thermodynamic Properties for Froth Bitumens and Clarified Aqueous Secondary Tailings, for 80 °C, all units in mN/m**

Aqueous Surface Tension	Bit./Aq. Interfacial Tension	Bitumen Surface Tension	A	S
50 to 60	2 to 12	22 to 28	24 to 50	10 to 36

Source: L.L. Schramm, unpublished results

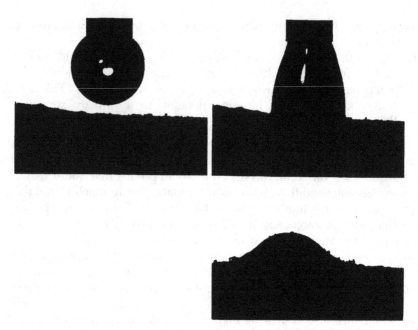

Figure 19.   Photographic sequence in which, viewing the images clockwise, an air bubble, on the tip of a capillary, is pushed down through an alkaline solution, at 80 °C, until it just touches a layer of bitumen that had been coated onto a silica surface. The bitumen spontaneously spreads over the surface of the bubble causing it to detach from the capillary and become engulfed. Note the obvious presence of solid particles in the bitumen on the surface in the lower photo. (Photomicrographs by L.L. Schramm, reproduced from Shaw et al. [55], copyright 1994, American Chemical Society.)

water and froth are later removed only with considerable difficulty, the question of how to produce higher quality froth is of interest. This is why so many oil sands patents are directed principally at the primary recovery aspect of the process (e.g. reference [6]). As a practical matter the solids floated do become part of the nature of the froths, and will be discussed later.

Processibility and Surfactant Concentrations.   Of the two natural surfactant classes identified earlier (see Figure 12), the carboxylate surfactant class has the greatest impact on process efficiency in most cases. Furthermore, optimal process conditions for best primary bitumen recovery correlate very well with a specific concentration of these surfactants in the aqueous phase of the conditioning slurry. This optimum concentration has been termed a "slurry-stage critical carboxylate surfactant concentration" (ζ_{CS}^0) [101], where the subscript specifies the carboxylate surfactant. Accounting for the addition of flood water, one can easily calculate the corresponding value of this concentration in the flooded slurry. This gives the value of the flooded-stage critical carboxylate surfactant concentration (C_{CS}^0) that would be measured were one to assay process streams such as the secondary tailings (batch extraction tests) or the primary separator middlings (continuous pilot or commercial process tests). The actual values of these parameters vary depending upon the specific process equipment used, but for the standard laboratory batch extraction unit (BEU) operated at 80 °C they are as follows:

Flooded Stage: $C_{CS}^0 = 1.2 \times 10^{-4}$ N
Slurry Stage: $\zeta_{CS}^0 = 7.5 \times 10^{-4}$ N

The existence of optimum concentrations for separation- and flotation-aid surfactants is quite common in mineral processing operations. In oil sand processing, reasons for reductions in process efficiency at higher than the optimum concentrations may be due to the formation of different adsorption layer orientations and/or multiple adsorption layers at interfaces. Either or both of these could explain the observed reductions in surface electric charge (Zeta potential) at the bitumen/aqueous and gas/aqueous interfaces (see Figure 14).

The other natural surfactant class (see Figure 12), the sulfonate surfactant class, has a significant impact on process efficiency under conditions where the carboxylate surfactant concentration is near zero. Otherwise preferential adsorption of the carboxylate surfactants dominates the properties of the interfaces. At near-zero concentrations of carboxylate surfactants, optimizing the process conditions for best primary bitumen recovery correlates very well with achieving a specific concentration of the sulfonate surfactants in the aqueous phase of the conditioning slurry. This optimum concentration is termed the "slurry-stage critical sulfonate surfactant concentration" (ζ_{SS}^0) [101], but note that the subscript now refers to sulfonate surfactant. Again, one can calculate a corresponding value of this concentration in the flooded slurry, which gives a flooded-stage critical sulfonate surfactant concentration (C_{SS}^0). For the standard laboratory batch extraction unit (BEU) operated at 80 °C these values are:

Flooded Stage: $C_{SS}^0 = 1.5 \times 10^{-4}$ N
Slurry Stage: $\zeta_{SS}^0 = 9.5 \times 10^{-4}$ N

With guidelines for the critical concentrations and recognizing the preferential adsorption behaviour (dominance) of the carboxylate surfactant, an extremely wide range of oil sand processing behaviours can be understood, as illustrated in a number of papers on the subject [70, 101–104, 116].

Figures 14, 15, and 20(a) show seven processibility curves that are typical of the different kinds of grades (rich through lean) and facies (estuarine through marine) origins that are encountered when processing Athabasca oil sands. In these cases the processibility is solely determined

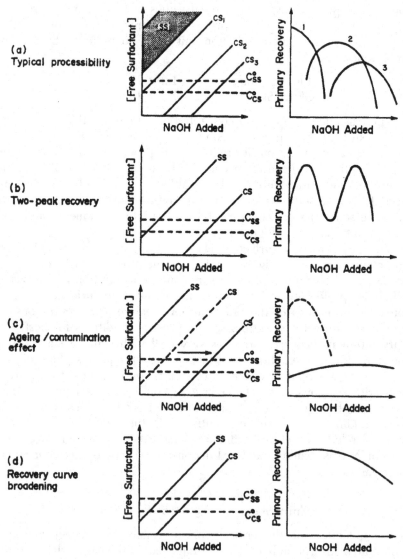

Figure 20.   Illustration of the processibilities and natural surfactant concentration variations in several kinds of oil sands. (From Schramm and Smith [104]. Copyright 1987.)

by the action of the natural carboxylate surfactants. Note the correlation between the primary bitumen recovery maxima and the point at which the free carboxylate surfactant concentration (C_{CS}) equals the critical concentration (C_{CS}^0).

Figure 20(b) illustrates a case where, at low NaOH addition, the concentration of carboxylate surfactants is near-zero and when the sulfonate surfactant concentration reaches C_{SS}^0, a first primary recovery maximum is observed. For the same oil sand processed at a moderately high NaOH addition level, the carboxylate surfactant concentration has increased and reaches C_{CS}^0, at which condition a second primary recovery maximum is observed. This "two-peak" processability curve is not an artifact, as it has been reproduced in multiple laboratory studies and also in continuous pilot-scale testing [104]. This particular kind of oil sand (a type of channel margin origin) first gained notice when a so-called "problem ore" in a commercial extraction plant was encountered, for which traditional feed-forward fines-content process control strategies failed.

Of course there exist oil sands which, when processed using moderate NaOH additions, yield surfactant concentrations that are intermediate between the two extremes just discussed. Figure 20(d) shows how rather broadened recovery curves result when both surfactant classes reach near critical concentrations at similar NaOH addition levels.

Ore Grades, Ageing, Blending of Ores and Contamination with Overburden.   The oil sand feed itself is an important process variable. Each oil sand type contains its own composition of compounds that can react with either the sodium hydroxide or directly with the natural surfactants. Figure 20(a) shows a series of processability curves governed by differing free surfactant concentration profiles. This illustration represents three grades of oil sand that yield progressively lower surfactant concentrations for given NaOH additions. Progressively higher fine mineral solids contents inhibit surfactant production by adsorbing surfactants as they are produced and by reacting with the NaOH itself, preventing it from generating the surfactants. The importance of such reactions extends beyond a call for more process aid to reach the critical surfactant concentrations: there is a concomitant decrease in obtainable primary oil recoveries.

Ageing.   Rather than three grades of separate oil sands, curves 1–3 in Figure 20(a) could also represent a single oil sand that "things have happened to". Curves 1–3 in this case could be three different "ages". Exposure of oil sand to ambient surface conditions causes a reduction in obtainable surfactant levels and hence oil recovery drops unless additional process aid is added. A number of ageing case studies have been reported by Schramm and Smith [117, 118] and by Wallace et al. [119]. Figure 21

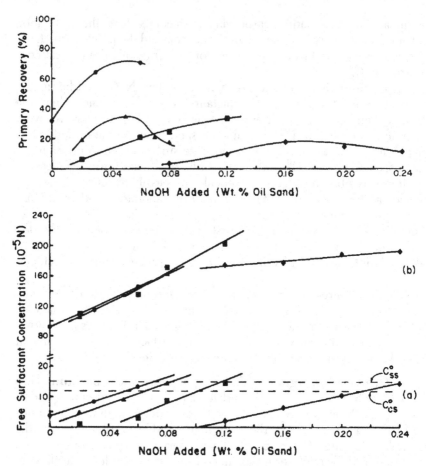

Figure 21. Processibility changes due to ageing of an oil sand. The curves correspond to processing the oil sand when freshly mined (●), and after 27 (▲), 54 (■), and 91 (◆) days of age. (From Schramm and Smith [117]. Copyright 1987, Alberta Oil Sands Technology and Research Authority.)

shows an example of the ageing of a moderate-size (20 kg) sample of oil sand over a period of three months. This figure illustrates the classic reduction in obtainable surfactant concentrations, the need for increased process aid addition, and the loss in maximum obtainable primary bitumen recoveries that accompany ageing. The "ageing effect" is due to the oxidation of minerals such as pyrite, which causes the release of polyvalent metal cations which precipitate the natural surfactants and also react directly to consume added NaOH during processing [118]. Figure 22 shows the reactions involved.

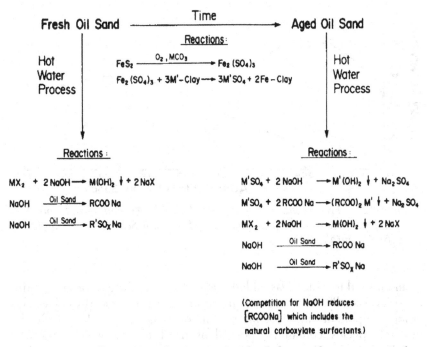

Figure 22.   Illustration of reactions involved during the processing of fresh and aged oil sands. (From Schramm and Smith [118]. Copyright 1987, Alberta Oil Sands Technology and Research Authority.)

It should be pointed out that whereas ageing phenomena present problems in laboratory and pilot-plant research and development studies, ageing has only occasionally been found to be a problem in commercial oil sand processing. In one instance a "small" quantity (several hundred thousand tonne) of oil sand was mined, set aside in a stockpile for about one year, then reclaimed and sent to an extraction plant. In this case, extremely poor recoveries were attributed to the ageing of the stockpile. In normal mining and processing operations oil sands are not left exposed for many months and even for dragline mining most of the oil sand in windrows does not age quickly. This was quantitatively demonstrated when Schramm and Smith [120] dissected a large commercial-scale windrow of about two months age. By determining the processibility of oil sand samples from different locations in the cross-section, it was found that only minor amounts of ageing had taken place.

Mineral Contamination.   Returning to Figure 20, curves 1–3 could also represent a single oil sand that has been contaminated with increasing levels of minerals from, for example, top reject material in the mine or simply a reject grade oil sand. Figure 23 shows a typical example,

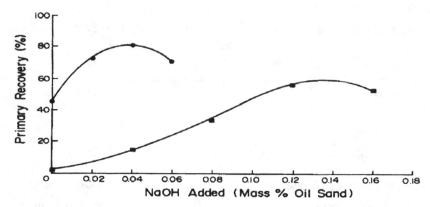

Figure 23.   Processibility of an oil sand by itself (●) and when blended with 2 mass % of overlying material (■). (From Schramm and Smith [116]. Copyright 1989.)

of an oil sand to which was added 2 mass % of overlying (non-ore grade) material. Usually the source of the problem can be traced to clays and other components in the contaminant overlying material; specifically, mineral species containing polyvalent-metal cations which consume NaOH and precipitate the surfactants. The problem can be traced to interfering polyvalent metal cation reactions, and alternative process aids can be used. Figure 24 shows an example in which an oil sand of a certain native processibility (▲) suffered greatly due to the addition of 2 mass % of overlying mineral material from the mine (◇). Almost exactly its original surfactant concentrations and processibility were restored (□) by an appropriate addition of chelating agent, EDTA [116]. Another use of EDTA to remediate poor processing is described in reference [118]. Figures 20(c) and 23 show how mineral contamination can have marked effects upon concentrations of carboxylate surfactants with correspondingly dramatic impacts on recovery. The kinds of mineral components that can have such dramatically negative effects on oil sand processibility include [70, 116]:

- polyvalent metal carbonates, e.g., $CaCO_3$, $MgCO_3$
- clays, e.g., Ca-illite, Ca-kaolinite
- polyvalent metal sulfates, e.g., $CaSO_4$, $FeSO_4$

Ore Blending.   An application of the influence of mineral components on processibility is the planned blending of different types of oil sands. Some fresh, rich grade oil sands, when processed with hot water only (no NaOH addition) yield natural surfactant concentrations that are above the critical concentration. As has been shown in earlier figures most oil sands yield surfactants at the critical surfactant concen-

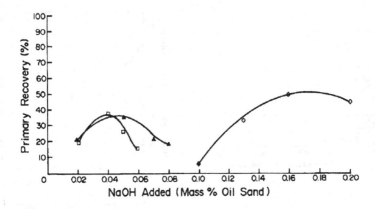

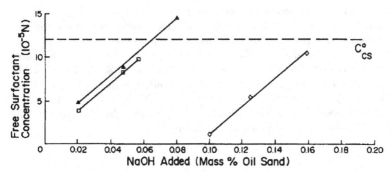

Figure 24. Processibility of an oil sand alone (▲), when blended with 2 mass % of overlying material (◇), and when blended with 2 mass % of overlying material and processed with the addition of EDTA (□). (From Schramm and Smith [116]. Copyright 1989.)

tration with either zero or a small NaOH addition. Some lean oil sands require significant NaOH additions and, as has been mentioned earlier, large NaOH additions dramatically increase the concentrations of other dissolved salts which has a depressing effect on primary bitumen recoveries. An alternative to large NaOH additions is to blend the very rich oil sands with the very lean oil sands in proportions designed to allow optimal processing of the final mixtures with little or no NaOH addition. The mixing rules for proper blending and the synergisms achievable are described in reference [103]. An example is shown in Figure 25. This approach has been patented [121] and used commercially in long-range mine operations planning.

Other Process Variables: Changing Slurry Water, Time, and Temperature. Although the "standard" hot water process

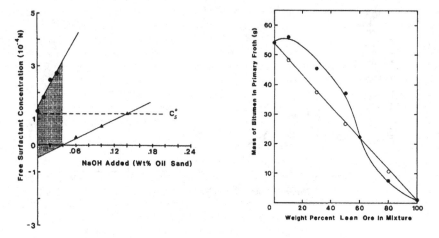

Figure 25.   Surfactant concentrations produced by processing rich (Left, Upper) and lean (Left, Lower) oil sands. The shaded area indicates the region of processing conditions within which blending could be synergistic. The right-hand figure shows maximum primary recoveries obtained from processing blends of the rich and lean ores (●). (Note that the recovery scale is in mass units and was intended to show relative rather than absolute recoveries.) The straight line (○) represents ideal mixing rule behaviour. (From Schramm et al. [103]. Copyright 1985, Alberta Oil Sands Technology and Research Authority.)

involves using NaOH as the process control variable there are, of course, other conditions of the process that can be varied. Building on the previous discussion of surfactant action in the process we can examine the effects of adjusting such variables as slurry water addition, slurrying time, and process temperature.

In an earlier section we discussed the difference between slurry-stage and flooded-stage critical surfactant concentrations. These allow one to directly predict the influence of changing slurry water addition levels in the process. To a first approximation, the standard processing conditions will generate a specific amount of natural surfactant in the slurry and their actual concentration will be determined by the total amount of slurry water present. This can be illustrated with an actual process example as given in Figure 26. Under standard conditions, the optimum processing condition was 0.02 wt% NaOH. One should be able to generate the same slurry-stage critical surfactant concentrations at a lower NaOH addition, say 0.01 wt% (where, under standard conditions $C_{CS} = 6.2 \times 10^{-4}$ N), by reducing the slurry water addition level. There is a limit to how much the slurry water addition levels can be reduced without impairing the ablation, mixing and dispersion processes (see the elementary process

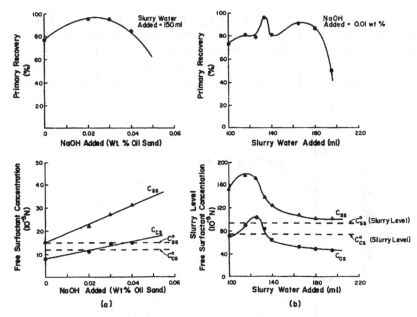

Figure 26. Example showing the effect of changing the slurry water addition levels. The left-hand graphs refer to standard processing conditions, including a slurry water addition level of 150 mL. The right-hand graphs refer to 0.01 mass % NaOH addition levels and varying slurry water addition levels. (Adapted from Schramm and Smith [101]. Copyright 1989, Alberta Oil Sands Technology and Research Authority.)

steps in Figure 3). Based on wet grinding research from other mineral processing industries, an estimate of the most effective water addition range for an average grade of oil sand processed under batch extraction conditions has been given [101] as 110 to 160 mL for 500 g of oil sand. In an ideal situation we should be able to cut the slurry water addition from 150 mL to about $(6.2 \times 10^{-4} \text{ N})(150 \text{ mL})/(7.5 \times 10^{-4} \text{ N}) = 124 \text{ mL}$, which is within the predicted range. At this condition C_{CS} should equal C_{CS}^0. From Figure 26, C_{CS} becomes $= 7.5 \times 10^{-4} \text{ N}$ and maximum primary recovery is achieved at 133 mL of slurry water which is quite close to the predicted value considering that reducing the slurry water addition level in reality changes the situation by more than just dilution factors.

It has been shown [101] that the simple dilution model can be used to fairly accurately predict how to increase slurry water addition levels as well as, for example, to counteract overdosing with NaOH in the slurry. It has also been shown [101] that the correlations between maximum primary bitumen recovery and surface electric charges (Zeta potentials) are also preserved when varying the slurry water addition levels.

While focussing on the slurrying stage we can address another important process variable – the slurry time (in the continuous process this would be the tumbler residence time). It takes some time for the natural surfactants to dissociate and diffuse into solution, even when no NaOH is added into the process, as illustrated in Figure 27. Also over time

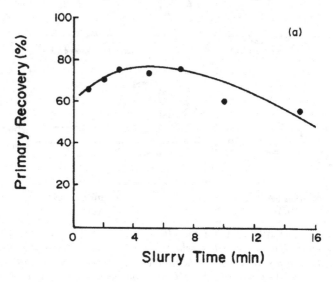

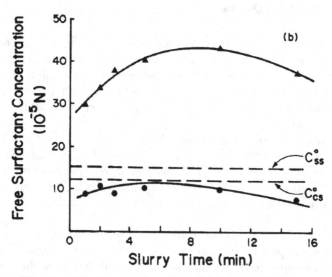

Figure 27.   Example showing the effect of changing the slurrying time in batch processing. (Adapted from Schramm and Smith [101]. Copyright 1989, Alberta Oil Sands Technology and Research Authority.)

the surfactants obviously react with other species since their concentration drops after reaching a maximum. See reference [101] for further details.

Finally, there has recently been much attention paid to the possibility of reducing commercial process costs by reducing the slurry temperature. Figure 28 shows some comparisons for the processing of an average grade of oil sand at 82 versus 55 °C. The latter temperature could be reached in a continuous process tumbler using only hot water (i.e., no steam addition). It can be seen that all of the trends are virtually identical, although it was found [101] that the lower temperature process is associated with a slightly lower critical surfactant concentration, a slightly lower optimum slurry water addition level, and a longer slurrying time requirement.

Alternative Process Aids. Numerous substitute process aids have been evaluated for possible use in the hot water flotation process. Indeed, this is probably the subject of the majority of patents directed at oil sand processing. It should be fairly obvious that if sodium hydroxide is a suitable process aid then other alkaline agents could be used as well. Research by many investigators has borne this out over the years. Other bases, including silicates and phosphates have been studied in some detail [122]. Some of these alternative alkalis have performance advantages over NaOH. The main reason NaOH has remained in commercial use over three decades has been economic.

A number of studies have shown that it is possible to directly add commercially available surfactants to the conditioning step of the process rather than rely on the addition of water alone or water plus an alkali such as NaOH. Some surfactants that have been successfully used include a variety of hydrocarbon- (ionic, nonionic, amphoteric), perfluoro-, and silicone-based commercial products [122]. The anionic hydrocarbon surfactants include tallow foots (a carboxylate-functional surfactant from the meat packing industry) [123], lignosulfonate (a sulfonate-functional surfactant from the pulp and paper industry) [124], sodium oleate [124, 125], and even the clothes-washing detergent Tide[®],[4] [124]. Although commercially available surfactants can be very effective as process aids, their cost has thus far posed a barrier to significant commercial use in hot water processing.

Another kind of process aid class is that of diluents, which would dissolve in the bitumen in order to reduce its viscosity and facilitate its separation and flotation. This is an essential additive in the processing of Utah oil sands, in which the bitumen is extremely viscous and for which kerosine has been recommended [42, 119]. A blend of kerosine and

---

[4] Tide is a trade-mark of Proctor & Gamble.

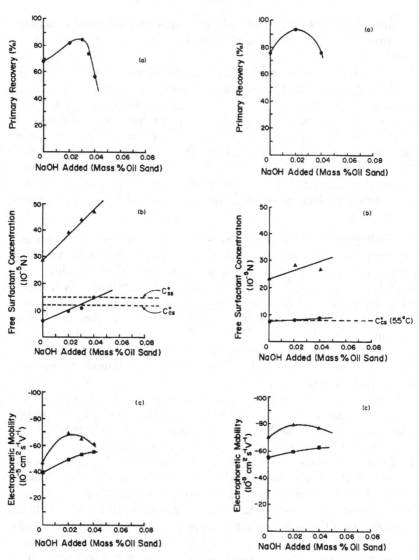

Figure 28.   Example showing the effect of changing the conditioning and flotation temperature in batch processing. The graphs show the processibility of an oil sand at 82 °C (Left) and at 55 °C (Right). (Adapted from Schramm and Smith [101]. Copyright 1989, Alberta Oil Sands Technology and Research Authority.)

methylisobutyl carbinol (MIBC, also known as methylamyl alcohol) is the process aid of choice for the so-called "OSLO" extraction process [126].

Finally, a number of other kinds of process aids, too numerous to cover here, have been evaluated. As an example, the reader may remember the discussion around Figure 24, in which it was shown that polyvalent metal cation problems could be counteracted by the addition of a chelating agent such as EDTA.

Beyond Primary Flotation

Reference was earlier made to the fact that under optimal processing conditions the bitumen droplets which float are for the most part spherical, consisting of a thin film of oil enveloping an air bubble. At the surface of the primary separation vessels, successfully floated bitumen globules form a froth layer. The packing density of aerated bitumen droplets at the froth interface is a function of droplet size distribution, the flux of material impinging on the boundary, and the bitumen droplet coalescence rate. The latter variables govern the time available for water drainage and the extent to which droplets can orient themselves relative to one another. Under normal vessel loading conditions, the rate of bitumen droplet coalescence is slow relative to the rate at which bitumen droplets collect at the interface. Paths for drainage become exceedingly tortuous and much of the occluded water is unable to drain (see reference [55]). The quality of this froth is determined not only by the relative amounts of water and solids present in the material, but also by the ease with which these constituents can be separated from the froth in downstream operations.

Calculations can be performed to show the expected relationships between froth quality, bitumen droplet aeration, and packing density for an ideal system of uniform spheres that congregate in a hexagonal close packed array with a volume fraction of 0.74. The ideal relationship between froth quality and aerated bitumen droplet density which follows (still assuming equal sized droplets) is shown in Figure 29. For comparison, Figure 29 also shows Danielson's [127] measured bitumen droplet densities and corresponding froth compositions for two oil sands of differing processibility. Although these froth compositions are somewhat higher than those predicted from the ideal system, the trend towards improved froth quality with increasing droplet density is still evident. Of course, in real systems the assumption of uniform droplet size is not completely adequate (small sized droplets would fill the interstices of larger droplets) and real droplets can deform in the froth, both of which give rise to increased bitumen concentration. The effect of varying packing density on froth bitumen content is shown in reference [55].

Based on microscopic studies, Swanson [128] proposed that the rising bitumen-coated gas bubbles collapse in the froth layer to yield water droplets dispersed in bitumen, as illustrated in Figure 30. This would produce dispersed gas bubbles with diameters just smaller than the original globules and also small water droplets, formed from the original

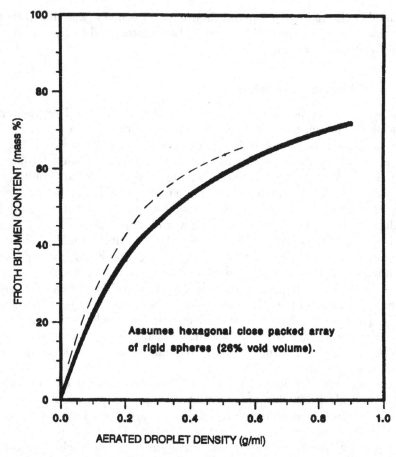

Figure 29. Relationship between froth quality and aerated bitumen droplet density showing model (solid curve) and experimental results (broken curve). (Adapted from Schramm [110]. Copyright 1994, American Chemical Society.)

thin films and containing sub-micron sized mineral particles. The membranous interfacial film was concluded to be formed from solids (mostly clays) having an organic coating, making them oil-wetting or bi-wetted. In addition, there will be larger sized water droplets that are entrained into the froth layer and are termed free water. This model is consistent with other microscopic investigations conducted by Chung et al. [129] and a magnetic resonance imaging study of froth structure conducted by Schramm and Axelson (reported in reference [55]). Given that froth contains an interstitial aqueous phase originating from the middlings of the separation vessel, one might expect that a certain amount

THE PROPOSED MECHANISM OF WATER-SAC FORMATION

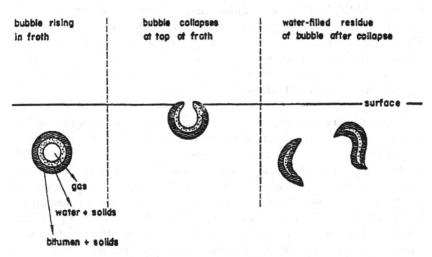

bubble rising in froth | bubble collapses at top of froth | water-filled residue of bubble after collapse

surface

gas
water + solids
bitumen + solids

Figure 30.   Swanson's mechanism of membranous emulsified water formation in froth, based on microscopic examinations. (From Schramm [110]. Copyright 1994, American Chemical Society.)

of dissolved natural surfactant and also dispersed water-wet clays and silica will be present as well. In addition, Schutte [130] found that froth is enriched in asphaltenes. Otherwise, there do not appear to be any published studies relating to surfactant occurrences in these bituminous froths.

Primary froth typically consists of 60% bitumen, 10% solids and 30% water. It also contains air and so is compressible and more viscous than bitumen [40]. To facilitate pumping, the froth is deaerated in towers by causing it to cascade through shed decks, flowing against the upward flow of steam. The froth from the secondary flotation is much more contaminated with water and solids than is the primary froth, and typically contains 15% bitumen, 20% solids and 65% water. Secondary froth is "cleaned" in stirred thickeners to remove some of the water and solids, then deaerated. Due to its higher quality, primary froth does not need the difficult cleaning and is more highly valued than secondary froth. Thus much emphasis is directed at optimizing the primary froth yield in the process (e.g., [48]). The primary and secondary froths, once deaerated, are combined into a single feed for a froth treatment process. This deaerated froth contains about 60% oil, 30% water and 10% solids, which is essentially the same as primary froth because secondary froth accounts for only about 5% of the total bitumen production. A review of froth structure and properties is given in reference [55].

A froth treatment process is used to remove water and fine solids from the combined froth [3, 10, 31, 44, 46]. The froth is first diluted with heated naphtha in about 1:1 volume ratio and then centrifuged in scroll centrifuges (at about 350 g) to remove coarse solids (greater than 44 μm). The diluted froth is then filtered and pumped to disc-nozzle centrifuges where higher g-forces (about 2500 g) are used to remove essentially all of the remaining solids and most of the water. After stripping off the naphtha the bitumen is upgraded into synthetic crude oil.

Froth Treatment.   As shown earlier, the primary and secondary froths comprise bitumen, air, water and solids. The two most often cited properties of concern with regard to processing behaviour are their high air content (compressable) and high viscosity [131, 132]. To make them easier to pump the froths are deaerated. Despite the presence of appreciable amounts of water and solids, the deaeration process brings the viscosity of the froths to very nearly that of bitumen itself [40]. The deaerated froth contains about 65% oil, 25% water and 10% solids. It also contains emulsions [128]. Emulsions of water-in-bitumen and of bitumen-in-water, both thought to be stabilized by asphaltenes and fine bi-wetted solids, have been found in interface layer emulsions in enhanced gravity separators [133].

The froth treatment process is used to remove water and fine solids from the deaerated froth. The froth is first diluted with heated naphtha in about 1:1 volume ratio. The diluted froth is then centrifuged in scroll centrifuges (at about 350 g) to remove coarse solids greater than 44 μm. The product from the scroll centrifuges is subsequently filtered and pumped to disc-nozzle centrifuges wherein higher g-forces (about 2500 g) are used to remove essentially all of the remaining solids and most of the water. In some cases lamella settlers are used rather than centrifuges [134].

The froth model described earlier, and shown in Figure 30, produces collapsed globules comprising a water (and solids) droplet surrounded by a membranous layer made up of asphaltenes and bi-wetted solids. When such froth is contacted with naphtha the time required to penetrate the bitumen-membrane coating is on the order of 30 minutes, whereas in a commercial process the elapsed time between naphtha addition and introduction into a settling vessel is less than one minute. Thus, the diluted froth process stream can contain these globules, probably in flocs, which have a bulk density intermediate between diluted bitumen and water. Such flocs then tend to accumulate in the separation vessel and form an interface layer (sometimes called rag-layer) emulsion, and could potentially form an effective barrier to gravity separation [134].

In both centrifuge and lamella settler operations highly emulsified samples which impair the separation processes have frequently been encountered. Such emulsions are apparently composed of the water-in-oil globules, dispersed in water, that is, an emulsion of water-in-oil-in-water (W/O/W). These emulsions appear gel-like and exhibit extremely high viscosities at very low shear rates (as high as 200,000 mPa·s at 80 °C). Under moderate shear however, the emulsions invert so that their viscosities are dramatically reduced (to about 10 to 20 mPa·s at 80 °C). Reference [55] contains an illustration of the effects of shear, including the shear-induced inversion from water-continuous (W/O/W) to oil-continuous (W/O) emulsion. Depending upon how a separation device is operated, such emulsions could accumulate into an emulsified layer thereby forming an effective barrier to gravity separation.

It has been shown that the emulsion structure-related problems encountered in the treatment of diluted froths have their origin in elements of the original froth structure, at least one component of which is made up of the natural surfactants released during the conditioning step. Another kind of surfactant chemistry comes into play in commercial froth treatment operations because a demulsifier is normally added to the diluted froth in order to improve centrifuge/settler performance. Commercial petroleum industry demulsifiers are formulated products, frequently based on polyglycols, polyglycol esters, ethoxylated alcohols, and ethoxylated resins in a heavy aromatic naphtha [135]. Dosage levels for emulsions of water-in-diluted froth are likely to be in the tens of parts per million.

Impacts of Natural Surfactants from Oil Sands on Water Quality Issues

Surfactants in Tailings. The tailings from the primary, secondary, and tertiary flotation processes are combined and hydraulically transported to a settling basin, as are the aqueous tailings from the froth treatment process. The coarse sand fraction settles out to form beaches and dykes, while the fines fraction (fine sand, silts, clays and unrecovered hydrocarbons) not retained within the sand, run off to a containment pond. Water is released during the subsequent settling and densification of the fines slurry to form mature fine tails (MFT) [136]. Over the first several years, the initial slurry of tailings fines (6–10% by wt) densifies to over 30% fines content, and the released water is recycled to the extraction plant. This represents over 70% of the water demand in the present processes. The full dissipation of the excess pore pressures of the fine tails to a fully consolidated clay (>70% solids by wt) by natural self-weight consolidation will be a very long process [136]. The tailings properties and characteristics have been reviewed by Mikula et al. [137]. Much effort has been expended over many years in looking for efficient and cost effective methods for managing the tailings (e.g., [6, 138]).

The sand deposits, fine tails zones and release water zones act as sinks for the extraction tailings and process waters associated with the bitumen recovery. In commercial-scale plants, operating with "zero discharge" of process-affected waters, water management of both present and future inventories is a complex issue, as can be seen in Figure 31(a) [139]. In the Syncrude operation, for example, there is, at present, an inventory of process waters in these various sinks of more than 700 Mm^3, Figure 31(b). It is projected that under present mine plans this stored volume will almost triple over the next 25 years of operation.

Over the projected life of a commercial-scale oil sands operation, such as Syncrude, the volumes of process-affected waters will be large. The quality and quantity of this water may affect water usage in the plant, as well as both onsite and offsite reclamation approaches [136]. As can be seen in Figure 31(a), much of the plant water requirements are drawn from the water released from the various sinks. To ensure optimized use of this water, factors that would adversely affect extraction efficiency or plant performance (scaling, corrosion) must be minimized.

The properties of the tailings waters are affected by the extraction process (Figure 32). The dissolved components, including the inorganic salts and dissolved organics associated with the water phase, are the most mobile, either through direct discharge or through long term seepage or release into the reclaimed landscape. The solids fraction in the extraction tailings includes sand, silts, clays and low levels of unrecovered hydrocarbons and can be expected to develop into traffic-able deposits. Such tailings deposits are suitable to terrestrial recla-mation strategies [136]. Fluid tailings will be incorporated into these deposits or will be placed in secure storage areas where they will retain their fluid character for a long time. The latter case is described as a "wet" reclamation option, and entails the formation of water-capped fine tails lakes or wetlands in which viable self-sustaining aquatic habitats will develop [140–142].

The properties of the water are important factors in the success of both "dry" and "wet" reclamation options, as is shown in Figure 33. The acute and chronic impacts of the process-affected waters on aquatic biota, wildlife, plants and humans exposed to them are being investigated. The common feature in these end-of-lease landscape units is the water. The inorganic salts or dissolved organic matter leached from the oil sand or added by process chemicals will have the potential to affect the plants used in terrestrial habitats, as well as the abundance and diversity of aquatic biota [136]. Issues such as survival, growth, fecundity and diversity of the reclaimed ecosystem are being addressed, and most of the short term detrimental biological effects of the process-affected waters to aquatic organisms have been shown to be associated with the organic acids in the dissolved organic fraction [142].

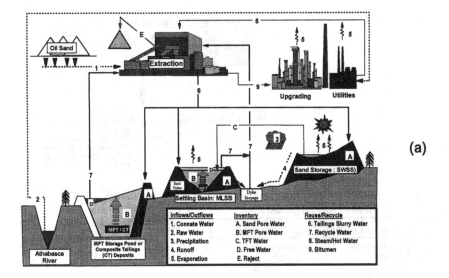

(a)

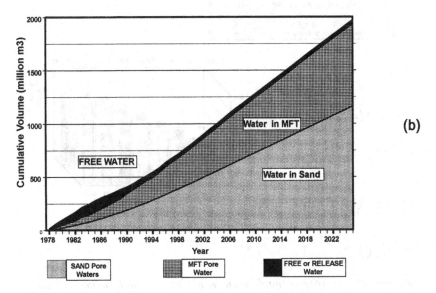

(b)

Figure 31. Simplified water balance at the Syncrude Canada operation (a), and estimates of the cumulative volumes of process-affected waters in various sinks, based on current operations (b).

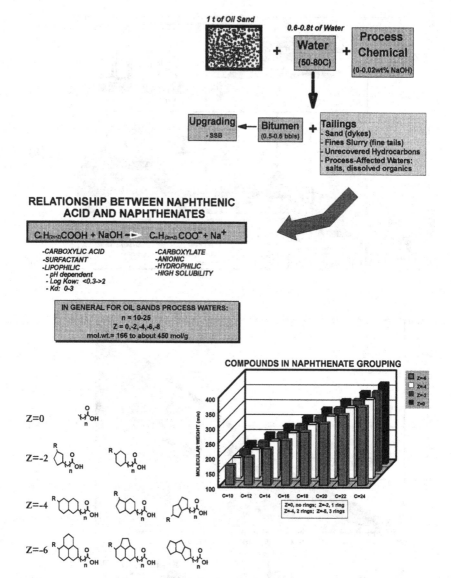

Figure 32.   Source of naphthenic acids in oil sand processing and examples of the types of structures and the range of compounds included in the naphthenic acid group.

While unrecovered bitumen accounts for the largest fraction of organics in the tailings materials, most of the dissolved organic matter is found as the polar components of the acid fraction. The major component of this has been identified as a mixture of carboxylic acids known as

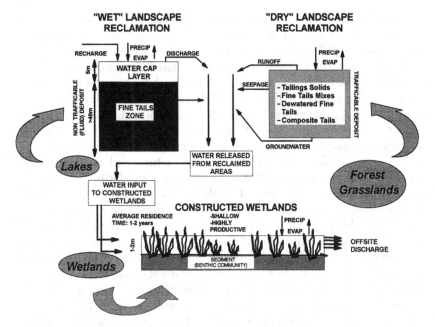

Figure 33.   Components of fine tailings reclamation.

naphthenic acids, some of which are surface active. Although, as discussed in earlier sections, the surface active sodium naphthenates are beneficial to bitumen recovery, the full range of extractable naphthenates presents both operational (corrosion) and environmental (toxicity) concerns [*140, 143, 144*]. This group of acids is represented by the general formula , RCOOH, where R is a naphthene. They are referred to as "petroleum acids" and include the carboxylic acids in crude oils [*143*]. Crude oils from various sources have been reported to contain between 0 and 3% organic acids. Generally, they are monobasic alkyl-substituted acyclic and cycloaliphatic carboxylic acids with little evidence of unsaturation and acid numbers in the 175–300 mg KOH/g range [*143*]. The commercial forms of the naphthenic acids are generally a complex mixture obtained by the caustic extraction of petroleum distillates from naphtha and kerosene fractions (200–370 °C). At temperatures above 500 °C, decomposition occurs. The manufacturing process produces a product that will be contaminated with other acidic components from the crude oils (phenolic and sulphur compounds). The use of commercially available naphthenic acids as standards is further complicated by differences in source and production [*145*]. Since the term, naphthenic acids, refers to a group of similar compounds, it is difficult to choose good representative "primary" standards for both process and environmental studies.

In the oil sands, the levels of organic acids in the bitumen have been reported to be in the 1–2 wt% range [146]. However, the concentrations of naphthenic acids measured in the extraction waters only represent a small fraction of the organic acids in the original bitumen. Based on the naphthenic acid concentrations determined in the extraction tailings, and expressing this relative to the original bitumen content of the oil sand, the calculated acid content is only about 0.1–0.2%, or about 10% of the total organic acid content expected in the bitumen [146, 147]. This may reflect the different analytical approaches and their selectivity.

Work undertaken to determine the structure and composition of naphthenic acids, using FABMS, has shown a similarity between the material present in conventional crude oils [93] and bitumen from the Athabasca oil sands deposits [94]. The naphthenic acid grouping is a mixture of a large number of saturated aliphatic and alicyclic carboxylic acids, represented by the general formula, $C_nH_{2n+z}O_2$, where n indicates the carbon number, and z indicates the number of hydrogens lost for each saturated ring structure in the molecule. For each ring in the structure, two hydrogens are removed, so by using the z-number, various series are indicated: straight chain $(z = 0)$, one-ring $(z = -2)$, two-ring $(z = -4)$, three-ring $(<z = -6)$ structures, etc. With a range of carbon numbers from about 10 to 30, and of z numbers from 0 to -6, as well as various possible isomers, the number of compounds within the naphthenic acid grouping can be quite large (Figure 32). At present, there is no evidence of significant unsaturation or aromatic character evident in the naphthenic acid groupings [94, 95]. A review of the sources, properties and environmental fates in the oil sands has recently been completed [147]. Questions about the characterization and pathways of this group of compounds still need to be examined.

As shown in Figure 32, the naphthenic acids will be released during processing of the oil sand. As shown earlier, the log K_{ow}'s of the naphthenic acids at the ambient pH of the Athabasca surface waters (pH 7.5–8) will be in the 0–1 range (Figure 13). This means that the bioaccumulation potential should be relatively low. In addition, the naphthenates have been shown to be relatively quickly degraded by microbes (naphthenate-degraders) under optimum processing conditions [136, 140, 142, 148, 149]. Under natural conditions, the rate of degradation of the naphthenic acids is slower, but does proceed quickly enough that within one year of isolation from fresh input of tailings waters, the acute toxicity of the water is removed [136]. Natural bioremediation of the toxicity associated with this group of compounds is the basis for many of the reclamation options that are now being pursued (Figure 33). The surfactant character of a fraction of the naphthenates will ensure that some attenuation of the naphthenic acids can be expected if process-affected waters intrude into the slightly alkaline conditions (pH 7–8.5) seen in most of the groundwater in the surface aquifers of the area. In Figure 34, the sorption distribution coefficients (K_d) for naphthenic acids on several possible substrates are shown. At equilibrium, K_d is given by

$$K_d = \frac{(\text{Naphthenic acids sorbed per g of solids})}{(\text{Naphthenic acids per g of water})}$$

In sand, the low K_d values (<0.5) would suggest little attenuation, while greater adsorption onto clay substrates takes place, since the K_d's were in the 1–3 range.

Even the contact between natural waters and exposed oil sands with erosion through the McMurray Formation will result in the release of low background levels into the waters of the area. In Figure 35(a), the results for water samples taken from the main stem Athabasca River (upstream, within and below the main oil sands deposit around Fort McMurray, in northeastern Alberta) and several tributaries in the deposit area, show low but measurable levels of naphthenic acids being observed. When compared to the levels of naphthenates found in the process waters and those influenced by them, the levels present in the process-affected waters are much higher. As one proceeds from the extraction tailings, to the tailings

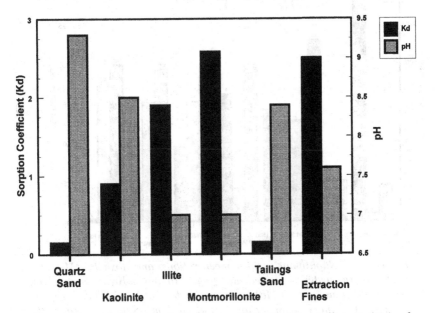

Figure 34. Change in the sorption distribution coefficient (K_d) of naphthenic acids with various substrates: sands (quartz, tailings coarse fraction) and clays (kaolinite, illite, montmorillonite, and oil sands fines). Naphthenic acids were determined using the FTIR method, and using a procedure based on ASTM Method E1195-87.

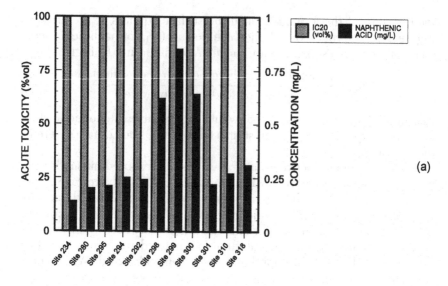

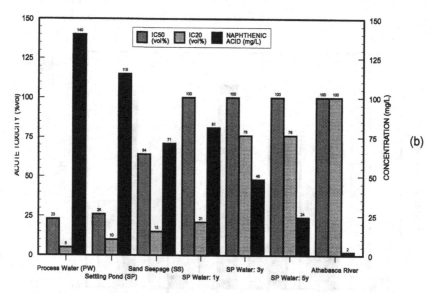

Figure 35. Naphthenic acid concentrations and toxicity (Microtox IC20) for (a) waters collected along the Athabasca River (from about 100 km upstream of Fort McMurray to the delta of Lake Athabasca), and (b) various waters at Syncrude's Mildred Lake Site. With time, the original toxicity and naphthenic acid levels in the fresh process waters (PW, SP) show a steady decrease when removed from fresh input of tailings (SS, 1, 3, and 5 years). Levels in the Athabasca River represent natural surface waters.

waters, to seepage waters, to influenced surface runoff and groundwaters on active site, there is almost a 2 orders of magnitude range (from > 130 to less than 1 mg/L), Figure 35(b).

Processing of oil sand using hot water digestion is more effective than the natural erosion seen in the Athabasca River drainage area, and, as expected, greater amounts of the naphthenic acids are released. In Figure 36, the naphthenic acid contents in tailings waters produced from the processing of six oil sand samples using the batch extraction unit (which uses a water:ore ratio of about 2, versus about 0.7 in commercial-scale operation) at various NaOH dosages (0–0.06 wt% of the oil sand) and waters (deionized and river water). Even with no NaOH addition (deionized and river waters), the pH's of the oil sand slurry were alkaline and concentrations of naphthenic acid were in the 30–50 mg/L range. With NaOH process aid addition, the pH values in the tailings waters were elevated, as were the naphthenic acid concentrations (60–125 mg/L). The addition of NaOH enhances the naphthenic acid release, but processing without caustic would also add substantial amounts of naphthenic acids to process-affected waters.

In the extraction process, the elevated pH's and abundance of sodium ion leads to greater solubilization of this group of compounds. In the produced waters, the acids will be in their carboxylate form, which in the oil sands extraction waters will be as sodium naphthenates (Figure 32). The extracted "petroleum acids" can be grouped into families based on the "c" and "z" numbers described earlier. However, there is no specific analytical method that has been shown to provide both qualitative and quantitative information on the individual compounds in the naphthenic acid group. As a result, most of the aquatic analyses have been carried out using the FTIR method described previously. This method is quite sensitive and suitable for the ranges seen in the process waters, but because of its lack of selectivity, interferences will be contributed by other organic acids that may be in the sample.

The levels of naphthenic acids in extraction waters are acutely toxic to many aquatic biota [*136, 147*]. Application of molecular toxicity methods, in which stress-inducible genes from *E. coli* are exposed to oil sands derived naphthenic acids, indicated that the main toxic response was indicative of cytotoxicity with osmotic stress and membrane disruption [*150*]. The results are consistent with toxic effects of surfactants. The application of the toxicity identification evaluation (TIE) protocol of the U.S. EPA (1991) to oil sands tailings waters was valuable in determining the fraction of the tailings water responsible for the reported acute toxicity [*151*]. The general physical and chemical fractionation scheme is shown in Figure 37(a), with the approximate percentages of the acute toxic response for each of the fractions. As shown in Figure 37(b), the acute toxic responses for several trophic levels (bacteria, invertebrates, and fish)

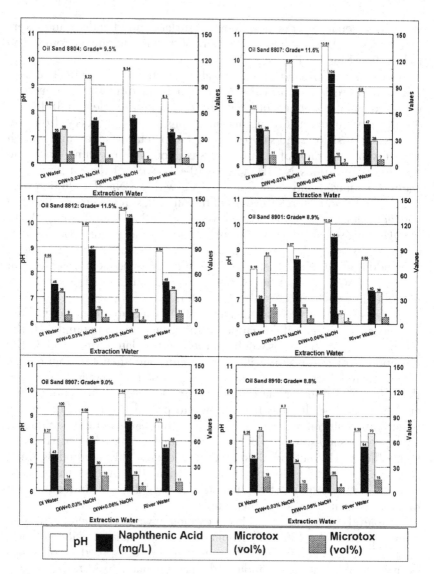

Figure 36. Relationship between acute toxicity (IC50 and IC20) and naphthenic acid concentrations (mg/L) in extraction waters from the batch extraction unit (BEU), with varying levels of NaOH added during processing.

show the same effect. Most of the acute toxicity was accounted for by the fraction precipitated out under acidic conditions or that removed by a reverse-phase solid absorbent (C_{18}) at all pH's. These observations on the loss of toxicity are consistent with organic acids with surfactant properties, such as the naphthenic acids. The TIE was conducted using the bacterial bioassay (Microtox), but the same general response was observed with other species, although the sensitivities were species dependent [150]. This indicates that the fresh tailings water had different relative toxicity, with the fish (rainbow trout, *Oncorhynchus mykiss*) being much more sensitive than the invertebrates *Daphnia magna* and slightly more sensitive than the bacteria (*Vibrio fischen*), Figure 37(b).

The naphthenic acids in fresh tailings water show a direct dose response as shown in the Microtox bioassay results in Figure 37(c), where fresh surface zone water from the Syncrude's Mildred Lake Settling Basin was diluted with deionized water. The initial tailings water, with a naphthenic acid concentration of 128 mg/L, was quite toxic (IC50 = 21% by vol.), and not until it had been diluted to about 10% was the IC50 greater than 100%.

However, with ageing, the toxic response per unit of naphthenic acid decreases. From Figure 38, the reduction in naphthenic acid content and of acute toxicity with time for tailings waters, stored under aerobic conditions, is evident for an array of biota. Natural bioremediation processes are proceeding in the oil sands extraction waters, once they are removed from fresh input of process waters and maintained under aerobic conditions. This biodegradation process is the basis on which the both the "wet" and "dry" landscape reclamation options operate. With time, waters will be slowly released from the various sinks (sand deposits, fine tails) in the end-lease landscape. If the rate of their release is less than the rate of the bioremediation processes, then the resulting waters should have the toxic components of the naphthenic acid group reduced to levels that will ensure the waters are neither acutely nor chronically toxic [136, 142].

The potential for the naphthenic acids to cause long term impacts is lessened by the properties of these compounds. It has been found that there are naturally occurring naphthenate-degrading bacteria present in most of the active water bodies at the Syncrude site, for example, with the highest concentrations (10^7 to 10^8 cells per g dry weight) being found at the sediment and fine tailings interfaces [152]. Naturally occurring populations of bacteria have been shown to degrade the naphthenates [148, 149]. With the use of surrogates of naphthenic acids that included a range of molecular weights, ring structures and positioning of substitutions, varying degrees of mineralization were reported [148]. When ^{14}C-labelled surrogates were used, rates of mineralization were also compound dependent and could be enhanced by nutrient addition [153]. Selective rates of removal of compounds within the naphthenic acid envelope can be expected. Differences in relative toxicity of the various compounds have also been indicated, so the application of specific concentration limits for environmental criteria of water quality will be difficult to establish.

Contribution of Various Fractions to Acute Toxicity

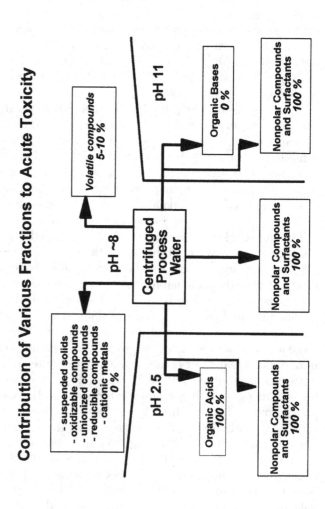

(a)

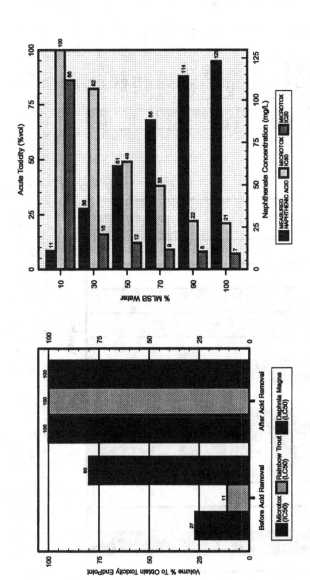

(c)

Figure 37. (a) Contribution to acute toxicity of Syncrude's fresh tailings pond water from MLSB based on the fractionation scheme and manipulations in the 1991 US EPA toxicity identification evaluation (TIE) method as reported in Verbeek et al. [151]. Toxicity was determined using the Microtoc® bacterial bioassay. (b) Change in acute toxicity of Syncrude's tailings water determined with bioassays using bacteria (Microtox®), rainbow trout, and Daphnia magna, before and after the removal of the acid fraction from the water. (c) Change in naphthenic acid content and acute toxicity (Microtox IC50) in tailings pond water from Syncrude's MLSB as it is diluted with deionized water.

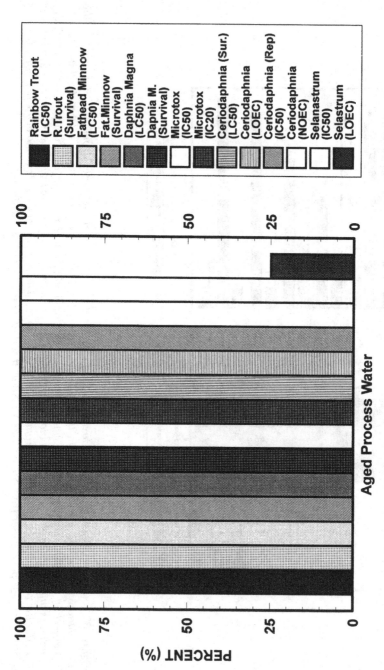

Figure 38. Results of a battery of bioassays (acute and chronic) conducted on extraction tailings water from Syncrude's hot water extraction process that was allowed to age naturally (isolated from fresh tailings input for 7 years). Initial LC50 to rainbow trout was <10%. (Note: the higher the percentage, the lower the toxic response.)

The levels of naphthenic acids in oil sands water will be relatively high in the fresh tailings, but over time the aerobic degradation of this material will limit its build up even in active process waters and eventually lead to its reduction. In an active oil sands settling pond, where tailings are added and released water is recycled back to extraction, the results of a four year period show little change in either the naphthenic acid content or the reported acute toxicity. Even though the MLSB water has been effectively recycled over eight times, no obvious change in the water quality with respect to toxicity and organic acids was measured, while during the same time an increase in about 25–40% of the ionic content was seen. The role of natural processes in limiting the toxic character of reclaimed sites is very important. Ongoing research projects are underway to assess the short and long term potential of bioremediation processes to treat process-affected waters and meet acceptable water quality criteria for preventing detrimental biological influences [136, 139, 141].

Surfactants and Corrosion.   Finally, in addition to the reclamation issues surrounding the levels and composition of the naphthenic acid group in oil sands waters, there is concern regarding the potential for corrosion. Naphthenic acids are a source of corrosion problems in the upgrading and refining stages of petroleum processing [93, 143, 156, 157]. They have been reported to initiate corrosion in the 200–400 °C range, when in the liquid phase. The corrosion rates associated with naphthenic acids are accelerated by velocity and turbulence, and can be mitigated by the presence of sulphur compounds and correct metallurgy [154].

Naphthenic acids have been identified as causing high-temperature, non-aqueous corrosion in refineries [155, 156, 157]. The exact mechanism is not well understood, but it seems to involve the formation of the metal naphthenates at the affected surface, and the mobilization of the naphthenate in the hydrocarbon phase because of its solubility. The naphthenic acids are only corrosive at temperatures above about 230 °C. The rate of naphthenic acid corrosion will increase with temperature since the process involves the mobilization of the metals rather than forming a scale or protective layer, as occurs with hydrogen sulphide based corrosion [155, 157]. When acting concurrently, the presence of naphthenic acids may increase the sulphidic corrosion. Because the vaporization point of the naphthenic acids is in the 250–350 °C range, the processing units that will be most at risk from the naphthenic acid corrosion will be those where distillation point oil products are handled or processed. In most refineries and in the upgrading facilities used in the oil sands industry, this will occur primarily in the atmospheric and vacuum towers, where the light and heavy gas oils are handled or treated [156].

In Figure 39, the results of the measurement of the naphthenic acids in various streams of Syncrude's commercial upgrading facility are

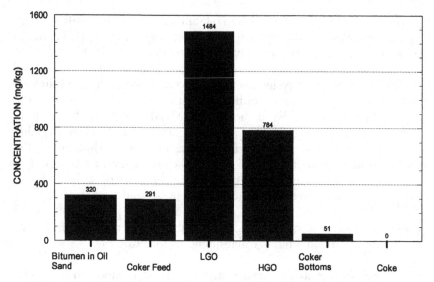

Figure 39. Naphthenic acid content (mg/kg of bitumen or carbon) in various hydrocarbon streams in Syncrude Canada's upgrading facility. Concentrations are based on leached naphthenates from a three-times extraction with 1N NaOH.

summarized. The reported values are based on the potential leachable naphthenic acids from the hydrocarbon streams, as determined in a three-time 1:1 extraction with 1N NaOH. With the log K_{ow} of the naphthenic acids being in the 0 range for such an aqueous separation (Figure 13), the extraction should remove most of the soluble naphthenic acids, but a more exhaustive extraction may be required. The results can be considered a relative gauge of where the naphthenic acids from the original oil sand bitumen are distributed in the upgrading process. It appears that the light (LVGO: 340–380 °C) and heavy (HVGO: 380–500 °C) gas oil fractions are enriched in the extractable naphthenic acids, while the coker bottoms (vacuum bottoms: > 500 °C) are depleted in this group of organic acids. At the higher temperatures, decomposition of the non-distilled naphthenic acids can be expected [157]. As a result, in an industry such as oil sands processing, the protection against corrosion associated with the naphthenic acids should focus on susceptible zones in the further refining and handling of the LVGO and HVGO cuts.

From an operational perspective, naphthenic acids are both beneficial and potentially a concern. As has been discussed earlier, the surface active properties play an important role in the efficiency of the bitumen separation methods from oil sands. These same surface active properties result in toxic responses to an array of biota that may affect the water management and reclamation options available to the oil sands industry. In addition, the corrosion resulting from the naphthenic acids poses an upgrading concern that must be controlled. At the same time, the research efforts are hampered by a lack of specific and sensitive analytical tools to better understand the role and pathway of impact of individual compounds within the naphthenic acid group of compounds. There is still considerable work required in the oil sands industry, to explain and predict the impact of this group of compounds. From the reclamation viewpoint, the relative ease and speed with which natural bioremediation, through microbial activity, proceeds appears to minimize or eliminate detrimental biological effects of the naphthenic acids and increases the confidence in end-of-lease reclamation options.

Summary

We have shown that in order to apply the hot water flotation process to Athabasca oil sands, one must deal with a number of phenomena originating from the discipline of colloid and interface science, particularly surfactant reactions and surface and interfacial surface activity leading to phase dispersions including suspensions, emulsions, and foams of various kinds. The initial oil sands slurry contains rocks and particles from which bitumen must be separated. Next, the bitumen must become aerated, preferably by encapsulating gas bubbles, and then floated, after which the floated bitumen globules coalesce to form a special kind of non-aqueous foam known as bituminous froth. The actions of natural surfactants originating in the bitumen form the physical chemical basis for several of these sub-processes. Surfactants from the process, particularly those in the naphthenate class, persist into the process tailings settling basin, giving rise to a number of concerns, including bioaccumulation, toxicity, and corrosion.

Acknowledgments

We thank the Petroleum Recovery Institute, Syncrude Canada Ltd., and the Alberta Oil Sands Technology and Research Authority for their support of, and permission to publish, this work. We also thank Vince Nowlan, Betty Fung, and Deib Birkholz for valuable advice and information.

List of Symbols and Abbreviations

A	Attachment coefficient
BEU	Lab-scale batch extraction unit for hot water conditioning and flotation processes

B/W/S	Bitumen, water and solids concentrations
C_{CS}	Aqueous concentration of carboxylate functional surfactant (flooded stage)
ζ_{CS}	Aqueous concentration of carboxylate functional surfactant (slurry stage)
C_{CS}^0	Critical aqueous concentration of carboxylate functional surfactant (flooded stage)
ζ_{CS}^0	Critical aqueous concentration of carboxylate functional surfactant (slurry stage)
EEC	Experimental extraction circuit, a 2.5 tonne per hour continuous hot water processing pilot plant operated by Syncrude Canada Ltd.
ESIMS	Electrospray ionization mass spectrometry
FABMS	Fast atom bombardment (negative ion-mode) mass spectrometry
FI-MS	Fluoride ion (chemical ionization) mass spectrometry
FTIR	Fourier transform infrared spectroscopy
K_d	Sorption distribution coefficient
K_{ow}	Octanol/water partition coefficient
PSV	Primary separation (flotation) vessel in a continuous hot water processing plant
S	Spreading coefficient
$\gamma_{Bit/Aq}$	Aqueous solution/bitumen interfacial tension
γ_{Bit}°	Bitumen surface tension
γ_{Aq}°	Aqueous solution surface tension

References

1. Walters, E.J. In *Oil Sands Fuel of the Future*; Hills, L.V., Ed.; Memoir 3, Canadian Society of Petroleum Geologists: Calgary, 1974; pp 240–263.
2. Demaison, G.J. In *The Oil Sands of Canada-Venezuela 1977*; Redford, D.A.; Winestock, A.G., Eds.; CIM Special Volume 17, Canadian Institute of Mining, Metallurgy and Petroleum: Calgary, 1977; pp 9–16.
3. Ruhl, W. *Tar (Extra Heavy) Sands and Oil Shales*; Ferdinand Enke: Stuttgart, 1982.
4. *Major Tar Sand and Heavy Oil Deposits of the United States*; Interstate Oil Compact Commission: Oklahoma City, OK, 1984.
5. Gutierrez, F.J. In *The Future of Heavy Crude Oils and Tar Sands*; Meyer, B.F.; Steele, C.T., Eds.; McGraw-Hill Inc.: New York, 1981; pp 107–117.
6. Rosing, K.E. In *The Future of Heavy Crude Oils and Tar Sands*; Meyer, B.F.; Steele, C.T., Eds.; McGraw-Hill Inc.: New York, 1981; pp 124–133.
7. Khalimov, E.M.; Muslimov, R.Kh.; Yudin, G.T. In *The Future of Heavy Crude Oils and Tar Sands*; Meyer, B.F.; Steele, C.T., Eds.; McGraw-Hill Inc.: New York, 1981; pp 134–138.
8. Valera, R. In *The Future of Heavy Crude Oils and Tar Sands*;

Meyer, B.F.; Steele, C.T., Eds.; McGraw-Hill Inc.: New York, 1981; pp 254–263.

9. Phizackerly, P.H.; Scott, L.O. In *Bitumens, Asphalts and Tar Sands*; Chilingarian, G.V.; Yen, T.F., Eds.; Elsevier: Amsterdam, 1978; pp 57–92.

10. Towson, D. In *Kirk-Othmer Encyclopedia of Chemical Technology*, 3rd ed.; Wiley: New York, 1983, 22, 601–627.

11. Carrigy, M.A. In *Guide to the Athabasca Oil Sands Area*; Carrigy, M.A.; Kramers, J.W., Eds.; Alberta Research Council: Edmonton, 1974; pp 173–185.

12. Fitzgerald, J.J. *Black Gold with Grit*; Gray's Publishing Co.: Sidney, B.C., 1978.

13. Spragins, F.K. In *Bitumens, Asphalts and Tar Sands*; Chilingarian, G.V.; Yen, T.F., Eds.; Elsevier: Amsterdam, 1978; pp 93–122.

14. Ells, S.C. *Bituminous Sands of Northern Alberta Occurrence and Economic Possibilities*; Mines Branch, Canada Department of Mines, Report No. 632, 1926.

15. Clark, K.A. Canadian Patent 289,058, April 23, 1929.

16. Blair, S.M. *Report on the Alberta Bituminous Sands*; Government of Alberta, King's Printer, 1951.

17. Hills, L.V., Ed. *Oil Sands Fuel of the Future*; Memoir 3, Canadian Society of Petroleum Geologists: Calgary, 1974.

18. Carrigy, M.A.; Kramers, J.W., Eds. *Guide to the Athabasca Oil Sands Area*; Information Series 65, Research Council of Alberta: Edmonton, 1973.

19. Stewart, G.A.; MacCallum, G.T. *Athabasca Oil Sands Guide Book*; Canadian Society of Petroleum Geologists: Calgary, 1978.

20. Chilingarian, G.V.; Yen, T.F., Eds. *Bitumens, Asphalts and Tar Sands*; Elsevier: Amsterdam, 1978.

21. Carrigy, M.A. *Geology of the McMurray Formation; Part III, General Geology of the McMurray Area*; Memoir 1, Research Council of Alberta: Edmonton, 1959.

22. Mossop, G.D. In *Facts and Principles of World Petroleum Occurrence*; Miall, A.D., Ed.; Memoir 6, Canadian Society of Petroleum Geologists: Calgary, 1980; pp 609–632.

23. Mossop, G.D.; Flach, P.D. *Sedimentology* **1983**, *30*, 493.

24. Stewart, G.A. In *K.A. Clark Volume*; Carrigy, M.A., Ed.; Alberta Research Council: Edmonton, 1963; pp 15–27.

25. Carrigy, M.A. *Sediment. Geol.* **1967**, *1*, 327–352.

26. Jardine, D. In *Oil Sands Fuel of the Future*; Hills, L.V., Ed.; Memoir 3, Canadian Society of Petroleum Geologists: Calgary, 1974, pp 50–67.

27. Pow, J.R.; Fairbanks, G.H.; Zamora, W.J.; In *K.A. Clark Volume*; Carrigy, M.A., Ed.; Alberta Research Council: Edmonton, 1963; pp 1–14.

28. O'Donnell, N.D.; Jodrey, J.M. *Proceedings of the SME of AIME Meeting*; Society of Mining Engineers of AIME: New York, 1984; preprint 84–421.

29. Mossop, G.D. *Science* (Washington, DC) **1980**, *207*, 145–152.

30. Carrigy, M.A. In *Proceedings of the 7th World Petroleum Congress*; Elsevier: Amsterdam, The Netherlands, 1967, *3*, 573–581.

31. Camp, F.W. *Tar Sands of Alberta, Canada,* 3rd ed.; Cameron Engineers Inc.: Denver, CO, 1976.
32. Carrigy, M.A. *Criteria for Differentiating the McMurray and Clearwater Formations in the Athabasca Oil Sands;* Bulletin 14, Research Council of Alberta: Edmonton, 1963.
33. Takamura, K. *Can. J. Chem. Eng.* **1982,** *60,* 538–545.
34. Vigrass, L.W. *Am. Assoc. Petrol. Geol. Bull.* **1968,** *52,* 1984–1999.
35. Ward, S.H.; Clark, K.A. *Determination of the Viscosities and Specific Gravities of the Oils in Samples of Athabaska Bituminous Sand;* Research Council of Alberta, Report No. 57, 1950.
36. Jacobs, F.A.; Donnelly, J.K.; Stanislav, J.; Svrcek, W.Y. *J. Can. Petrol. Technol.* **1980,** *19,* 46–50.
37. Svrcek, W.Y.; Mehrotra, A.K. *J. Can. Petrol. Technol.* **1982,** *21,* 31–38.
38. Jacobs, F.A. M.Sc. Thesis, University of Calgary: Calgary, Alberta, 1978.
39. Dealy, J.M. *Can. J. Chem. Eng.* **1979,** *57,* 677–683.
40. Schramm, L.L.; Kwak, J.C.T. *J. Can. Petrol. Technol.* **1988,** *27,* 26–35.
41. Carrigy, M.A. *J. Sedimentary Petrol.* **1962,** *32,* 312–325.
42. Yang, Y.J.; Bukka, K.; Miller, J.D. *Energy Processing Canada* **1989,** *82,* 14–21.
43. Miller, J.D.; Misra, M. *Fuel Proc. Technol.* **1982,** *6,* 27–59.
44. Erskine, H.L. In *Handbook of Synfuels Technology;* Meyers, R.A., Ed.; McGraw-Hill: New York, 1984; pp 5-1 to 5-79.
45. Sanford, E.C. *Can. J. Chem. Eng.* **1983,** *61,* 554–567.
46. Schutte, R.; Ashworth, R.W. In *Ullmans Encyklopadie der Technischen Chemie;* Verlag Chemie GmbH: Weinheim, Germany, 1979; Vol. 17.
47. Innes, E.D.; Fear, J.V.D. *Proceedings of the 7th World Petroleum Congress;* Elsevier: Amsterdam, The Netherlands, 1967; Vol. 3, pp 633–650.
48. Schramm, L.L.; Smith, R.G. Canadian Patent 1,188,644, June 11, 1985; U.S. Patent 4,462,892, July 31, 1984.
49. Perrini, E.M. *Oil From Shale and Tar Sands;* Noyes Data Corp.: Park Ridge, NJ, 1975.
50. Ranney, M.W. *Oil Shale and Tar Sands Technology;* Noyes Data Corp.: Park Ridge, NJ, 1979.
51. Schramm, L.L. *J. Can. Petrol. Technol.* **1989,** *28,* 73–80.
52. Schramm, L.L. Canadian Patent 1,232,854, February 16, 1988; U.S. Patent 4,637,417, January 20, 1987.
53. Adam, D.G. In *Proceedings, Advances in Petroleum Recovery and Upgrading Technology;* AOSTRA: Edmonton, Alberta, June 6–7, 1985; session 2, paper 1, 20 pp.
54. Sanford, E.C.; Seyer, F.A. *CIM Bull.* **1979,** *72(803),* 164–169.
55. Shaw, R.C.; Czarnecki, J.; Schramm, L.L.; Axelson, D. In *Foams, Fundamentals and Applications in the Petroleum Industry;* Schramm, L.L., Ed.; American Chemical Society: Washington, DC, 1994; pp 423–459.
56. Shaw, R.C.; Schramm, L.L.; Czarnecki, J. In *Suspensions, Fundamentals and Applications in the Petroleum Industry;* Schramm, L.L., Ed., American Chemical Society: Washington, DC, 1996; pp 639–675.

57. Cymbalisty, L.M. *Proceedings of the 30th Canadian Chemical Engineering Conference*; Edmonton, Alberta, 1980, 4, 1168–1182.
58. Seitzer, W.H. *Proceedings ACS Division of Petroleum Chemistry*; San Francisco, April 2–5, 1968; pp F19–F24.
59. Malmberg, E.W.; Bean, R.M. *Proceedings ACS Division of Petroleum Chemistry*; San Francisco, April 2–5, 1968; pp F25–F37.
60. Malmberg, E.W.; Bean, R.M. *Proceedings ACS Division of Petroleum Chemistry*; San Francisco, April 2–5, 1968; pp F38–F49.
61. Clark, K.A.; Pasternack, D.S. *Ind. and Eng. Chem.* **1932**, *24*, 1410–1416.
62. Bowman, C.W. U.S. Patent 3,623,971, November 30, 1971.
63. Floyd, P.H.; Schenk, R.C.; Erskine, H.L.; Fear, J.V.D. Canadian Patent 841,581, May 12, 1970; U.S. Patent 3,401,110, September 10, 1968.
64. Sanford, E. U.S. Patent 4,201,656, May 6, 1980.
65. Houlihan, R. In *The Future of Heavy Crude and Tar Sands*; Meyer, R.F.; Wynn, J.C; Olson, J.C., Eds.; United Nations Institute for Training and Research: New York, 1982; pp 1076–1086.
66. Smith, R.G.; Schramm, L.L. *Fuel Proc. Technol.* **1989**, *23*, 215–231.
67. Bowman, C.W. *Proc. 7th World Petroleum Congress*, **1967**, *3*, 583.
68. Baptista, M.V.; Bowman, C.W. *Proceedings, 19th Canadian Chemical Engineering Conference*, Edmonton, AB, 1969.
69. Leja, J.; Bowman, C.W. *Can. J. Chem. Eng.* **1968**, *46*, 479.
70. Smith, R.G.; Schramm, L.L. *Fuel Proc. Technol.* **1992**, *30*, 1–14.
71. Schramm, L.L.; Smith, R.G.; Axelson, D., unpublished results, 1987.
72. Misra, M.; Aguilar, R.; Miller, J.D. *Sep. Sci. Technol.* **1981**, *16*, 1523–1544.
73. Cross, J., Ed., *Anionic Surfactants – Chemical Analysis*; Dekker: New York, 1977.
74. Cross, J., Ed., *Nonionic Surfactants – Chemical Analysis*; Dekker: New York, 1986.
75. Porter, M.R., Ed., *Recent Developments in the Analysis of Surfactants*; Elsevier: Essex, 1991.
76. Schmitt, T.M., *Analysis of Surfactants*; Dekker: New York, 1992.
77. Cullum, D.C., Ed., *Introduction to Surfactant Analysis*; Blackie: U.K., 1994.
78. Rosen, M.J.; Goldsmith, H.A., *Systematic Analysis of Surface-Active Agents*, 2nd. ed., Wiley: New York, 1972.
79. Wang, L.K.; Kao, S.F.; Wang, M.H.; Kao, J.F. *Ind. Eng. Chem. Prod. Res. Dev.* **1978**, *17*, 186.
80. Schwartz, A.M.; Perry, J.W.; Berch, J. *Surface-Active Agents and Detergents*, Vol. 2, Kreiger: N.Y., 1977.
81. Llenado, R.A.; Jamieson, R.A. *Anal. Chem.* **1981**, *53*, 174R.
82. Wang, L.K.; Kao, S.F.; Wang, M.H.; Kao, J.F. *Ind. Eng. Chem. Prod. Res. Dev.* **1978**, *17*, 186.
83. Llenado, R.A.; Jamieson, R.A. *Anal. Chem.* **1981**, *53*, 174R.
84. Epton, S.R. *Trans. Faraday Soc.* **1948**, *44*, 226.
85. Epton, S.R. *Nature* **1947**, *160*, 795.
86. Glazer, J.; Smith, T.D. *Nature* **1952**, *169*, 497.
87. Reid, V.W.; Longman, G.F.; Heinerth, E. *Tenside* **1967**, *4*, 292–304.

88. Li, Z-P.; Rosen, M.J. *Anal. Chem.* **1981**, *53*, 1516–1519.
89. Birch, B.J.; Clarke, D.E. *Anal. Chim. Acta* **1973**, *67*, 387–393.
90. Vytras, K. *Mikrochimica Acta [Wien]*, **1984**, *111*, 139–148.
91. Oei, H.H.Y.; Toro, D.C. *J. Soc. Cosmet. Chem.* **1991**, *42*, 309–316.
92. *ASTM Designation D 4251–89*, American Society for Testing and Materials, Philadelphia, PA.
93. Fan, T. *Energy & Fuel* **1991**, *5*, 371–375.
94. Morales, A.; Hrudey, S.E.; Fedorak, P.M. *Mass Spectrometric Characterization of Naphthenic Acids in Oil Sands Wastewaters, Analysis, Biodegradation, and Environmental Significance*; Report, Alberta Dept. of Energy: Edmonton, AB, 1993.
95. St. John, W.P.; Rughani, J.; Green, S.; McGinnis, G. *J. Chromatography* **1998**, *807*, 241–251.
96. Morales-Izquierdo, A. *Analysis of Naphthenic Acids in Oil Sands Wastewater Samples by Electrospray Ionization Mass Spectrometry*; Report, Syncrude Canada Ltd.: Fort McMurray, AB, 1998.
97. MacKinnon, M., unpublished results, 1990.
98. Ribo, J.M. *Chemosphere* **1988**, *17(4)*, 709–715.
99. *ASTM Designation E 1147–92*, American Society for Testing and Materials, Philadelphia, PA.
100. Schramm, L.L.; Smith, R.G. *Colloids and Surfaces* **1985**, *14*, 67–85.
101. Schramm, L.L.; Smith, R.G. *AOSTRA J. Research* **1989**, *5*, 87–107.
102. Schramm, L.L.; Smith, R.G.; Stone, J.A. *AOSTRA J. Research* **1984**, *1*, 5–14.
103. Schramm, L.L.; Smith, R.G.; Stone, J.A. *AOSTRA J. Research* **1985**, *1*, 147–161.
104. Schramm, L.L.; Smith, R.G. *Can. J. Chem. Eng.* **1987**, *65*, 799–811.
105. Takamura, K.; Chow, R.S. *Colloids and Surfaces* **1985**, *15*, 35–48.
106. Schramm, L.L.; Smith, R.G. *Colloids and Surfaces* **1984**, *11*, 247–263.
107. Hupka, J.; Miller, J.D. *Int. J. Mineral Processing* **1991**, *31*, 217–231.
108. Drelich, J.; Hupka, J.; Miller, J.D.; Hanson, F.V. *AOSTRA J. Res.* **1992**, *8*, 139–147.
109. Schramm, L.L. (Ed.), *Emulsions, Fundamentals and Applications in the Petroleum Industry*; American Chemical Society: Washington, DC, 1992.
110. Schramm, L.L. (Ed.), *Foams, Fundamentals and Applications in the Petroleum Industry*; American Chemical Society: Washington, DC, 1994.
111. Schramm, L.L.; Smith, R.G. Canadian Patent 1,265,463, February 6, 1990.
112. Schramm, L.L.; Smith, R.G., unpublished results, 1985.
113. Houlihan, R.N., unpublished results, 1976.
114. Robinson, J.W.; Woods, W.W. *J. Soc. Chem. Ind.* **1948**, *67*, 361–365.
115. Harkins, W.D. *J. Chem. Phys.* **1941**, *9*, 552–568.
116. Smith, R.G.; Schramm, L.L. *Fuel Proc. Technol.* **1989**, *23*, 215–231.
117. Schramm, L.L.; Smith, R.G. *AOSTRA J. Research* **1987**, *3*, 195–214.
118. Schramm, L.L.; Smith, R.G. *AOSTRA J. Research* **1987**, *3*, 215–224.
119. Wallace, D.; Henry, D.; Takamura, K. *Fuel Sci. & Technol. Int.* **1989**, *7*, 699–725.

120. Schramm, L.L.; Smith, R.G., unpublished results, 1986.
121. Schramm, L.L.; Smith, R.G. Canadian Patent 1,214,421, November 25, 1986; U.S. Patent 4,474,616, October 2, 1984.
122. Bichard, J.A., *Oil Sands Composition and Behaviour Research*; AOSTRA: Edmonton, AB, 1987.
123. Schramm, L.L.; Smith, R.G., unpublished results, 1985.
124. Sanford, E. Canadian Patent 1,100,074, April 28, 1981.
125. Schramm, L.L.; Smith, R.G., unpublished results, 1986.
126. Sury, K.N.; Paul, R.; Dereniwski, T.M.; Schultz, D.G. "Next generation oil sands technology: The new OSLO processes". In *Proceedings, Oil Sands Our Petroleum Future Conference*, Edmonton, AB, 1993.
127. Danielson, L., unpublished results, 1989.
128. Swanson, W.D., unpublished results, 1965, 1966.
129. Chung, K.H.; Ng, S.; Sanford, E.C. *AOSTRA J. Res.* **1991**, 7, 183–193.
130. Schutte, R., unpublished results, 1974.
131. Wood, R. Canadian Patent 1,137,906, December 21, 1982.
132. Kizior, T.E. Canadian Patent 1,072,474, February 26, 1980.
133. Schramm, L.L.; Hackman, L.P., unpublished results, 1986.
134. Shelfantook, W.E.; Hyndman, A.W.; Hackman, L.P. U.S. Patent 4,859,317, August 22, 1989.
135. Grace, R. In *Emulsions, Fundamentals and Applications in the Petroleum Industry*; Schramm, L.L., Ed.; American Chemical Society: Washington, DC, 1992; pp 313–339.
136. Fine Tails Fundamentals Consortium, *Advances in Oil Sands Tailings Research*, Vol. 1, Alberta Dept. of Energy: Edmonton, AB, 1995.
137. Mikula, R.J.; Kasperski, K.L.; Burns, R.D.; MacKinnon, M.D. In *Suspensions, Fundamentals and Applications in the Petroleum Industry*; Schramm, L.L., Ed.; American Chemical Society: Washington, DC, 1996; pp 677–723.
138. Morgenstern, N.R.; Scott, J.D. In *Proceedings of Geoenvironment 2000 Conference*; Geotechnical Special Publication, 46, ASCE, 1995, 1663–1683.
139. Rogers, M.E.; Ferguson, D.; MacKinnon, M. In *Proceedings of NACE International Conference and Exposition: Corrosion 96*; North American Corrosion Engineers, 1996, Paper No. 568, 26pp.
140. Boerger, H.; MacKinnon, M.: Van Meer, T.; Verbeek, A. In *Proceedings of the 2nd International Conference on Environmental Issues and Management of Waste in Energy and Mineral Production*; Singhal, R.K.; Mehrotra, A.K.; Fytas, K.; Collins, J., Eds.; Balkema: Rotterdam, 1992, pp 1248–1261.
141. Gulley, J.; MacKinnon, M. In *Proceedings of Fine Tailings Symposium in Oil Sands: Our Petroleum Future Conference*; Liu, J.K., Ed., AOSTRA: Edmonton, AB, 1993, Paper F23, 24pp.
142. Nelson, L.R.; Gulley, J.R.; MacKinnon, M. In *Proceedings of the 6th UNITAR Conference on Heavy Crude and Tar Sands*; UNITAR, 1995, pp 705–718.
143. Brient, J.A.; Wessner, P.J.; Doyle, M.N. In *Kirk-Othmer Encyclopedia of Chemical Technology*, 4th Edn, 1995, 16, 1017–1029.

144. Zenon Environmental Inc. *Novel Methods for Characterization and Treatment of Toxic Oil Sands Wastewater, Phase II- Step1*. Report for Environmental Protection Service, Environment Canada, Contract #52 SS.KE145–3–0413, 1986.
145. Hsu, C.S.; Dechert, G.J.; Robbins, W.K.; Fukuda, E.; Roussis, S.G., In *Proceedings of Symposium on Acidity in Crude Oil, 215th National ACS Meeting*, Dallas, TX, March 29–April 3, 1998, pp 127–130.
146. Strausz, O.P. In *AOSTRA Technical Handbook of Oil Sands Bitumens and Heavy Oils*; Hepler, L.G.; Hsi, C., Eds; Alberta Oil Sands Technology and Research Authority: Edmonton, AB, 1989.
147. Canadian Oil Sands Network for Research and Development *Naphthenic Acids, Background Information Discussion Report*; Alberta Department of Energy: Edmonton, AB, 65pp.
148. Nix, P.G.; Martin, R.W. *Envir. Tox. and Chem.* **1992**, *7*, 171–188.
149. Herman, D.C.; Fedorak, P.M.; MacKinnon, M.; Costerton, J.W. *Canadian J. Microbiology* **1994**, *40*, 467–477.
150. EnviroTest, *Cellular Toxicity Associated with Syncrude Tailings Pond Water*; Syncrude Canada Ltd.: Fort McMurray, AB, 1996, 46pp.
151. Verbeek, A.G.; Mackay, W.C.; MacKinnon, M. *Can. Tech. Rep. Fisheries & Aquatic Sci.* **1994**, *1989*, 196–207.
152. Sobolewski, A. *Microbial Distributions in Process-affected Aquatic Eco-systems*; report by Microbial Technologies, for Syncrude Canada Ltd.: Fort McMurray, AB, 1998.
153. Lai, J.W.S; Pinto, L.J.; Kiehlmann, E.; Bendell-Young, L.I.; Moore, M. *Envir. Tox. and Chem.* **1996**, *15*, 1482–1491.
154. Babaian-Kibala, E.; Petersen, P.R.; Humphries, M.J. In *Proceedings of Symposium on Acidity in Crude Oil, 215th National ACS Meeting*, Dallas, TX, March 29–April 3, 1998, pp 106–110.
155. Johnson, J.H. et al. (Eds.) *Metals Handbook*; 9th Edn, Vol. 13, ASM International, 1987.
156. Babaian-Kibala, E. *Oil & Gas Journal* **1994**, *Feb.*, 31–40.
157. Slavcheva, E.; Shone, B.; Turnbull, A. *Corrosion J.* (in press).

RECEIVED for review July 21, 1998. ACCEPTED revised manuscript October 21, 1998.

ENVIRONMENTAL, HEALTH, AND SAFETY APPLICATIONS

11

Surfactant Enhanced Aquifer Remediation

Varadarajan Dwarakanath[2] and Gary A. Pope[*1]

[1] Center for Petroleum and Geosystems Engineering, The University of
Texas at Austin, Austin, TX 78712
[2] Duke Engineering and Services, Inc., Austin, TX

*The use of surfactants to remediate groundwater contaminated
by nonaqueous phase liquids has been under investigation and
field testing since at least the 1980s. Surfactant enhanced aquifer
remediation (SEAR) is especially important for dense nonaqueous phase liquids such as chlorinated solvents because they are
difficult to remediate and because there are few good alternatives
to SEAR to remove these contaminants from groundwater. The
technology has continued to improve and recent field demonstrations at superfund sites have shown that under certain conditions very favorable results can be obtained with SEAR. Some of
these advances can be attributed to the adaptation of technology
developed for surfactant enhanced oil recovery over the past 30
years. The emphasis on phase behavior for screening and
evaluating surfactants is especially noteworthy and important.
In this chapter, we first briefly review the phase behavior of
surfactants when mixed with organic liquids of interest, and then
give a detailed example of a study done at the University of Texas
to further evaluate surfactant candidates in soil column tests in
the laboratory.*

Introduction

The contamination of groundwater by nonaqueous phase liquids (NAPLs)
is a cause for concern throughout the world. NAPLs can be classified by
their density as those lighter than water (LNAPLs) and denser than water
(DNAPLs) [1]. NAPLs migrate into aquifers because of gravity and
capillary forces and may be trapped in the form of immobile blobs or
ganglia [2] or when present in sufficient volume DNAPLs may form pools
above aquitards. DNAPL contamination is especially significant as typical
DNAPLs like tetrachloroethylene (PCE), trichloroethylene (TCE), and

---

*Corresponding author.

1,1,1-trichloroethane (TCA) exhibit (1) low viscosity which enables easy invasion into the subsurface, (2) high volatilities which enable easy contamination of the vadose zone, (3) low absolute solubilities which limit pump and treat methods, (4) high solubilities with respect to drinking water standards and (5) low biodegradability [1]. For these reasons, DNAPL can persist in the soil for many decades and present a long term threat to groundwater quality [3]. Large dissolved plumes can continue to form and migrate for decades unless the DNAPL source is removed from the aquifer. Surfactant flooding is one of the few remediation methods that has the potential to do this quickly and effectively.

The conventional remediation method of pump and treat involves pumping of contaminated water followed by treatment at the surface by air stripping, steam stripping, activated carbon filtration and various other means. Pump and treat is limited by parameters such as flow rates, NAPL composition and mass transfer rates, and the surface area available for mass transfer of NAPL constituents from the NAPL to the water [3–5]. Furthermore, dilution effects [6] caused by hydrodynamic dispersion and mixing of contaminated and uncontaminated water adversely affect pump and treat remediation operations. Even in those cases where a pool has formed and has been located and DNAPL pumped out, large amounts of residual NAPL remain to dissolve and pollute the water [7]. The need for cost effective alternatives to pump and treat is compelling.

Surfactants can be used to vastly increase the solubility of the NAPL constituents in water [8–9]. Surfactants also lower the interfacial tension at the water–NAPL interface which, if sufficiently low, will result in mobilizing the NAPL [8]. However, mobilization of DNAPL in aquifers is not always desirable because of concerns of downward migration [1].

Surfactants have been studied for remediating gasoline, polychlorinated biphenyls, dichlorobenzene and automatic transmission fluid in many laboratory studies [10–13]. NAPL recoveries varied between 33 and 85% in these early laboratory studies. Peters et al. [14] and Bourbonais et al. [15] have reported recoveries of total petroleum hydrocarbons between 60 and 90%. However, they identified problems such as mobilization of fines, difficulty in removing surfactant residue from soil and surfactant precipitation. Pennell et al. [16] and Shiau et al. [17] have shown that between 90 and 99% of pure PCE can be removed from Ottawa sand during surfactant column experiments. Encouraging field results have been reported by Fountain et al. [9, 18]. Below we summarize several of our own column experiments which showed 99% removal of various NAPLs from Ottawa sand and up to 99.9% removal of DNAPL from a Hill AFB field soil. The latter laboratory data served as the basis for the design of the surfactant remediation field tests at the DNAPL site at Hill AFB [19–21] where 99% of the TCE-rich DNAPL was removed from the swept volume in about three weeks of surfactant flooding followed by water flooding.

Surfactant Enhanced Aquifer Remediation

Surfactant enhanced aquifer remediation is a complex subject involving many considerations of surfactant chemistry, soil physics, engineering testing and design, hydrogeology, site characterization, predictive multi-phase flow and transport modeling, waste water treatment, economics and regulatory compliance. Surfactant selection and testing by itself is a complex subject with many pitfalls and many widely misunderstood aspects, and yet without a suitable surfactant the process cannot be used in an efficient, reliable and predicable way let alone one that is cost effective. Thus, we focus in this introductory treatment of the subject almost exclusively on surfactant selection and laboratory testing.

A suitable surfactant must first of all be biodegradable at the end of the remediation process and otherwise acceptable environmentally for use in the groundwater. A very wide variety of anionic and nonionic surfactants are biodegradable and we will assume in the following that this is the case as we describe the appropriate laboratory testing to select the best surfactants to remediate a given nonaqueous phase liquid (NAPL). We want a surfactant that will solubilize large amounts of the NAPL and transport easily through the soil with low adsorption and without significant loss of hydraulic conductivity. These criteria as well as all of the key properties of the surfactant such as how it changes the interfacial tension between the NAPL and water are directly linked to its phase behavior, so any screening study as well as any final evaluation of the surfactant should emphasize this aspect of its behavior.

Surfactant Phase Behavior. The process of selecting surfac-tants for remediating NAPL-contaminated aquifers requires an in-depth understanding of the behavior of surfactants in the presence of the NAPL, i.e. surfactant phase behavior. This section presents a brief review of surfactant phase behavior, the various surfactant selection criteria, and finally the technical approach followed for surfactant selection.

Surfactants are amphiphilic agents that show dual behavior; i.e. they are both water and oil soluble. A typical structure of an anionic surfactant is shown in Figure 1 with a water-soluble head and an NAPL-soluble tail. Surfactants are also characterized by their ability to exist in the form of aggregates or micelles above a certain concentration called the critical micelle concentration. The amphiphilic nature also induces surfactants to aggregate at the water–NAPL interface, which brings about a decrease in the interfacial tension. Surfactant micelles also have the ability to solubilize significant volumes of NAPL components such as trichloroethy-lene (TCE), tetrachloroethylene (PCE), and 1,2-dichloroethane (DCA) among many others.

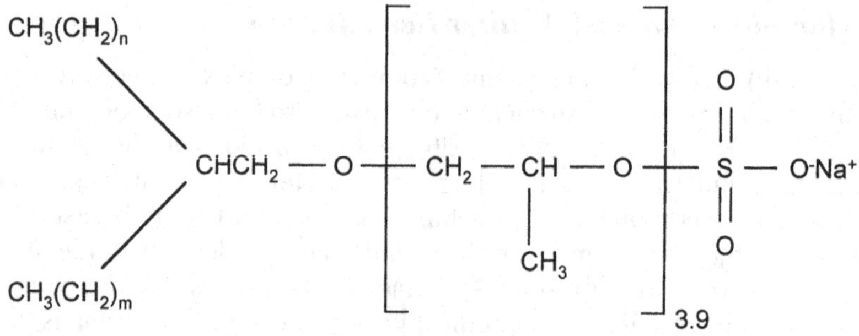

$$CH_3(CH_2)_n \diagdown CHCH_2 - O - \left[CH_2 - \underset{\underset{CH_3}{|}}{CH} - O - \right]_{3.9} \underset{\underset{O}{||}}{\overset{\overset{O}{||}}{S}} - O^-Na^+$$

CH₃(CH₂)ₘ

m + n = 10 - 11 and m. n = 0 - 11

Figure 1.   IsalChem 145 (PO)$_{3.9}$ sodium ether surfactant.

Furthermore, the reduction of interfacial tension between the water and NAPL caused by the addition of surfactant reduces the effect of the capillary forces that entrap the NAPL. Both these mechanisms greatly enhance the ability of surfactants to recover trapped NAPLs from contaminated soils. Hence, surfactant enhanced aquifer remediation (SEAR) is a very promising technology for remediating NAPL-contaminated soils.

Classical surfactant phase behavior involves the formation of microemulsions consisting of the NAPL-components, surfactant, and water as a thermodynamically stable phase. When the NAPL components are solubilized in the center of a micelle it is called a Winsor Type I microemulsion. This happens when the surfactant resides in the aqueous phase. When the surfactant is strongly hydrophobic and resides in the NAPL and has solubilized water in the center of a micelle, it is called an inverted or Winsor Type II microemulsion. A third separate microemulsion exists in the transition between the Winsor Type I and Type II. This is called the Winsor Type III microemulsion. In the case of anionic surfactants, the addition of electrolyte or an increase in the hydrophobe tail length can cause the transition between Winsor Type I to Type III to Type II [22]. Cosolvents such as alcohols typically promote this same transition from Type I to Type III, although the shift can be in either direction depending on the cosolvent and other variables. Nonionics are more sensitive to temperature than anionics and the Type I to Type III transition can be induced by increasing the temperature as well as by changing the structure of the surfactant, e.g. by decreasing the number of ethylene oxide units. See also the discussion presented in Chapter 6 of this book.

Winsor Type III systems have usually been favored by the petroleum industry for surfactant enhanced oil recovery as they are associated with ultra-low interfacial tension and the most efficient displacement of crude oil by increased capillary number. Winsor Type I systems are generally associated with higher interfacial tension and are preferred when solubilization dominated recovery of NAPL is required although mobilization sometimes occurs with Type I systems when the capillary number exceeds its critical value, which is typically on the order of 10^{-5}. Winsor Type II behavior is undesirable as it causes a loss of surfactant due to partitioning into the NAPL and should be avoided. An illustration of the Winsor Type I, Type III and Type II microemulsions with anionic surfactants is shown in Figure 2. As seen in Figure 2, at low electrolyte concentrations the surfactant and the solubilized NAPL form a Winsor Type I microemulsion. However, at higher electrolyte concentrations, the surfactant and solubilized water form a Winsor Type II microemulsion in the NAPL-rich phase.

Although the above discussion indicates the potential of surfactants to remove trapped contaminants in the form of NAPL from soils, there still remain a number of important concerns and problems with their use [23]. For example, some investigators have reported problems with surfactants such as pore plugging [11, 24] and the reduction of permeability [25]. More recently, the loss of hydraulic conductivity has been attributed to

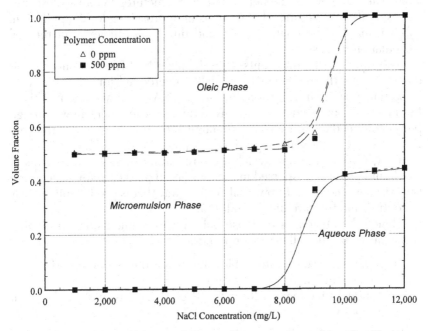

Figure 2. Volume fraction diagram for 4% sodium dihexyl sulfosuccinate, 8% IPA and TCE, with and without xanthan gum polymer.

the adsorption of surfactant by the organic material and clays in the soil [26]. But the most likely reasons for the loss of hydraulic conductivity and pore plugging are the formation of liquid crystals, gels or macroemulsions rather than classical microemulsions, which have lower viscosity and generally better transport behavior [27]. The loss of permeability is undesirable during remediation as it will necessitate lower injection rates and lengthen project life, and in extreme cases will prevent continuation of the flood.

When surfactant, water, electrolytes and NAPL are vigorously mixed in the laboratory, an unstable macroemulsion consisting of NAPL, surfactant micelles, and water containing the electrolytes is formed. This macroemulsion separates into a thermodynamically stable microemulsion of low or moderate viscosity under favorable conditions. This phenomenon is termed coalescence, and the time required for coalescence into equilibrium phases is a measure of good surfactant phase behavior. Phase separation into equilibrium phases in less than 24 hours is an indication of a good surfactant, whereas with poor surfactants the macroemulsion may never coalesce into equilibrium phases. The addition of a cosolvent such as isopropanol will promote faster equilibration, reduce the viscosity of the microemulsion and generally improve phase behavior. Without cosolvent, most surfactants at low temperature never form microemulsions, but rather gels and other undesirable structures. Rapid coalescence corresponds to high rates of mass transfer approaching local equilibrium solubilization of the contaminants, and thus high efficiency in the soil remediation process.

This ability of surfactants to solubilize NAPL components and coalesce into thermodynamically stable microemulsions is of critical importance in their selection for use in NAPL remediation. This is because microemulsions are highly fluid and can easily flow through permeable media and transport contaminants with them under the very low hydraulic gradients typical of shallow, unconfined aquifers. By contrast, macroemulsions, which are physical mixtures or dispersions of one liquid phase interspersed in another liquid phase are unstable, often very viscous, and do not flow easily or in a way that is predictable or easy to control, and thus are highly undesirable.

Hence the objectives of phase behavior experiments are to identify surfactants with the following properties:

- rapid coalescence into stable microemulsions of low viscosity
- high contaminant solubilization
- minimal tendency to form liquid crystals, gels and emulsions at groundwater temperature

Since there is a wide variety of surfactants and contaminants of interest as well as wide variations in groundwater composition and temperature, many phase behavior tests are typically required. This can be done rapidly in practice by making up many combinations of surfactant concentration, cosolvent concentration, NAPL concentration, electrolyte concentration, temperature and so forth in small pipettes and observing the coalescence of these mixtures over a period of a few days and then selecting the most promising ones for further study. Details and large amounts of data of this type can be found in Baran et al. [28–33], Shotts [34] and Dwarakanath [35]. New surfactants can be tailored to meet specific goals, e.g. by adding various numbers of ethylene oxide or propylene oxide units to the surfactant. Examples can be found in Baran et al. [28], Ooi [36] and Weerasooriya et al. [37]. Regardless, the next step in the selection and evaluation procedure is soil column testing. We next describe our soil column procedures and give illustrative results of one such study from our laboratory [35].

Soil Column Procedures. The soil column procedures were similar to those of Pennell et al. [16, 38] and Jin [39] with several important additions and refinements. These include the addition of pressure gradient measurements, the use of cosolvent with the injected surfactant, the use of partitioning tracers for performance assessment of the surfactant floods, the use of xanthan gum polymer to improve the efficiency of the surfactant, the use of field soil and field DNAPL in soil column experiments, and finally the measurement of surfactant adsorption by the field soil. The Ottawa sand used for the column experiments was rinsed with 4 N hydrochloric acid for four hours. This was followed by rinsing with deionized water until all traces of acid were removed. The sand was dried out in an oven at 62 °C for 24 hours. After washing and drying, more than 99% of the sand was observed to be between 600 μm and 53 μm (#40 mesh to the #270) mesh range with an average size of 180 μm (#80 mesh). The column was secured to a vibrating jig and the inlet piece was removed. The sand was placed in a funnel and affixed to the jig. The apparatus was activated and the soil column was allowed to fill at a rate of approximately 1 cm (of height in column) per minute for the 2.21 cm diameter columns and 0.5 cm (of height in column) per minute for the 4.8 cm diameter column. After the column was full, it was removed from the jig and the inlet cap was tightened on to the end of the column. Two 250 μm (#60 mesh) and one 99 μm (#150 mesh) stainless steel screens were affixed to the end pieces to hold the sand in place in the column.

Field soil samples from Hill AFB were obtained in SOLINST cores or as loose soil samples in glass jars. The Hill soil was composed of gravel, cobbles and sand grains. The bulk of the sand grains in the Hill field soil varied between 0.02 mm and 4 mm. The mean diameter varied with the depth of the soil samples and was usually between 0.05 mm and 0.42 mm.

The average clay content of the Hill soil was between 1 and 2%. The fractional organic content of the Hill field soil was 0.006 by weight.

The glass columns used to pack the Hill soil were 4.8 cm diameter and either 15 cm or 30 cm long with an adjustable plunger to provide confinement. Two 250 μm and one 99 μm stainless steel screens were used to hold the sand in place. The contents of the SOLINST core were slowly emptied into the glass column using a steel piston. When loose soil samples from glass jars were used, a spatula was used to pour the soil into the column. The soil in the glass column was compacted by tapping the glass column slowly. Large cobbles and stones (greater than 2 cm diameter) were removed to ensure that no large spaces remained in the soil column. Once the desired soil column length was obtained, the end pieces were adjusted into place and the column was ready for water saturation.

In the case of columns with Ottawa sand, a vacuum was pulled to remove most of the air from the column. The column was flooded with carbon dioxide for about 30 minutes. This was followed by flushing with de-aired water for about 5 to 8 pore volumes. The pore volume measurement was made by using the difference in weight between dry and water saturated soil columns. When field soil was used, the column was flushed with several pore volumes of de-aired groundwater with a back pressure of about 0.021 MPa to 0.034 MPa (3 to 5 psi). The permeability was measured by flowing the water through the column at various flow rates and measuring the induced pressure drop using a 0 to 0.034 MPa (5 psid) differential pressure transducer.

An initial tracer test was performed at 100% water saturation to estimate the pore volume and determine the homogeneity of the soil pack in columns. Then 1.1 to 1.5 pore volumes of DNAPL were injected from the bottom at an interstitial velocity of 6 to 7 m/d. The flow direction was then reversed and about 5 pore volumes of water were injected from the top of the column at an interstitial velocity of 6 to 7 m/d until residual DNAPL saturation was reached. In order to reach residual jet fuel saturations, jet fuel was injected from the top followed by a waterflood from the bottom of the soil column. In each case, the flow direction corresponds to the gravity stable direction. Residual NAPL saturations were calculated based on the difference in weight of the uncontaminated soil column and the contaminated soil column. The residual saturation estimate based on weighing the soil column will be referred to as the mass balance estimate of residual NAPL saturation. Residual saturations were also estimated from partitioning tracer tests [40]. The surfactant solutions were injected into the column from the top for the DNAPLs and from the bottom in case of the LNAPLs since these are the favorable directions with respect to gravity. A list of the column experiments discussed in this chapter is given in Table 1.

Table 1.   Surfactant Flood Column Conditions for all the Soil Columns

Experiment	NAPL	Length (cm)	Diameter (cm)	Soil	End Pieces	Velocity (m/d)	Residual Saturation (%)
DW#2	PCE	30.50	2.21	Ottawa	Steel	0.6	19.9
DW#3	PCE	30.50	2.21	Ottawa	Steel	0.6	17.7
DW#4	TCE	30.50	2.21	Ottawa	Steel	1.2	17.7
DK1	TCE	13.50	4.80	Ottawa	Teflon	0.5	21.0
DW#5	JP4	15.00	4.80	Ottawa	Teflon	1.2	16.7
JP4#2	JP4	30.50	2.21	Ottawa	Steel	1.7	15.3
DK4	TCE	42.7	2.50	Ottawa	Teflon	1.8	17.5
POLYTCE#1	TCE	26.8	4.80	Ottawa	Teflon	1.8	19.6
POLYTCE#2	TCE	22.8	4.80	Ottawa	Teflon	1.6	18.5
POLYTCE#3	TCE	75.0	2.21	Ottawa	Steel	1.3	16.4
HILLOU2#7	DNAPL	21.6	4.80	Field	Teflon	0.5	25.8
HILLOU2#8	DNAPL	9.8	4.80	Field	Teflon	0.3	8.1

The effluent samples during the surfactant flood and partitioning tracer column experiments were collected in a fraction collector using volumetrically calibrated test tubes. The samples were immediately transferred into glass vials with aluminum lined caps and stored at 4 °C to ensure preservation. The analysis of the samples was usually completed within 48 hours of sample collection. The viscosities of the various injectate surfactant solutions were measured using a Couette viscometer. The interfacial tensions were measured using a spinning drop tensiometer. A description of these instruments and the measurement techniques are given in Dwarakanath [35].

Column Results and Discussion.   As shown in Figure 3, the solubility of the contaminants in the DNAPL from Hill Air Force Base is increased from about 1100 mg/l in groundwater to about 625,000 mg/l by a surfactant solution containing 8% by weight sodium dihexyl sulfosuccinate, 4% by weight isopropanol and 7000 mg/l NaCl at 12.2 °C. All samples coalesced to microemulsions in less than 15 hours, and samples close to optimal salinity coalesced in less than 4 hours. Optimal salinity is used here in the classical sense to mean the salinity at which the microemulsion contains equal volumes of oil and water [22]. These are very fast coalescence rates compared to most surfactant mixtures and this is due in part to the nature of the surfactant and in part to the presence of the IPA cosolvent. No evidence of gels or liquid crystals was observed based on visual observations. Surfactant mixtures that take longer than

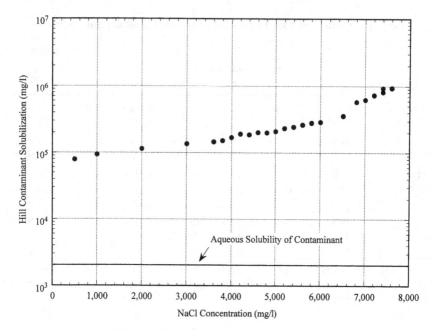

Figure 3.   Solubility enhancement of Hill contaminant by 8% sodium dihexyl sulfosuccinate, 8% IPA and sodium chloride.

24 hours to coalesce to equilibrium microemulsions typically do not perform well in surfactant column experiments.

The addition of xanthan gum polymer did not affect phase behavior. This was evidenced by the close match between the volume fraction diagrams for a surfactant solution (4% sodium dihexyl sulfosuccinate, 8% IPA) with 500 mg/l xanthan gum polymer and without polymer as shown in Figure 2. As shown in Figure 4, the viscosity of the surfactant solution is increased by the addition of polymer. All samples with polymer were observed to coalesce to microemulsions in less than 20 hours, which is still fast enough to be acceptable based upon subsequent column floods.

Based on phase behavior experiments, surfactants were selected for remediating the NAPL-contaminated soil columns. A summary of the initial conditions of all the soil columns is given in Table 1. The compositions of the surfactant solutions used in the soil column experiments are given in Table 2. The contaminant solubilization and interfacial tension data are given in Table 3. The surfactant solutions used in the soil column experiments produce equilibrium microemulsions with contaminant solubilization that ranges from 39,000 to 1,000,000 mg/l and corresponding interfacial tensions between the microemulsion and NAPL that varied from 0.20 to 0.01 mN/m (0.20 to 0.01 dynes/cm). Table 3 gives the Winsor phase type for each case.

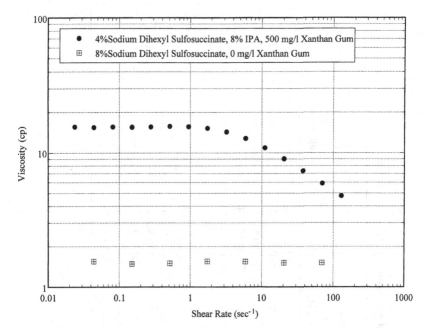

Figure 4. Viscosity enhancement of surfactant with xanthan gum in surfactant solution.

The contaminant concentration in the effluent during the surfactant flood provided useful information regarding the rate of remediation, mass of NAPL recovered, mass transfer characteristics between surfactant micelles and trapped NAPL, and most importantly the final contaminant concentration as a result of surfactant flooding.

In mobilization dominated experiments, greater than 80% of the contaminant was recovered as free phase NAPL followed by a rapid decline to low concentrations of contaminant in the effluent. We illustrate this behavior for four mobilization experiments using Ottawa sand. This is observed in Figure 5 for column experiment POLYTCE#3, in which high concentrations on the order of 500,000 mg/l TCE were observed during the first pore volume of the surfactant flood. Similar results are shown in Figure 6 where peak effluent contaminant concentrations between 500,000 and 1,000,000 mg/l were observed at about one pore volume for three other mobilization experiments.

We next illustrate three column experiments using Ottawa sand where little or no mobilization occurred because of the relatively high interfacial tension. Mobilization actually depends on the trapping number [41], which is a dimensionless parameter defined as the magnitude of the vector sum of the buoyancy and viscous forces acting on the trapped

Table 2. Properties of Injected Surfactant Solutions for all Soil Columns

Experiment	NAPL	Surfactant Solution (wt.%)					xg (mg/l)	Electrolyte (mg/l)	Mobilization or Solubilization	Pore Volumes of Surfactant Injected
		C5	C6	C8	SBA	IPA				
DW#2	PCE	2	0	2	0	0	0	1300 $CaCl_2$	Mobilization	14.7
DW#3	PCE	0	4	0	0	0	0	25,000 NaCl	Solubilization	12.5
DW#4	TCE	0	8	0	0	0	0	2000 NaCl	Solubilization	8.7
DK1	TCE	0	4	0	0	0	0	10,800 NaCl	Mobilization	4.8
DW#5	JP4	0	0.6	1.4	2	0	0	6000 NaCl		3.0
JP4#2	JP4	0	2	2	8	0	0	11,700 NaCl		5.6
								1300 $CaCl_2$		
DK-4	TCE	0	4	0	0	8	0	4000 NaCl	Solubilization	13.9
POLYTCE#1	TCE	0	4	0	0	8	500	4000 NaCl	Solubilization	11.9
POLYTCE#2	TCE	0	4	0	0	8	0	9350 NaCl	Mobilization	1.0
POLYTCE#3	TCE	0	4	0	0	8	0	9350 NaCl	Mobilization	1.0
HILLOU2#7	DNAPL	0	4	0	0	4	500	11,250 NaCl	Mobilization	2.1
HILLOU2#8	DNAPL	0	8	0	0	8	500	5850 NaCl	Mobilization	1.9

C5 = Sodium diamyl sulfosuccinate   C6 = Sodium dihexyl sulfosuccinate
C8 = Sodium dioctyl sulfosuccinate   SBA = Secondary butyl alcohol
xg = xanthan gum

Table 3. Surfactant Solution Properties

Experiment	NAPL	Winsor Type	Interfacial Tension (mN/m)	Solubilization (mg/l)	Viscosity (mPa·s)
DW#2	PCE	I	0.01	1,000,000	36.4
DW#3	PCE	II	0.14	58,000	1.2
DW#4	TCE	I	0.20	52,000	1.5
DK1	TCE	III	—	163,000	1.2
DW#5	JP4	I	—	45,000	5.8
JP4#2	JP4	III	—	230,000	1.9
DK4	TCE	I	0.19	39,000	1.2
POLYTCE#1	TCE	I	0.19	39,000	11.8
POLYTCE#2	TCE	III	0.02	516,000	1.2
POLYTCE#3	TCE	III	0.02	516,000	1.2
HILLOU2#7	DNAPL	III	0.01	600,000	11.8
HILLOU2#8	DNAPL	III	0.01	425,000	11.6

Shear rate = 0.945 sec^{-1}

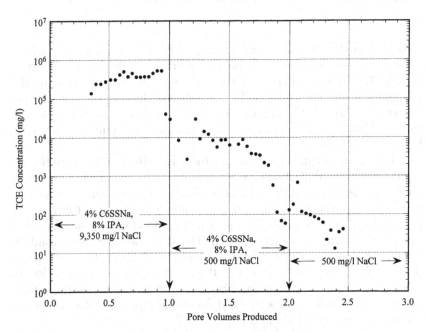

Figure 5. TCE concentration in effluent during surfactant flood and post surfactant waterflood for experiment POLYTCE#3.

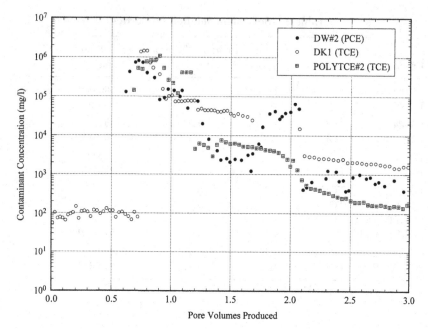

Figure 6.   Effluent contaminant concentration during surfactant flood and post surfactant waterflood for three mobilization experiments.

NAPL divided by the interfacial tension. The interfacial tension was high in these experiments due to the low salinity (Table 2). As shown in Figure 7, a steady plateau of the contaminant concentration in the effluent was observed for 5 to 10 pore volumes followed by a decline to low concentrations. The effluent contaminant concentration was close to the equilibrium solubility of the contaminant as determined by the phase behavior experiments.

These experiments also illustrate the ease with which the preferred type of flood (solubilization or mobilization) can be conducted by simply changing the salinity of the injected surfactant mixture. Since the salinity can be adjusted continuously and inexpensively in the field, this is a very practical approach to controlling a surfactant flood when anionic surfactants such as sulfosuccinates are used. This is also a significant advantage of using anionic surfactants over nonionic surfactants. The final water flood to flush the surfactant out of the aquifer should always be low in salinity similar to the groundwater both to restore the electrolytes to their original state and because surfactant floods are more robust and efficient if a negative salinity gradient is imposed [27].

In experiment DW#3, the peak effluent PCE concentration was 45,000 mg/l compared to the equilibrium solubilization of 58,000 mg/l.

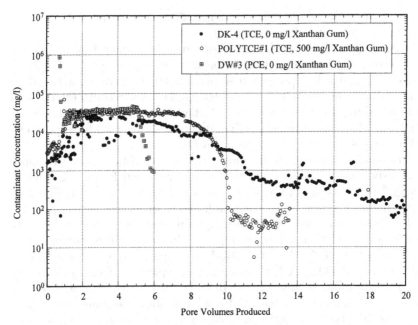

Figure 7. Effluent contaminant concentration during surfactant flood and post surfactant waterflood for three solubilization experiments.

The effluent PCE concentration did not change when the flow rate was increased from 0.6 m/day to 2.4 m/day during the surfactant flood. This suggests a close approach to local equilibrium in the soil pores. Similarly in experiment DK-4, the peak effluent TCE concentration was 29,000 mg/l compared to the equilibrium solubilization of 39,000 mg/l. In experiment POLYTCE#1, the peak effluent TCE concentration was about 37,000 mg/l compared to the equilibrium solubilization of 39,000 mg/l. These values are the same well within the uncertainty of the GC analysis, which is about 10%. This very close approach to equilibrium behavior with polymer is thought to be due to the increased viscosity of the solution containing polymer, which promotes more uniform transport of the surfactant solution and thus increases the contact with the trapped DNAPL and causes a more efficient mass transfer of TCE between the trapped DNAPL and the surfactant micelles. Thus, even in a linear column experiment, with nearly uniform packing, polymer appears to increase the efficiency of the flooding process. In an aquifer, additional benefits of the polymer will be manifested in less bypassing of low hydraulic conductivity zones and other effects promoting better sweep of the aquifer, which will reduce the amount of surfactant and time needed to remediate the aquifer and thus reduce the cost of the remediation. The cost of the polymer itself is very small since only about 0.05 wt. % is needed to generate adequate viscosity.

The higher efficiency due to the presence of polymer translates into a faster remediation rate. This can be inferred by comparing Figure 7, which shows that 10.5 pore volumes of surfactant were required to reduce the TCE concentration to less than 100 mg/l in experiment POLYTCE#1 compared to 20 pore volumes of surfactant in experiment DK-4. A more uniform TCE concentration in the effluent was observed for experiment POLYTCE#1 compared to experiment DK-4.

In general, a comparison of mobilization experiments and solubilization experiments showed that the volume required for NAPL removal ranged from about 2 to 4 pore volumes of surfactant for mobilization experiments whereas 11 to 20 pore volumes of surfactant were required in solubilization experiments. For example, more than 10 pore volumes of surfactant were needed to recover greater than 95% of the contaminant in experiments POLYTCE#1 and DK-4 as compared to just 1 pore volume in experiment POLYTCE#3. However, in typical field sites the average NAPL saturation is much lower than 18–20% and hence will require fewer pore volumes of surfactant for solubilization dominated recoveries.

The final contaminant concentration in the effluent varied between 10 mg/l and 100 mg/l in most column experiments that used plastic end pieces and tubing. Both nylon tubing and Teflon end pieces used in several columns absorb TCE and then cause a persistent tailing in the effluent as it desorbs. When both nylon tubing and Teflon end pieces were eliminated in experiment POLYTCE#3, the effluent TCE concentration declined to less than 1 mg/l. Once the surfactant flood has removed all of the DNAPL from the soil as in this experiment, the contaminant concentrations in the water can become very low, and plastic must be eliminated from the entire apparatus to prevent an artificially high measurement.

Hydraulic Gradient Measurements and Interpretation. The pressure gradient is one of the most important measurements that must be made in all column experiments. In the groundwater literature, the pressure gradient is typically expressed as a hydraulic gradient in meters of water head per unit distance in meters. The hydraulic gradients for these column results are all reported in units of m/m.

In a typical mobilization experiment, the hydraulic gradient increases until the oil bank reaches the outflow end piece of the column, i.e. until the oil bank breakthrough time. Many examples of this behavior can be found in the EOR literature [42]. In typical solubilization experiments, the hydraulic gradient increases until surfactant breakthrough. This is followed by a slow decline to lower values. This behavior is shown in Figure 8 where the hydraulic gradient during the surfactant flood and

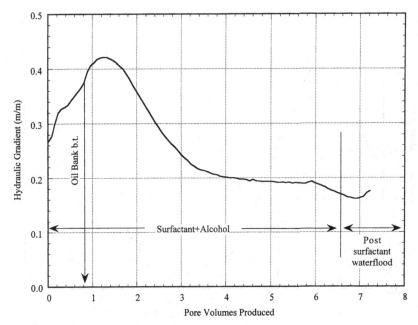

Figure 8. Hydraulic gradients across the soil column during surfactant flood and post surfactant waterflood for experiment JP4#2.

post surfactant waterflood for experiment JP4#2 is plotted. The final hydraulic gradient declined to less than 0.2 after surfactant flooding. The hydraulic gradient for a surfactant/polymer flood is shown in Figure 9. The hydraulic gradient is higher in this flood since the viscosity of the surfactant solution was increased by the xanthan gum polymer but eventually declined to about 0.3 m/m. Considering the velocities of these experiments, these are acceptable hydraulic gradients. The final permeability to water was very nearly the same as the initial value in these experiments.

An illustration of a column flood with a very high and unacceptable hydraulic gradient is shown in Figure 10. The hydraulic gradient increased continuously during surfactant injection and eventually reached more than 34, at which time a waterflood was started and it decreased rapidly. Tracers were injected into the soil column after the surfactant flood as another means to understand the cause of the high hydraulic gradient. As shown in Figure 11, the final tracers break through much earlier than the initial tracers, which indicates bypassing of many of the soil pores because of plugging of the soil. Surfactants which show this type of behavior are unsuitable for use in remediation. In this case, the problem can be attributed to the dioctyl sulfosuccinate that was mixed

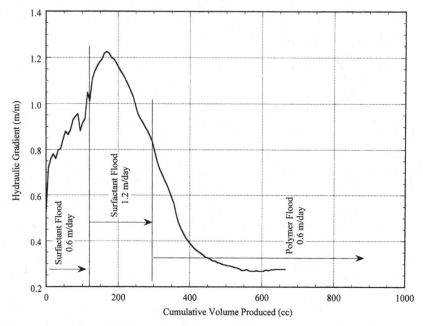

Figure 9.   Hydraulic gradients across the soil column during surfactant flood and post surfactant polymer flood for experiment HILLOU2#7.

with the dihexyl sulfosuccinate (Table 2). The longer hydrophobe produces structures with JP4 that cause high viscosity and pore plugging. With other NAPLs, this same surfactant mixture is acceptable. The gels may also be eliminated by adding sufficient amounts of light alcohols or by increasing the temperature.

A comparison of the hydraulic gradients for all the experiments is given in Table 4. Values are given under the initial 100% water saturated conditions, while injecting water at residual NAPL saturation before the surfactant floods and during the final waterflood after surfactant flooding. The permeabilities at each of these conditions were calculated from these hydraulic gradients and are given in Table 5. The end point water relative permeability was calculated by dividing the water permeability at residual NAPL saturation by the initial water permeability and is typically about 0.3 for the Ottawa sand experiments.

In experiment DW#2, the permeability was decreased from an initial value of 15.4 Darcy to a final waterflood value of 1.5 Darcy. Column DW#5 was plugged as a result of surfactant flooding. In both experiments, a surfactant mixture with sodium dioctyl sulfosuccinate was used without enough cosolvent to prevent gels. Sodium dihexyl sulfosuccinate is less prone to forming gels and less alcohol is needed to eliminate gel when it

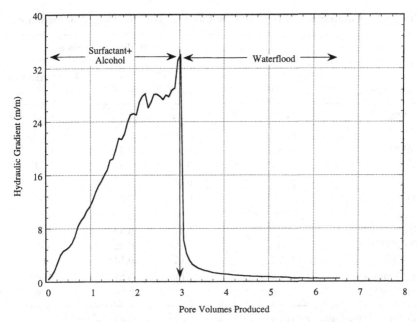

Figure 10. Hydraulic gradients across the soil column during the surfactant flood and post surfactant waterflood for experiment DW#5.

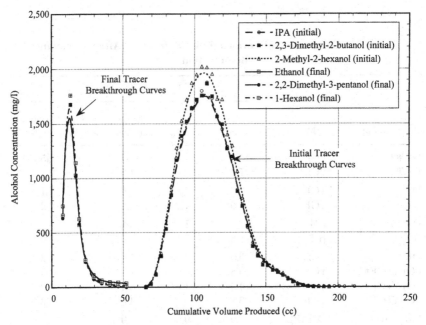

Figure 11. Comparison of initial tracer and post surfactant tracer breakthrough curves for experiment DW#5.

Table 4.   Comparison of Gradients Before and After Surfactant Remediation

		Hydraulic Gradient		
Experiment	NAPL	Initial	at S_{or}	Post Surfactant Water Chase
DW#2	PCE	0.04	0.16	0.35
DW#3	PCE	0.07	0.21	0.08
DW#4	TCE	0.07	0.21	0.08
DK1	TCE	0.1	0.39	0.21
DW#5	JP4	0.07	0.29	—
JP4#2	JP4	0.06	0.24	0.06
DK4	TCE	0.07	0.19	0.11
POLYTCE#1	TCE	0.13	0.43	0.16
POLYTCE#2	TCE	0.10	0.32	0.10
POLYTCE#3	TCE	0.07	0.14	0.07
HILLOU2#7	DNAPL	0.08	0.51	0.09
HILLOU2#8	DNAPL	—	0.16	0.06

All gradient readings have been adjusted for an interstitial velocity of 1.2 m/d

Table 5.   Comparison of Permeabilities Before and After Surfactant Remediation

		Permeability (Darcy)			
Experiment	NAPL	Initial	at S_{or}	End of Surfactant	Post Surfactant, Water Chase
DW#2	PCE	15.3	3.3	—	1.5
DW#3	PCE	7.3	2.6	1.9	6.9
DW#4	TCE	8.3	2.5	1.6	7.1
DK1	TCE	4.5	1.2	1.8	2.2
DW#5	JP4	7.1	1.8	0.1	—
JP4#2	JP4	8.8	2.2	5.8	8.6
DK4	TCE	6.5	2.0	1.1	3.2
POLYTCE#1	TCE	5.8	1.8	—	4.9
POLYTCE#2	TCE	4.4	1.4	3.4	4.2
POLYTCE#3	TCE	6.8	3.2	4.8	6.2
HILLOU2#7	DNAPL	5.9	0.9	1.9	4.5
HILLOU2#8	DNAPL	—	3.0	3.8	7.7

does form. Low pressure gradients are difficult to measure accurately with pressure transducers and result in a large uncertainty in the hydraulic gradients and permeabilities, but for experiments DW#3, DW#4, JP4#2, POLYTCE#2 and POLYTCE#3, it is clear that a negligible reduction in permeability was observed. Based on these results, it can be concluded that when sodium dihexyl sulfosuccinate is used with these soils, there will be no significant loss of permeability. When used with xanthan gum polymer as in experiments POLYTCE#1 and HILLOU2#7, the final permeability was somewhat less than the initial value. This may have been due to insufficient removal of the polymer by the final waterflood, but in any case it is still in the acceptable range of behavior and does not offset the value the polymer has for other reasons such as illustrated and discussed below.

The viscosities of the aqueous surfactant solutions were measured before injection into the columns. It can be seen in Table 3 that the solutions injected into DW#3, DW#4, DK1, JP4#2, POLYTCE#2 and POLYTCE#3 have low viscosities ($\leqslant 1.5$ cp), whereas the solutions injected into DW#2 and DW#5 have higher viscosities. In POLYTCE#1, HILLOU2#7 and HILLOU2#8, the viscosities of the surfactant solutions were about 11.7 cp at a shear rate of 0.945 sec^{-1}. This is due to the addition of 500 mg/l xanthan gum polymer. The variation of viscosity with shear rate for several of the surfactant solutions is plotted in Figure 12. High viscosity and non-Newtonian behavior was observed for surfactants used in experiments DW#2 and DW#5 despite the absence of polymer. In both these experiments, plugging and permeability reduction was observed.

Based on the comparison of the viscosities of the surfactant solutions, it can be inferred that the more viscous solutions with non-Newtonian behavior are more likely to cause higher hydraulic gradients and loss of hydraulic conductivity during surfactant flooding. After observing high gradients in DW#5, the amount of secondary butyl alcohol was increased to 8 wt% in the surfactant solution in experiment JP4#2. As a result, lower hydraulic gradients and negligible permeability reduction were observed in experiment JP4#2. The addition of alcohol helps in melting the gels and lowers the viscosity and density of the surfactant solution. Light alcohols such as IPA lower the contaminant solubilization slightly but the benefit of adding alcohol cosolvent far out weighs this slight disadvantage. In experiments DK-4, POLYTCE#1, POLYTCE#2, POLYTCE#3 and HILLOU2#8, 8 wt% IPA was used to prevent gel formation.

In the case of DNAPLs, another potential benefit of adding light alcohols such as ethanol and IPA which have low densities is the lower density of both the injected surfactant solution and the microemulsion that forms in-situ as the injected surfactant solution solubilizes the dense contaminants from the DNAPL. Without alcohol, the difference between

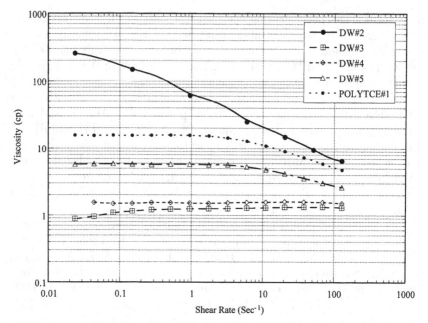

Figure 12.   Variation of viscosity with shear rate for surfactant solutions used in column experiments.

the microemulsion density and the groundwater density may be significant enough to cause downward movement of the microemulsion. The addition of large amounts of alcohol decreases the density contrast between the microemulsion and groundwater and minimizes or eliminates downward movement of the microemulsion. When sufficient amounts of alcohol are added to reduce the density to that of water, the microemulsion becomes neutrally buoyant and experiments by Kostarelos [43] have shown that even without an aquitard downward migration of the dense contaminants can be completely prevented. The neutral buoyancy approach to surfactant remediation is developed and illustrated by Shook et al. [44].

Anionic surfactants such as the sulfonates used in this study typically have lower adsorption than nonionic surfactants on silica minerals assuming neutral to high pH. This is because the negative sulfonate head of the surfactant is repelled by the net negative charge of silica and other typical minerals that make up alluvium aquifers at typical values of groundwater pH. Surfactant sorption experiments can be conducted in either batch experiments or soil column experiments. When the surfactant adsorption is very low and the surfactant concentration high as in our experiments, batch surfactant adsorption is very difficult to measure accurately because of the very small changes in surfactant concentration. In addition these experiments do not quantify surfactant behavior under dynamic conditions in the presence of both field DNAPL and soil, so column experiments were preferred in this work.

Surfactant retention (adsorption plus any other phenomena that might cause retention) was measured by injecting a surfactant solution containing ^{14}C radiolabeled sodium dihexyl sulfosuccinate into a column packed with contaminated Hill field soil. Tritiated water was used as a conservative tracer. Figure 13 shows the effluent concentration from the column for both these labeled molecules as measured by liquid scintillation counting. The almost identical breakthrough and elution of these molecules indicate very low adsorption of the surfactant on this soil.

A retardation factor was calculated from these data and yielded a distribution coefficient (K_d) of 0.0017, which corresponds to an adsorption of 0.16 mg of surfactant per gram of soil, but these values are not significantly different from zero taking the experimental error into account. Regardless, these values are lower than those reported by Rouse et al. [45] for other surfactants on similar soils. We attribute this in part to the very favorable characteristics of dihexyl sulfosuccinate and in part to the use of the IPA cosolvent. High values of surfactant retention

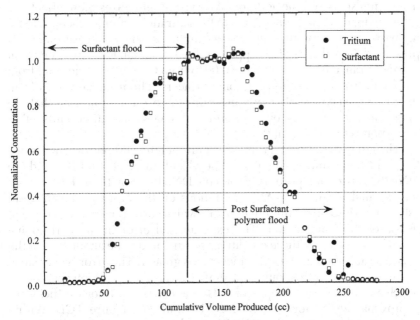

Figure 13. Surfactant and tritium concentrations in the effluent during surfactant and post surfactant polymer flooding for experiment HILLOU2#8.

Table 6.   Final NAPL Saturation After Surfactant Remediation

| Experiment | NAPL | Final NAPL Saturation | | Final NAPL Conc. (mg/l) |
		Mass Balance	Tracers	
DW#2	PCE	0.0046	—	<100
DW#3	PCE	−0.0009	0.005	<100
DW#4	TCE	0.0000	0.005	<50
DK1	TCE	0.004	0.016	<10
DW#5	JP4	—	—	—
JP4#2	JP4	0.023	0.035	—
DK4	TCE	0.01	0.000	<10
POLYTCE#1	TCE	−0.0023	0.0001	<10
POLYTCE#2	TCE	0.0038	0.0010	<10
POLYTCE#3	TCE	−0.0024	0.0002	<1
HILLOU2#7	DNAPL	−0.0038	0.0002	—
HILLOU2#8	DNAPL	—	0.0015	<15

can be caused by undesirable phases such as gels and liquid crystals or by macroemulsions and the addition of cosolvent minimizes these effects.

A comparison of the residual NAPL saturations after surfactant flooding based on both partitioning tracers and mass balance measurements is presented in Table 6. The apparent final residual DNAPL saturation estimates range from −0.0038 to 0.0046 from the mass balance and from 0.0001 to 0.005 from the partitioning tracers. The negative values from the mass balance are obviously experimental error and indicate that the mass balance error is on the order of 0.004 saturation units. This corresponds to only about 0.2 mL of DNAPL.

The last five column experiments represent the best performance since the surfactant solution contained dihexyl sulfosuccinate and IPA and the DNAPL was either pure TCE or Hill DNAPL. For these best experiments, final residual DNAPL saturations on the order of 0.0002 were estimated based on partitioning tracers, which corresponds to a removal of 99.9% of the contaminant. The experimental error in the partitioning tracer method under these conditions is estimated to be about 10% of the residual saturation or 0.0002, whichever is greater. The error is very small due to the use of a partitioning tracer with a very high partition coefficient, which in this case was n-heptanol. n-Heptanol has a partition coefficient of approximately 140 for the Hill DNAPL and 90 for pure TCE. Within experimental error, all the DNAPL has been removed from these soil columns. This is consistent with the final TCE concentration in the water for the experiment in which a stainless steel column was used. In that column experiment (POLYTCE#3), the final TCE concentration was less than 1 mg/l, which was the detection limit on the GC used in this work.

Conclusions and Recommendations.   Laboratory soil column experiments have demonstrated that with a good surfactant formulation based upon good phase behavior, up to 99.9% of the DNAPL can be recovered from a soil column in as little as 1.0 to 2.0 pore volumes of surfactant flooding. Phase behavior experiments are a very important step to select good surfactants with high contaminant solubilization, fast coalescence rates and minimal liquid crystal forming tendencies. The use of xanthan gum polymer as a viscosifier helped in improving the rate of contaminant removal in solubilization experiments.

The measurement of hydraulic gradients during a surfactant flood is essential for quantifying surfactant transport in a porous medium. Surfactants that are more likely to form gels and liquid crystals will cause high hydraulic gradients during surfactant flooding and should be avoided. Acceptable surfactants will show low hydraulic gradients during the surfactant flood. Low hydraulic gradients and negligible permeability reduction with both Ottawa sand and Hill field soil were observed when sodium dihexyl sulfosuccinate was used with isopropanol as a cosolvent in soil column experiments.

Based on the results presented in this chapter the following guidelines are suggested for designing surfactant floods for remediating soils contaminated by NAPLs:

1. Phase behavior experiments should be performed to identify surfactants with high contaminant solubilization, fast coalescence rates to classical microemulsions and minimal liquid crystal/gel/ macroemulsion forming tendencies over the expected range of conditions of temperature, electrolyte, surfactant, cosolvent and contaminant concentrations. The viscosity of both the aqueous surfactant solution that is injected and the microemulsion that forms when it mixes with and solubilizes the NAPL should be low, i.e. not much greater than the corresponding water (or polymer solution if polymer is added to the water).

2. Pressure gradients should be measured during column experiments to verify that the hydraulic gradient during and after the surfactant flood is not too high for the aquifer conditions of interest and surfactant solutions that show unacceptable hydraulic gradients should be not be used to flood aquifers even if the recovery of the NAPL is high. The final water permeability of the soil after surfactant flooding should be close to its initial value.

3. The partitioning tracer test is a valuable and accurate complement to mass balance for estimating the residual NAPL saturation before and after surfactant flooding.

4. Adding polymer to a surfactant solution to increase its viscosity has the potential to greatly increase the efficiency of surfactant remediation and should be tested in the soil of interest to evaluate its performance and benefit for applications where the aquifer permeability is sufficiently high and a sufficiently high hydraulic gradient can be imposed during the flood of the aquifer.

Acknowledgments

We would like to thank Dino Kostarelos, Meng Lim, Kiam Ooi, Bruce Rouse, Doug Shotts and Vinitha Weerasooriya for their assistance with the laboratory experiments and we would especially like to acknowledge the late Professor W.H. Wade for his many contributions to the science of surfactants including the scientific application of surfactants to groundwater cleanup. Surfactant remediation research at the University of Texas during the past several years has been funded by the Advanced Technology Program of the state of Texas, the U.S. Environmental Protection Agency, the Air Force Center for Environmental Excellence, the Department of Energy and the Army Corps of Engineers Waterways Experiment Station.

References

1. Pankow, J.F.; Cherry, J.A. *Dense Chlorinated Solvents*; Waterloo Press: Portland, OR, 1996.
2. Kueper, B.H. 1989. The behavior of dense, non-aqueous phase liquid contaminants in heterogeneous porous media. Ph.D. dissertation, University of Waterloo, Waterloo, Ont.
3. Mackay, D.M.; Cherry, J.A. *Environ. Sci. Technol.* **1989**, *23(6)*, 630–635.
4. Powers S.E.; Loureiro, C.O.; Abriola, L.M.; Weber Jr., W.J. *Water Resour. Res.* **1991**, *27(4)*, 463–477.
5. Gellar, J.T.; Hunt, J.R. *Water Resour. Res.* **1993**, *29(4)*, 833–845.
6. Jackson, R.E.; Mariner, P.E. *Ground Water* **1995**, *33(3)*, 407–414.
7. Oolman, T.; Godard, S.T.; Pope, G.A.; Jin, M.; Kirchner, K. *GWMR* **1995**, *15(4)*, 125–137.
8. Fountain, J.C.; Klimek, A.; Beikirch, M.G.; Middleton, T.M. *J. Haz. Mater.* **1991**, *28(3)*, 295.
9. Fountain, J.C.; Starr, R.C.; Middleton, T.; Beikirch, M.; Taylor C.; Hodge, D. *Ground Water* **1996**, *34(5)*, 910–916.
10. Ellis, W.D.; Morgan, D.R.; Ranjithan, S.R. 1986. Treatment of Contaminated Soils with Aqueous Surfactants. EPA/600/2-85/129, Hazardous Waste Engineering Research Laboratory.
11. Ziegenfuss, P.S. 1987. The Potential Use of Surfactant and Cosolvent Soil Washing as Adjutant for In-Situ Aquifer Restoration. M.S. thesis, Rice University, Houston, Texas.

12. Abdul, A.S.; Gibson, T.L.; Rai, D.N. *Ground Water* **1990**, *28(6)*, 920.

13. Ang, C.C.; Abdul, A.S. *GWMR* **1991**, *11(2)*, 121.

14. Peters, R.W.; Montemagno, C.D.; Shem, L.; Lewis, B. *Hazardous Waste and Hazardous Materials* **1992**, *9(2)*, 113.

15. Bourbonais, K.A.; Compeau, G.C.; MacClellan, L.K. In *Surfactant Enhanced Subsurface Remediation Emerging Technologies*, Sabatini, D.A.; Knox, R.C.; Harwell, J.H. Eds; 1995; ACS Symposium Series 594, American Chemical Society Washington, DC, 1995; 161–177.

16. Pennell, K.D.; Jin, M.; Abriola, L.M.; Pop, G.A. *J. Cont. Hydrol.* **1994**, *16(1)*, 35.

17. Shiau, B.; Rouse, J.D.; Sabatini, D.A.; Harwell J.H. In *Surfactant Enhanced Subsurface Remediation Emerging Technologies*, Sabatini, D.A.; Knox, R.C.; Harwell J.H. Eds; ACS Symposium Series 594, American Chemical Society, Washington, DC, 1995; 65–81.

18. Fountain, J.C.; Waddel-Sheets, C.; Lagowski, A.; Taylor, C.; Frazier, D.; Byrne, M. In *Surfactant-Enhanced Subsurface Remediation Emerging Technologies*, Sabatini, D.A.; Knox, R.C.; Harwell, J.H. Eds; ACS Symposium Series 594, American Chemical Society, Washington, DC, 1995; 177–190.

19. Brown, C.L.; Delshad, M.; Dwarakanath, V.; McKinney, D.C.; Pope, G.A. Design of a Field-Scale Surfactant Enhanced Remediation of a DNAPL Contaminated Aquifer. Presented at the I&EC Special Symposium, American Chemical Society, Birmingham, AL, September 9–12, 1996.

20. Brown C.L.; Delshad, M.; Dwarakanath, V.; McKinney, D.M.; Pope, G.A.; Wade, W.H.; Jackson, R.E.; Londergan, J.T.; Meinardus, H.W. In *Innovative Subsurface Remediation: Field Testing of Physical, Chemical, and Characterization Technologies*, Brusseau, M.; Annable, M.; Gierke, J. Eds; ACS Symposium Series 725, American Chemical, Washington, DC, 1999.

21. INTERA (now DE&S) Final Report on The Demonstration of Surfactant Enhanced Aquifer Remediation of Chlorinated Solvent DNAPL at Operable Unit 2, Hill AFB, Utah. Prepared for AFCEE Technology Transfer Division Brooks AFB, San Antonio, TX, 1998.

22. Bourrel, M.; Schechter, R.S. *Microemulsions and Related Systems*; Marcel Dekker Inc: New York, 1988.

23. West, C.C.; Harwell, J.H. *Environ. Sci. Technol.* **1992**, *26(12)*, 2324–2330.

24. Nash, J.H. Field Studies of In-Situ Washing; EPA/600/2-87/110, U.S. Environmental Protection Agency, Cincinnati, OH, 1987.

25. Allred, B.; Brown, G.O. In *Surfactant-Enhanced Subsurface Remediation Emerging Technologies*, Sabatini, D.A.; Knox, R.C.; Harwell, J.H. Eds; ACS Symposium Series 594, American Chemical Society, Washington DC, 1995; 216–230.

26. Renshaw C.E.; Zynda G.D.; Fountain J.C. *Water Resour. Res.* **1997**, *33(3)*, 371–378.

27. Pope, G.A.; Wade W.H. In *Surfactant-Enhanced Subsurface Remediation Emerging Technologies*, Sabatini, D.A.; Knox, R.C.; Harwell, J.H.

Eds; ACS Symposium Series 594, American Chemical Society, Washington DC, 1995; 142–160.

28. Baran, J.R. Jr.; Pope, G.A.; Wade, W.H.; Weerasooriya, V. *Langmuir* **1994**, *10(4)*, 1146.

29. Baran, J.R. Jr.; Pope, G.A.; Wade, W.H.; Weerasooriya, V.; Yapa A. *J. Colloid Interface Sci.* **1994**, *168(1)*, 67.

30. Baran, J.R. Jr.; Pope, G.A.; Wade, W.H.; Weerasooriya, V.; Yapa A. *Environ. Sci. Technol.* **1994**, *28*, 1361–1366.

31. Baran, J.R. Jr.; Pope, G.A.; Wade, W.H.; Weerasooriya, V. *J. Dispersion Sci. Technol.* **1996**, *17(2)*, 131–138.

32. Baran, J.R. Jr., Pope, G.A.; Schultz, C.; Wade, W.H.; Weerasooriya, V.; Yapa, A. In *Surfactants in Solution*, Chattopadhyay, A.K.; Mittal, K.L. Eds; Marcel Dekker Inc: New York, 1996; 393–411.

33. Baran, J.R. Jr.; Pope, G.A.; Wade, W.H.; Weerasooriya, V. *Environ. Sci. Technol.* **1996**, *30(7)*, 2143–2147.

34. Shotts, D.R., 1996. Surfactant Remediation of Soils Contaminated with Chlorinated Solvents. M.S. thesis, University of Texas, Austin.

35. Dwarakanath, V., 1997. Characterization and Remediation of Aquifers Contaminated by Nonaqueous Phase Liquids Using Partitioning Tracers and Surfactants. Ph.D. dissertation, University of Texas, Austin.

36. Ooi, K., 1998. Laboratory Evaluation of Surfactant Remediation of Nonaqueous Phase Liquids. M.S. thesis, University of Texas, Austin.

37. Weerasooriya,V.; Yeh, S.L.; Pope, G.A. Integrated Demonstration of Surfactant-Enhanced Aquifer Remediation (SEAR) with Surfactant Regeneration and Reuse. *Proceedings of the ACS Surfactant-Based Separations: Recent Advances Symposium*, Dallas, Texas, March 29–30, 1998.

38. Pennell, K.D.; Abriola, L.M.; Weber Jr., W.J. *Environ. Sci. Technol.* **1993**, *27(12)*, 2332–2340.

39. Jin, M., 1995. Surfactant Enhanced Remediation and Interwell Partitioning Tracer Test for Characterization of NAPL Contaminated Aquifers. Ph.D. dissertation, University of Texas, Austin.

40. Jin, M.; Delshad, M.; Dwarakanath, V.; McKinney, D.C.; Pope, G.A.; Sepehrnoori, K.; Tilburg, C.; Jackson, R.E. *Water Resour. Res.* **1995**, *31(5)*, 1201–1212.

41. Pennell, K.D.; Pope, G.A.; Abriola, L.M. *Environ. Sci. Technol.* **1996**, *30(4)*, 1328–1335.

42. Lake, L.W. *Enhanced Oil Recovery*; Prentice Hall: Englewood Cliffs, NJ; 1989.

43. Kostarelos, K., 1998. Surfactant Enhanced Recovery of DNAPL using Neutral Buoyancy. Ph.D. dissertation, University of Texas, Austin.

44. Shook, G.M.; Pope, G.A.; Kostarelos, K. *J. Contaminant Hydrology* **1998**, *34(4)*, 363–382

45. Rouse J.D.; Sabatini, D.A. *Environ. Sci. Technol.* **1993**, *27(10)*: 2072–2078.

RECEIVED for review November 11, 1998. ACCEPTED revised manuscript January 7, 1999.

Use of Surfactants for Environmental Applications

Merv Fingas

Emergencies Science Division, River Road Environmental Technology Centre, Environment Canada, 3439 River Road, Ottawa, Ontario, Canada K1A 0H3

This chapter discusses dispersants, surface-washing agents, and emulsion breakers and inhibitors, all of which are oil-spill-treating agents that contain surfactants. These spill-treating agents are used on oil spills for specific purposes. Dispersants are used to remove oil from a water surface, surface-washing agents to remove oil from beaches or similar surfaces and emulsion breakers are used to break or prevent the formation of water-in-oil emulsions. The major issues related to the use of these agents are effectiveness and toxicity, which are discussed for the three types of agents. On the subject of effectiveness, the results of past and present field trials and laboratory tests are reviewed and the factors influencing effectiveness are discussed. For dispersants, toxicity is discussed both in terms of the dispersed oil and of the agent itself. The use of these agents and methods of application are also reviewed.

Introduction

Major oil spills can grab the headlines around the world. These incidents have created a global awareness of oil spills and the potential damage they can cause to the environment. The risk of oil spills will not diminish in the near future, however, preventative measures have and will continue to reduce the frequency and amount of spills. Better and faster cleanup measures will continue to reduce the impact of spills.

The most effective cleanup and containment methods vary from spill to spill and often vary from site to site in a single given spill. The efficiency of cleanup equipment and techniques may also change with time, as weather conditions fluctuate and the character of the spilled oil is altered. Consequently, a wide range of cleanup techniques and equipment is often considered.

The first priority in any spill is to stop the source of leakage. The second priority is to contain the spill so that further environmental damage does not occur. Spills on water can be contained using the many commercial spill containment booms. Booms will contain floating liquids up to a relative speed of $\frac{3}{4}$ knot. This is an important limitation because in many situations this velocity is exceeded. Attempts to contain oil on some open waters and across tidal bays will be futile. Tidal currents often exceed 2 knots and can be as much as 8 knots. Placement of booms requires extensive manpower and time.

Once the petroleum or solvent is contained, the conventional means of recovery is to recover oil using devices known as "skimmers". Skimmers are available in a wide variety of configurations and sizes. They too require extensive manpower to deploy and operate. Often skimmers are unable to recover more than about 10% of the volume of the oil spill. Often the oil will reach shoreline, from where the oil will be removed if it can cause potential for further water recontamination or damage to the shoreline ecosystem. The cost of shoreline cleanup, the amount of manpower and the time taken to do manual cleanup is often staggering. In the case of the *Exxon Valdez* spill in 1989, the total cost of cleanup was $2,000,000,000, most of which was shoreline cleanup and secondarily on-water recovery.

Many surfactant mixtures for treating oil spills have been promoted in the past two decades to overcome the extensive problems and costs of physical recovery. Of particular interest to this volume are dispersants, surface-washing agents, and emulsion breakers and inhibitors. All of these are formulations containing surfactants as active ingredients. Dispersants, in particular, promise to reduce the efforts and costs of cleaning up oil spills.

Surfactants have varying solubilities in water and varying actions toward oil and water. The parameter used to characterize surfactants is the hydrophilic–lipophilic balance (HLB) [1]. HLB is determined using theoretical equations that relate the length of the water-soluble portion of the surfactant to the oil-soluble portion of the surfactant. A surfactant with an HLB between 1 and 8 promotes the formation of water-in-oil emulsions and one with an HLB between 12 and 20 promotes the formation of oil-in-water emulsions. A surfactant with an HLB between 8 and 12 may promote either type of emulsion, but generally promotes oil-in-water emulsions. Dispersants have an HLB in this range.

Dispersants are formulated to "disperse" oil slicks into the sea or another water body. Surface-washing agents, or beach cleaners as they are sometimes called, are surfactant formulations designed to remove oil from surfaces such as beaches. Emulsion breakers and inhibitors are intended to break water-in-oil emulsions or to prevent their formation.

Although many of these treating agents have been promoted, few are still being produced. More than 100 dispersants have been tested for toxicity and effectiveness by Environment Canada, but only three remain on the department's list of accepted products [2]. The compendium of oil-spill-treating agents prepared by the American Petroleum Institute in 1972 lists 69 dispersants and 43 surface-washing agents, most of which are also listed as dispersants [3]. Only two of these are commercially available today, each produced in a different formulation. More than 100 surface-washing agents have been sold in the North American market, but only about six of these are still commercially available. It is estimated that approximately 600 spill-treating agents have been developed worldwide, of which only about 200 were ever tested in the lab or field, even in a limited way. The abundance of products makes it difficult for potential buyers and environmentalists to discriminate between effective products and those that are ineffective or could actually cause more damage than if the oil was left without intervention.

The current trend for governments is to establish stricter criteria for accepting spill-treating agents. In the United States, the 1993 National Contingency Plan (NCP) listed 48 dispersants, 2 surface-collecting agents, 42 biological additives, and 10 miscellaneous products [4], many of which contained surfactants. The current list contains only four dispersants, largely as a result of more stringent testing standards, but also as a result of attrition. As already noted, the Canadian government has similarly tested many dispersants but only three remain on the list of approved products. Many agents were listed in a recent review of oil-spill-treating agents in general, but only a few of these are still available 5 years after the report was published [5].

Effectiveness remains the major problem with most treating agents. Effectiveness is generally a function of the type of oil and its composition. Crude oil and refined oil products vary widely and include whole categories of materials with a wide range of molecular sizes, such as asphaltenes, alkanes, aromatics, and resins. What is often effective for small asphaltene compounds in an oil may be ineffective for large asphaltenes. What is effective for an aromatic compound may not be effective for a polar compound. This leaves little scope for a spill-control agent that is universally applicable and effective. Other major factors that influence the effectiveness of chemical agents are environmental parameters such as temperature and sea energy. These can be highly dominating and will sometimes overwhelm most other factors. The influence of both oil composition and environmental parameters on effectiveness must be determined to establish the usefulness of a spill-treating agent.

Spill-treating agents have been tested for both toxicity and effectiveness. Toxicity testing is very important. Most vendors of treating agents have not tested their products for aquatic toxicity, although a few have tested their products for mammalian toxicity to meet transportation requirements. Many products tested by Environment Canada have shown high and unacceptable aquatic toxicities. Even natural products can show high and unacceptable aquatic toxicities when tested. The aquatic toxicity of treating agents will be reported in this chapter when available, but the aquatic toxicity methodology itself will not be discussed.

Dispersants have been extensively reviewed in the past, especially in a major review by the National Academy of Sciences, completed in 1988 and published in 1989 [6]. In this chapter, some of the data from that report will be summarized in tables under Toxicity in the section on dispersants and the information will be discussed and updated. No comparable reviews have been published for other spill-treating agents, probably because these agents have only been developed recently and have a narrower range of application than dispersants.

Dispersants

Introduction. Much controversy has been generated in the past three decades over the use of dispersants and there are still strong proponents and opponents of their use. Controversy has often been based on outdated and unsubstantiated information or poorly documented and contradictory reports from the actual use of dispersants in the field. The difficulty arose largely from some dispersants used in the late 1960s and early 1970s that were either ineffective and resulted in wasted effort or were highly toxic and severely damaged the marine environment [6]. Thus, the two major issues associated with the use of dispersants are their effectiveness and the toxicity of the oil that is dispersed into the water column. Both these topics will be discussed extensively in this section.

Dispersants still generate much discussion and many studies since the birth of the oil spill industry many years ago after the 1968 *Torrey Canyon* incident [6]. There still exists a strong polarization between dispersant proponents and opponents. Documentation on actual field use is poor. Interviews with operators who have used dispersants often result in contradictory opinions on whether the dispersant worked in that situation or not. Large-scale biological experiments have failed to convince environmentalists that the use of dispersants is safe in all conditions, although the evidence is becoming increasingly clear that dispersants cause little, if any, ecological damage above that by untreated oil in many situations, particularly in offshore regions [6].

Dispersant effectiveness is defined as the amount of oil that the dispersant puts into the water column versus that which remains on the surface. There are many factors that influence dispersant effectiveness: sea energy, oil composition, state of oil weathering, rate of dispersant application, dispersant type, temperature, salinity of the water, etc. The most important factor for dispersant effectiveness is the composition of the oil, followed very closely by sea energy and amount of dispersant applied.

Certain oil components such as resins, asphaltenes and larger aromatics or waxes are barely dispersible, if at all [6]. Oils that contain mostly the latter components will disperse poorly even with dispersant application. On the other hand, oils that contain mostly saturates, such as diesel fuel, disperse both naturally and with the addition of dispersant. The additional amount of diesel dispersed using dispersants, over that naturally dispersed, depends primarily on the amount of sea energy present; however, dispersant will often be unnecessary. Laboratory studies have found a trade-off interrelationship between the two factors of amount of dispersant applied (dose) and the sea energy. That is, less sea energy implies that a higher dose of dispersant is needed to yield the same amount. There are other interrelationships as well, such as with salinity and temperature.

Effectiveness of dispersants is relatively easy to measure in the laboratory, however, there are many nuances in testing procedures [6]. One concern is that these tests are representative of real conditions. Since it is impossible to mimic all conditions directly, it is important to both consider the important factors such as sea energy and salinity while considering the laboratory tests as a form of screening or representative value, rather than a direct representation of what can be obtained in the field. Field "measurements" of dispersant effectiveness are also fraught with difficulty because it is very difficult to measure the concentration of oil in the water column over wide distances in appreciably small times, because there are no commonly available oil slick thickness measures with which to assess the amount of oil remaining on the surface and because of the fact that the sub-surface oil often moves differently than the surface slick. Any field measurement at this time is best viewed as an estimate. Actual dispersant effectiveness is very difficult to assess for the same reasons. Effectiveness is indicated by the presence of a coffee-coloured dispersed-oil plume in the water column. This is visible from ships and aircraft. Lack of the coffee-coloured plume indicates no or very little effectiveness.

The second issue of dispersants of importance is that of the toxicity of the dispersant and the toxicity of the dispersed oil droplets. Toxicity became a large issue in the early 1970s when application of toxic products resulted in substantial loss of sea life [6]. Subsequent generations of dispersants have had substantially less (often one hundredth) the toxic effect that early generations of dispersants had [6]. A standard measure of toxicity is the acute toxicity to standard species such as the rainbow trout. The LC_{50} of a substance is the "Lethal Concentration to 50% of a test population" usually given in mg/L, nearly equivalent to parts per million.

The specification is also given with a time period. For larger test organisms such as fish, this is often 96 hours. The smaller the LC_{50} number, the more toxic the product. The toxicity of the dispersant alone used during the early 1970s ranged from about 5 to 50 mg/L measured as an LC_{50} to the Rainbow Trout over 96 hours. The toxicity of current dispersants varies between 200 and 500 mg/L. Current dispersants contain a mixture of surfactants and a solvent that is less toxic than that in former generations.

The toxicity of the oil itself is now higher than that of the dispersants. The toxicity of diesel and light crude oil typically ranges from 20 to 50 mg/L. This is irrespective of whether the oil is chemically or naturally dispersed. No increase in toxicity has been observed to dispersed oil as a result of the addition of dispersants. However, the natural or chemical dispersion of oil in shallow waters can result in a mixture toxic to sea life by increasing the amount of hydrocarbons that enter the water. A spill of diesel fuel off the Atlantic coast of North America, into a shallow bay, resulted in the deaths of thousands of lobsters and other sea life. This occurred without dispersants. Similar toxicity could result from the use of dispersants in waters where the mixing was insufficient so that levels of the dispersed oil were toxic to sea life.

In some quarters, the use of dispersants remains a controversial issue. This is generally reflected that, in most jurisdictions, special permission is required to use dispersants. In other jurisdictions, dispersants are not allowed.

Formulations.   Dispersants are oil-spill-treating agents formulated to disperse oil into water in the form of fine droplets. Typically, the hydrophilic–lipophilic balance (HLB) of dispersants ranges from 9 to 11. Ionic surfactants can be rated using an expanded scale and have HLBs ranging from 25 to 40. Ionic surfactants are strong water-in-oil emulsifiers, very soluble in water and relatively insoluble in oil, which generally work from the water onto any oil present. Such products disappear rapidly in the water column and are not effective on oil. Because they are readily available at a reasonable price, however, many ionic surfactants are proposed for use as dispersants. These agents are better classified as surface-washing agents. Some dispersants contain ionic surfactants in small proportions, yielding a total HLB more toward 15 than 10. Studies on the specific effect of this on effectiveness or mode of action, have not been done.

A typical dispersant formulation consists of a pair of non-ionic surfactants in proportions to yield an average HLB of 10, and some proportion of ionic surfactants. Studies have been done on this mixture, one of which used statistical procedures in an attempt to determine the best mixture of the three ingredients [7]. A performance improvement was claimed by adjusting the three ingredients. Several patents are held

Table 1. Contents of Dispersants (patent information)

Type	Surfactants and Solvents
Hydrocarbon-based-1	Sorbitan monooleate
	Ethoxylated monooleate
	Na dioctyl sulfosuccinate
	Solvent – hydrocarbon and butyl cellosolve
Hydrocarbon-based-2	Sorbitan monooleate
	Ethoxylated sorbitan monooleate
	Ethoxylated sorbitan trioleate
	Na tridecyl sulfosuccinate
	Solvent – hydrocarbon and butanols
Hydrocarbon-based-3	Mixtures of polyethylene glycol monooleate
	Solvent – hydrocarbon
Aqueous-based-1	Tall oil esters (35%), ethyl dioxitol (47%)
	Sorbitan monolaurate (7%), water (10%)
	Calcium sulfonate (1%)

on dispersants [6, 8, 9]. The typical ingredients, from patents, are listed in Table 1. Some dispersants approved for use in Canada, the United States, and Europe are listed in Table 2.

Effectiveness of Dispersants. Dispersant effectiveness is defined as the amount of oil that the dispersant puts into the water column compared to the amount of oil that remains on the surface. In the field, effectiveness is indicated by the formation of a white to coffee-coloured plume of dispersed oil in the water column which is visible from ships and aircraft. This is shown in Figure 1. Many factors influence dispersant effectiveness, including oil composition, sea energy, state of oil weathering, the type of dispersant used and the amount applied, temperature, and salinity of the water. The most important of these is the composition of the oil, followed closely by sea energy and the amount of dispersant applied.

Certain components of oil, such as resins, asphaltenes, and larger aromatics or waxes, are barely dispersible, if at all [6]. Oils that are made up primarily of these components will only disperse poorly even when dispersants are applied. On the other hand, oils that contain mostly saturates, such as diesel fuel, will disperse both naturally and when dispersants are added. The additional amount of diesel dispersed when dispersants are used compared to the amount that would disperse naturally depends primarily on the amount of dispersant entering the oil. Laboratory studies have found a trade-off between the amount of dispersant applied, or the dose, and the sea energy. In general, less sea energy implies that a higher dose of dispersant is needed to yield the same

Table 2.   Approved Dispersants in Various Countries

Product	Manufacturer/Origin	Canada	United States	Britain	France
Corexit 9500	Exxon, Houston	✓	✓	✓	✓
Corexit 9527	Exxon, Houston	✓	✓	✓	
NEOS	Neos, Japan		✓		
Mare Clean 20	Taiho, Japan		✓		
Dasic Slickgone xx	Britain			✓	
Dispolene 3xx	France			✓	✓
Enersperse xx	Britain			✓	
Finasol xx	Belgium			✓	
Gamlen xx	Britain			✓	✓
Inipol xx	France			✓	✓
Shell Dispersant xx	Britain			✓	
Wellchem Welaid	Britain			✓	
Bioreco	France				✓
Emulgal	Israel				✓
Oceana	France				✓
PTI	France				✓

Examples only – lists may not be complete
xx indicates that several different sub-types are approved

Figure 1.   Effectiveness of dispersant is shown by the formation of a white to coffee-coloured dispersion in the water column as shown in this test tank experiment.

degree of dispersion as obtained when the sea energy is high. There are other interrelationships as noted above.

While it is easier to measure the effectiveness of dispersants in the laboratory than in the field, there are no standard testing procedures [6] and tests may not be representative of actual conditions. Therefore, important factors that influence effectiveness, such as sea energy and salinity, may not be accurately reflected in laboratory tests. Results obtained from laboratory testing should therefore be viewed as representative values only and not necessarily reflecting what would take place in actual conditions.

When testing dispersant effectiveness in the field, it is very difficult to measure the concentration of oil in the water column over large areas and at frequent enough time periods. It is also difficult to determine how much oil is left on the water surface as there are no methods available for measuring the thickness of an oil slick and the oil at the subsurface often moves differently than the oil on the surface. Field effectiveness trials, laboratory effectiveness tests, and factors influencing the effectiveness of a dispersant will be discussed in the following sections.

Field Trials. Many field trials have been conducted to assess the effectiveness of dispersants. The objectives of these tests were:

1. To quantify the effectiveness of dispersants on a given oil in a given application situation.
2. To demonstrate the effectiveness of dispersants and/or application techniques.
3. To measure concentrations of oil in the subsurface as a result of dispersant use.
4. To determine dispersibility conditions and relationships between factors.
5. To quantify application factors, such as effect of application rate and droplet size.

As can be seen in Table 3, in the past few years offshore trials have been conducted in the North Sea primarily by Great Britain and also by Norway [10–38]. In the 1980s, similar trials were also conducted in France and North America. Several papers have assessed the techniques used to measure effectiveness in these tests. There is no general consensus that effectiveness and other parameters can actually be measured in the field using some of the current methodologies.

The effectiveness determined during these trials varies significantly. Recent results, which may be more reliable, claim that dispersants removed about 10 to 40% of the oil to the subsurface. This is based on questionable analytical methodology. Ideal methodology may result in

Table 3. Dispersant Field Tests

Location/Identifier	Ref.	Year	Number	Oil Type	Amount, m³	Dispersant	Application Method	Rate, D:0	Sea State	Effectiveness Claimed (% unless noted otherwise)
North Sea – Great Britain	[11]	1997	136	Forties crude – weathered	50	Corexit 9500	airplane	1:19	3-4	good
			135	Forties crude – weathered	50	Slickgone NS	airplane	1:19	3-4	good
			134	Alaska North Slope – weathered	30	Corexit 9500	airplane	1:30		good
North Sea – Norway	[12]	1995	133	Troll – weathered	15	Corexit 9500	helicopter	1:20	1-2	good
			132	Troll – weathered	15	Corexit 9500	ship	1:20	1-2	good
			131	Troll – weathered	15	Corexit 9500	control	control	1-2	good
North Sea – Great Britain	[13]	1995	130	Medium Fuel Oil	continuous 50 L/min	OSR 5	boat	1:20	> 2	33
			129	Medium Fuel Oil	continuous 50 L/min	Corexit 9527	boat	1:20	> 2	32
			128	Medium Fuel Oil	continuous 50 L/min	Slickgone NS	boat	1:20	> 2	23
			127	Medium Fuel Oil	continuous 50 L/min	BP 1100X	boat	1:20	> 2	10
			126	Medium Fuel Oil	continuous 50 L/min	LA 1834	boat	1:20	> 2	9
			125	Medium Fuel Oil	continuous 50 L/min	control	boat	1:20	> 2	4
			124	Forties crude	continuous 50 L/min	Slickgone NS	boat	1:20	> 2	16
			123	Forties crude	continuous 50 L/min	control	boat	1:20	> 2	6
			122	Medium Fuel Oil	continuous 50 L/min	Slickgone NS	boat	1:20	< 2	10.5
			121	Medium Fuel Oil	continuous 50 L/min	control	boat	1:20	< 2	1.5
			120	MFO emulsion	continuous 50 L/min	Corexit 9527	boat	1:20	< 2	4
			119	MFO emulsion	continuous 50 L/min	Slickgone NS	boat	1:20	< 2	6
			118	MFO emulsion	continuous 50 L/min	control	boat	1:20	< 2	0.2
North Sea – Norway	[14]	1994	117	Sture Blend Crude	20	Corexit 9500	helicopter	1:12	4-5	good
			116	Sture Blend Crude	20	control then Corexit 9500	helicopter – next day	control then 1:20	4-5	good
North Sea – Great Britain	[15]	1993	115	MF/CO (Medium Fuel Oil/Gas Oil)	20	Dasic Slickgone NS	airplane	1:10	2-3	good
			114	MF/CO	20	control			2-3	
North Sea – Great Britain	[16]	1993	113	MF/CO (Medium Fuel Oil/Gas Oil)	continuous 50 L/min	OSR 5	boat	1:20	5-6	29.5
	[17]		112	MF/CO	continuous 50 L/min	Slickgone NS	boat	1:20	5-6	17
	[18]		111	MF/CO	continuous 50 L/min	BP 1100X	boat	1:20	5-6	10
			110	MF/CO	continuous 50 L/min	control	boat	control	5-6	1.6

Location [ref]	Year	No.	Oil	°C	Dispersant	Application	Ratio		
North Sea – Great Britain [19]	1992	109	Forties crude	12.3	LA 1834 then Dasic LTSW	airplane	1:100, 1:28	4–5	good
		108	Forties crude	12.3	LA 1834	airplane	1:100	4–5	
		107	Forties crude	12.3	control			4–5	
Beaufort Sea – Canada [20]	1986	106	topped Federated crude	2.5	control			2–3	
		105	topped Federated crude	2.5	Corexit CRX-8	helicopter	1:1	2–3	poor
		104	topped Federated crude	2.5	BP MA700	helicopter	1:1	2–3	poor
		103	topped Federated crude	2.5	BP MA700	helicopter	1:10	2–3	—
		102	topped Federated crude	2.5	control			1–2	
North Sea – Haltenbanken Norway [21]	1985	101	topped Statfjord crude	12.5	Alcopol	premixed	250 ppm	1–2	
		101	topped Federated crude	2.5	control			1–2	
		100	topped Statfjord crude	12.5	control			1–2	
		99	topped Statfjord crude	12.5	Finasol	premixed, 3 m below surface	1:50	1–2	—
Brest, Protecmar VI France [22, 23]	1985	98	topped Statfjord crude	12.5	control			1–2	—
		97	fuel oil	part of below	Dispolene 355	ship-aerosol	1:9	1	—
		96	fuel oil	part of below	Dispolene 355	ship-spray	1:9	1	—
		95	fuel oil	28	Dispolene 355	helicopter	1:9	1	—
		94	fuel oil	5	control	control		1	—
Norway [24]	1984	93	Statfjord	10	Corexit 9527	airplane	1:50	2	—
		92	Statfjord	12	Corexit 9527	premixed	1:33	2	—
		91	Statfjord	10	Corexit 9527	airplane	1:80	2	—
		90	Statfjord	10	control	control		2	—
		89	Statfjord	10	Corexit 9527	airplane, Islander	1:75	1	—
		88	Statfjord	10	control	control		1	—
Halifax, Canada [25–27]	1983	87	ASMB	2.5	control	control		1	
		86	ASMB	2.5	BP MA700	helicopter	1:10	2–3	7
		85	ASMB	2.5	control	control		2–3	
		84	ASMB	2.5	Corexit 9550	helicopter	1:10	1	10–41
		83	ASMB	2.5	control	control		1	
		82	ASMB	2.5	Corexit 9527	helicopter	1:20	1	1.3
Holland [28]	1983	81	Statfjord	2	Finasol OSR-5	airplane	1:10–30	1–2	2
		80	Statfjord	2	Finasol OSR-5	airplane	1:10–30	1–2	2
		79	light fuel	2	control	control		2–3	2
		78	Statfjord	2	Finasol OSR-5	premixed	1:20	2–3	100
		77	light fuel	2	Finasol OSR-5	airplane	1:10–30	1	2
		76	Statfjord	2	Finasol OSR-5	airplane	1:10–30	2	2
		75	Statfjord	2	control	control		2	2
		74	light fuel	2	control	control		1–2	2
		73	Statfjord	2	control	control		1–2	2

Table 3. (*Cont.*)

Location/ Identifier	Ref.	Year	Number	Oil Type	Amount, m³	Dispersant	Application Method	Rate, D:0	Sea State	Effectiveness Claimed (% unless noted otherwise)
Mediterranean Protecmar V – France	[25, 29]	1982	72	light fuel	5	control	control	—	2	
			71	light fuel	2	premixed	premixed	1:20	1–2	40–50
			70	light fuel	4	Dispolene 325	helicopter	1:2.9	1–2	—
			69	light fuel	3.5	Dispolene 325	ship	1:2.6	1–2	—
			68	light fuel	5	Dispolene 325	airplane, CL215	1:2.8	2	—
			67	light fuel	5	Dispolene 325	ship	1:2.8	2	—
			66	light fuel	5	Dispolene 325	airplane, CL215	1:2.4	3	—
			65	light fuel	3	10% Dispolene	ship	1:2	3	—
North Sea, Britain	[25, 30]	1982	64	Arabian	20	Corexit 9527	airplane, Islander	1:4	1	—
			63	Arabian	20	Corexit 9527	airplane, Islander	1:2	1	—
			62	Arabian	20	control	control	—	1	—
Norway	[31]	1982	61	Statfjord	0.2	10% Corexit 9527	ship	1:13	2–3	2
			60	Statfjord	0.2	10% Corexit 9527	ship	1:18	2–3	22
			59	Statfjord	0.2	10% Corexit 9527	ship	1:17	2–3	1.9
			58	Statfjord	0.2	control	control	—	2–3	2.6
			57	Statfjord	0.2	10% Corexit 9527	ship	1:10	2–3	1.7
			56	Statfjord	0.2	10% Corexit 9527	ship	1:10	2–3	6
			55	Statfjord	0.2	control	control	—	2–3	0.6
Newfoundland	[32]	1981	54	ASMB	2.5	Corexit 9527	airplane, DC-6	1:10	1	—
			53	ASMB	2.5	control	control	—	1	—
Mediterranean, Protecmar III – France	[23, 25, 33]	1981	52	light fuel	6.5	control	control	—	1–2	—
			51	light fuel	6.5	Shell	airplane, CL215	1:3	2–3	—
			50	light fuel	6.5	Dispolene 325	airplane, CL215	1:3	1–2	50
Mediterranean, Protecmar II – France	[23, 25]	1980	46–49			Corexit 9527	airplane, CL215	CL215		
			45			Corexit 9527	airplane, CL215			
			44			Finasol OSR-5	airplane			
			43			BP 1100WD	various			
Mediterranean, Protecmar I – France	[23]	1979	42	light fuel	1–5.5	BP 1100X	ship, helicopter	—	1–3	—
			33–41	light fuel	3 each	BP 1100X	ship, helicopter	—	1–3	—
			32			Corexit 9527	airplane, CL215			
			31			Corexit 9527	airplane, CL215			
			30			Finasol OSR-5	airplane			
			29			BP 1100WD	various			

Location	Ref	Year	No.	Oil	Amount	Dispersant	Application	Ratio		
Long Beach, USA	[34]	1979	27	Prudhoe Bay	1.6	2% Corexit 9527	ship	1:11	2–3	62
			26	Prudhoe Bay	1.6	2% Corexit 9527	ship	1:11	2–3	11
			25	Prudhoe Bay	3.2	conc.	airplane, DC-4	1:27	2–3	60
			24	Prudhoe Bay	1.6	control	control	—	2–3	1
			23	Prudhoe Bay	1.6	conc.	airplane, DC-4	1:25	2–3	45
			22	Prudhoe Bay	3.2	conc.	airplane, DC-4	1:20	2–3	78
Victoria, BC, Canada	[35]	1978	21	Prudhoe Bay	1.6	2% Corexit 9527	ship	1:67	2–3	5
			20	Prudhoe Bay	1.6	2% Corexit 9527	ship	1:67	2–3	8
			19	Prudhoe Bay	1.6	control	control	—	2–3	0.5
Southern California, USA	[36]	1978	18	North Slope	0.2	10%, 9527	ship, WSL	1:1	1	—
			17	North Slope	0.4	10%, 9527	ship, WSL	1:1	1	—
			16	North Slope	0.2	10%, 9527	ship, WSL	1:1	2	—
			15	North Slope	0.6	Several, demonstration	Several, demonstration	—	1–2	—
			14	North Slope	0.8	BP 1100WD	ship, WSL	>1:5	1–2	—
			13	North Slope	0.8	Corexit 9527	ship	>1:5	0–1	—
			12	North Slope	0.8	Corexit 9527	ship	>1:5	1–2	—
			11	North Slope	3.2	Corexit 9527	airplane, Cessna	>1:5	1–2	—
			10	North Slope	0.8	BP 1100WD	ship, WSL	>1:5	0–1	—
			9	North Slope	1.7	Recovery +	helicopter	>1:5	0–1	—
			8	North Slope	3.2	Corexit 9527	airplane, Cessna	>1:5	0–1	—
			7	North Slope	1.7	control later Corexit 9527	control then helicopter	>1:5	0–1	—
Wallops Island, USA	[25, 37]	1978	6	La Rosa	1.7	Corexit 9527	helicopter	1:11	1	50
			5	Murban	1.7	Corexit 9527	helicopter	1:11	1	100
			4	La Rosa	1.7	Corexit 9527	helicopter	1:5	1	—
			3	Murban	1.7	Corexit 9527	helicopter	1:5	1	—
North Sea, Britain	[25, 38]	1976	2	Kuwait		10% BP 1100	ship, WSL	1:20	2–3	100
			1	Ekofisk	0.5	10% BP 1100	ship, WSL	—	1	—

larger or smaller values; the results are not predictable at this time. The validity of the older results is even more questionable because of both the analytical methodology, which is now known to be incorrect, and data treatment methods [10, 39]. It is interesting that the percentage values that have been assigned average 16%, both in the older and more recent field trials.

All tests relied heavily on developing a mass balance of oil in the water column and that left on the surface [10]. In early tests, samples from under the oil plume were analysed in a laboratory using colorimetric methods. Colorimetric methods are not accurate for this type of analysis and are no longer used. Firstly, the concentrations to be measured were near or well below the threshold of the technique and secondly, a significant amount of hydrocarbons was lost between the sampling and the laboratory that could not be accounted for. Fluorometry has recently been used, but this method is also unreliable as it measures only a small and varying portion of the oil (middle aromatics) and does not discriminate between dissolved components and oil that actually dispersed. Furthermore, it is difficult to calibrate fluorometers for whole oil dispersions in the laboratory without using accurate techniques such as extraction and gas-chromatographic analysis. It is uncertain whether the aromatic ratio of the oil changes as a result of the dispersion process.

In early tests, it was not recognized that the plume of dispersed oil forms near the heavy oil in the tail of the slick and that this plume often moves off in a separate trajectory from the slick [10]. Many researchers "measured" the hydrocarbon concentrations beneath the slick and then integrated this over the whole slick area. As the area of the plume is always far less than this area, the amount of hydrocarbons in the water column was greatly exaggerated. Since the colorimetric techniques used at the time always yield some value of hydrocarbons, the effectiveness values were significantly increased. When effectiveness values from past tests were recalculated using only the area where the plume was known to be, those values decreased by factors as much as 2 to 5 [10].

Although no applications of dispersants on freshwater spills have been found, one field test was carried out in freshwater [40]. While effectiveness was not measured specifically, it was found that the dispersants appeared to reduce the long-term impact of the spill. The ASTM standards on the use of dispersants in freshwater such as lakes and rivers suggest that they not be used in freshwater primarily because most lakes and rivers are used as sources of drinking water [41].

In summary, testing in the field is difficult because effectiveness values depend on establishing a mass balance between oil in the water column and on the surface. Because this mass balance is difficult to achieve, results are questionable.

Laboratory Tests. Many different types of procedures and apparatus for testing dispersants are described in the literature. Fifty different tests or procedures are described in one paper [42]. Only a handful of these are in common use, however, including the Labofina or rotating flask test, the Mackay or MNS test, the swirling flask test, and the IFP (French Institute for Petroleum) test method.

Several investigators have reported results of apparatus comparison tests conducted in early years [42]. In the 11 papers reviewed in this reference, all authors have concluded that the results of the different tests do not correlate well, but some conclude that some of the rankings are preserved in different tests. Generally, the more different types of oil tested, the less the results correlate. It has been shown that laboratory tests can be designed to give a comparable value of oil dispersion if the parameters of turbulent energy, oil-to-water ratio, and settling time are set at similar values [42].

The most common laboratory apparatuses are listed in Table 4. In the literature, different protocols are sometimes described for the same apparatus. Previous comparisons of the different apparatus have generally been limited to two or three apparatuses [43–45]. Fingas and coworkers compared the Labofina, Mackay, oscillating hoop, and swirling flask apparatus for 10 oils and 3 dispersants and concluded that the numerical results correlated poorly and that rank of effectiveness correlated only weakly [46]. Results obtained using the oscillating hoop test in particular correlated poorly with other results.

Work has been done recently on determining the reason for the poor correlation between test results [47–49]. It was concluded that the differences in energy levels and the way the energy was applied to the oil/water mixture result in effectiveness values that are unique. In the past, investigators followed the specified test procedure when using an apparatus and did not vary any of the conditions. Fingas and coworkers found that, by adjusting the oil-to-water ratio and settling time, equivalent effectiveness values could be achieved using five different apparatuses [48]. It was found that energy was important but appeared to simply give higher values along the same line, that is, the relative ranking of dispersant/oil combinations was preserved. Clayton et al. reviewed the entire spectrum of laboratory testing and came to a similar conclusion that oil-to-water ratio, mechanisms of action, and settling times were not taken into account [50].

Clayton and Marsden tested the swirling flask test, which was the standard EPA test at the time, and the French IFP dilution test using Prudhoe Bay crude and South Louisiana crude and three commercial dispersants, Corexit 9527, Corexit CRX-8, and Enersperse 700 [51]. Results obtained with the swirling flask test were found to be comparable to the EPA modified tests, but results obtained with the other two tests

Table 4. Apparatus for Laboratory Testing of Dispersant Effectiveness

Test Name	Alternative Names(s)	Energy Source	Water Volume (L)	Prime Use	Where Used
LABOFINA	Warren Springs rolling flask	vessel rotation	0.25	regulatory general	Britain
Mackay	MNS Mackay–Nadeau–Steelman	air stream	6	regulatory general	Norway
Swirling flask		vessel movement	0.12	regulatory general	United States Canada, others
High-energy		moving vessel	5	experimental general	Canada
EXDET		wrist-action shaker	0.25	experimental	Exxon
IFP	French standard	oscillating hoop	16	regulatory general	France
SET	Simulated Environmental Test Tank	circulating pump	119	regulatory	not used at present
Cascading weir	Mackay flume	fall over weir	constant flow (0.5 L/s)	experimental	not used at present
Flowing column		fall down tube	1 – flowing	experimental	not used at present
Concentric tube	Bobra	water flow	constant flow (~0.05 L/s)	experimental	not used at present
Oscillating hoop		oscillating hoop	35	experimental	rarely used
Wave-plate tank	South African BP Sunbury	moving plates	30	regulatory general	not used at present
Spinning drop	Interfacial	water movement	<0.05	experimental	not used at present
Blender		propeller	1.5	experimental	not used at present

Table 5. Intercomparison of Laboratory and Field Effectiveness Results

| | | Effectiveness Results in Percent | | | | | |
| | | Field Test [16–17] | SF GC | SF CA | IFP | WSL Lab 1 | WSL Lab 2 | EXDET |
Oil type	Dispersant							
Medium fuel oil	Corexit 9527	26	54	50	91	42	42	67
Medium fuel oil	Slickgone NS	17	49	46	94	29	23	50
Medium fuel oil	LA 1834/Sur	4	2	2	50	16	11	38
Forties crude	Slickgone NS	16	47	65	95	28	25	60
Forties crude	LA 1834/Sur	5	2	2	61	15	12	53
Correlation with field test (R^2)			0.89	0.7	0.54	0.87	0.94	0.41
Ratio lab test/field test			0.4	0.35	0.19	0.56	0.62	0.27

SF: Swirling Flask, GC: analysis by Gas Chromatography, CA: Colorimetric Analysis, IFP: French Institute for Petroleum test, WSL: Warren Springs Laboratory Test

were less comparable. The IFP test resulted in different test values than the other three tests. Based on simplicity and cost, the swirling flask test was recommended over the other tests [51].

The only test developed recently is an internal Exxon effectiveness test known as EXDET [52]. The trend in recent years has been to either use existing tests or modify existing tests to accommodate new findings.

In an inter-laboratory evaluation of dispersant effectiveness tests, there was some agreement between test results on fresh oils, but very poor agreement between results of tests on oils that were more weathered or had any amount of water content [53, 54]. Some of these laboratory data were compared to the field data by Lunel and coworkers and the results are shown in Table 5 [53–55]. While the data correlate somewhat to the field data, with the wide spread in effectiveness numbers and the few data points, this correlation should not be overstated. Another interesting point is that the effectiveness values obtained in the field are lower than the data obtained in the laboratory, indicating that the energy levels may be much higher in laboratory tests than those in the field conditions described here. This is contrary to what was thought in previous years.

The results of a number of dispersant effectiveness tests taken from published laboratory results are shown in Table 6 [47–49]. As can be seen, the correlation among tests varies from high to low. This may be due to errors associated with the measurement, such as errors in measurement of volumes, variances in energy of the apparatus, etc. It can also be seen that the ranking of effectiveness is generally consistent, that is those oils and dispersants that show the highest or lowest effectiveness do so in all tests.

Table 6.   Dispersant Effectiveness Testing Results

Oil	Dispersant	Swirling Flask Colorimetric	Swirling Flask GC-analysis	Warren Spring Lab Test	Mackay–Nadeau–Steelman Test
		Effectiveness in percent			
Adgo	Corexit 9500	30	29		
Adgo	Corexit 9527	57		78	64
Adgo	Corexit CRX-8	51		77	87
Adgo	Dasic	11			
Adgo	Enersperse 700	71		76	93
AEA medium fuel oil	Corexit 9500	49	45	42	
AEA medium fuel oil	Corexit 9527	50	54		
Alaska North Slope	Corexit 9500				
Alaska North Slope	Corexit 9527	9			
Amauligak	Corexit 9500	47			
Amauligak	Corexit 9527	55		86	44
Amauligak	Corexit CRX-8			73	85
Amauligak	Dasic	26			
Amauligak	Enersperse 700	54		59	73
Arabian Light	Corexit 9500	21			
Arabian Light	Corexit 9527	24			
Arabian Light	Corexit CRX-8	17			
Arabian Light	Dasic	24			
Arabian Light	Enersperse 700	12			
Arabian Medium	Corexit 9500	19			
Arabian Medium	Corexit 9527	12			
Arabian Medium	Corexit CRX-8	14			
Arabian Medium	Dasic LTS	8			
Arabian Medium	Enersperse 700	10			
ASMB	Corexit 9500	37	43		
ASMB	Corexit 9527	39		31	39
ASMB	Corexit CRX-8	36		34	61
ASMB	Dasic	22			
ASMB	Enersperse 700	32		62	76
Atkinson	Corexit 9527	27		57	17
Atkinson	Corexit CRX-8	34		47	19
Atkinson	Dasic	26			
Atkinson	Enersperse 700	48		57	76
Avalon	Corexit 9527	8			
Avalon	Corexit CRX-8	10			
Avalon J-34	Corexit 9527	19			
Avalon J-34	Corexit CRX-8	7			
Avalon J-34	Dasic	7			
Avalon J-34	Enersperse 700	7			
Avalon Zone 4	Corexit 9527	21			
Avalon Zone 4	Corexit CRX-8	17			
Avalon Zone 4	Dasic	12			
Avalon Zone 4	Enersperse 700	17			
Barrow Island	Corexit 9500	49			
Barrow Island	Corexit 9527	42			
Barrow Island	Corexit CRX-8	51			
Barrow Island	Dasic LTS	13			
Barrow Island	Enersperse 700	48			
BCF-24	Corexit 9500	16	12		
BCF-24	Corexit 9527	16			
BCF-24	Corexit CRX-8	9			

Table 6. (*Cont.*)

Oil	Dispersant	Effectiveness in percent			
		Swirling Flask Colorimetric	Swirling Flask GC-analysis	Warren Spring Lab Test	Mackay–Nadeau–Steelman Test
BCF-24	Dasic LTS	1			
BCF-24	Enersperse 700	4			
Belridge Heavy	Corexit 9500	4	4		
Belridge Heavy	Enersperse 700	2	2		
Bent Horn	Corexit 9500	15	25		
Bent Horn	Corexit 9527	21		29	29
Bent Horn	Corexit CRX-8	15		27	51
Bent Horn	Dasic	16			
Bent Horn	Enersperse 700	14		19	42
Bent Horn Dyed	Corexit 9527	12			
Bent Horn Dyed	Corexit CRX-8	14			
Bent Horn Dyed	Enersperse 700	17			
Beta	Corexit 9500	1	1		
Beta	Corexit 9527	1			
Beta	Corexit CRX-8	0			
Beta	Dasic LTS	0			
Beta	Enersperse 700	0			
Bunker C	Corexit 9527	2			
Bunker C	Corexit CRX-8	4			
Bunker C	Dasic	2			
Bunker C	Enersperse 700	1			
Bunker C Light (IFO-200)	Corexit 9527	4			
Bunker C Light (IFO-200)	Corexit 9527	1			
Bunker C Light (IFO-200)	Corexit CRX-8	1			
Bunker C Light (IFO-200)	Dasic	1			
Bunker C Light (IFO-200)	Enersperse 700	1			
California Crude (11.0)	Corexit 9527	1			
California Crude (11.0)	Corexit CRX-8	2			
California Crude (11.0)	Dasic	0			
California Crude (11.0)	Enersperse 700	1			
California Crude (15.0)	Corexit 9527	2			
California Crude (15.0)	Corexit CRX-8	1			
California Crude (15.0)	Dasic	1			
California Crude (15.0)	Enersperse 700	1			
Carpenteria	Corexit 9500	11			
Carpenteria	Corexit 9527	2			
Carpenteria	Corexit CRX-8	0			
Carpenteria	Dasic LTS	0			
Carpenteria	Enersperse 700	2			
Cat Cracking Feed	Corexit 9500	9			
Cat Cracking Feed	Corexit 9527	7			
Cat Cracking Feed	Corexit CRX-8	10			
Cat Cracking Feed	Dasic LTS	3			
Cat Cracking Feed	Enersperse 700	7			
Cohasset	Corexit 9527	76			
Cohasset	Corexit CRX-8	36			
Cohasset	Dasic LTS	5			
Cohasset	Enersperse 700	35			
Cold Lake Bitumen	Corexit 9527	2			
Cold Lake Bitumen	Corexit CRX-8	2			
Cold Lake Bitumen	Dasic	1			

Table 6.   (*Cont.*)

Oil	Dispersant	Effectiveness in percent			
		Swirling Flask Colorimetric	Swirling Flask GC-analysis	Warren Spring Lab Test	Mackay–Nadeau–Steelman Test
Cold Lake Bitumen	Enersperse 700	1			
Dos Cuadras	Corexit 9500	9			
Dos Cuadras	Corexit 9527	4			
Dos Cuadras	Corexit CRX-8	5			
Dos Cuadras	Dasic LTS	1			
Dos Cuadras	Enersperse 700	7			
Empire	Corexit 9500	26	31		
Empire	Corexit 9527	12			
Empire	Corexit CRX-8	16			
Empire	Dasic LTS	11			
Empire	Enersperse 700	9			
Endicott	Corexit 9500	9			
Endicott	Corexit 9527	12			
Endicott	Corexit CRX-8	16			
Endicott	Dasic	6			
Endicott	Enersperse 700	8			
Endicott v 12.5% w 11.7%	Corexit 9527	3			
Endicott v 12.5% w 11.7%	Corexit CRX-8	3			
Endicott v 12.5% w 11.7%	Dasic	1			
Endicott v 12.5% w 11.7%	Enersperse 700	2			
Endicott v 8.38% w 7.46%	Corexit 9527	3			
Endicott v 8.38% w 7.46%	Corexit CRX-8	5			
Endicott v 8.38% w 7.46%	Dasic	1			
Endicott v 8.38% w 7.46%	Enersperse 700	5			
Eugene Island	Corexit 9500	27			
Eugene Island	Corexit 9527	5			
Extreme pressure gear oil	Corexit 9527	29			
Extreme pressure gear oil	Corexit CRX-8	40			
FCC Heavy Cycle Dyed	Corexit 9527	87			
FCC Heavy Cycle Dyed	Corexit CRX-8	37			
FCC Heavy Cycle Dyed	Dasic LTS	9			
FCC Heavy Cycle Dyed	Enersperse 700	11			
Federated	Corexit 9500	55	60		
Federated	Corexit 9527	50		51	35
Federated	Corexit CRX-8	56		35	76
Federated	Dasic	20	19		
Federated	Enersperse 700	39		70	76
Federated (15.5% weathered)	Corexit 9500	38	38		
Federated (28.5% weathered)	Corexit 9500	19	22		
Federated (41.8% weathered)	Corexit 9500	16	18		
Forties Blend	Corexit 9500	55	60		
Forties Blend	Corexit 9527	57	59		
Green Canyon block 109	Corexit 9500	19			
Green Canyon block 109	Corexit 9527	5			
Gullfaks	Corexit 9500	27			
Gullfaks	Corexit 9527	20			
Hibernia	Corexit 9527	10		23	6
Hibernia	Corexit CRX-8	11		19	9

Table 6. (*Cont.*)

		Effectiveness in percent			
Oil	Dispersant	Swirling Flask Colorimetric	Swirling Flask GC-analysis	Warren Spring Lab Test	Mackay–Nadeau–Steelman Test
Hibernia	Dasic	7			
Hibernia	Enersperse 700	7		23	14
Hondo	Corexit 9527	3			
Hondo	Corexit CRX-8	0			
Hondo	Dasic LTS	0			
Hondo	Enersperse 700	1	4		
Hondo (12.4% weathered)	Corexit 9500	0	0		
Hondo (12.4% weathered)	Corexit 9527	0	0		
Hout	Corexit 9500	15	20		
Hout	Corexit 9527	5	2		
Iranian Heavy	Corexit 9500	30	14		
Iranian Heavy	Corexit 9527	8			
Iranian Heavy	Corexit CRX-8	8			
Iranian Heavy	Dasic LTS	6			
Iranian Heavy	Enersperse 700	12			
Issungnak	Corexit 9527	51		22	41
Issungnak	Corexit CRX-8	41		76	100
Issungnak	Dasic LTS	31			
Issungnak	Enersperse 700	51		60	100
Kuwait	Corexit 9527	5			
Kuwait	Corexit CRX-8	4			
Lago	Corexit 9500	1			
Lago	Corexit 9527	2		29	16
Lago	Corexit CRX-8	4		19	19
Lago	Dasic LTS	0			
Lago	Enersperse 700	5		24	27
Lago	Shell VDC	4			
Lago Medio	Corexit 9527	7		8	16
Lago Medio	Corexit CRX-8	10		15	19
Lago Medio	Dasic LTS	5			
Lago Medio	Enersperse 700	11		23	27
Louisiana	Corexit 9500	19	33		
Louisiana	Corexit 9527	8	13		
Louisiana	Corexit CRX-8	15			
Louisiana	Dasic LTS	17			
Louisiana	Enersperse 700	7	14		
Lucula	Corexit 9500	22			
Lucula	Corexit 9527	3			
Lucula	Corexit CRX-8	4			
Lucula	Dasic LTS	3			
Lucula	Enersperse 700	4			
Lucula	Shell VDC	4			
Main Pass block 37	Corexit 9500	32			
Main Pass block 37	Corexit 9527	20			
Main Pass block 306	Corexit 9500	36			
Main Pass block 306	Corexit 9527	26			
Malongo	Corexit 9500	15			
Malongo	Corexit 9527	3			
Malongo	Corexit CRX-8	3			
Malongo	Dasic LTS	2			
Malongo	Enersperse 700	5			

Table 6. (*Cont.*)

Oil	Dispersant	Effectiveness in percent			
		Swirling Flask Colorimetric	Swirling Flask GC-analysis	Warren Spring Lab Test	Mackay–Nadeau–Steelman Test
Malongo	Shell VDC	6			
Mayan	Corexit 9527	2			
Mayan	Corexit CRX-8	1			
Mayan	Dasic LTS	0			
Mayan	Enersperse 700	6			
Mississippi Canyon block 194	Corexit 9500	30			
Mississippi Canyon block 194	Corexit 9527	15			
Mousse Mix	Corexit 9527	7		27	30
Mousse Mix	Corexit 9550	0			
Mousse Mix	Corexit CRX-8	15		18	26
Mousse Mix	Dasic	6			
Mousse Mix	Enersperse 700	9		23	43
Norman Wells	Corexit 9500	33			
Norman Wells	Corexit 9527	31		65	47
Norman Wells	Corexit CRX-8	53		70	65
Norman Wells	Dasic LTS	19			
Norman Wells	Enersperse 700	63		74	89
Oseberg	Corexit 9500	16			
Oseberg	Corexit 9527	29			
Oseberg	Corexit CRX-8	34			
Oseberg	Dasic LTS	12			
Oseberg	Enersperse 700	19			
Panuk	Corexit 9527	84		89	100
Panuk	Corexit CRX-8	78		85	100
Panuk	Dasic	42			
Panuk	Dasic LTS	26			
Panuk	Enersperse 700	88		87	100
Pitas Point	Corexit 9500	45	40		
Pitas Point	Corexit 9527		40		
Pitas Point	Corexit CRX-8	13			
Pitas Point	Dasic LTS	8			
Pitas Point 36% Weathered	Corexit 9500		65		
Pitas Point 36% Weathered	Corexit 9527		40		
Pitas Point 36% Weathered	Corexit CRX-8	83	65		
Pitas Point 36% Weathered	Dasic LTS	73	40		
Point Arguello Comingled	Corexit 9500	3	4		
Point Arguello Comingled	Corexit 9527	3	1		
Point Arguello Comingled	Dasic LTS	1			
Point Arguello Comingled	Enersperse 700	2			
Point Arguello Comingled	Corexit CRX-8	1			
Point Arguello Heavy	Corexit 9500	2	1		
Point Arguello Heavy	Enersperse 700	2	0		
Point Arguello Heavy	Dasic LTS	2			
Point Arguello Heavy	Corexit CRX-8	2			
Point Arguello Heavy	Corexit 9527	2	2		

Table 6. (*Cont.*)

Oil	Dispersant	Swirling Flask Colorimetric	Swirling Flask GC-analysis	Warren Spring Lab Test	Mackay–Nadeau–Steelman Test
Point Arguello Light	Corexit 9500	14	15		
Point Arguello Light	Corexit 9527	5	10		
Point Arguello Light	Dasic LTS	2	3		
Point Arguello Light	Enersperse 700	6	6		
Port Hueneme	Corexit 9500	0			
Port Hueneme	Corexit 9527	0			
Port Hueneme	Corexit CRX-8	0			
Port Hueneme	Dasic LTS	0			
Port Hueneme	Enersperse 700	1			
Prudhoe Bay	Corexit 9500	10			
Prudhoe Bay	Corexit 9527	29		47	27
Prudhoe Bay	Corexit CRX-8	22		38	23
Prudhoe Bay	Dasic LTS	13			
Prudhoe Bay	Enersperse 700	35		48	37
Prudhoe Bay weathered 14.5%	Experimental	7			
Prudhoe Bay weathered 14.5%	Corexit 9527	4			
Prudhoe Bay weathered 14.5%	Corexit CRX-8	9			
Prudhoe Bay weathered 14.5%	Dasic LTS	2			
Prudhoe Bay weathered 14.5%	Enersperse 700	4			
Prudhoe Bay weathered 7.55%	Experimental	30			
Prudhoe Bay weathered 7.55%	Corexit 9527	10			
Prudhoe Bay weathered 7.55%	Corexit CRX-8	13			
Prudhoe Bay weathered 7.55%	Dasic LTS	12			
Prudhoe Bay weathered 7.55%	Enersperse 700	9			
Rangely Crude	Corexit 9527	7			
Rangely Crude	Corexit CRX-8	11			
Rangely Crude	Enersperse 700	9			
Rangely Crude	Dasic LTS	13			
Santa Clara	Corexit 9527	2			
Santa Clara	Corexit CRX-8	0			
Santa Clara	Dasic LTS	0			
Santa Clara	Enersperse 700	3			
Sockeye	Corexit 9500	20			
Sockeye	Corexit 9527	7			
Sockeye	Corexit CRX-8	12			
Sockeye	Dasic LTS	1			
Sockeye	Enersperse 700	7			
South Louisiana Crude	Experimental	72			
South Louisiana Crude	Corexit 9527	53			
South Louisiana Crude	Corexit 9527	8			
South Louisiana Crude	Corexit CRX-8	56			

Table 6.   *(Cont.)*

Oil	Dispersant	Swirling Flask Colorimetric	Swirling Flask GC-analysis	Warren Spring Lab Test	Mackay–Nadeau–Steelman Test
		Effectiveness in percent			
South Louisiana Crude	Corexit CRX-8	15			
South Louisiana Crude	Dasic	28			
South Louisiana Crude	Dasic LTS	11			
South Louisiana Crude	Enersperse 700	31			
South Louisiana Crude	Enersperse 700	7			
South Pass Block 60	Corexit 9500	30			
South Pass Block 60	Corexit 9527	44			
South Pass Block 60	Corexit CRX-8	40			
South Pass Block 60	Dasic LTS	15			
South Pass Block 60	Enersperse 700	9			
South Pass Block 93	Corexit 9527	25			
South Pass Block 93	Corexit CRX-8	26			
South Pass Block 93	Dasic LTS	25			
South Pass Block 93	Enersperse 700	23			
South Timbalier Block 130	Corexit 9500	29			
South Timbalier Block 130	Corexit 9527	10			
South Timbalier Block 130	Corexit CRX-8	35			
South Timbalier Block 130	Dasic LTS	20			
South Timbalier Block 130	Enersperse 700	19			
Statfjord	Corexit 9500	25	40		
Statfjord	Corexit 9527	37			
Statfjord	Corexit CRX-8	44			
Statfjord	Dasic LTS	17			
Statfjord	Enersperse 700	14			
Sumatran Minas Light	Corexit 9527	0			
Sumatran Minas Light	Corexit CRX-8	0			
Sumatran Minas Light	Dasic LTS	0			
Sumatran Minas Light	Enersperse 700	0			
Synthetic Crude	Experimental	93			
Synthetic Crude	Corexit 9527	67		78	83
Synthetic Crude	Corexit CRX-8	56		40	91
Synthetic Crude	Dasic	23			
Synthetic Crude	Enersperse 700	65		76	88
Taching	Corexit 9527	0			
Taching	Corexit CRX-8	0			
Taching	Dasic LTS	0			
Taching	Enersperse 700	0			
Takula	Corexit 9500	11			
Takula	Corexit 9527	3			
Takula	Corexit CRX-8	5			
Takula	Dasic LTS	2			
Takula	Enersperse 700	6			
Takula	Shell VDC	9			
Tarsuit	Corexit 9527	53			
Tarsuit	Corexit CRX-8	6			
Terra Nova	Experimental	40			
Terra Nova	Corexit 9527	29			
Terra Nova	Corexit CRX-8	22			
Terra Nova	Dasic	19			
Terra Nova	Enersperse 700	21			
Thevenard Island	Corexit 9500		89		

Table 6. (*Cont.*)

Oil	Dispersant	Swirling Flask Colorimetric	Swirling Flask GC-analysis	Warren Spring Lab Test	Mackay–Nadeau– Steelman Test
		Effectiveness in percent			
Thevenard Island	Corexit 9527	55	56		
Thevenard Island	Corexit CRX-8	38			
Thevenard Island	Dasic LTS	24			
Thevenard Island	Enersperse 700	31			
Transmountain Blend	Experimental	25			
Transmountain Blend	Corexit 9527	15			
Transmountain Blend	Corexit CRX-8	13			
Transmountain Blend	Dasic	12			
Transmountain Blend	Enersperse 700	17			
Udang	Corexit 9527	0			
Udang	Corexit CRX-8	0			
Udang	Dasic LTS	0			
Udang	Enersperse 700	0			
Used Motor Oil	Experimental	42			
Used Motor Oil	Corexit 9527	42			
Used Motor Oil	Corexit CRX-8	39			
Used Motor Oil	Dasic	30			
Used Motor Oil	Enersperse 700	47			
Uvilik	Corexit 9527	31			
Uvilik	Corexit CRX-8	10			
Waxy Light Heavy	Corexit 9527	7			
Waxy Light Heavy	Corexit CRX-8	6			
Waxy Light Heavy	Dasic LTS	0			
Waxy Light Heavy	Enersperse 700	42			
West Delta Block 97	Corexit 9500		90		
West Delta Block 97	Corexit 9527		51		
West Delta Block 97	Dasic LTS		16		
West Texas Intermediate	Corexit 9527	32			
West Texas Intermediate	Corexit CRX-8	88			
West Texas Intermediate	Dasic LTS	10			
West Texas Intermediate	Enersperse 700	40			
West Texas Sour	Corexit 9500	24			
West Texas Sour	Corexit 9527	37			
West Texas Sour	Corexit CRX-8	25			
West Texas Sour	Dasic LTS	10			
West Texas Sour	Enersperse 700	27			
Zaire	Corexit 9500	1			
Zaire	Corexit 9527	3			
Zaire	Corexit CRX-8	3			
Zaire	Dasic LTS	1			
Zaire	Enersperse 700	3			
Zaire	Shell VDC	2			

Physical Studies.   Traditionally, the effectiveness of a dispersant was viewed as simply a result of interfacial phenomena, that is the lowering of the surface tension of the oil by the use of surfactants [56]. It is now apparent that many factors influence the effectiveness of dispersants, the most important of which are sea energy, the composition of the oil, the type of dispersant and the amount applied, temperature, and salinity of the water [48].

Given a certain type of oil and salinity, the important considerations are the sea energy and the amount of dispersant. Experiments have focussed on defining effectiveness for oils that might be spilled frequently. An energy–dispersant amount diagram for Alberta Sweet Mixed Blend, a common oil in North America, is shown in Figure 2. The diagram is based on experimental data [48]. Energy is indicated by the rotational rate of the shaker unit, which shows that there is a predictable relationship between the three factors of effectiveness, energy, and dispersant quantity. While it can be seen in Figure 2 that energy is a very important factor, the same dispersant effectiveness can be achieved at several different energy–dispersant combinations.

Determining the relationship between laboratory data as given here and data from larger scale experiments, tank experiments, and field experiments [57–59] is an ongoing issue. Several workers have shown that mass balance is not preserved in larger scale and field experiments, because of analytical difficulties in measuring oil in the water column and remaining on the water surface. This implies a greater reliance on laboratory experiments until better analytical methods are developed for larger scale experiments.

Energy.   Varying levels of energy have often been cited as the reason for the inconsistent results of dispersant effectiveness tests in the field and in the laboratory. These inconsistent results have not been studied due to difficulties in measuring energy levels in the field or in the

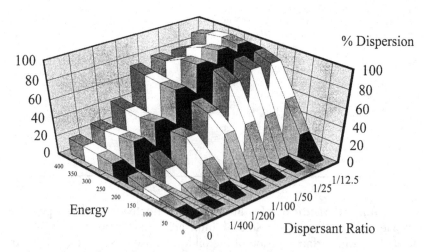

Figure 2.   Relationship between effectiveness (% dispersion), energy and dispersant amount (ratio), ASMB and Corexit 9527.

laboratory. Energy, as it relates to the sea and oil spills, is a complex topic [48, 60–63]. There are different ways of defining energy as it relates to oil dispersion. One means of describing energy is to measure the wavelength and the amount of turbulence in the near surface. Sea energy has also been described in terms of steepness of waves and period [48].

Several series of experiments examining the relationship of energy to dispersion were conducted by Environment Canada. The results are summarized in Figure 3. Chemical dispersion increases with energy (measured in these experiments as revolutions of the experimental apparatus in a given time period) in a linear fashion until a maximum is reached. For light oils, this maximum is about 80% and for heavier oils, it is about 65%. The dispersion curve is very steep; that is, only a small amount of energy causes a significant change in dispersion. There is an energy threshold below which little dispersion occurs. Chemical dispersion curves for different oils are parallel (have similar slopes).

Natural dispersion is similar to chemical dispersion except that the increase in dispersion with energy is much less. It is difficult to determine the energy level at which natural dispersion occurs, but it occurs at a higher energy level than chemical dispersion.

The major question raised by these experiments is how the energy in these tests relates to that at sea. There are a few observations that show that most typical sea energies occur at the low end of the energy range shown in Figure 3. The effectiveness is given as percent, that is, percent of the oil in the water column. The x-axis is given in relative energy. The

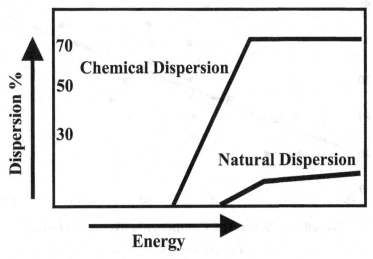

Figure 3.   Relationship of energy to amount of dispersion.

technology to measure energy both at sea and in the laboratory is complex and still being developed, and has therefore not yet been fully applied.

Composition of Oil.   In the past, it was thought that viscosity was the only quality of an oil that influenced the effectiveness of a dispersant [*64*]. It soon became apparent, however, that the chemical constituents of oil had a major influence on the effectiveness of dispersants. Studies correlating effectiveness and oil composition revealed that the most important factor was the amount of saturates in the oil [*48, 65*]. It was also found that the effectiveness of dispersants decreases with increasing amounts of resins and asphaltenes in the oil. Furthermore, it was found that effectiveness could be predicted using a simple model of saturates, less the other components of the oil, including resins, asphaltenes, and aromatics. This simple model is illustrated in Figure 4 which shows that the fit is poor and additional information is required to accurately describe dispersant effectiveness as a function of the composition of the oil. The figure does illustrate, however, that the composition of the oil is an important factor in the effectiveness of a dispersant.

Amount of Dispersant.   While several workers have noted the decline in the amount of dispersion with decreasing dosages of dispersant,

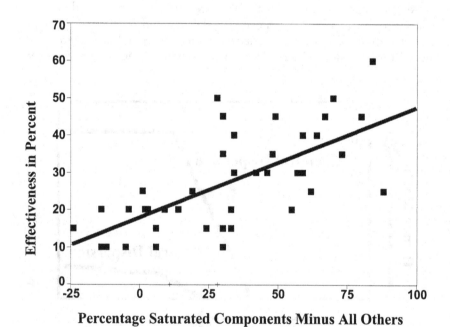

Percentage Saturated Components Minus All Others

Figure 4.   Simple model of dispersant effectiveness as a function of oil composition.

few have actually measured this decline [48, 50]. The relationship of the amount of dispersion that occurs and the dosage of dispersant used as conducted in a series of laboratory experiments is shown in Figure 5, taken from Fingas et al. [48]. The dosage is given as the ratio of dispersant to oil, $1/x$ where x is the amount of dispersant relative to one volume of oil. Thus 1/5 represents a dosage of 1/5 the volume of dispersant versus the volume of oil or a ratio of 1:5 dispersant:oil.

Temperature.   Several workers have noted that dispersant effectiveness declines with temperature [48, 50, 66]. It was noted, however, that it is difficult to distinguish between the effect of viscosity and other factors and the effect of temperature. A typical decline in effectiveness with temperature is shown in Figure 6, taken from Fingas et al. [48].

Salinity.   Several workers have noted that conventional dispersants did not function well in waters of low salinity. Belk et al. found that effectiveness decreases as salinity decreases and that effectiveness is minimal in absolute freshwater [67]. One freshwater dispersant showed limited effectiveness in "hard" water with a high ionic content. Brandvik and coworkers tested the effectiveness of dispersants at both low temperature and low salinity conditions and found that most dispersants dropped by as much as a factor of 100 and typically about 1/5 in going from the salinity of 33‰ to 5‰ [66]. Both series of tests were conducted

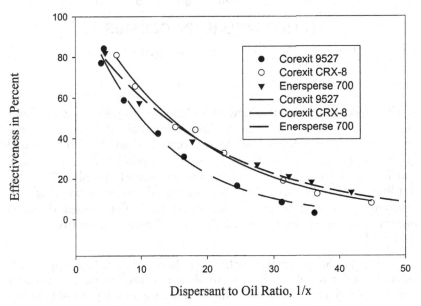

Figure 5.   Variation in effectiveness with amount of dispersant.

EFFECTIVENESS %

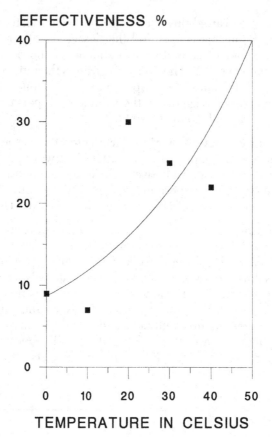

TEMPERATURE IN CELSIUS

Figure 6.   Variation in effectiveness with temperature.

at 0 °C. It was concluded that new formulations of dispersants would have to be developed for use in the Arctic, because of both the lower temperatures and salinities of Arctic waters. Similarly, Eriksson and Moet and coworkers tested dispersants in modified Warren Springs apparatus and found that effectiveness decreased sharply with time (settling time) and temperature, and decreased somewhat when oil was weathered [68, 69]. Most authors found that effectiveness declined sharply with salinity similar to that shown in Figure 7, taken from Fingas et al. [48].

Particle Size.   In the past, researchers believed that dispersants created smaller droplets which were obviously more stable in the water column [50]. Several researchers found that droplet sizes did not change with the amount of dispersion used, but that more dispersant simply

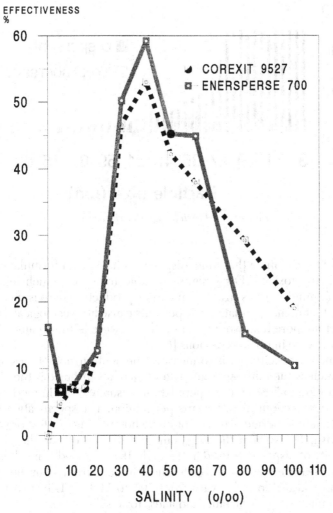

Figure 7. *Variation in effectiveness with salinity.*

created a larger amount of droplets of relatively the same size [*48, 50, 70*]. A typical distribution of droplet sizes, with and without the use of dispersant, is shown in Figure 8, which is adapted from Lunel [*70*].

Toxicity. The second important issue when discussing dispersants is toxicity, both of the dispersant itself and of the dispersed oil droplets. Toxicity became an important issue in the late 1960s and early 1970s when application of toxic products resulted in substantial loss of sea life [*71*]. For example, the use of dispersants during the *Torrey Canyon* episode in Great Britain in 1968 caused massive damage to intertidal and

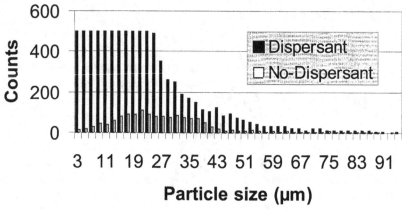

Figure 8.   Particle size distribution.

sub-tidal life [6]. Since that time, dispersants have been formulated with lesser aquatic toxicity. Dispersants available today are much less toxic (often one hundredth as toxic) than earlier products. There is increasing evidence that in many situations dispersants cause little ecological damage or at least no more than would occur if the oil were left untreated. This is particularly true in offshore regions [6].

A standard toxicity test is to measure the acute toxicity to a standard species such as the rainbow trout. The LC_{50} of a substance is the "Lethal Concentration to 50% of a test population", usually given in mg/L, which is approximately equivalent to parts per million. The specification is also given with a time period, which is often 96 hours for larger test organisms such as fish. The smaller the LC_{50} number, the more toxic the product. The toxicity of dispersants used in the early 1970s ranged from about 5 to 50 mg/L measured as an LC_{50} to the rainbow trout over 96 hours. Dispersants available today vary from 200 to 500 mg/L in toxicity and contain a mixture of surfactants and a less toxic solvent.

Today, the oil itself is more toxic than the dispersants, with the LC_{50} of diesel and light crude oil typically ranging from 20 to 50 mg/L, for either chemically or naturally dispersed oil. No increase in toxicity has been observed to dispersed oil as a result of the addition of dispersants. However, the natural or chemical dispersion of oil in shallow waters can result in a mixture that is toxic to sea life. For example, a spill in 1996 from the *North Cape* in a shallow bay on the Atlantic coast caused massive loss of benthic life without the use of dispersants [72]. Another significant factor in terms of the impact of this spill was the proximity to shore which caused a high concentration of hydrocarbons in the water. Similar toxicity could also result if dispersants are not mixed in the correct ratio and the resulting dispersed oil could be toxic to sea life.

Dispersants have been reviewed extensively in terms of toxicity, particularly in a major review by the National Academy of Sciences published in 1989 [6]. Although data and references will not be repeated here, some of the major data will be summarized in tables. The major issues have changed since this review was published. First, the concern over the exposure regimes has subsided. In the last decade, there was concern that the time–dose applied to test organisms was not relevant to the regime that the same organisms would be exposed to in an actual dispersant application. New methods for testing aquatic toxicity have enabled more realistic dosing. Furthermore, many of the methods used in the past were questionable. Toxicity testing is more accurate today due to new analytical techniques.

Results of older aquatic toxicity studies of the ubiquitous dispersant Corexit 9527, produced by Exxon, are shown in Table 7, as summarized by the National Academy of Sciences (NAS) [6]. The results show that the effects are highly dependent on the species and its life-stage, as would be expected. The 96-hour acute toxicity to many fish averages 100 mg/L (approximately equivalent to parts-per-million) and approximately 10 mg/L to more sensitive life forms. Many of these results are obtained from screening tests, which are aquatic toxicity tests designed to determine if a dispersant meets minimum acceptability criteria. The toxicity of several dispersants is given in Table 8. These data are derived from both Environment Canada's testing program and the National Academy of Sciences study [2, 6].

Since the NAS study, several researchers have studied the aquatic toxicity of oil, dispersants, and dispersed oil. Some of the typical studies will be summarized here.

Gulec and Holdway studied the toxicity of oil and the dispersant Corexit 9527 to the amphipod, *Allorchestes compressa* (73). They found that the mean ($m = 4$) acute 96-hour LC_{50} for *A. compressa* exposed to Corexit 9527 was 3 mg/L, to dispersed crude oil was 16.2 mg/L, and for the water-accommodated fraction of Bass Strait crude oil was 311,000 mg/L. Sub-lethal effects were also measured for a 30-minute exposure. The EC_{50} (threshold for sub-lethal effects) was found to be 50.2 mg/L for Corexit 9527, 64.4 mg/L for Corexit 9527, 65.4 mg/L for dispersed crude oil, and 190,000 mg/L for the water-accommodated fraction of Bass Strait crude oil.

Singer and coworkers also found that the dispersed fraction was more toxic than the oil or dispersant alone [74–76]. This group tested the toxicity of Prudhoe Bay oil and Corexit 9527 to red abalone (*Halliotis rufescens*), a kelp forest mysid (*Holesimysis costata*) and the topsmelt (*Atherinops affinis*). The dispersed oil showed much higher toxicological responses; in fact, no responses were noted with the WAF (water-accommodated fraction) itself. The toxicity of the dispersant and dis-

Table 7. Aquatic Toxicity Studies on Corexit 9527

Species	Specific Species	Type of Threshold	Dispersant Toxicity Concentration (mg/L)	Oil Only Concentration (mg/L)	Dispersed Oil Effects
Freshwater					
Algae	*Chlamydomonas*	LC_{50} – 48 hours	575		
Ciliate protozoa		IEC	100–150		
Trout	*Salmo gairdneri*	LC_{50} – 96 hours	140–233		
Zebra fish	*Brachydanio*	LC_{50} – 48 hours	550		
Saltwater					
Amphipods	several	LC_{50} – 96 hours	104–170		Lethality + Prudhoe Bay crude
Arctic fish	*Myoxocephalus*	LC_{50} – 96 hours	<40		
Barnacle larvae	*Belanus*	EC_{50} – 144 hours	10–100		
Calanoid copepods		LC_{50} – 24 hours	100–1000		
Clams	*Protothaca staminea*	LC_{50} – 72–100 hours	100		
Clams	*Protothaca staminea*	LC_{50} – 240 hours	>10		
Clams	several	+ Venezuelan crude			
Cod eggs	*Gadus morhua*	EC_{50} – 504 hours	<1		Behavioural effects
Cod embryo	*Gadus morhua*	EC_{50} – 504 hours	10–100		Lethality + Ekofisk crude
Coelenterate larvae		EC_{50} – 96 hours	1–10		Lethality + Ekofisk crude
Copepods	*Acartia*	LC_{50} – 48 hours	22–35		
Copepods	*Pseudocalanus minatus*	LC_{50} – 48 hours	7.1–11		
Copepods	various	+ Venezuelan and Ekofisk crudes			Lethality

Flounder larvae	*Platichthys fleus*	+ Ekofisk crude			Lethality
Gastropods	*Patella vulgata*	+ crudes and refined oils			Lethality
Isopods	several	+ Prudhoe Bay crude			Lethality
Lobster	*Homarus americanus*	+ Louisiana crude			Lethality
Plaice eggs	*Pleuronectes*	EC_{50} – 240 hours	10–100		
Salmon	Coho	LC_{50} – 96 hours		19	9 mg/L toxicity
Scallops	*Argopectan irradians*	LC_{50} – 6 hours	1900–2500		Behavioural effects + Kuwait crude
Sea urchins	several	EC_{50} – 96 hours	10–100, 1–10		
Sea urchins	*Paracentrotus*	EC_{50} – 0.15 hours	0.03–0.05		
Sea urchins	*Paracentrotus*	EC_{50} – 0.5 hours	0.003–0.0003		
Shrimp larvae	*Artemia*	EC_{50} – 48 hours	40, 42–72	984–1560	Lethality + Lago medio
Shrimp	several	+ Kuwait and Louisiana crudes		15	Lethality, 1970–2880
Spot eggs	*Leiostomus xanthurus*	+ Ixtoc I crude			Lethality
Stickleback	*Gasterosteus*	LC_{50} – 96 hours	28		
Marine Plants					
Fucus	*Fucus seratus*	+ Ekofish crude			effects noted on zygotes
Seagrass	*Thalassia testudinum*	+ Prudhoe Bay crude			lethality, leaf coloration
Seagrass	*Thalassia halodule*	+ Murban crude			mortality

Table 8.   Aquatic Toxicity of Common Dispersants

	Aquatic Toxicity Ranges		
Dispersant Trade Name	LC_{50} to Rainbow Trout (96 hours)	LC_{50} to *Daphnia* *magna* (48 hours)	LC_{50} to *Artemia* (48 hours)
Corexit 9500	354^1		
Corexit 9527	108^1, $26-293^2$	$31-42^1$	$52-104^2$
Corexit 7664	851^1, $490-857^2$		$99-500^2$
BP 1100X	$42-150^1$, $210-$ 2400^2		
Enersperse 700	$50-150^1$	$40-60^1$	

[1] Environment Canada method (2)
[2] NAS (6)

persed oil ranged from 28.6 to 74.7 mg/L for the mysid, 10.5 to 16.8 mg/L for the topsmelt, and 17.8 to 32.7 mg/L for the abalone.

Similarly, Midlaugh and Whiting reported on tests to embryonic inland silversides, *Menidia beryllina* [77]. Effects were ranked in terms of toxicity observed at the percentage of the No. 2 fuel oil water-accommodated fractions, with and without Corexit 7664 and Corexit 9527. Dispersants were also tested alone. Embryos exposed to the No. 2 fuel oil in 20‰ salinity water showed responses only at the 100% WAF concentration. Corexit 7664 alone elicited response at 10% WAF and when combined with fuel oil at 1% WAF. Corexit 9527 and fuel oil elicited responses at 10% WAF.

The acute toxicity of physically and chemically dispersed crude oil to the estuarine mysid, *Mysidopsis bahia*, and the kelp forest mysid, *Holesimysis costata*, was evaluated in continuous and spiked exposure conditions [78]. The continuous exposure LC_{50} for *M. bahia* was about 0.65 mg/L for chemically dispersed oil and was similar for the physically dispersed oil. Continuous exposure LC_{50} for *H. costata* were 0.17 mg/L for the chemically dispersed oil and 0.10 mg/L for the physically dispersed oil. No toxicity for physically dispersed oil was observed for either species in the spiked-exposure tests but the chemically dispersed oil showed a toxicity of 13 to 25 mg/L for *M. bahia* and 1 to 5 mg/L for *H. costata*. It was concluded that only spiked-exposure tests should be used because this is the type of exposure occurring during oil spills and dispersant usage.

Fucik and coworkers tested the toxicity of dispersant and oil (Western and Central Gulf oils) and dispersant (Corexit 9527) to several indigenous species from the Gulf of Mexico [79]. The 96-hour LC_{50} of oil alone to shrimp (both *Penaeus aztecus* and *Penaeus setiferus*) was around 12 mg/L and of oil and dispersant was from 14 to 15 mg/L. The LC_{50} of dispersant alone for the blue crab (*Callinectes sapidus*) ranged from 78 to 80 mg/L and for dispersant and oil was 20 to 91 mg/L. The LC_{50} of dispersant alone for the eastern oyster (*Crassostrea virginica*) was 5 mg/L and for dispersant and oil was 4 to 11 mg/L. The LC_{50} of dispersant alone for inland silverside larvae (*Menidia beryllina*) ranged from 43 to 47 mg/L and for dispersant and oil was 59 to 100 mg/L. The LC_{50} of dispersant alone for inland silverside embryos (*Menidia beryllina*) was over 100 mg/L and the same for dispersant and oil. The LC_{50} of dispersant alone for Atlantic menhaden (*Brevoortia tyrannus*) was 42 mg/L and for dispersant and oil was 22 to 65 mg/L. The LC_{50} of dispersant alone for the Spot (*Leiostomus xanthurus*) was 27 mg/L and for dispersant and oil was 50 to 68 mg/L. The LC_{50} of dispersant alone for the red drum (*Sciaenops ocellatus*) ranged from 52 to over 100 mg/L and for dispersant and oil was over 100 mg/L.

Burridge and Shir evaluated the toxicity of oil and dispersants to marine algae [80]. They found that sometimes germination was enhanced and sometimes there was little effect, depending on dispersant/oil combinations. The 48-hour EC_{50} for dispersant alone was 10,500 µL/L in the case of Corexit 7664, 27,000 µL/L for Corexit 8667, 0.7 µL/L for Corexit 9500, and 30 µL/L for Corexit 9527. The EC_{50} for the dispersant/ oil mixture (Bass Strait crude) was 130 µL/L for the crude alone, 4000 µL/L for Corexit 7664, 2500 µL/L for Corexit 8667, 20 µL/L for Corexit 9500, and 200 µL/L for Corexit 9527.

Wolfe et al. studied dispersants to a primary producer, algae, *Isochrysis galbana*, and a primary consumer, a rotifer, *Brachionus plicatilis* (81). Results showed that the uptake of naphthalene increased significantly in the presence of dispersant in algae and also in the rotifer via trophic transfer.

Unsal studied aquatic toxicity to the prawn, *Palaemon elegans*, and found the 24-hour LC_{50} for oil alone (Turkish crude) was 83.5 mL/L, for dispersant (Spillwash) was 0.0112 mL/L, and for the oil dispersant mixture was 1.1 mL/L [82].

Many researchers studied the effects of dispersants on whole ecosystems and not on individual species [83]. The ASTM guides on dispersants are prepared for whole systems [84]. The advantages of this approach are that often users think in terms of their ecosystem as a whole and also because of the many links between trophic levels in a given ecosystem. The disadvantages are that studies may not be applicable to other ecosystems and certain species may not be studied in a given system, simply because they were not considered.

Several researchers studied tropical ecosystems focussing on corals, seagrasses, and mangroves [85–88]. Knap, Ballou, and coworkers reviewed the fate and effects of oil on tropical ecosystems. It was noted that different researchers found different effects of oil and dispersants together and oil alone. Some found that oil and dispersed oil had severe and irreversible effects on coral systems and others found exactly the opposite. The application of dispersants to a tropical ecosystem was studied by applying oil to separate areas containing seagrass, mangrove, and coral habitats. One site was treated with a dispersant, Corexit 9527, and the sites were monitored over a 2-year period. It was found that in the short-term, the chemically dispersed oil caused the number of inverte-brates, including corals, to decline but the effects disappeared over the long term. Fresh, untreated oil had several long-term effects on the survival of the mangrove and associated fauna, but only minor effects on seagrasses, corals, and associated organisms.

Thorhaug and coworkers tested the toxicity of dispersants to corals, mangrove, seagrasses, and selected Jamaican fish species [89, 90]. It was found that the mortality of mangroves exposed for 10 hours to 1250 ppm of dispersed oil varied from 0 to 80% with 20% being a typical value. The other species were exposed to 125 ppm of dispersed oil for 6 hours. Mortality for the fish ranged from 0 to 100%, with 100% being the most typical value. The corals and seagrasses showed similar mortality trends at the same concentrations.

When the fate and effects of dispersants and oil were studied in a freshwater lake, it was found that dispersants reduced the overall impact of the oil by reducing the adhesion of the oil to grasses and reeds around the lake [91].

The effects of oil, dispersant and oil, and dispersant were examined at a salt marsh in Nova Scotia [92]. After 4 years, it was concluded that the effects of the oil ranged from minimal but persistently negative in the creek-edge and high-marsh zones, to negative but short-lived in the mid-marsh zone. The effects of the dispersant ranged from slightly positive in the creek-edge to acutely negative but short-lived in the mid- and high-marshes. The effects of the oil and dispersant together ranged from slightly positive in the creek-edge to slightly negative in the mid- and high-marsh zones.

In a project carried out in an intertidal zone in Maine, oil was released into a cove and its fate monitored over a 2-year period [93]. It was concluded that the dispersed oil had less impact than undispersed oil.

A major project to evaluate the fate, effects of, and countermeasures for dealing with oil spilled in the Arctic was conducted over several years beginning in 1980 [94]. Dispersants were applied in one bay and the fate of the oil and effects on the ecosystem monitored for several years. The dispersed oil narcotized benthic life in the first day, however, within one week, the bay recovered. The oil was deposited in sediment to a degree, but was largely carried away and traces could be found in sediment up to 2 km away, the furthest extent of sampling. The fate of oil in this bay was compared to that of an adjacent bay in which the same amount of oil was released and recovered to some extent with skimmers and the remainder left. The remaining oil also diminished with time and had little impact.

Another group of researchers assessed the use of dispersants in cold water in a large outdoor test vessel [95]. Several series of tests were conducted using Forties crude oil and a dispersant composed of Brij 92 and Brij 96 in salt water at $-1.6\,^{\circ}$C. The effectiveness of the dispersant was greatly reduced compared to when used in warmer water. The biodegradation of the dispersant-treated oil was reduced compared to that of the control and the population of heterotrophs initially decreased in all tanks, but soon recovered.

Use of Dispersants.   The use of dispersants remains a controversial issue. In most jurisdictions, special permission is required to use dispersants. In other jurisdictions, their use is banned. In Canada, special permission is required from Environment Canada, through the Regional Environmental Emergencies Team (REET) [96]. Similarly, in the United States, special permission to use dispersants is required from the U.S. Environmental Protection Agency (U.S. EPA) [97] and in waters close to shore, permission is also required from the State. In both Canada and the U.S., products must pass standard test procedures for toxicity and effectiveness before they can be used. Only about 5 dispersants of approximately 30 proposed products are approved for use (Table 2).

Countries that allow dispersants to be used do so only outside of a specific distance from shore. The toxicity associated with the use of dispersants is caused primarily by the dispersed droplets. As there is not enough capacity for dilution closer to shore, in most countries where dispersants are allowed, they can only be used at a specific distance, i.e., 1 to 10 km, from shore. Many jurisdictions also specify that the water must be at least 10 m deep before dispersants can be used which means that dispersants are generally used in offshore ocean waters. Dispersants are seldom used in freshwater. Many countries restrict such use since freshwater is often a source of drinking water [71]. Formulations designed for use in saltwater are not effective in freshwater, although there are a few dispersants specifically formulated for freshwater.

The use of dispersants varies around the world and their use has been declining steadily. In Canada, dispersants were used freely in the late 1970s and early 1980s, however, almost all stockpiles and equipment have now been sold. It is thought that dispersants were last used in Canada in about 1984. In the United States, dispersants have been used about twice per decade in the last 30 years. Dispersants have rarely been used in North America in the past 10 years [71].

Dispersants are not used in Europe, except for Great Britain and occasionally in France and potentially in Norway, Italy and Spain. Britain has the longest history of dispersant use and has well documented procedures for use [98]. Dispersants are occasionally used in Africa, especially in South Africa and Nigeria. In the Middle East, dispersants are used around the Arabian Gulf. India permits the use of dispersants and in the Far East they are used in Singapore, Malaysia, and Indonesia. Japan permitted dispersant use in the past. Australia allows the use of dispersants. Countries that have used dispersants in the past are listed in Table 9 (data adapted from reference 71).

An analysis of the potential for the use of dispersants in United States waters was conducted by Kucklick and Aurand [99]. Data was obtained on 138 refined and 69 crude oil spills. Using the given criteria, it was found that dispersants could have been used at approximately 25% of the large spills of crude oil and 7% of the large spills of refined oil that have occurred in the past 20 years. This compares to about 45% of the same sample of crude oil spills and 25% of spills of refined products at which burning could have been conducted. This would indicate that dispersants are rarely used in the United States on spills at which conditions are appropriate for their use.

In Canada and the United States, dispersant use remains a trade-off between toxicity to aquatic life and saving birds and shoreline species. Unfortunately, dispersants are never 100% effective and both surface and aquatic life may be affected by a spill even if it is treated. It has been shown that oil that is treated with dispersants but does not disperse is less adhesive to surfaces such as shoreline or birds, than oil that is untreated, although there are not many situations where this would be a benefit.

Information available in the literature on the use of dispersants over the last 30 years is provided in Table 10. The table was largely based on a recent survey [39, 100–109]. Some large scale uses were probably not documented or noted in the literature, while many uses, even of a smaller nature, are not documented in North America. The rate of reporting can be estimated from the use situation in Canada. It is known by the author that dispersants were used on a small scale at 23 spills on the east coast of Canada in the 1980s, although not one of these uses appears in the literature cited in Table 10. The applications listed in Table 10 therefore probably represent the larger applications and only a few smaller applications.

From the information in Table 10, the following conclusions can be drawn about dispersant use.

1. Given that the table is as complete as it can be, either dispersants are used infrequently or their use is poorly documented. Information on their use is certainly sketchy in the literature.
2. The large scale use of dispersants appears to be declining.
3. Only a few countries have used dispersants on a large scale in the past.

Table 9. Dispersant Usage as a Cleanup Option
(data adapted from reference 71)

Primary Option	Secondary Option		Last Resort Option
Angola	Albania	Kuwait	Finland
Bahamas	Algeria	Latvia	Ireland
Brazil	Anguilla	Malaysia	Jamaica
Brunei	Antigua/Barbuda	Malta	Rep. Korea
Bulgaria	Argentina	Mexico	Lithuania
Cameroon	Aruba	Monaco	Mauritius
China	Australia	Montserrat	Portugal
Congo-Brazz.	Bahrain	Morocco	Samoa (West.)
Croatia	Bangladesh	Neth. Antilles	Sweden
Cyprus	Barbados	New Zealand	Venezuela
Dominican Rep.	Belgium	Nigeria	
Ecuador	Bermuda	Norway	**Unlikely Option**
Egypt	Canada	Oman	Gambia
Eritrea	Cayman Is.	Philippines	Georgia
Falkland Is.	Chile	Poland	Guatemala
Fiji	Colombia	Qatar	Guinea
Gabon	Costa Rica	Russian Fed.	Liberia
Grenada	Côte d'Ivoire	Senegal	Marshall Is.
India	Cuba	Sierra Leone	Micronesia
Kenya	Denmark	Spain	Palau
Kiribati	Estonia	Tanzania	Panama
Lebanon	France	Trinidad/Tobago	Peru
Mozambique	French Antilles	Tunisia	St. Kitts/Nevis
Namibia	French Guiana	Turkey	St. Vincent/Grenadines
Pakistan	Germany	United States	Sudan
Papua New Guinea	Ghana	Vietnam	Syria
Saudi Arabia	Greece		Ukraine
Seychelles	Guinea Bissau		Uruguay
Singapore	Guyana		Yemen
South Africa	Hong Kong		
Sri Lanka	Iceland		**Prohibition**
Taiwan	Indonesia		Fr. Polynesia
Thailand	Israel		Greenland
Turks & Caicos Islands	Italy		Netherlands
United Arab Emirates	Japan		Romania
United Kingdom	Jordan		

Table 10. Use of Dispersants on Spills – 1966–1998

Country	Date	Location	Name	Volume of Oil	Oil Type	Dispersant Used	Effectiveness	Reported Effects	References Cited
United States	1998	Offshore	Undersea pipeline	115–355 tons	light crude	Corexit 9527 – 11.4 tons	effective	no reports	[101] OSIR 29 Jan 98
United States	1998	Offshore	Red Seagull	64 tons	light crude	Corexit 9500	effective	no reports	[101] OSIR 29 Jan 98
United States	1998	Offshore	Undersea pipeline	71 tons	medium crude	Corexit 9527	effective	no reports	[101] OSIR 22 Jan 98
United Kingdom	1997	Offshore	Production Platform	685 tons	North Sea crude	Corexit 9500	somewhat	no reports	[101] OSIR 16 Oct 97
Singapore	1997	Offshore	Evoikos/Orapin Global	25–30 k tons	heavy fuel oil	various	poor	no reports	[101] OSIR 16 Oct 97
Japan	1997	Offshore	Nakhodka	27,000 tons	medium fuel oil	various – limited amounts	ineffective	no reports	[101] OSIR 17 Jul 97
Korea	1997	Nearshore	No. 3 O Sung	500–1200 tons	Bunker C	various – 112 tons	ineffective	no reports	[101] OSIR 24 Apr 97
Uruguay	1997	Offshore	San Jorge	4,500 tons	medium crude	various	ineffective	no reports	[101] OSIR 13 Feb 97
United Kingdom	1996	Estuary	Sea Empress	19 million gal	Forties crude oil	Corexit, Dasic – 444 tons	unknown	unknown	[102]; [105]
South Korea	1996	Offshore	Hang Chang No. 8	100,000 gal	fuel oil	unknown	unknown	unknown	[101] OSIR, issue XIX[25]
Greece	1996	Estuary (dock)	Kriti Sea	>10,000 gal	Arabian light crude?	unknown	some oil hit beaches	unknown	[101] OSIR, issue XIX/31
South Korea	1996	Estuary (at pier)	Barge Yung Jung No. 1	unknown	Bunker C and marine diesel	3200 gallons – dispersant	no reports	unknown	[101] OSIR, issue XIX/32
Singapore	1996	Nearshore	unknown	unknown	unknown	Chemkleen and Corexit 9500	effective	unknown	[101] OSIR, issue XIX/32
Australia	1995	Nearshore	Iron Baron	88,000 gal	heavy fuel	BP AB & Androx 6120	unknown	unknown	[102], [101] OSIR 13 Jul 95
Korea	1995	Nearshore	Sea Prince	220,000–412,000	Bunker C	unknown	unknown	unknown	[101] OSIR 3 Aug 95
Korea	1995	Nearshore	Honam Sapphire	294,000 gal	Arabian heavy	unknown	unknown	unknown	[102]
United States	1995	Offshore	West Cameron 198	500–700 bbls	light natural gas condensate	unknown	unknown	unknown	[102]
Thailand	1994	Offshore	Visahakit	106,000 gal	diesel oil	unknown	unknown	unknown	[102]
South Africa	1994	Offshore	Sunken vessel	Leaking oil	unknown	Chemserve OSE 750	unknown	unknown	[102]
Singapore	1994	Offshore	Honam Pearl	3000–30,000 gal	slop oil	Shell VDC & Corexit 9527	unknown	unknown	[102]
South Africa	1994	Nearshore	Apollo Sea	763,000 gal	fuel oil, gasoline (49,000 gal)	Chemserve OSE 750	unknown	unknown	[103]

Country	Year	Location	Name	Amount	Oil type	Dispersant			Reference
India	1994	Embay-ment?	*Maharishi Dayanand*	30,000 gal	crude oil	unknown	unknown	unknown	[102]
United Kingdom	1993	Nearshore	*Braer*	25 million gal	Gulfaks crude	Dasic-LTSW (95 tons), Dispolene-345 (15 tons),	ineffective	no reports	[102]
India	1993	Offshore	*Maersk Navigator*	unknown	unknown	unknown	unknown	unknown	[102]
Argentina	1993	Bay/Estuary	Two tankers	10,000 gal/6000 gal	crude oil	unknown	unknown	unknown	[102]
India	1993	Offshore	Pipeline	1.5–1.8 million gal	crude oil	unknown	unknown	unknown	[102]
Japan	1993	Bay/Estuary	*Taiko Maru*	100,000 gal	heavy oil	unknown	unknown	unknown	[102]
Korea	1993	Bay/Estuary	Barge *Keumdong No. 5*	376,000 gal	heavy fuel oil	unknown	unknown	unknown	[102]
Korea	1993	Offshore	*Sambo No. 11*	1200 gal	bilge	unknown	unknown	unknown	[102]
France	1993	Nearshore	*Lyria*	800,000 gal	Iranian/Saudi heavy crude	unknown	no reports	no reports	[103]
South Africa	1993	Nearshore	Fishing trawler	32,000 gal	diesel and blended fuel	unknown	unknown	unknown	[103]
Greece	1992	Offshore	*Geroi Cheromorya*	400,000 gal	light crude	unknown	unknown	unknown	[102]
United States	1992	Offshore	Cook Inlet spill	8000 gal	medium crude	Corexit 9527	ineffective	no reports	[101] OSIR 9 Jan 92
United Kingdom	1992	Nearshore	Russian factory	8000 gal	diesel oil	Enersperse 1583	unknown	unknown	[102]
Australia	1992	Estuary?	*Era*	87,000 gal	light crude	unknown	unknown	unknown	[102]
Greece	1992	Offshore	Unknown slick	500 × 3.2 km	unknown	unknown	unknown	unknown	[102]
Egypt	1992	Nearshore?	*Soheir*	21,000 gal	fuel oil	unknown	unknown	unknown	[102]
Argentina	1992	Estuary	*President Arturo Umberto Ulia*	184,000 gal	crude	unknown	unknown	unknown	[102]
Nevis (France)	1991	Nearshore	Barge *Vista Bella*	550,000 gal	No. 6 fuel oil	Finasol OSR-7, OSR-52	variable reports	unknown	[104]; [106]
Australia	1991	Nearshore	*Sanko Harvest*	150,000–180,000 gal	fuel oil	BP A-B, Arderox B1 20	unknown	unknown	[104]
Japan	1991	Estuary	*Scan Alliance*	150 × 50 mile slick	heavy fuel oil	unknown	unknown	unknown	[102]
Australia	1991	Offshore	*Kirki*	6 million gal	light crude oil	unknown	unknown	unknown	[102]

Table 10. (*Cont.*)

Country	Date	Location	Name	Volume of Oil	Oil Type	Dispersant Used	Effectiveness	Reported Effects	References Cited
Japan	1991	Offshore	*Aiko Maro*	34,000 gal	diesel	unknown	unknown	unknown	[102]
United Kingdom	1991	Estuary	Pipeline in Thames River	unknown	crude oil	unknown	unknown	unknown	[102]
Japan	1991	Offshore	South Korean container ship	17,000 gal	light fuel oil	unknown	unknown	unknown	[102]
United Kingdom	1990	Estuary	*British Resolution*	minor	crude	unknown	no information	no reports	[104]
United Kingdom	1990	Nearshore	*Rosebay*	1100 tonnes	Iranian light crude	110 tonnes of concentrate	75% dispersed	no reports	[31]
United Kingdom	1990	Nearshore	*Kondor*	not specified	diesel and lubes	unknown (6 tons)	appeared successful	no reports	[104]
Greece	1990	Bay	*Nalkratis*	1200 gal	Iraqi crude oil	unknown	unknown	no reports	[104]
United States	1990	Offshore	*Mega Borg*	45–150 tons (12,000–40,000 gal)	light crude	Corexit 9527	effective	minimal	[104]
Greece	1990	Nearshore	*Happy Leader*	large amounts	crude oil	unknown	unknown	no reports	[104]; [106]
Greece	1990	Offshore	unknown	unknown	unknown	unknown	partially successful	minimal	[104]
Spain	1990	Offshore	*Sea Spirit/ Hesperus*	11,400 tons (3 million gal)	fuel oil	unknown	appeared successful	no reports	[104]; [106]
Hong Kong	1990	Nearshore	Barge *Hoi Fung*	100 tons	waste oil	unknown (26,000 gal)	appeared successful	no reports	[104]
United Kingdom	1990	Estuary	Barge *Portfield*	large amounts	medium fuel oil	unknown (15–20 tons)	unknown	unknown	[104]
UAE	1989	Nearshore	*Tropical Lion*	unknown	unknown	unknown	no information	no information	[104]
Japan	1989	Offshore	*Otake Maru/ Taiho Maru*	unknown	diesel fuel	unknown (700 L)	no information	no information	[104]
United States	1989	Nearshore	*Exxon Valdez*	11 million gal (258,000 bbls)	Alaska North Slope crude	Corexit 9527	ineffective	no reports	[30]
United Kingdom	1989	Nearshore	*Phillips Oklahoma/ Fiona*	800–900 tons	British Maureen crude	Dasic Slickgone LTSW (70 tons)	appeared successful	no reports	[104]

Country	Year	Location	Name	Amount	Oil type	Dispersant	Effectiveness	Reports	References
Japan	1989	Nearshore	*Mansion Trader*	minor	fuel oil	unknown	unknown	no reports	[104]
Uruguay	1989	Bay	*Presidente Rivera*	90,000 gal (340 tons)	fuel oil	unknown	unknown	no reports	[104]
United Kingdom	1989	Estuary	*Texaco Westminster*	15,000 gal (57 tons)	light fuel oil	Enersperse	unknown	no reports	[104]
United Kingdom	1988	Offshore	Piper Alpha Platform	unknown	crude	unknown	successful on small patch	unknown	[104]
United States	1987	Offshore	*PacBaroness*	30 bbls/day	possibly diesel	Corexit 9527	appeared successful	no reports	[104]; [106]
Panama	1986	Nearshore	Texaco Refinery	20,000–30,000 bbls	mixture, Mexican & Venezuelan	Corexit 9527 (72 drums)	effective ?	no reports	[104]; [106]
Singapore	1986	Nearshore	Bunkering barge	<100 metric tons	unknown	locally manufactured	unknown	unknown	[104]
United States	1984	Offshore	*Puerto Rican*	100,000 bbls	lube oil & additives	Corexit 9527 (2000 gal)	initially effective	no reports	[104]; [106]
South Africa	1983	Offshore	*Castillo de Bellver*	160,000–190,000 tons	unknown	Solvent-based & concentrates	unknown	no evidence	[104]
United Kingdom	1983	Estuary	*Sivand*	6000 tons	Nigerian Forcados crude	BP 1100, Dasic, Disp. 34S (28,500 gal)	estimated 1/6 to 1/2	no reports	[104]; [106]
Ireland	1979	Bay	*Betelgeuse*	3–5 tons per day after fire	Saudi Arabian crude	BP 1100 WD	effective	no reports	[104]; [106]
Denmark	1979	Nearshore	*Thuntank 3*	2800 bbls (400 tons)	heavy fuel oil	unknown	start only	no reports	[104]
South Korea	1979	Bay	*Continental Friendship*	unknown	Bunker C	unknown	no information	no information	[104]
Egypt	1979	Bay (Suez Canal)	*Skyron II*	14,000 bbls (2264 tons)	crude	unknown	no information	no information	[104]
Greece (Crete)	1979	Nearshore	*Messiniaki Frontis*	35,000–70,000 bbls (5000–10,000 tons)	Libyan crude	unknown	no information	no information	[104]
Mexico	1979	Offshore	*Ixtoc I – platform*	20,000 bbls/day (Total 3,750,000 bbls)	crude oil	Corexit 9527	questionable	no reports	[104]

Table 10. (*Cont.*)

Country	Date	Location	Name	Volume of Oil	Oil Type	Dispersant Used	Effectiveness	Reported Effects	References Cited
United States	1978	Nearshore	Barge *Pennsylvania*	881 bbls (126 tons)/143 bbls (20 tons)	No. 6 fuel oil/No. 2 fuel oil	Corexit 9527 (2000 gal)	effective	unknown	[104]; [106]
United Kingdom	1978	Nearshore	*Christos Bitas*	14,500–22,000 bbls (2000–3000 tons)	Iranian crude	several types (70,000 gal)	poor	no reports	[104]
United Kingdom	1978	Offshore	*Eleni V*	56,000 bbls	heavy fuel oil	6800 bbls of BP1100D and 10% Dasic LTD	ineffective	no reports	[6]
Saudi Arabia	1978	Nearshore	Hasbah 6 platform	105,000 bbls	heavy crude oil	unknown	ineffective	no reports	[6]
France	1978	Nearshore	*Amoco Cadiz*	220,000 tons	heavy crude oil	various (2500 tons)	ineffective	no reports	[6]
Spain	1976	Nearshore	*Urquiola*	100,000 tons	heavy oil	various (2400 tons)	ineffective	some reports	[39]
United Kingdom	1975	Offshore	*Olympic Alliance*	2000 tons	Iranian light crude	BP 1100X & Dasic LT 2 (200 tons)	some noted	no reports	[104]
Portugal	1975	Nearshore	*Jakob Maersk*	88,000 tons	heavy oil	various (110 tons)	ineffective	no reports	[39]
Singapore	1975	Nearshore	*Showa Maru*	15,000 tons	heavy oil	various (500 tons)	ineffective	no reports	[39]
Spain	1971	Bay	*Polycommander*	100,000 bbls (14,500 tons)	crude	Corexit 7664	effective	no evidence	[104]
United States	1970	Nearshore	*Delian Apollon*	unknown	No. 6 fuel oil	Corexit 8666 and 7664	no reports	no reports	[104]
United States	1970	Offshore	Chevron Main Pass Block 41	35,000–65,000 bbls (5000–9300 tons)	COM crude	mostly Corexit 7664 (2000 drums)	unknown	no evidence	[104]; [106]
Canada	1970	Nearshore	*Arrow*	5000 tons	Bunker C	various – 12 tons	ineffective	no reports	[39]
United Kingdom	1970	Nearshore	*Pacific Glory*	6300 tons	heavy oil	various	mixed reports	mixed reports	[39]
Saudi Arabia	1970	Bay (Tarut)	Pipeline Spill	unknown	light Arabian crude	unknown	unknown	no evidence	[104]
South Africa	1969	Offshore	*World Glory*	322,000 bbls (46,000 tons)	Kuwait crude	unkown up to 20,000 gallons/day for 20 days	no reports	no reports	[104]; [106]
United States	1969	Nearshore	Barge *Florida*	175,000 gal (550 tons)	No. 2 fuel oil	unknown	not effective	severe shore impacts	[109]

Country	Year	Location	Ship/Well	Amount	Oil type	Treatment	Effectiveness	Effects	Refs
United States	1969	Nearshore	Well A-21, Santa Barbara	77,000 bbls (12,000 tons)	Santa Barbara crude	ARA Gold Crew Bilge Cleaner (37,500 gal)	no estimates	no reports	[104]
Bahamas	1968	1 km offshore on coral reef	General Colocotronis	19,370 bbls (2600 tons)	Bunker C	Corexit 7664, Magnus Oil Spill Disperser, Ameroid	reportedly worked	none?	[104]; [106]
Puerto Rico	1968	Nearshore	Golden Eagle	12,000 tons	heavy oil	various – 60 tons	ineffective	no reports	[39]
South Africa	1968	Nearshore	Esso Essen	105,000 bbls (115,000 tons)	Arabian heavy	Corexit 7664 (15.6 m^3)	unknown	no benthic effects	[104]
United Kingdom	1967	Nearshore	Torrey Canyon	105,000 bbls	Kuwait crude oil	10,000 tons of toxic degreasers/detergents	variable reports	intertidal mortalities	[6]
Germany	1966	Offshore – Elbe	Anne Mildred Brovis	118,900 bbls	Iranian crude	Moltoclar, Asca Super, A-11, Slix/Navee, Gamlen, BP-1002, Ameroid-Drew	mixed reports	none?	[104]; [71]

4. Most applications of dispersants do not involve assessment of their effectiveness.
5. Dispersants do not appear to have been used on freshwater in the past.

Dispersants are used so infrequently in some locations that stockpiles are sometimes in place for as long as 20 years before use. In Great Britain, the government studied the aging of dispersants by testing effectiveness [110]. The laboratory performed both long-term tests and short-term accelerated tests in various containers and at 20 and 30 °C. The tests did not show dispersant deterioration as indicated by the effectiveness values.

Application of Dispersants.   To be effective on an oil spill, dispersant must be applied as soon as possible after the spill before the slick thins out too much or the oil weathers excessively. Thin slicks have been found to disperse poorly as do highly weathered oils. Dispersants were first applied on oil spills using boat-mounted spray systems. Early in the 1970s, however, it was realized that such systems, usually with a spray width of about 10 m, did not provide adequate coverage of a spill. Large systems for larger boats and small ships were developed, but the dispersant had to be mixed with seawater, requiring extra equipment to control dilution and application rates. Use of smaller vessels for applying dispersant is not common today.

Aerial spraying, which is done from small and large fixed-wing aircraft as well as from helicopters, is the most popular application method today. This method allows the dispersant to be applied neat which is thought to be best because, when diluted in water, dispersants may not repartition to the oil phase and could be lost to the water column. The other benefit of aerial spray systems is that they have the potential to cover a large area. The aerial coverage that can be achieved from various types of aircraft as well as from different sized vessels is shown in Table 11 (data summarized from reference 111). As can be seen, the best coverage is obtained when spraying from a Hercules transport aircraft.

It is important to note that dispersant must be applied with a system designed specifically for that purpose. For example, the spray volume of pesticide spraying equipment is generally 10 to 50 times less than a dispersant spray system. In addition, pesticide spraying equipment is designed to apply pesticide as a fine spray or mist with droplet sizes of about 50 to 200 μm. It was thought that dispersants are best applied in larger droplets, 400 to 700 μm in size, so that the droplets will not blow away and an adequate amount of dispersant will be deposited onto the oil slick.

The essential elements in applying dispersant are to supply enough dispersant to a given area in droplets of the correct size and to ensure that

Table 11. Dispersant Spray Equipment

Platform	Typical Characteristics		
	Dispersant Load (L)	Coverage per hour (Ha)	Coverage per day (Ha)[a]
Small boat	1000	10	80
Small ship	3000	20	160
Supply ship	10,000	30	240
Small helicopter	700	170	280
Large helicopter	2000	280	800
Agriculture spray plane	400	170	270
DC-3	4500	540	2400
DC-4	8000	840	4800
DC-6	11,000	1010	7330
C-130 (Hercules)	13,000	1010	8670

[a] Presuming a working day of 8 hours and typical sorties 50 km from base, and a target dosage of 15 L/Ha

the dispersant comes into direct contact with the oil. Several tests of aerial application systems have been carried out in the past [6] and a new set of studies has recently been conducted. Giammona and coworkers report on tests of aerial application of dispersants to determine the effectiveness of applying dispersants at different altitudes, in varying flows, and several other parameters [112, 113]. It was found that the important factors affecting the effective application of dispersants are: altitude, wind speed and direction, major droplet characteristics, rate of flow of dispersant, boom configuration, deposition, and the width and patterns of the spray swath. Based on several tests of equipment, including the ADDSPAK (an aerial application package designed specifically for dispersants) in a Hercules and a DC-3, a U.S. Air Force MASS (systems for general spray purposes) system in a Hercules and an Air Tractor, and a DC-4 spray system, it was concluded that, while altitude did not change the amount of dispersant applied, it did affect the spray pattern to a certain extent. The deposited droplets were sometimes in the desired range of 400 to 700 μm, although many of the flights did not result in droplets in that size range. It was also found that many of the flights did not achieve the targeted 5 gpa (gallons per acre) dosage on the ground.

Past studies on the desired droplet size were based on the premise that droplets ranging from 400 to 700 μm in size were best for an effective application [114, 115]. The impact of droplets of this size range on oil spills has not been determined. Oil spills spread rapidly and the oil droplets in the slick are often in this size range or smaller within hours. It is known that slicks often spread out to a 100 μm thickness within hours of the spill [116]. Earlier researchers of aerial trials have suggested that the mean droplet diameter should be about half of the slick thickness [117].

Another important consideration when applying dispersants is the concept of "windows-of-opportunity", that is appropriate time periods after a spill during which conditions are correct for the use of dispersants. One group of researchers estimated such windows-of-opportunity for dispersant use, based primarily on viscosity data and laboratory tests [118]. It was found that the prime window-of-opportunity for the use of dispersants on Alaska North Slope oil was during the first 26 hours after a spill. It was estimated that a reduced effectiveness window would last from 26 to 120 hours after the spill. After this time, the oil would not be considered "treatable". The most effective time for dispersing Bonnie Light crude oil was during the first 2 hours after the spill, and the time period for reduced dispersibility was 2 to 4 hours. As above, the oil would be considered untreatable after 4 hours. The formation of water-in-oil emulsions was a factor and it was suggested that the treatment time could be increased by using emulsion-breaking agents.

Various methods of spraying dispersant are shown in Figures 9 to 12.

Figure 9.   Close-up of a nozzle system on a boat adapted for spraying dispersant.

Figure 10. DC-4 aircraft spraying dispersant during IXTOC incident in 1979.

Figure 11.   The dispersant spray system, ADDSPAK, in operation from a Hercules aircraft.

Figure 12.   A helicopter bucket spray system in operation.

Surface-Washing Agents or Beach Cleaners

Introduction. Surface-washing agents or beach cleaners are formulations of surfactants designed to remove oil from solid surfaces such as shorelines and beaches. Since they are intended to remove oil rather than to disperse it, surface-washing agents contain surfactants with higher hydrophilic–lipophilic balance (HLB) than those in dispersants. Most surface-washing agents are formulated not to disperse oil into the water column, but to release oil from the surface where it floats. Surface-washing agents are a recent phenomenon and are still in the development stage. Agents have only been classified as surface-washing agents rather than dispersants in the past 5 years, with most of the newer products promoted after the *Exxon Valdez* spill in 1989. Before that, dispersants were assessed on shorelines, with mixed results [119]. In the oil spill industry, the new specially designed products are still called dispersants by some.

As with dispersants, effectiveness and toxicity are the main issues with surface-washing agents, although the level of concern is not as great. There are several reasons for this. Firstly, surface-washing agents have not been used on a large scale anywhere in the world. Unlike dispersants, they are not a universally applicable agent, but are used in specific cases of supratidal or intertidal oiling. Secondly, no adverse incidents have occurred using surface-washing agents, such as the killing of aquatic life when dispersants were used after the *Torrey Canyon* spill [71]. Finally, many surface-washing agents are relatively effective and much less toxic than dispersants. The ability to remove oil from a surface appears to be easier than dispersing it from the sea surface. Furthermore, the surfactants used in surface-washing agents have inherently less fish toxicity than those used for dispersants.

There is some concern about whether surface-washing agents can result in appreciable amounts of dispersed oil. Some products currently listed as surface-washing agents do disperse the oil given moderate agitation or sea energies. If this occurs, the situation can be similar to that with dispersants.

At this time, the only product approved by Environment Canada as a surface-washing agent is Corexit 9580 from Exxon [96]. The U.S. Environmental Protection Agency has approved the following seven agents: Corexit 9580, Corexit 7664, Topsall #30, CN-110, Premier 99, Simple Green, and Aquaclean [97].

Formulations. Little information is available on specific formulations for surface-washing agents because the formulations vary extensively and many are not patented. Several basic types of formulations are:

1. Non-ionic or anionic surfactants with HLBs of more than 11 in a low-aromatic hydrocarbon solvent
2. D-Limonene in various solvents
3. Surfactants mixed with various solvents
4. Surfactants in glycol-type solvents similar to dispersants
5. Detergents with little or no solvent
6. Solvent mixtures

Several papers have been written on the development of surface-washing agents [119–123]. Many of the agents were developed after the *Exxon Valdez* spill in 1989. The following three products were tested on oiled shorelines resulting from the *Exxon Valdez* spill: Corexit 7664, Corexit 9580, and PES-51. Most products functioned and Corexit 9580 appeared to be very successful.

Application.   Surface-washing agents are applied directly on the beached oils and left to penetrate for at least 15 to 30 minutes [124, 125]. The oil is then flushed with water to remove it and direct it to a cleanup area. From there, the oil is generally removed with a conventional skimmer system. Since the surface-washing agents are typically applied to a small expanse of oil at the upper or intertidal zone, they are applied manually using hand-held or backpack sprayers. This is shown in Figure 13. It would be difficult to apply the agent using airborne spray

Figure 13.   Application of a surface-washing agent to an oiled beach during an experiment.

systems and much product would be lost. On beaches, the product must be applied during low tide and the oil removed before the tide rises and the oil is no longer accessible. No extensive research or testing of application methods for surface-washing agents have yet been done.

Effectiveness. *Field Trials.* Several tests of the effectiveness of surface-washing agents have been conducted at actual spills. The results of some of these tests are listed in Table 12 [*119, 122, 126–135*]. Effectiveness was not quantified in any of these field tests, however, in every case, except where dispersants were used in earlier years, the tests were declared to be successful. The earlier dispersant trials showed variable effectiveness and, where penetration was measured, showed that dispersants promoted penetration of the oil into the sub-sediments [*135*].

Little and Baker reported on field and laboratory studies of the use of dispersants in nearshore areas or on shorelines [*135*]. Tests showed that some dispersant treatments can increase penetration of oil into sediment and that the oil may be retained in the subsurface. The nature of the shoreline or sediment was the main factor determining whether the penetration was enhanced by dispersant. On some shorelines it was shown that natural removal can be enhanced by dispersant usage. It was also found that dispersant-enhanced toxicity of oil could pose a problem and suggested that work be done on defining an effective minimum dispersant-to-oil ratio.

While field evaluation methods have not been fully developed for surface-washing agents, field screening kits for evaluating both effectiveness and toxicity have been developed and tested. Clayton and coworkers reported on the development of test kits for evaluating the effectiveness and aquatic toxicity of surface-washing agents [*136–139*]. The test was evaluated using natural substrates including gravel, rock fragments, and eelgrass. It was concluded from laboratory tests that the field test would be an appropriate indicator of effectiveness in the field. Four field-applicable methodologies for testing the aquatic toxicity of surface-washing agents were tested, including the Microtox unit, echinoderm fertilization, byssal thread attachment in mussels, and righting and water-escaping ability in periwinkle snails. While all methodologies were able to detect differences in toxicity, the Microtox and echinoderm fertilization showed greater sensitivity and/or precision.

Laboratory Testing. Laboratory tests for surface-washing agents were first developed by Environment Canada [*140*]. After evaluating about 25 testing methods including troughs, surfaces and coupons in flasks, the trough was found to be the most appropriate. A close-up of the sloped-trough test is shown in Figure 14. A heavy oil such as Bunker C was placed on a small metal trough, agent applied, and then the oil was

Table 12. Use of Surface-Washing Agents and Major Field Tests

Country	Date	Location	Name	Volume of Oil	Oil Type	Agent Used	Effectiveness	References Cited
United States	1998	Alaska	*Exxon Valdez*	test only	North Slope	PES 51	not known	[126]
United States	1997	Maine	*Julie N.*	test only	Bunker C	Corexit 9580	50% removed	[127]
Uruguay	Mar-97	shoreline	*San Jorge*	test only		Corexit 9580	successful	[128] OSIR 6 Mar 97
Uruguay	Mar-97	shoreline	*San Jorge*	test only		Enviroclean	successful	[128] OSIR 6 Mar 97
United States	06-Oct-96	Maine	*Julie N.*	test only	Bunker C	Corexit 9580	varied	[128] OSIR, 3 Oct 96, 17 Oct 96
New Zealand	late 96	Wellington	*Sydney Express/ Maria Luisa*	8 tonnes	Diesel	OSD 9	successful	[128] OSIR, 5 Jun 97
United States	1994	Puerto Rico	*Morris J. Berman*	test only	Bunker C	Corexit 9580	successful	[129] [130]
United States	1994	Puerto Rico	*Morris J. Berman*	test only	Bunker C	PES 51	successful	[129] [130]
United States	1994	Texas	San Jacinto River	small amount	Crude	Corexit 9580	successful	[131]
United States	1994	Louisiana	oil marsh	small amount	Crude	Corexit 9580	successful	[131] [132]
United States	1993	Alaska	*Exxon Valdez*	test only	North Slope	PES 51	successful	[122] [134]
Great Britain	1987	Folkestone	test	test only	Fuel Oils and emulsion	dispersants	variable	[119]
Great Britain	1985–88	Wales	test	test only	Fuel Oils and crude	dispersants	variable	[135]
United States	1970	Florida	*Delian Apollan*	test only	Bunker C	Corexit 8666	variable	[129]

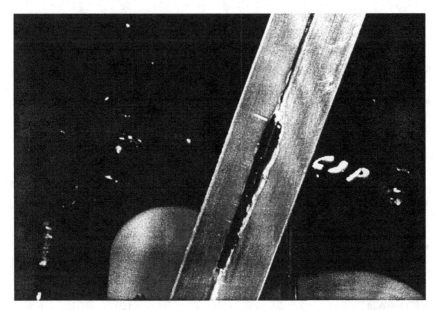

Figure 14.   Close-up of the sloped-trough test for evaluating surface-washing agents.

flushed away with water. Quantitation was by weight. The U.S. EPA subsequently evaluated a number of test methods and then evaluated several products with a trough test similar to that used by Environment Canada [141, 142]. The French government laboratory developed a small coupon test to screen products for acceptability [143].

A variety of agents, including dispersants, have been extensively tested by Environment Canada using the trough test [144, 145]. The results of some of these tests are shown in Table 13. Included in this table are effectiveness results from the trough test for both freshwater and salt water and effectiveness as a dispersant using the swirling flask test and Alberta Sweet Mixed Blend crude oil. These test results show that products which are effective as a dispersant are not effective as a surface-washing agent and vice versa. This effect, which was noted in previous studies, is thought to be due to the difference in HLB needed for a dispersant (HLB ~ 10) and for a surface-washing agent (HLB > 10) [47]. The table also includes household products and other products that are not intended for use on oil spills.

Toxicity.   The acute lethal toxicity of many surface-washing agents is shown in Table 13 [47, 49]. Unlike dispersants, the aquatic toxicity of surface-washing agents varies from nontoxic (> 1000 mg/L) to

Table 13.   Test Results for Surface-Washing Agents

Product	Percentage Removed from Trough		Aquatic Toxicity LC$_{50}$ to Rainbow Trout, 96 hours	Effectiveness as a Dispersant (%)
	Freshwater	Saltwater		
Corexit 9580	69	53	>10,000	0
D-Limonene	51	52	35	0
Penmul R-740	49	44	24	9
Limonene '0'	38	43	35	0
Formula 2067	41	39	11	0
Ecologic 5M10MB10	24	38	62	0
Citrikleen XPC	37	36	34	2
Formula 861	32	32	24	0
Oriclean	—	32	70	0
Corexit 7664	25	27	850	2
Pronatur Extra	19	25	9	0
Bioorganic	—	23	18	0
BP 1100 X AB	28	23	2900	0
BP 1100WD	30	21	120	6
Tesoro Pes 51	23	21	14	0
Ecologic BF-104	35	20	62	0
Champion JS10-232	27	20	1060	0
Corexit 7664/Isopar	17	20	1500	1
Biosurf	15	20	42	0
Champion JS10-242	27	19	380	<5
Tesoro Pes 41	22	19	9	0
Oil Gon	20	19	134	0
Pronatur	23	17	75	0
Re-Entry	17	17	8	0
Biocat 145	14	17	104	0
Sea Spray	26	16	420	0
Palmolive	14	16	13	9
Per 4m	14	16	566	0
Topsall	—	14	354	0
Breaker-4	17	13	340	0
Nokomis 3	13	13	110	0
Ecologic 5M5B4	11	13	46	0
Ortec	0	13	123	0
Sunlight	16	12	13	9
Inprove	14	12	78	0
Citrikleen 1855	14	12	55	0
Citrikleen FC1160	10	12	75	0
Con-Lei	8	12	70	0
Alcopol 60	—	11	62	18
Ecologic 10M10B10	19	11	23	0
Pyprr	12	11	650	0

Table 13. *(Cont.)*

Product	Percentage Removed from Trough		Aquatic Toxicity LC$_{50}$ to Rainbow Trout, 96 hours	Effectiveness as a Dispersant (%)
	Freshwater	Saltwater		
Bioversal	8	11	120	0
Oil Spill Eater	5	11	135	0
Icoshine	12	10	40	0
Ecologic BF-102	25	9	46	0
Envirosperse OSD	0	9	108	<5
Siallon Emulsifier	6	8	375	0
Ecologic BF-103	7	7	71	0
IDX 20	6	7	140	0
Mr Clean	13	6	30	0
Gran Control	5	6	75	0
Corexit CRX8	14	5	20	45
Formula 730	3	5	33	0
Corexit 9527	13	3	108	33
Tornado	8	3	1350	0
Balchip 215	8	3	157	0
Firezyme	4	3	521	0
Equisolve	0	3	60	0
Jansolve	25	2	57	0
Citrikleen 1850	24	2	18	11
Value 100	4	2	4250	0
Biosolve	2	2	9	0
Lestoil	9	1	51	0
Enersperse 700	1	1	50	51
Oil Dissolver	5	0	40	0
Petrotech	0	0	1460	0
Inipol EAP-22	0	0	17	9

highly toxic (<50 mg/L). Toxicity does not correlate with effectiveness. In fact, the most effective product noted in Table 13, Corexit 9580, is also the least toxic.

Shigenaka et al. found no adverse biological effects of Corexit 9580 during an application to a saltwater marsh [146]. Pezeshki and coworkers studied the effects of Corexit 9580 on seagrasses and also found no adverse effects [132, 133]. Similarly, Teas et al. studied the use of Corexit 9580 on mangroves and found benefits and no toxicity [147]. Hoff et al. reviewed PES-51, which consists primarily of D-Limonene and found its aquatic toxicity relatively high [122].

Emulsion Breakers and Inhibitors

Introduction.   As with dispersants and surface-washing agents, effectiveness and toxicity are the main issues with emulsion preventers and breakers. The level of concern is not as high, however, as these agents are not used on a large scale anywhere in the world. The issues overall are not as great as with dispersants for several reasons. Emulsion breakers are not universally applicable agents as dispersants are, but are used specifically with oils that would form emulsions. Also, no adverse effects have resulted from their use. There is little concern over the use of emulsion breakers in closed systems, that is when little water is present. There is concern, however, that some dispersion occurs when emulsion breakers are used and they could thus result in near-shore toxicity in a similar manner to dispersants and surface-washing agents.

At this time, Environment Canada has approved only one product as an emulsion-breaking agent, Vytac DM from Cartier Chemical [96]. The U.S. Environmental Protection Agency has not approved any emulsion breakers for use [97].

Formulations.   Emulsion breakers and inhibitors are formulated to break water-in-oil emulsions or to prevent them from forming. Some information is available on specific formulations of these agents, but the formulations vary extensively and many are not specifically patented. Several basic types of formulations are:

1. Sodium salts of sulfosuccinates
2. Polypropylene/polyethylene glycol block copolymers
3. Surfactants mixed with various solvents
4. Surfactants generally with an HLB of more than 15
5. Unspecified polymeric materials

Only three products, Gamelin EB439, Vytac DM, and Breaxit OEB-9, are specifically marketed for oil spills at this time [148]. Another product, Alcopol 60, has also been used extensively in field trials. Many products of this type are marketed for use in breaking emulsions that occur in petroleum production, but most have never been applied to oil spills [149]. Interestingly, it was found that a bacterial metabolite, later identified as acetoin, de-emulsified oil-in-water emulsions [150].

Effectiveness.   The effectiveness of emulsion breakers and inhibitors is measured as the minimum dose required to break a stable emulsion or prevent one from forming. There is as much concern about the effectiveness of these agents as there is with the effectiveness of dispersants.

Field Trials. Although no large-scale use of emulsion breakers has been reported, several large-scale field trials have been conducted [19, 21, 151–153]. These tests are summarized in Table 14. Some of these tests were done in conjunction with dispersant tests and conducted in a similar manner. The end-points and measurement techniques were generally not up to the current standards. In most cases, the amount of water in the oil was used, which is not a reliable end point because some water can be retained even though the emulsion is no longer stable. Researchers concluded that emulsion breakers can be used effectively on the open sea, although these results have not been quantified.

Countermeasures Tests. Emulsion breakers are used primarily before other spill countermeasures, such as skimming and burning, can be applied. In most cases, skimming and burning cannot be done if the oil is emulsified. Some products have been tested specifically in this regard. Lewis and coworkers conducted a series of tests with pumps (gear, lobe, and progressive cavity) to assess the capability of Alcopol 60 to break emulsions in closed systems [154]. The oil used was a heavy Bunker C blended with 31.5% (by volume) of diesel fuel. They found that a dose of 300 to 400 ppm by total weight of the incoming fluid was sufficient to break the emulsions passing through the pumps. Buist and coworkers tested several combinations of the oilfield emulsion breaker, EXO 0894, to break emulsions of Alaska North Slope oil before burning [155]. It was found that 500 to 5000 ppm of EXO 0894 was sufficient to break emulsions of less than 65% water, so that these would burn. Emulsions containing more water would not burn. These laboratory-scale tests also found that at least one hour of mixing time was often required after spraying with the emulsion breaker before the emulsion would break.

Physical Basis of Effectiveness. The most important characteristic of a water-in-oil emulsion is its "stability" because this must be known before the properties of the emulsion can be identified [156]. Properties change very significantly for each type of emulsion. The stability of emulsions has only been defined recently [157]. Therefore, studies were difficult because the end points of analysis were not defined. The following four "states" or ways that water can exist in oil have been defined: stable emulsions, mesostable emulsions, unstable emulsions (or simply water and oil), and entrained water [157]. These four "states" are identified by visual appearance as well as by rheological measures.

Recent studies have shown that certain emulsions can be classed as stable, characterized by their persistence over several months [156]. The viscosity of these stable emulsions actually increases over time until their viscosity is at least three orders-of-magnitude greater than the starting oil. The stability of these emulsions is derived from the strong visco-elastic interface caused by asphaltenes, perhaps combined with resins. Increased

Table 14. Field Trials of Emulsion Breakers

Location/Identifier	Reference	Year	Oil Type	Amount m³	Emulsion Breaker	Application Method	Rate A:0	Wind/Sea State	Effectiveness
North Sea – Great Britain	[19]	1992	Forties crude	12.3	LA 1834 then Dasic LTSW	airplane	1:100	4–5	good
			Forties crude	12.3	LA 1834	airplane	1:100	4–5	good
			Forties crude	12.3	Control			4–5	
North Sea – Great Britain	[151]	1991	MFO (medium fuel oil)	9.9	LA 1834 in 50% solvent	airplane	1:207	2–3	good
			MFO	10.2	Control			2–3	
			MF/GO (Medium Fuel Oil/ Gas Oil)	18.3	LA 1834 in 50% solvent	airplane	1:172	2–3	good
			MF/GO	14.7	Control			2–3	
			Forties crude	14	LA 1834 in 50% solvent	airplane	1:74	2–3	good
			Forties crude	15	Control			2–3	
			Forties crude	16	Control			2–3	
			MF/GO emulsion	28.5	LA 1834 in 50% solvent	airplane	1:75	2–3	good
			MF/GO emulsion	26.5	Control			2–3	
Canada – Atlantic	[152, 153]	1987	Bunker/crude mix	0.8	Demoussifier	pretreated	0:00	1–2	good
			Bunker/crude mix	0.8	Demoussifier	sprayer	0:00	1–2	good
			Bunker/crude mix	0.8	Control	control		1–2	
			Bunker/crude mix	0.8	Demoussifier	sprayer	250 ppm	1–2	good
			Bunker/crude mix	0.8	Demoussifier	sprayer	1000 ppm	1–2	good
North Sea – Haltenbanken	[21]	1985	topped Statfjord crude	12.5	Alcopol	premixed	250 ppm	1–2	good
			topped Federated crude	2.5	Control		—	1–2	

viscosity may be caused by increasing alignment of asphaltenes at the oil–water interface. These stable emulsions may be more difficult to break than less stable emulsions. These emulsions have been monitored for as long as 3 years in the laboratory [156].

Mesostable emulsions are probably the most commonly formed emulsions in the field. These emulsions are actually oil/water mixtures and have properties between stable and unstable emulsions [156]. Mesostable emulsions may lack enough asphaltenes to render them completely stable or contain too many de-stabilizing materials such as smaller aromatics. The viscosity of the oil may be high enough to stabilize some water droplets for a period of time. Mesostable emulsions may degrade to form layers of oil and stable emulsions.

Unstable emulsions are those that break down into water and oil rapidly after mixing, generally within a few hours. Some water, usually less than about 10%, may be retained by the oil, especially if the oil is viscous. Applying emulsion breakers to these oils will result in apparent success, but the emulsions will break down in time without the addition of these agents. Unstable emulsions do not exhibit an increase in viscosity with time and their viscosity is less than about 20 times greater than that of the starting oil.

Forced oscillation rheometry studies are the most accurate way of determining the type of emulsion. The visco-elastic properties are the simplest way to discriminate between the four types of water-in-oil states. The presence of significant elasticity clearly defines whether a stable emulsion has been formed. The viscosity by itself can be an indicator of the stability of an emulsion, although it is not necessarily conclusive, unless the viscosity of the starting oil is known. Colour is an indicator, but may not be definitive. Most stable emulsions are brown or reddish [157]. Some mesostable emulsions are brown in colour and unstable emulsions are always the colour of the starting oil. Water content is not an indicator of stability and is error-prone because "excess" water may be present. It should be noted that the water content of stable emulsions is greater than 70% and that unstable emulsions or entrained water-in-oil generally contain less than 50% water. Water content of an emulsion after a period of about one week is more reliable than the water content of the emulsion when it has first formed because the oil and water will separate in a less stable emulsion.

The differences between the four types of water-in-oil states are summarized in Table 15 [156]. It can be seen in the table that precise analysis by viscosity will provide information on the stability of the emulsion and that this can also be used as a test of emulsion breakers.

Laboratory Tests. Specific laboratory tests for emulsion breakers have only been developed recently [158]. Similarly, the identification

Table 15. Typical Properties for the Water-in-Oil States

	Stable Emulsions	Mesostable Emulsions	Entrained Water	Unstable
Appearance on day of formation	brown solid	brown viscous liquid	black with large droplets	like oil
Water content on first day, %	80	62	42	5
Appearance after one week	brown solid	broken, 2 or 3 phases	separated oil and water	like oil
Water content after one week, %	79	38	15	2
Stable time, days	>30	<3	<0.5	not
Starting Oil				
Density, g/mL	0.85–0.97	0.84–0.98	0.97–0.99	0.8–1.03
Viscosity, (mPa·s)	15–10,000	6–23,000	2000–60,000	2–5.1×10^6
Saturates, %	25–65	25–65	19–32	23–80
Aromatics, %	20–55	25–40	30–55	5–12
Resins, %	5–30	6–30	15–30	0–32
Asphaltenes, %	3–20	3–17	3–22	0–32
Asphaltenes/Resins	0.74	0.47	0.62	0.45
Properties on day of formation				
Average ratio of viscosity increase	1100	45	13	1
Properties after one week				
Average ratio of viscosity increase	1500	30	2	1

of the different emulsion stabilities has not been applied until recently. Testing has shown that mesostable emulsions and entrained water especially, can show apparent breakage with very little product. Since these results are artifacts of the lack of discrimination between starting emulsion states, the results where emulsion stability was not considered are questionable to say the least.

In the past, many emulsion tests used modified dispersant tests. Several of the older tests of emulsion breakers were reviewed for the petroleum industry [149, 159]. While most of the older tests reported positive results, the stability of the target emulsion was not measured or considered. In one series of tests, however, the "elasticity" of the emulsions was measured before and after treatment and this served as a test for effectiveness measurements [160]. These scattered, early tests did show, however, that the amount of emulsion-breaking agent required was very low compared to dispersants [159]. Many of the earlier investigators found that effectiveness was achieved at ratios as low as 1:1000 of agent to emulsion.

Testing was conducted with both specialty products and with dispersants. Brown et al. tested the emulsification behaviour of oils treated with dispersants (Corexit 9550 or 9527). They found that some oils took up water more readily after treatment with dispersants, e.g., Drift River crude, while others took up less, e.g., North Slope oil [161]. Tests after the *Exxon Valdez* spill showed that the dispersant Corexit 9527 was not effective in breaking the emulsions created on the seas [162]. This was also the case in subsequent tests in the laboratory with mesostable or stable emulsions, but a de-watering effect was noted with entrained water states [163].

Several more recent tests have been conducted. Krawezkh and co-workers studied demulsification using oil-soluble Exxon and Nalco products and concluded that interfacial tension is a significant factor and that strictly oil-soluble products might not be highly effective [164]. Several Norwegian tests were conducted and various aspects of the process reported [165, 166]. In one of the tests reported on, the researchers used a small test tube to test emulsions made from the weathered residue of a crude and the emulsion breaker, Alcopol 60. The researchers found that there was no difference in the doses of 250 and 500 ppm and that heat accelerated the breaking of the emulsion.

Four demulsifiers were studied by Bhardwaj and Hartland using water-in-oil emulsions made from Velden crude oil and water at 0.5 and 10% salinities [167]. The demulsification process was followed by monitoring the reduction in surface tension as well as the percentage of water separated. It was found that the reduction in surface tension was by itself not correlated with water reduction but the surface tension of water must be reduced by at least 25 dynes/cm before demulsification occurred.

The products tested were a high propylene oxide polymer of molecular weight 7500, a block copolymer of molecular weight 3700, a mixture of ethylene oxide–polypropylene oxide polymer of molecular weight 3100, a nonylphenol resin surfactant of molecular weight 2000, and a nonylbutyl phenol surfactant of molecular weight 400.

Fiocco and coworkers demonstrated the use of two dispersant apparatuses to test demulsifiers, the "WRASET", which uses the wrist-action tester or EXDET test, and the "ROFLET", which uses the rolling flask or Warren Springs test [168]. To demonstrate the test, an emulsion with Kuwait crude oil was used with the Exxon products, Breaxit OEB-9, 7877, 8150, and 8160, and the surfactant dioctyl sulfosuccinate. Effectiveness was measured by visually observing the amount of water/oil separation. A Sture blend crude was used to evaluate the same demulsifiers plus Ameroid 372101, Breaxit 4018, 711, 7125, 7128, and 7652, as well as Shell LA 1834, Alcopol 60, and a "demoussifier". Ameroid was found to be the best product, followed closely by Breaxit 4018, 7111, 4018, and then LA 1834.

Studies in recent years have emphasized that the physics and chemistry of water-in-oil emulsions dominate the development of effectiveness tests. Emulsions vary in stability, depending on the type of oil and the degree of weathering. These factors complicate the development of a test. Emulsions with low stability will break easily with chemical emulsion breakers [158]. Broken emulsions will form a foam-like material, called "rag", which retains water that is not part of the stable emulsion.

Analytical methods used to determine the final stability of the broken or unbroken emulsion have varied. It has been shown that measurements of water content yielded unreliable results [158, 163]. Viscosity measurements show correlation to emulsion stability and provide a more reliable measure of emulsion stability but special instrumentation is required [169, 170]. Simple viscometers without controlled shear or stress rates yield unreliable results.

An additional consideration is the action required of the product. It has been shown that some products will inhibit emulsification better than they will break an already formed emulsion [158]. It would therefore seem appropriate to have two types of tests for each of these functions. In addition, some emulsion breakers are used on the open sea, which is called an open system, and others are used where little water is present in conjunction with skimmers, tanks, and pumps, which is called a closed system. Thus a total of four types of tests would be appropriate to test all facets of water-in-oil emulsion-treating agents.

Environment Canada has evaluated two treating agents in tests that are designed to measure each of the four effectiveness regimes (an example is shown in Figure 15) and results are shown in Table 16 [158]. The table shows that different results were obtained with the same agents

Table 16. Summary Results of Emulsion Breaking and Inhibiting Tests

Action	System	O:W Ratio	Shaker/Device	Results (min. operative dose)
Breaking	Open	1:5000	New Brunswick	Vytac 1:300 Alcopol 1:200
	Closed	1:200	Burrell	Vytac 1:250 Alcopol 1:280
Inhibition	Open	1:25	Rotary	Vytac 1:6000 Alcopol 1:2000
	Closed	1:04	Blender	Vytac 1:7000 Alcopol 1:2500

Figure 15. One type of rotary shaker used to evaluate emulsion breakers.

in the four different tests. Much less agent is required to inhibit the formation of a water-in-oil emulsion than to break such an emulsion.

Toxicity. The sub-lethal or long-term toxicity of emulsion breakers has not been reported in the literature. The aquatic toxicity of Vytac-DM and Alcopol 60 was measured for a 96-hour exposure of the rainbow trout [158]. The LC_{50} for Vytac-DM was $> 10,000$ mg/L and for Alcopol 60 was 62 mg/L. This indicates that other products may show an equally wide range, from relatively non-toxic to relatively toxic.

Acknowledgments

The author acknowledges the United States Minerals Management Service and Environment Canada who supported much of his work in dispersants in the past 10 years. The author also acknowledges Jennifer Charles who assisted in editing the final version.

References

1. Becher, P. *Emulsions: Theory and Practice*, Robert E. Krieger Publishing Company, Huntington, NY, 1977.

2. *Guideline on the Use and Acceptability of Oil Spill Dispersants – Second Edition*, Environment Canada, Report Number EPS 1-EP-84–1, Ottawa, Ont., 1984, p 31.

3. *Oil Spill Treating Agents: A Compendium*, The American Petroleum Institute, API Report No. 4150, Washington, DC, 1972.

4. IT Corporation, *Use of Chemical Dispersants for Marine Oil Spills*, Environmental Protection Agency Report, EPA/600/R-93/195, Edison, NJ, 1993.

5. Walker, A.H.; Michel, J.; Canevari, G.; Kucklick, J.; Scholz, D.; Benson, C.A.; Overton, E.; Shane, B. *Chemical Oil Spill Treating Agents: Herding Agents, Emulsion Treating Agents, Solidifiers, Elasticity Modifiers, Shoreline Cleaning Agents, Shoreline Pre-treatment Agents and Oxidation Agents*, MSRC Technical Report 93-015, Marine Spill Research Corporation, Washington, DC, 1993.

6. *Using Oil Spill Dispersants on the Sea*, Marine Board, National Research Council, National Academy Press, Washington, DC, 1989.

7. Brandvik, P.J.; Daling, P.S. *Proceedings of the Thirteenth Arctic and Marine Oilspill Program (AMOP) Technical Seminar*, Environment Canada, Ottawa, Ont., 1990, pp 243–254.

8. World Intellectual Property Organization, "Chemical Dispersant for Oil Spills", patent applied for by Exxon, 1993.

9. Exxon Research Engineering, U.S. Patent 5,618,468, "Chemical Dispersant for Oil Spills", April 8, 1997.

10. Fingas, M.F. *Oil Dispersants: New Ecological Approaches, ASTM STP 1018*, L. Michael Flaherty, Ed., American Society for Testing and Materials, Philadelphia, PA, 1989, pp 157–178.

11. OSIR, *Oil Spill Intelligence Report*, Cutter Information Corporation, Arlington, MA, October 16, 1997.

12. Brandvik, P.J.; Strom-Kristiansen, T.; Lewis, A.; Daling, P.S.; Reed, M.; Rye, H.; Jensen, H. *Proceedings of the Nineteenth Arctic and Marine Oilspill Program (AMOP) Technical Seminar*, Environment Canada, Ottawa, Ont., 1996, pp 1395–1415.

13. Lunel, T.; Davies, L. *Proceedings of the Nineteenth Arctic and Marine Oilspill Program (AMOP) Technical Seminar*, Environment Canada, Ottawa, Ont., 1996, pp 1355–1392.

14. Lewis, A.; Daling, P.S.; Strom-Kistiansen, T.; Brandvik, P.J. *Proceedings of the Eighteenth Arctic and Marine Oilspill Program (AMOP) Technical Seminar*, Environment Canada, Ottawa, Ont., 1995, pp 453–469.

15. Lunel, T. *Proceedings of the Seventeenth Arctic and Marine Oilspill*

12. Fingas *Use of Surfactants for Environmental Applications* 529

Program (AMOP) Technical Seminar, Environment Canada, Ottawa, Ont., 1994, pp 1015–1021.
16. Lunel, T. *Proceedings of the Seventeenth Arctic and Marine Oilspill Program (AMOP) Technical Seminar*, Environment Canada, Ottawa, Ont., 1994, pp 952–977.
17. Lunel, T. *Proceedings of the 1995 International Oil Spill Conference*, American Petroleum Institute, Washington, DC, 1995, pp 147–155.
18. Lunel, T.; Davies, L.; Brandvik, P.J. *Proceedings of the Eighteenth Arctic and Marine Oilspill Program (AMOP) Technical Seminar*, Environment Canada, Ottawa, Ont., 1995, pp 603–627.
19. Lunel, T.; Lewis, A. *Proceedings of the Sixteenth Arctic and Marine Oilspill Program (AMOP) Technical Seminar*, Environment Canada, Ottawa, Ont., 1993, pp 955–972.
20. Swiss, J.J.; Vanderkooy, N.; Gill, S.D.; Goodman, R.H.; Brown, H.M. *Proceedings of the Tenth Arctic and Marine Oilspill Program (AMOP) Technical Seminar*, Environment Canada, Ottawa, Ont., 1987, pp 307–328.
21. Sørstrøm, S.E. *Proceedings of the International Seminar on Chemical and Natural Dispersion of Oil on the Sea*, Centre for Industrial Research, Trondheim, Norway, 1986.
22. Bocard, C. *Bulletin de Cedre*, No. 22, France, 1985, pp 6–10.
23. Bocard, C.; Castaing, G.; Dureux, J.; Gatellier, C.; Croquette, J.; Merlin, F. *Proceedings of the 1987 Oil Spill Conference*, American Petroleum Institute, Washington, DC, 1987, pp 225–229.
24. Lichtenthaler, R.G.; Daling, P.S. *Proceedings of the 1985 Oil Spill Conference*, American Petroleum Institute, Washington, DC, 1985, pp 471–478.
25. Nichols, J.A.; Parker, H.D. *Proceedings of the 1985 Oil Spill Conference*, American Petroleum Institute, Washington, DC, 1985, pp 421–427.
26. Swiss, J.J.; Gill, S.D. *Proceedings of the Seventh Arctic and Marine Oilspill Program (AMOP) Technical Seminar*, Environment Canada, Ottawa, Ont., 1984, pp 443–453.
27. Gill, S.D.; Goodman, R.H.; Swiss, J.J. *Proceedings of the 1985 Oil Spill Conference*, American Petroleum Institute, Washington, DC, 1985, pp 479–482.
28. Delvigne, G.A.L. *Sea Measurements on Natural and Chemical Dispersion of Oil*, Report No. M1933-1, Delft Hydraulic Laboratory, The Netherlands, 1983.
29. Bocard, C.; Ducreaux, J.; Gatellier, C. *Protecmar IV*, Report IFP 31478, Institut Français de Petrole, France, 1983.
30. Cormack, D. *The Use of Aircraft for Dispersant Treatment of Oil Slicks at Sea*, Marine Pollution Control Unit, Department of Transport, London, England, 1983.
31. Lichtenthaler, R.G.; Daling, P.S. *Oil Pollution Control: Research and Development Program*, PFO Projects No. 1406 and 1470, ISBN 82-7224-198-6, Trondheim, Norway, 1983.
32. Gill, S.D.; Ross, C.W. *Proceedings of the Fifth Arctic and Marine Oilspill*

Program (AMOP) Technical Seminar, Environment Canada, Ottawa, Ont., 1982, pp 255–263.

33. Bocard, C.; Gatellier, C. *Protecmar III*, Report IFP 30482, Institut Français du Petrole, France, 1982.

34. McAuliffe, C.D.; Steelman, B.L.; Leck, W.R.; Fitzgerald, D.E.; Ray, R.P.; Barker, C.D. *Proceedings of the 1981 Oil Spill Conference*, American Petroleum Institute, Washington, DC, 1981, pp 269–282.

35. Greene, D.R.; Buckley, J.; Humphrey, B. *Fate of Chemically Dispersed Oil in the Sea: A Report on Two Field Experiments*, Report No. EPS 4-EC-82-5, Environment Canada, Ottawa, Ont., 1982.

36. Smith, D.D.; Holliday, G.H. *Proceedings of the 1979 Oil Spill Conference*, American Petroleum Institute, Washington, DC, 1979, pp 475–482.

37. McAuliffe, C.D.; Johnson, J.C.; Greene, S.H.; Canevari, G.P.; Searl, T.D. *Environ. Sci. Technol.* **1980**, *14(12)*, pp 1509–1518.

38. Cormack, D.; Nichols, J.A. *Proceedings of the 1977 Oil Spill Conference*, American Petroleum Institute, Washington, DC, 1977, pp 381–385.

39. Fingas, M.F. *Proceedings of a Dispersant Workshop*, December, 1989, Reston, VA, sponsored by National Oceanic and Atmospheric Administration, Washington, DC, 1989, p 18

40. Brown, H.M.; Goodman, R.H. in *Oil Dispersants: New Ecological Approaches, ASTM STP 1018*, L. Michael Flaherty, Ed., American Society for Testing and Materials, Philadelphia, PA, 1989, pp 31–40.

41. ASTM, *Guide for Ecological Considerations for the Use of Oilspill Dispersants in Freshwater and Other Inland Environments, Lakes and Large Water Bodies, STP F 1210-89*, American Society for Testing and Materials, Philadelphia, PA, 1989.

42. Fingas, M.F. *Proceedings of the EPA Workshop on Dispersant Laboratory Testing*, Edison, NJ, 1991, p 46.

43. Byford, D.C.; Green, P.J. in *Oil Spill Chemical Dispersants: Research, Experience and Recommendations*, Tom E. Allen, Ed. STP 840, American Society for Testing and Materials, Philadelphia, PA, 1984, pp 69–86.

44. Meeks, D.G. *Proceedings of the 1981 Oil Spill Conference*, American Petroleum Institute, Washington, DC, 1981, pp 19–25.

45. Rewick, R.T.; Sabo, K.A.; Gates, J.; Smith, J.H.; McCarthy, Jr., L.T., *Proceedings of the 1981 Oil Spill Conference*, American Petroleum Institute, Washington, DC, 1981, pp 5–10.

46. Fingas, M.F.; Bobra, M.A.; Velicogna, D. *Proceedings of the 1987 Oil Spill Conference*, American Petroleum Institute, Washington, DC, 1987, pp 241–246.

47. Fingas, M.F.; Kyle, D.A.; Wang, Z.; Handfield, D.; Ianuzzi, D.; Ackerman, F. *The Use of Chemicals in Oil Spill Response, ASTM STP 1252*, Peter Lane, Ed., American Society for Testing and Materials, Philadelphia, PA, 1995, pp 3–40.

48. Fingas, M.F.; Kyle, D.A.; Tennyson, E. *The Use of Chemicals in Oil Spill Response, ASTM STP 1252*, Peter Lane, Ed., American Society for Testing and Materials, Philadelphia, PA, 1995, pp 92–132.

49. Fingas, M.F.; Kyle, D.A.; Wang, Z.; Ackerman, F.; Mullin, J. *Proceedings*

of the Seventeenth Arctic and Marine Oilspill Program (AMOP) Technical Seminar, Environment Canada, Ottawa, Ont., 1994, pp 905–941.

50. Clayton, J.R.; Payne, J.R.; Farlow, J.S.; Sarwar, C. *Oil Spill Dispersants: Mechanisms of Action and Laboratory Tests*; CRC Press: Boca Raton, FL, 1993.

51. Clayton, J.R.; Marsden, P. *Chemical Oil Spill Dispersants: Evaluation of Three Laboratory Procedures for Estimating Performance*, Environmental Protection Agency Report PB92-222041, Cincinnati, OH, 1992.

52. Becker, K.W.; Coker, L.G.; Walsh, M.A. *Proceedings of the Oceans 91 Conference*, IEEE Service Centre, Piscataway, NJ, 1991.

53. Nordvik, A.B.; Hudon, T.J. *Interlaboratory Calibration Testing of Dispersant Effectiveness, Phase I*, MSRC Technical Report Series Report 93-003.1, Marine Spill Research Corporation, Washington, DC, 1993.

54. Nordvik, A.B.; Osborn, H.G. *Interlaboratory Calibration Testing of Dispersant Effectiveness, Phase II*, MSRC Technical Report Series Report 935-003.2, Marine Spill Research Corporation, Washington, DC, 1993.

55. Lunel, T.; Baldwin, G.; Merlin, F. *Proceedings of the Eighteenth Arctic and Marine Oilspill Program (AMOP) Technical Seminar*, Environment Canada, Ottawa, Ont., 1995, pp 629–651.

56. Liu, H.; Zhan, Y.; Wang, W.; Wang, Z. *Proceedings of the Eighteenth Arctic and Marine Oilspill Program (AMOP) Technical Seminar*, Environment Canada, Ottawa, Ont., 1995, pp 355–365.

57. Brown, H.M.; Goodman, R.H.; Canevari, G.P. *Proceedings of the 1987 International Oil Spill Conference*, American Petroleum Institute, Washington, DC, 1987, pp 305–312.

58. Payne, J.R.; Reilley, T.J.; Martrano, R.J.; Lindblom, G.P.; Kennicutt II, M.C.; Brooks, J.M. *Proceedings of the 1993 International Oil Spill Conference*, American Petroleum Institute, Washington, DC, 1993, pp 791–792.

59. Lewis, A.; Daling, P.S.; Fiocco, R.J.; Nordvik, A.B. *Proceedings of the Seventeenth Arctic and Marine Oilspill Program (AMOP) Technical Seminar*, Environment Canada, Ottawa, Ont., 1994, pp 979–1010.

60. Fingas, M.F.; Kyle, D.A.; Tennyson, E. *Proceedings of the Sixteenth Arctic and Marine Oilspill Program (AMOP) Technical Seminar*, Environment Canada, Ottawa, Ont., 1993, pp 861–876.

61. Fingas, M.F.; Kyle, D.A.; Holmes, J.B.; Tennyson, E.J. *Proceedings of the 1993 International Oil Spill Conference*, American Petroleum Institute, Washington, DC, 1993, pp 567–572.

62. Fingas, M.F.; Kyle, D.A.; Tennyson, E.J. *Proceedings of the Fifteenth Arctic and Marine Oilspill Program (AMOP) Technical Seminar*, Environment Canada, Ottawa, Ont., 1992, pp 135–142.

63. Fingas, M.F.; Kyle, D.A.; Bier, I.D.; Lukose, A.; Tennyson, E.J. *Proceedings of the Fourteenth Arctic and Marine Oilspill Program (AMOP) Technical Seminar*, Environment Canada, Ottawa, Ont., 1991, pp 87–106.

64. Canevari, G.P. *Oil Spill Chemical Dispersants: Research, Experience and*

Recommendations, STP 840, Tom E. Allen, Ed., American Society for Testing and Materials, Philadelphia, PA, 1984, pp 87–93.

65. Fingas, M.F.; Bier, I.; Bobra, M.A.; Callaghan, S. *Proceedings of the 1991 International Oil Spill Conference*, American Petroleum Institute, Washington, DC, 1991, pp 419–426.

66. Brandvik, P.J. *Proceedings of the Fifteenth Arctic and Marine Oilspill Program (AMOP) Technical Seminar*, Environment Canada, Ottawa, Ont., 1992, pp 123–134.

67. Belk, J.L.; Elliot, D.J.; Flaherty, L.M. *Proceedings of the 1989 International Oil Spill Conference*, American Petroleum Institute, Washington, DC, 1989, pp 333–336.

68. Eriksson, F. *Proceedings of Seminar on Combating Marine Oil Spills in Ice and Cold Conditions*, National Board of Waters and the Environment, Helsinki, Finland, 1992, pp 99–109.

69. Moet, A.; Bakr, M.Y.; Abdelmonim, M.; Abdelwahab, O. *Proceedings of Division of Petroleum Chemistry, 210th Meeting*, American Chemical Society, Chicago, IL, 1995, pp 564–566.

70. Lunel, T. *The Use of Chemicals in Oil Spill Response*, P. Lane, Ed., ASTM, STP 1252, Philadelphia, PA, 1995, pp 240–285.

71. Etkin, D.S. *Proceedings of the Twenty-first Arctic and Marine Oilspill Program (AMOP) Technical Seminar*, Environment Canada, Ottawa, Ont., 1998, pp 281–304.

72. Spaulding, M. *Proceedings of the Nineteenth Arctic and Marine Oilspill Program (AMOP) Technical Seminar*, Environment Canada, Ottawa, Ont, 1996, pp 1487–1497.

73. Gulec, I.; Holdway, D.A. *Proceedings of the 1997 International Oil Spill Conference*, American Petroleum Institute, Washington, DC, 1997, pp 1010–1011.

74. Singer, M.M.; George, S.; Jacobsen, S.; Wectman, L.L.; Tjeerdema, R.S.; Aurand, D.; Blondina, G.; Sowby, M.L. *Proceedings of the 1997 International Oil Spill Conference*, American Petroleum Institute, Washington, DC, 1997, pp 1020–1021.

75. Singer, M.M.; George, S.; Benner, P.; Jacobsen, S.; Tjeerdema, R.S.; Sowby, M.L. *Environ. Toxicol. Chem.*, **1993**, *12*, 1855–1863.

76. Singer, M.M.; George, S.; Benner, P.; Jacobsen, S.; Lee, I.; Weetmore, L.L.; Tjeerdema, R.S.; Sowby, M.L. *Ecotoxicol. Environ. Safety*, **1996**, *35*, 183–189.

77. Midlaugh, D.P.; Whiting, D.D. *Arch. Environm. Contam. Toxicol.* **1995**, *29*, 535–539.

78. Gragin, G.E.; Clark, J.R.; Pace, C.B. *Comparison of Physically and Chemically Dispersed Crude Oil Toxicity Under Continuous and Spiked Exposure Scenarios*, MSRC Technical Report Series Report 94-015, Marine Spill Research Corporation, Washington, DC, 1994.

79. Fucik, K.W.; Carr, K.A.; Balcom, B.J. *Dispersed Oil Toxicity Tests with Biological Species Indigenous to the Gulf of Mexico*, Minerals Management Report MMS94-0021, Minerals Management Service, New Orleans, LA, 1994.

80. Burridge, T.R.; Shir, M-J. *Mar. Poll. Bull.*, **1995**, *31(4–12)*, 446–452.

81. Wolfe, M.F.; Schwartz, G.J.B.; Singaram, S.; Mielbrecht, E.E.; Tjeerdema, R.S.; Sowby, M.L. *Proceedings of the Twentieth Arctic and Marine Oilspill Program (AMOP) Technical Seminar*, Environment Canada, Ottawa, Ont., 1997, pp 1215–1226.
82. Unsal, M. *Toxicol. Environ. Chem.*, **1991**, *31–32*, 451–459.
83. Baker, J.M.; Cruthes, J.H.; Little, D.I.; Oldham, J.H.; Wilson, C.M. *Oil Spill Chemical Dispersants: Research, Experience and Recommendations*, STP 840, T.E. Allen, Ed., American Society for Testing and Materials, Philadelphia, PA, 1984, pp 239–279.
84. American Society for Testing and Materials (ASTM) guides:

   F 929-85 (1993) Guide for Ecological Considerations for the Use of Chemical Dispersants in Oil Spill Response – Marine Mammals

   F 930-85 (1993) Guide for Ecological Considerations for the Use of Chemical Dispersants in Oil Spill Response – Rocky Shores

   F 931-85 (1993) Guide for Ecological Considerations for the Use of Chemical Dispersants in Oil Spill Response – Seagrasses

   F 932-85 (1993) Guide for Ecological Considerations for the Use of Chemical Dispersants in Oil Spill Response – Coral Reefs

   F 971-85 (1993) Guide for Ecological Considerations for the Use of Chemical Dispersants In Oil Spill Response – Mangroves

   F 972-85 (1993) Guide for Ecological Considerations for the Use of Chemical Dispersants in Oil Spill Response – Nearshore Subtidal

   F 973-85 (1993) Guide for Ecological Considerations for the Use of Chemical Dispersants in Oil Spill Response – Tidal Flats

   F 990-85 (1993) Guide for Ecological Considerations for the Use of Chemical Dispersants in Oil Spill Response – Sandy Beaches

   F 999-85 (1993) Guide for Ecological Considerations for the Use of Chemical Dispersants in Oil Spill Response – Gravel or Cobble Beaches

   F 1008-85 (1993) Guide for Ecological Considerations for the Use of Chemical Dispersants in Oil Spill Response – Bird Habitats

   F 1012-85 (1993) Guide for Ecological Considerations for the Use of Chemical Dispersants in Oil Spill Response – The Arctic

   F 1209-89 Guide for Ecological Considerations for the Use of Oilspill Dispersants in Freshwater and Other Inland Environments – Ponds and Sloughs

   F 1210-89 Guide for Ecological Considerations for the Use of Oilspill Dispersants in Freshwater and Other Inland Environments – Lakes and Large Water Bodies

   F 1231-89 Guide for Ecological Considerations for the Use of Oilspill Dispersants in Freshwater and Other Inland Environments – Rivers and Creeks

   F 1279-90 Guide for Ecological Considerations for the Use of Oilspill Dispersants in Freshwater and Other Inland Environments – Permeable Surfaces

   F 1280-90 Guide for Ecological Considerations for the Use of Oilspill Dispersants in Freshwater and Other Inland Environments – Impermeable Surfaces,

American Society for Testing and Materials, Philadelphia, PA, 1985–1990.

85. Knap, A.H.; Dodge, R.E.; Baca, B.J.; Sleeter, T.D.; Snedaker, S.D. *Proceedings of the Second International Oil Spill Research and Development Forum*, International Maritime Organization, London, England, 1995, pp 262–270.

86. Ballou, T.G.; Hess, S.G.; Getter, C.D.; Knap, A.; Dodge, R.; Sleeter, T. *Proceedings of the 1987 International Oil Spill Conference*, American Petroleum Institute, Washington, DC, 1987, p 634.

87. Ballou, T.G.; Hess, S.C.; Dodge, R.E.; Knap, A.H.; Sleeter, T.D. *Proceedings of the 1989 International Oil Spill Conference*, American Petroleum Institute, Washington, DC, 1989, pp 447–454.

88. Baca, B.J.; Snedaker, S.C.; Dodge, R.E.; Knap, A.H.; Sleeter, T.D. *Proceedings of Oceans 96 MTS/IEEE*, Fort Lauderdale, FL, 1996, pp 469–485.

89. Thorhaug, A.; Marcus, J.H. *Proceedings of the 1985 International Oil Spill Conference*, American Petroleum Institute, Washington, DC, 1985, pp 597–501.

90. Thorhaug, A.; Carby, B.; Reese, R.; Sidrak, G.; Anderson, M.; Aiken, K.; Walker, W.; Rodriquez, M. *Proceedings of the 1991 International Oil Spill Conference*, American Petroleum Institute, Washington, DC, 1991, pp 415–417.

91. Brown, H.M.; Goudey, J.S.; Foght, J.M.; Cheng, S.K.; Dale, M.; Hoddinott, J.; Quaife, L.R.; Westlake, D.W.S.*Oil and Chemical Pollution*, **1990**, 6, pp 37–54.

92. MacKinnon, D.S.; Lane, P.A. *Saltmarsh Revisited: the Long-term Effects of Oil and Dispersant on Saltmarsh Vegetation*, Environmental Studies Research Funds Report No. 122, Ottawa, Ont., 1993.

93. Gilfillan, E.S.; Page, D.S.; Hanson, S.A.; Foster, J.; Holtham, J.; Vallas, D. et al. *Proceedings of the 1985 International Oil Spill Conference*, American Petroleum Institute, Washington, DC, 1985, pp 553–559.

94. Sergy, G.A., Ed. *Arctic, 40*, Supplement I, Environment Canada, Edmonton, Alberta, 1987, p 279.

95. Siron, R.; Pelletier, E.; Delille, D.; Roy, S. *Mar. Environ. Res.* **1993**, 35, 273–302.

96. *Environment Canada Standard List of Approved Treating Agents*, Environment Canada, Ottawa, Ont, 1998.

97. *Environmental Protection Agency – National Contingency Plan Product Schedule*, United States Environmental Protection Agency (U.S. EPA), May, 1998.

98. MAFF, *The Approval and Use of Oil Dispersants in the UK*, Ministry of Agriculture, Fisheries and Food, 1997.

99. Kucklick, J.H.; Aurand, D.V. *An Analysis of Historical Opportunities for Dispersant and In-situ Burning Use in the Coastal Waters of the United States, Except Alaska*, MSRC Technical Report 95–005, Marine Spill Research Corporation, Washington, DC, 1995.

100. Lewis, A.; Aurand, D. *Putting Dispersants to Work: Overcoming Obsta-*

cles, American Petroleum Institute Technical Report IOSC-004, Washington, DC, 1997.

101. OSIR, *Oil Spill Intelligence Report*, Cutter Information Corporation, Arlington, MA. (Issue numbers and dates listed in Table 10.)

102. Fiocco, R., Communication to A. Lewis and D. Aurand, 1996.

103. Welsh, J. *International Oil Spill Statistics: 1994*, Oil Spill Intelligence Report, Cutter Information Corporation, Arlington, MA, 1994.

104. Exxon, *Exxon Research and Engineering Dispersant Application Guidelines*, Exxon Publishing, Florham Park, NJ, 1994.

105. Lunel, T.; Rusin, J.; Bailey, N.; Halliwell, C.; Davies, L. *Proceedings of the 1997 International Oil Spill Conference*, American Petroleum Institute, Washington, DC, 1997, pp 185–194.

106. National Oceanic and Atmospheric Administration (NOAA), Oil Spill Case Histories 1967–1991: Summaries of Significant U.S. and International Spills, Report No. HMRAD 92–11, NOAA Hazardous Materials Response and Assessment Division, Seattle, WA, 1992.

107. Davidson, A. *In the Wake of the Exxon Valdez*; Douglas and McIntyre: Vancouver, BC, 1990.

108. IPIECA, *Dispersants and Their Role in Oil Spill Response*, IPIECA Report Series Vol. V, International Petroleum Industry Environmental Conservation Association, London, England, 1993.

109. U.S. Environmental Protection Agency, *A Review of Ecological Assessment Case Studies from a Risk Assessment Perspective*, EPA/630/R-92/005, U.S. Environmental Protection Agency, Washington, DC, 1993.

110. Albone, D.J.; Kibblewhite, M.G.; Sansom, L.E.; Morris, P.R. *The Storage Stability of Oil Spill Dispersants*, Warren Spring Laboratory Report LR670 (CS), Hertfordshire, UK, 1988.

111. Exxon, *Oil Spill Response Field Manual*, Exxon Production Research Company, Houston, TX, 1992.

112. Giammona, C.; Binkley, K.; Fay, R.; Denoux, G.; Champ, M.; Geyer, R.; Bouse, F.; Kirk, I.; Gardisser, D.; Jamail, R. *Aerial Dispersant Application: Field Testing Research Program, Alpine, Texas*, MSRC Technical Report 94-019, Marine Spill Research Corporation, Washington, DC, 1994.

113. Geyer, R.; Fay, R.; Giammona, C.; Binkley, K.; Denoux, G.; Jamail, R. *Aerial Dispersant Application: Assessment of Sampling Methods and Operational Altitudes, Volume 1*, MSRC Technical Report 93-009.1, Marine Spill Research Corporation, Washington, DC, 1993.

114. *ASTM Guide for Oil Spill Dispersant Application Equipment; Boom and Nozzle Systems, F-1413-92*, American Society for Testing and Materials, West Conshohocken, PA, 1992.

115. *ASTM Standard Test Method for Determination of Deposition of Aerially-Applied Oil Spill Dispersants, F-1738-96*, American Society for Testing and Materials, West Conshohocken, PA, 1996.

116. Fingas, M.F.; Brown, C.E. *Proceedings of the Twenty-first Arctic and Marine Oilspill Program (AMOP) Technical Seminar*, Environment Canada, Ottawa, Ont., 1998, pp 819–840.

117. Smedley, J.B. *Proceedings of the 1981 International Oil Spill Conference*, American Petroleum Institute, Washington, DC, 1981, pp 253–257.

118. Nordvik, A. *Spill Sci. Technol. Bull.*, **1995**, 3(1), 17–46.

119. Morris, P.R.; Thomas, D.H. *Evaluation of Oil Spill Dispersant Concentrates for Beach Cleaning – 1987 Trials*, Warren Spring Laboratory Report LR 624(OP), Hertfordshire, UK, 1987.

120. Fiocco, R.J.; Canevari, G.P.; Wilkinson, J.B.; Jahns, H.O.; Bock, J.; Robbins, M.; Markarian, R.K. *Proceedings of the 1991 International Oil Spill Conference*, American Petroleum Institute, Washington, DC, 1991, pp 395–400.

121. Canevari, G.P.; Fiocco, R.J.; Lessard, R.R.; Fingas, M.F. *The Use of Chemicals in Oil Spill Response, ASTM STP 1252*, Peter Lane, Ed., American Society for Testing and Materials, Philadelphia, PA, 1995, pp 227–239.

122. Hoff, R.; Shigenaka, G.; Yender, R.; Payton, D. *Chemistry and Environmental Effects of the Shoreline Cleaner PES-51*, HAZMAT Report No. 942, National Oceanic and Atmospheric Administration, Seattle, WA, 1994.

123. Clayton, J.R.; Tsang, S.-F.; Frank, V.; Marsden, P.; Chau, N.; Harrington, J. *Chemical Surface Washing Agents for Oil Spills*, U.S. Environmental Protection Agency Report, EPA/600/SR-93/113, 1993.

124. Fiocco, R.J.; Lessard, R.R.; Canevari, G.P. *Proceedings of 1996 Petro-Safe Conference*, Houston, TX, 1996, pp 276–280.

125. ASTM, *Standard Guide for the Use of Chemical Shoreline Cleaning Agents: Environmental and Operational Considerations, STP F 1872-98*, American Society for Testing and Materials, Philadelphia, PA, 1998.

126. ADEC, Private communication with Alaska Department of Environmental Conservation, 1998.

127. Michel, J.; Lehmann, S.M.; Henry, Jr., C.B. *Proceedings of the Twenty-first Arctic and Marine Oilspill Program (AMOP)* Technical Seminar, Environment Canada, Ottawa, Ont., 1998, pp 841–856.

128. OSIR, *Oil Spill Intelligence Report*, Cutter Information Corporation, Arlington, MA. (Issue numbers and dates listed in Table 12.)

129. Clayton, J.R.; Michel, J.; Snyder, B.J.; Lees, D.C. *Utility of Current Shoreline Cleaning Agent Tests in Field Testing*, MSRC Technical Report Series Report 95-004, Marine Spill Research Corporation, Washington, DC, 1995.

130. Michel, J.; Benggio, B.L. *Proceedings of the 1995 International Oil Spill Conference*, American Petroleum Institute, Washington, DC, 1995, pp 197–202.

131. Tomblin, T.G. *San Jacinto River Incident, Report to Federal On-Scene Commander, Use of Corexit 9580 for Shoreline Cleanup in Mitchell Bay*, Exxon Report, 1994.

132. Pezeshki, S.R.; DeLaune, R.D.; Nyman, J.A.; Lessard, R.R.; Canevari, G.P. *Proceedings of the 1995 International Oil Spill Conference*, American Petroleum Institute, Washington, DC, 1995, pp 203–209.

133. Pezeshki, S.R.; DeLaune, R.D.; Nyman, J.A. *Investigation of Corexit 9580 for Removing Oil from Marsh Grass*, Technical Report submitted to Exxon Research and Engineering from Louisiana State University, Baton Rouge, LA, 1994.

134. Rog, S.; Owens, D.; Pearson, L.; Tumeo, M.; Braddock, J.; Venator, T. *Proceedings of the Seventeenth Arctic and Marine Oilspill Program (AMOP) Technical Seminar*, Environment Canada, Ottawa, Ont., 1994, pp 607–620.

135. Little, D.I.; Baker, J.M. *Ecological Impacts of the Oil Industry*, B. Dicks, Ed.; John Wiley and Sons: Chichester, England, 1989, pp 169–201.

136. Clayton, J.R.; Stransky, B.C.; Schwartz, M.J.; Lees, D.C.; Michel, J.; Snyder, B.J.; Adkins, A.C. *Development of Protocols for Testing Cleaning Effectiveness and Toxicity of Shoreline Cleaning Agents (SCAs) in the Field*, MSRC Technical Report Series Report 95-020.1, Marine Spill Research Corporation, Washington, DC, 1995.

137. Clayton, J.R.; Stransky, B.C.; Adkins, A.C.; Lees, D.C.; Michel, J.; Schwartz, M.J.; Snyder, B.J.; Reilly, T.J. *Proceedings of the Nineteenth Arctic and Marine Oilspill Program (AMOP) Technical Seminar*, Environment Canada, Ottawa, Ont., 1996, pp 423–451.

138. Clayton, J.R.; Stransky, B.C.; Adkins, A.C.; Lees, D.C.; Michel, J.; Schwartz, M.J.; Snyder, B.J.; Reilly, T.J. *Proceedings of the Eighteenth Arctic and Marine Oilspill Program (AMOP) Technical Seminar*, Environment Canada, Ottawa, Ont., 1995, pp 423–451.

139. Clayton, J.R.; Stransky, B.C.; Schwartz, M.J.; Lees, D.C.; Michel, J.; Snyder, B.J.; Adkins, A.C. *Development of Protocols for Testing Cleaning Effectiveness and Toxicity of Shoreline Cleaning Agents (SCAs) in the Field: Data Report*, MSRC Technical Report Series Report 95-020.2, Marine Spill Research Corporation, Washington, DC, 1995.

140. Fingas, M.F.; Stoodley, G.; Harris, G.; Hsia, A. *Proceedings of the Workshop on the Cleanup of Beaches in Prince William Sound Following the EXXON VALDEZ Spill*, Anchorage, Alaska, November, 1989, sponsored by National Oceanic and Atmospheric Administration, Seattle, WA, 1989, p 5.

141. Sullivan, D.; Sahatjian, K.A. *Proceedings of the 1993 International Oil Spill Conference*, American Petroleum Institute, Washington, DC, 1993, pp 511–514.

142. Clayton, J.R.; Renard, E.P. *Proceedings of the Seventeenth Arctic and Marine Oilspill Program (AMOP) Technical Seminar*, Environment Canada, Ottawa, Ont., 1994, pp 877–907.

143. Merlin, F.X.; Le Guerroue, P. *Proceedings of the Seventeenth Arctic and Marine Oilspill Program (AMOP) Technical Seminar*, Environment Canada, Ottawa, Ont., 1994, pp 943–950.

144. Fingas, M.F.; Kyle, D.A.; Laroche, N.D.; Fieldhouse, B.G.; Sergy, G.; Stoodley, R.G. *Spill Technology Newsletter, Vol. 18, No. 3*, Environment Canada, Ottawa, Ont., 1993, pp 1–14.

145. Fingas, M.F.; Kyle, D.A.; Laroche, N.D.; Fieldhouse, B.G.; Sergy, G.; Stoodley, R.G. "The Effectiveness Testing of Spill Treating Agents," *The Use of Chemicals in Oil Spill Response, ASTM STP1252*, Peter Lane, Ed., American Society for Testing and Materials, Philadelphia, PA, pp 286–298.

146. Shigenaka, G.; Vicente, V.P.; McGehee, M.A.; Henry, C.B. *Proceedings of*

the *1995 International Oil Spill Conference*, American Petroleum Institute, Washington, DC, 1995, pp 177–184.

147. Teas, H.J.; Lessard, R.R.; Canevari, G.P.; Brown, C.D.; Glenn, R. *Proceedings of the 1993 International Oil Spill Conference*, American Petroleum Institute, Washington, DC, 1993, pp 147–151.

148. Walker, A. H.; Michel, J.; Canevari, G.; Kucklick, J.; Scholz, D.; Benson, C.A.; Overton, E.; Shane, B. *Chemical Oil Spill Treating Agents: Herding Agents, Emulsion Treating Agents, Solidifiers, Elasticity Modifiers, Shoreline Cleaning Agents, Shoreline Pre-treatment Agents and Oxidation Agents*, MSRC Technical Report Series Report 93-015, Marine Spill Research Corporation, Washington, DC, 1993.

149. Ross, S.L.; Canevari, G.P.; Consultchem, *State-of-the-Art Review: Emulsion Breaking Chemicals*, Canadian Petroleum Association Report, Calgary, Alberta, 1992.

150. Janiyani, K.L.; Purohit, H.J.; Shanker, R.; Khanna, P. *World J. Microbiol. Biotechnol.*, **1994**, *10*, 452–456.

151. McDonagh, M.; Colomb-Heiliger, K. *Proceedings of the Fifteenth Arctic and Marine Oilspill Program (AMOP) Technical Seminar*, Environment Canada, Ottawa, Ont., 1992, pp 107–121.

152. Seakem Oceanography, *Field Test of Two Spill Treating Agents*, Environment Canada Manuscript Report EE-124, Ottawa, Ont., 1990.

153. Fingas, M.F. *Spill Technology Newsletter, Vol. 13, No. 2*, Environment Canada, Ottawa, Ont., 1988, pp 40–46.

154. Lewis, A.; Singsaas, I.; Johannessen, B.O.; Jensen, H.; Lorenzo, T.; Nordvik, A.B. *Large Scale Testing of the Effect of Demulsifier Addition to Improve Oil Recovery Efficiency*, MSRC Technical Report Series Report 95-033, Marine Spill Response Corporation, Washington, DC, 1995.

155. Buist, I.A.; Glover, N.; McKenzie, B.; Ranger, R. *Proceedings of the 1995 International Oil Spill Conference*, American Petroleum Institute, Washington, DC, 1995, pp 139–146.

156. Fingas, M.F.; Fieldhouse, B.; Mullin, J.V. *Proceedings of the Twenty-First Arctic and Marine Oilspill Program (AMOP) Technical Seminar*, Environment Canada, Ottawa, Ont., 1998, pp 1–25.

157. Fingas, M.F.; Fieldhouse, B.; Gamble, L.; Mullin, J.V. *Proceedings of the Eighteenth Arctic and Marine Oilspill Program (AMOP) Technical Seminar*, Environment Canada, Ottawa, Ont., 1995, pp 21–42.

158. Fingas, M.F.; Fieldhouse, B. *Proceedings of the Seventeenth Arctic and Marine Oilspill Program (AMOP) Technical Seminar*, Environment Canada, Ottawa, Ont., 1994, pp 213–244.

159. Buist, I.A.; Ross, S.L. *Oil Chem. Poll.*, **1987**, *3*, 485–503.

160. Bobra, M.A.; Kawamura, P.; Fingas M.F.; Velicogna, D. *Mesoscale Application and Testing of an Oil Spill Demulsifying Agent and Elastol*, Environment Canada Manuscript Report EE-105, Ottawa, Ont., 1998, p 41.

161. Brown, H.M.; Weiss, D.K.; Goodman, R.H. *Proceedings of the Thirteenth Arctic and Marine Oilspill Program (AMOP) Technical Seminar*, Environment Canada, Ottawa, Ont., 1990, pp 255–264.

162. Fingas, M.F., Personal observations during the application of Corexit 9527 on emulsions off Seward Alaska, April, 1989.
163. Fingas, M.; Fieldhouse, B.; Bier, I.; Conrod, D.; Tennyson, E. *Proceedings of the Workshop on Emulsions*, Marine Spill Response Corporation, Washington, DC, 1993, p 9.
164. Krawezhk, M.A.; Wasan, D.T.; Shetty, C.S. *Ind. Eng. Chem. Res.* **1991**, *30*, 367–375.
165. Knudsen, O.O.; Brandvik, P.J.; Lewis, A. *Proceedings of the Seventeenth Arctic and Marine Oilspill Program (AMOP) Technical Seminar*, Environment Canada, Ottawa, Ont., 1994, pp 1023–1034.
166. Strøm-Kristiansen, T.; Daling, P.S.; Nordvik, A.B. *Demulsification by Use of Heat and Emulsion Breaker, Phase I*, MSRC Technical Report Series Report 93-026, Marine Spill Response Corporation, Washington, DC, 1993.
167. Bhardwaj, A.; Hartland, S. *J. Dispersion Sci. Technol.* **1993**, *14(5)*, 541–557.
168. Fiocco, R.J.; Becker, K.W.; Walsh, M.A.; Hokstad, J.N.; Daling, P.S.; Lewis, A. *Proceedings of the 1995 International Oil Spill Conference*, American Petroleum Institute, Washington, DC, pp 165–170.
169. Fingas, M.F.; Fieldhouse, B.; Bier, I.; Conrod, D.; Tennyson, E. *The Use of Chemicals in Oil Spill Response, ASTM STP 1252*, Peter Lane, Ed., American Society for Testing and Materials, Philadelphia, PA, 1995, pp 41–54.
170. Fingas, M.F.; Fieldhouse, B.; Mullin, J.V. *Proceedings of the Twentieth Arctic and Marine Oilspill Program (AMOP) Technical Seminar*, Environment Canada, Ottawa, Ont., 1997, pp 21–42.

RECEIVED for review June 29, 1998. ACCEPTED revised manuscript October 22, 1998.

Toxicity and Persistence of Surfactants Used in the Petroleum Industry

Larry N. Britton

Condea Vista Company, Austin, Texas

The environmental risks associated witht the use of surfactants in the petroleum industry are reviewed from the perspectives of persistence and toxicity. While, in general, the use of surfactants in the petroleum industry should not cause undue concern, new products are continually being introduced and the regulatory system continues to evolve. This chapter provides an assessment of the current state of knowledge and the processes involved in environmental risk assessment.

Introduction

Environmental risks associated specifically with the use of surfactants in the petroleum industry generally have not been an issue. This is because surfactants are perceived as being less environmentally hazardous compared to certain other chemicals used in the petroleum industry. Also, surfactants should not reach sensitive receptors unless there is an unexpected release, such as a spill or inappropriate discharge of process waters containing the surfactants. Consequently, there are few published studies on the fate and effects of surfactants in the petroleum industry. However, the times are changing. Environmental data are required on all chemicals listed as part of the pre-manufacturing notification (PMN) compliance with the Toxic Substances Control Act (TSCA), and even Material Safety Data Sheets (MSDS) currently have a section on environmental safety. There is a need to understand the environmental properties of large volume chemicals, and chemical manufacturers are responding by collecting more data. Likewise the users of these chemicals must understand the environmental risks associated with their use which in the case of the petroleum industry can be unique.

Fortunately, there is a great deal of data on environmental properties of surfactants used in cleaning applications. It is simply a matter of extrapolating this information for the same classes of surfactants used in the petroleum industry. The caveat is the use of this data to make a risk assessment may not be relevant to the unique conditions of surfactant usage in the petroleum industry. For example, the biodegradation rate of a surfactant in a subsurface petroleum reservoir is not likely to be the same as that in a stream. One could argue that biodegradation of the surfactant in the subsurface petroleum reservoir is not particularly important, or wanted, if there is never the chance that the surfactant will encounter a sensitive receptor population. However, biodegradation rates (as well as toxicity) become profoundly important if the tanker truck carrying the surfactant to the oil field crashes and spills its load into a stream.

This chapter reviews the persistence (i.e., fate) and toxicity (i.e., effects) of classes of surfactants used in the petroleum industry albeit the data were obtained from the viewpoint of the same surfactant classes as cleaning agents. Although down-the-drain cleaning agents and surfactant usage in the oil field are seemingly disparate topics, the central issue is the environmental consequences or the *risks* associated with the use and disposal of these surfactants. The process known as risk assessment is the product of fate and effects evaluations in the unique context of use and disposal. It is important that the reader understand the concept of risk assessment in order to judge environmental consequences. The chapter starts with this concept and is followed by categorization of surfactants used in the petroleum industry and discussions on biodegradation and toxicity.

The Concept of Risk Assessment

Risk assessment can be summarized in the following two relationships and using the terminology PEC (predicted environmental concentration) and PNEC (predicted no-effect concentration):

If PEC/PNEC × safety factor ≥ 1, then there is a risk

If PEC/PNEC × safety factor < 1, then risk is acceptable

Risk assessment is a systematic process that starts with estimating the concentrations (PEC) of the chemicals of interest in the relevant environmental compartments such as surface waters, sediments, soils, groundwater, etc. The estimate is generated by either direct measurement or by modeling. Since biodegradation is one of the most important processes that affect concentrations, it is important to know biodegradation rates from monitoring data, laboratory data or modeling. If the predicted environmental concentration is less than the concentration that causes toxic effects (PNEC), a margin of safety exists. Safety factors of 10,

100, or 1000 often are added for uncertainty and to determine if the margin of safety between PEC and PNEC differs by one, two or three orders of magnitude. The size of the safety factor is inversely related to the quantity and quality of fate and effects data. This uncertainty factor also applies to the quandary of how to ascertain risk in an environmental compartment when it is not known which are the most sensitive species.

A risk assessment is possible even without actual biodegradation or toxicity data. The modeling of biodegradation rates based on chemical structure can be done with BIOWIN, a program for estimating stream biodegradation rates from the U.S EPA's Office of Pollution Prevention & Toxics. Toxicity endpoints can also be modeled using quantitative structure activity relationships (QSAR) such as that used to predict general and polar narcosis modes of surfactant toxicity based on calculated octanol/water partitioning coefficients [1]. The U.S. EPA's ECOSAR program calculates toxicity endpoints using physical and chemical data generated from SMILES notation (Simplified Molecular Input Line Entry System) or CAS number of the chemical. The underlying assumptions and conservatism in calculating toxicity endpoints and the PEC from biodegradation rates and other fate determiners tends to decrease the accuracy of a "computed" risk assessment. Sometimes there is no substitute for actual testing. This is certainly the case when the environmental compartment is unique and does not fit the usual rivers and streams models.

Standardized biodegradation and toxicity testing have taken the uncertainty out of how to conduct laboratory testing. Organizations such as the Organization for Economic Cooperation and Development (OECD), the American Society for Testing and Materials (ASTM), the U.S. EPA, the International Organization for Standardization (ISO), Environmental Centre for Ecotoxicology and Toxicology of Chemicals (ECETOC), and the Japanese Ministry of International Trade and Industry (MITI) are active in developing and issuing relevant biodegradation and toxicity testing methodology. The standardized test methods tend to employ the same species such as daphnids (*Daphnia* sp., *Ceriodaphnia* sp.) and fathead minnows for freshwater tests, earthworms for soil, sheepshead and inland silversides minnows and mysid shrimp for marine tests, and chironomids, oligochaetes, and water scuds for sediments.

Surfactants used in the Petroleum Industry

The unique amphiphilic property of surfactants has made them useful in oil field applications. Their abilities to solubilize, emulsify/demulsify, foam, alter interfacial tension, viscosity and friction have made them ingredients in a variety of fluids. However, on a volume basis, the greatest potential usage would be surfactant waterflooding for enhanced oil recovery. The 1970s and 1980s were active periods for research on surfactants for EOR, and a large number of patents were issued (see references 2 and 3 for examples of early patents).

For the purpose of understanding and evaluating fate and effects of surfactants in EOR it is necessary to categorize individual surfactants into surfactant classes. Since the chemistries of surfactant classes generally determine the toxicity and biodegradation properties of individual surfactants, this simplification seems reasonable. The following surfactant classes are representative of the large number of surfactant candidates for EOR:

- alcohol ether sulfates
- alcohol ethoxylates (alkoxylates)
- alkyl aryl sulfonates and petroleum sulfonates
- alkyl phenol ethoxylates
- dialkyl sulfosuccinates
- quaternary ammonium (cationic) surfactants

These surfactant classes are a subset of those used in cleaning applications, and there are classes that are noticeably absent from the list above, such as alcohol sulfates, salts of fatty acids (soaps), alkanolamides, aliphatic sulfonates, betaines (amphoterics), methyl ester sulfonates and the relatively new alkyl polyglycosides. All of these have seen various applications in the petroleum industry, but perhaps lesser importance in EOR. Cost, availability and performance are the obvious issues in determining the potential use of surfactants in EOR.

The sections below review the toxicity and biodegradation data for this list of surfactant classes, although the information is not from the perspective of the petroleum industry. With relatively few exceptions (e.g., reference 4) the studies on environmental properties of surfactants deal with releases of these compounds into sewer and septic systems and ultimately to surface waters and soil. Nonetheless the data are valid for surfactants used in the petroleum industry if they enter these same environmental compartments. The data may not apply to unique subsurface conditions in EOR or other oil field applications.

Alcohol Ether Sulfates (AES)

AES can be represented by the following structure:

$$R\text{-}O\text{-}(CH_2\text{-}CH_2\text{-}O\text{-})_n \, SO_3^-$$

where R represents a linear or branched alkyl moiety of primary and secondary alcohols or the alkylbenzene part of alkylphenol. Although not shown above, AES may contain repeating units of butylene oxide, propylene oxide (PO) or mixtures of PO and ethylene oxide (EO).

Environmental data for AES surfactants have been reviewed extensively by the U.S. Soap and Detergent Association [5], by the Dutch Soap and Detergent Association [6] for a governmental risk assessment [7], and by Painter [8]. AES biodegradation is also reviewed in Swisher's classic book on surfactant biodegradation [9] and by Steber and Berger [10].

AES Biodegradation.   Aerobic biodegradation data show that AES surfactants are readily and completely biodegradable. Biological half-lives (i.e., time for biological removal of 50% of compound) for the group of AES surfactants in laboratory tests and river water die-away tests are generally less than 10 days. This supports sparse monitoring data [7] and predictions that AES from detergents should be less than 50 μg/L in river water below sewage treatment plants [11, 12].

Anaerobic biodegradation expectedly proceeds slower than under aerobic conditions, and compared to alcohol sulfates, anaerobic biodegradation of AES is not as fast [6, 8].

Surfactant structure is a determining factor in biodegradability, and data on AES biodegradation illustrate this principle. The following biodegradation rate relationships have been observed:

linear > branched
primary alcohol > secondary alcohol > alkylphenol
EO > PO
increasing alkoxamer number increases ultimate biodegradation rate

The length of the alcohol moiety does not appear to affect biodegradation rate, at least in the commonly encountered lengths of C_8 to C_{20}.

The pathway of AES biodegradation can be illustrated using a linear primary alcohol ethoxy sulfate as an example.

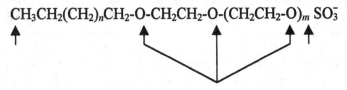

$$CH_3CH_2(CH_2)_nCH_2\text{-O-}CH_2CH_2\text{-O-}(CH_2CH_2\text{-O})_m\,SO_3^-$$

The points of enzymatic attack are shown by the arrows, and three routes or mechanisms have been observed:

1. Omega hydroxylation of the terminal methyl group of the alkyl chain, followed by further oxidation to the carboxyl function, followed by beta oxidative shortening of the alkyl group by two-carbon units.
2. Etherase cleavage of the ethoxylate moiety at any of the ethoxamers to produce glycol ether sulfates with varying EO number. The polyethylene glycol sulfate is further oxidized to the carboxy-

Table 1.   Range of Toxicity Values for AES[a]

Taxonomic Group	Acute Toxicity, EC_{50} (mg/L)	$NOEC^b$ (mg/L)
Bacteria	100–18,000 (growth inhibition)	1.5–2.2 (thymidine & glucose metabolism)
Algae	4–>50 (growth inhibition)	—
Aquatic invertebrates	1–50 (LC_{50})	≤0.27 (reproduction)
Fish	1.1–80 (LC_{50})	<0.1 (growth)
Other	1–6 g/kg (rat oral LD_{50})	1000–5000 ppm (dietary effect with rats)

[a] For all AES structures; specific studies listed in references 6 & 8
[b] No Observed Effect Concentration

late and ultimately cleaved apart in C_2 units with accompanying desulfation.
3. Sulfatase-mediated hydrolytic cleavage of the sulfate moiety to create the non-ionic alcohol ethoxylate, which is further degraded as described in a section below.

Any and all of the three mechanisms are observed in culture experiments; however, the first step appears to be the etherase cleavage. Ultimately, the final products are CO_2, H_2O, SO_4^-, and biomass.

AES Toxicity.   The effect of surfactant structure on toxicity has obvious importance. With AES, there is the tendency of decreased toxicity with increasing EO numbers, at least when comparing AES with the same hydrophobe. Also, increasing alkyl chain length in the hydrophobe will generally increase toxicity. These trends are understandable when one considers that the toxicity mechanism of surfactants, namely membrane disruption and protein denaturation, is a function of the surface-active properties of surfactants. Therefore, the alteration of surface-active properties via structure changes should affect toxicity.

Table 1 on toxicities is a "broad-brush" treatment of the toxicity data and does not consider the effects on toxicity values due to AES structure, differential sensitivity of test species within taxonomic groups or the variables of the test methods. The Dutch government surveyed the complex toxicity literature and developed rules for determining a risk-based, maximum permissible concentration (MPC) for AES in their country's surface waters. In 1995 they established the MPC at 0.4 mg/L [7].

Alcohol Ethoxylates (Alkoxylates)

$$R\text{-}O\text{-}(CH_2\text{-}CH_2\text{-}O)_n H$$

where R represents a linear or branched alkyl moiety of a primary or secondary alcohol.

Although the ethoxylate structure is shown, alcohol alkoxylates can contain units of ethylene oxide, propylene oxide, butylene oxide or mixtures. Alkyl phenol ethoxylates technically are alcohol ethoxylates; however, they are discussed separately below.

This class of non-ionic surfactant is widely used in cleaning products and in agricultural, cosmetic, textile, paper and other process applications. Consequently, there are numerous studies on their environmental properties. The U.S. Soap and Detergent Association has published a book on the environmental and human safety of alcohol ethoxylates [13], and there are other reviews and data compilations on AE toxicity and biodegradation [9, 14–16].

AE Biodegradation. All of the standardized laboratory tests plus die-away tests using river water and sediments demonstrate the ready and complete aerobic biodegradation of AE. Die-away tests with river water show biological half lives of 0.5–6 days. Monitoring of non-ionic surfactants through sewage treatment plants typically shows >98% removal (sorption plus biodegradation) during the six or less hours of hydraulic flow residence time. Half-lives of 2.8–8.6 days for $C_{12}AE_{8-9}$ mineralization (i.e., complete biodegradation) were observed in pond sediments that received laundromat wastewater [17]. These rapid biodegradation rates plus limited river monitoring data lend credence to the predictions that AE concentrations in rivers receiving treated sewage should not exceed low ppb levels. In the Dutch government risk assessment the calculated AE value was 0.5 ppb in its river waters below sewage outfalls [7]. Fate studies on AE in soils are lacking; however, the ubiquity of AE degraders should ensure complete biodegradation provided that bioavailability of the AE is not a problem because of AE sorption. Complete anaerobic biodegradation is also predicted from the biodegradation pathways, and this has been demonstrated in laboratory tests [18] albeit at a slower rate than aerobic degradation.

The biodegradation pathways for AE have been studied extensively using linear, primary alcohol ethoxylates as the model [9, 19–22]. Figure 1 depicts the three possibilities for initial point of attack:

1. Intramolecular scission of the hydrophile and hydrophobe to produce an alcohol and polyethylene glycol (PEG).
2. Omega alkyl oxidation, which is hydroxylation and further oxida-

tion of the terminal methyl group of the alkyl chain to the carboxylic acid.
3. Omega glycol oxidation, which is the oxidation of the terminal EO to give a carboxylic acid.

The hydrophobe is further oxidized by beta oxidation to CO_2, H_2O and biomass apparently more rapidly than the PEG or carboxylated PEG. The latter undergo oxidative cleavage of C_2 units. It is important to note that the pathways in Figure 1 are the sum of microbial processes on AE and not necessarily the capability of single microbial species. A proposed anaerobic pathway for AE biodegradation is also shown [23].

The structure of the AE molecule has significant influence on the pathway and on rates and extent of biodegradation. The structure of the alcohol hydrophobe is a determining factor, and there is considerable variation in commercial AE with respect to alcohol branching. Like the AES, biodegradation of linear hydrophobes is most rapid. Therefore, highly branched alcohols such as those synthesized by hydroformylation of polymerized propylene or butylene would exhibit slower biodegradation kinetics compared to a linear variety of the same carbon number. Single methyl branching is believed to have little or no effect. Therefore, AE synthesized from the "oxo-alcohol" process (which produces about 50:50 linear and methyl- or 2-alkyl-branched structures) show equivalent biodegradation compared to the completely linear derivatives [24]. No significant differences in biodegradation rates and extent can be detected in EO chain lengths until approximately 20 EO units whereupon ultimate biodegradation rate is reduced. The substitution of oxypropylene (PO) for EO, such as in EO–PO block polymers or PO capping, reduces biodegradation rate and extent [25]. Biodegradation has been found to be inversely proportional to the amount of PO in the surfactant.

AE Toxicity. The large range of values in the toxicity ranges table (Table 2) reflects differences in test methodology, but the main reasons are differential sensitivity of species and the effect of structure on AE toxicity. The latter refers to alkyl chain length and EO number. Generally, increasing the EO chain length or decreasing the alkyl chain length will decrease the aquatic toxicity. This really reflects changing the lipophilicity. That is, changes in structure that decrease lipohilicity will decrease aquatic toxicity. Table 3 illustrates this point with the toxicity of primary alcohols on fathead minnows.

The water solubilities are an indirect, and inversely proportional, measurement of lipophilicity. As alkyl chain length increases, the water solubility decreases (increase in lipohilicity) and the toxicity increases. With AE, the lipohilicity can be altered by both alkyl chain length and EO number. Quantitative structure activity relationship (QSAR) models have

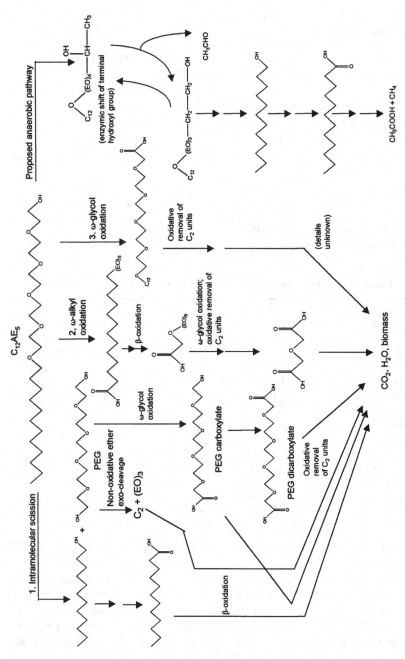

Figure 1.   Biodegradative pathways of linear alcohol ethoxylate.

Table 2.   Range of Toxicity Values for AE[a]

Taxonomic Group	Acute Toxicity, EC$_{50}$ (mg/L)	NOEC[b] (mg/L)
Bacteria	0.2–>2250 (Microtox™ EC$_{50}$) (ref.26)	<1000 (nitrification)
Algae	0.1–95 (avg. = ~3) (growth inhibition)	—
Aquatic invertebrates	0.6–>3300 (avg. = ~2.6) (LC$_{50}$)	0.1–10 (avg. = ~0.5) (reproduction)
Fish	0.5–100 (avg. = ~2.3) LC$_{50}$)	0.23–1 (growth, reproduction)
Other	21 (duckweed)	<100 (grass growth) <1000 (cowpea tissue damage)

[a] For all AE; from references 13–15
[b] No Observed Effects Concentration

Table 3.   Acute Aquatic Toxicities of Primary Alcohols (Reference 27)

Chemical	Fathead minnow 96 hr LC$_{50}$ (mg/L)	Water Solubility (mg/L)
Methanol	28,200	Miscible
Ethanol	14,700	Miscible
2-Propanol	10,000	Miscible
1-Butanol	1740	74,700
1-Hexanol	97.2	6270
1-Octanol	13.4	587
1-Nonanol	5.7	158
1-Decanol	2.3	34
1-Undecanol	1.04	8.5
1-Dodecanol	1.01	1.9
1-Tridecanol	No mortality in sat. solution	0.33

been developed to relate aquatic toxicity to alkyl chain length and EO number [28].

The differential sensitivity of aquatic species to surfactants (as illustrated in the toxicity ranges table) can lead to misinterpretations of the overall ecotoxicity of surfactants. Recent novel approaches have used small-scale ecosystems which are surveyed for the effects on a broader range of aquatic organisms. This stategy, called mesocosm testing, has been used for a $C_{12-13}AE_{6.5}$ surfactant, and tests produced a "mesocosm NOEC" of 0.28 mg/L [29].

Alkyl Aryl Sulfonates and Petroleum Sulfonates

Although these surfactant names have different product meanings to people in the surfactants field, both have common chemistries. Alkyl aryl sulfonates, by definition, have a branched or linear alkyl group attached to a sulfonated aromatic structure (benzene, substituted benzene, naphthalene, etc.). This definition could also be applied to petroleum sulfonates. Subtle distinctions in history, feedstocks and structure help to distinguish the two names. Alkyl aryl sulfonates history probably begins in the 1930s with kerylbenzene sulfonates, which were synthesized by alkylating benzene with chlorinated kerosene, and the generic name "alkylarylsulfonate" was applied [30]. An improved detergent alkylarylsulfonate was introduced after World War II. The alkyl group was tetrapropylene, and the surfactant was known as docecylbenzene sulfonate or DDBS, TPBS or ABS, also known as "hard" alkylate because of its poor biodegradability. In 1965 the linear alkyl version was introduced to give good biodegradability, and this started a worldwide conversion to "soft" alkylate which is almost complete.

Petroleum sulfonates are distinguished somewhat from alkyl aryl sulfonates by often containing more than one alkyl group (e.g., dialkyl benzene sulfonates), by higher molecular weight and oil solubility thereby making them useful in motor oils, and by the feedstock. Petroleum sulfonates are commonly associated with white oil manufacturing, and are formed by the oleum (SO_3 plus sulfuric acid) sulfonation of streams such as raffinate from lubricating oil streams or bottoms from other operations. Thus, they are a by-product of petroleum refining and are often called natural sulfonates. These petroleum sulfonates are chemically very ill-defined. Synthetic petroleum sulfonates such as alkyl orthoxylene sulfonates are produced to replace the natural products, usually when a more defined chemical structure is needed. The concept of producing petroleum sulfonates from crude oil, specifically for EOR, has been investigated. Like the natural petroleum sulfonates, these sulfonates from crude oil would be a "mixed bag" of chemicals.

The dilemma in discussing environmental acceptability of petroleum sulfonates is (1) the dearth of environmental data on this class and (2) the diversity of chemical structures which makes biodegradation and toxicity information difficult to interpret. A logical approach is to look at the environmental properties of a prototypical structure that represents petroleum sulfonates, This structure would be an alkyl aryl sulfonate, and the best candidate is linear alkylbenzene sulfonate. Linear alkylbenzene sulfonate (LAS) has been the focus of more environmental studies than any other surfactant, and it is a good surrogate for understanding the fate and effects of alkyl aryl sulfonates and petroleum sulfonates.

The structure below depicts the 3-phenyl isomer of a C_{12}-LAS; however, the phenyl portion can be attached at any of the internal alkyl carbons.

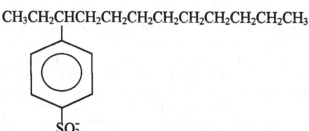

CH₃CH₂CHCH₂CH₂CH₂CH₂CH₂CH₂CH₂CH₂CH₃

The environmental properties of LAS have been the subject of many reviews. The discussions below on LAS fate and effects are largely included in these reviews [8–10, 31–34].

LAS Biodegradation.   A wide variety of laboratory tests have demonstrated that LAS is readily and completely degradable under aerobic conditions. Interestingly, the laboratory tests show longer biological half-lives than rates observed in rivers, streams and activated sewage sludge treatment facilities [35]. For example, half-lives in laboratory tests tend to be 1–4 days whereas field-measured values are just several hours. The stringency of many laboratory tests accounts for the discrepancy. All potential environmental compartments that could receive LAS have been tested, and no evidence of accumulation due to lack of biodegradability has been found. Monitoring of activated sludge type wastewater treatment plants in the U.S. [36, 37] and in Europe [38] showed 99 + % removal. Monitoring of the heavily impacted Mississippi River [39] and certain German rivers [40] indicated that LAS biodegradation prevents accumulation.

LAS does not degrade in laboratory anaerobic tests. There is the fear that this lack of anaerobic biodegradability will result in accumulation in anaerobic environmental compartments, even though environmental monitoring does not show accumulation. The lack of anaerobic biodegradability is a property of aromatic sulfonates, and presumably all aliphatic sulfonates. The reason is that microbial desulfonation mechanisms apparently rely primarily on broad-substrate range oxygenase enzymes that utilize O_2 to break the C—S bond to produce C-OH [41, 42]. Anaerobic desulfonation of LAS has not been demonstrated, although anaerobic desulfonation of 2-(4-sulfophenyl) butyrate and 4-tolylsulfonate has been observed [43].

The pathway of LAS biodegradation is shown in Figure 2. No single cultural isolate has been shown to mineralize LAS. The best characterized consortium of LAS degraders contained four members, three of which

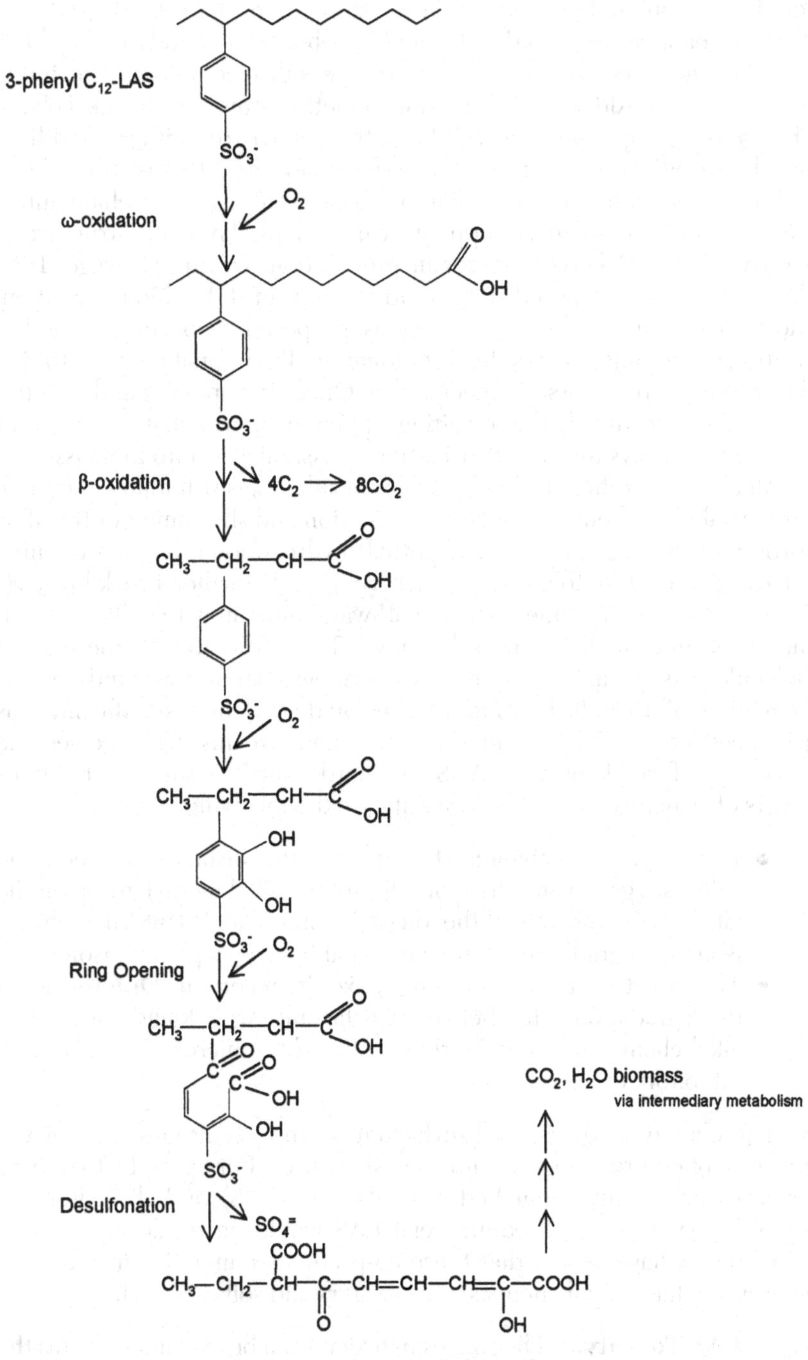

Figure 2. LAS catabolic pathway.

could carry out initial biodegradative steps or primary biodegradation [44]. The pathway compiled by Cain [45], Schöberl [46] and Swisher [9] is the most accepted, although knowledge is still incomplete. The initial attack is omega oxidation of the terminal methyl groups of the alkyl chain. The methyl group that is most distal to the phenyl group is attacked first, and the resulting oxidation produces an alkanoic acid that is shorted two carbons at a time via beta oxidation. The resulting short-chain inter-mediate is often called sulfophenyl carboxylate. The ring structure is hydroxylated next by oxygenases in preparation for ring cleavage. Ring cleavage is the proposed rate-limiting step in LAS biodegradation. Dioxygenase-catalyzed ring cleavage is proposed to occur at the 1–2 position of the ring followed by desulfonation. Desulfonation prior to ring cleavage is another possible mechanism. Once ring opening and desulfo-nation have occurred, the resulting aliphatic intermediates can enter common pathways for further oxidation or assimilation into biomass.

All alkyl aryl sulfonates and petroleum sulfonates will follow the same basic catabolic scheme: (a) terminal oxidation and shortening of the alkyl portion(s); (b) ring (phenyl or naphthyl) hydroxylation; (c) ring opening (cleavage) and desulfonation (or vice versa); (d) further breakdown via common intermediary metabolism pathways, ultimately to CO_2 or assim-ilation of intermediates into biomass. The structure of the parent molecule has great importance on biodegradation rate and extent. Branching of the alkyl group retards biodegradation significantly, as evidenced pre-1965 by foaming in rivers and streams that received the tetrapropyl form known as ABS or "hard" alkylate surfactant. Other effects of structure on biodegradability are the following:

- For a given homologue, the greater the distance between the sulfonate group and the more distant terminal methyl group on the alkyl chain the faster the degradation. That is, internal phenyl isomers degrade slower than external (e.g. 2-, 3-phenyl) isomers.
- The effect of alkyl chain length size is uncertain. Differences in biodegradation rates between surfactants with longer or shorter alkyl chains may actually reflect solubility (microbial uptake) and inhibitory effects.

Biodegradation of commercial surfactant mixtures is also affected by the presence of co-products with differing structures. For example, there was concern that methyl branched LAS (i.e., iso-LAS) and dialkyltetralin sulfonates (DATS) in all commercial LAS would be recalcitrant. How-ever, studies have shown that these co-products mineralize in receiving environmental compartments such as waters and soils [47, 48].

LAS Toxicity. The ranges in toxicity can be explained by: (a) the differential sensitivity of test species; marine organisms tend to be more

Table 4. **Range of Toxicity Values for LAS**[a]

Taxonomic Group	Acute Toxicity, EC_{50} (mg/L)	$NOEC^b$ (mg/L)
Bacteria	20–100 (growth inhibition)	0.5–172 (nitrification, respiration, structure)
Algae	0.1–170 (growth inhibition)	0.25–54 (photosynthesis, community composition)
Aquatic invertebrates	0.4–154 (LC_{50})	0.04–10 (growth, reproduction)
Fish	0.2–100 (LC_{50})	0.1–50 (growth, reproduction)
Other	10–1000 (plant foliar damage)	> 100 (effects from LAS in soil to food crops)
	> 500 (earthworm)	1 (photosynthesis)
	500–2000 mg/kg (rat oral toxicity)	> 5000 (dietary effects on rats)

[a] For C_{10}–C_{14} alkyl chain length LAS (detergent range); studies cited in refs. 8, 31, 32, 34
[b] No Observed Effects Concentration

sensitive; (b) substantial differences in toxicity between C_{10}-LAS (less toxic) and C_{14}-LAS (more toxic); (c) test methodology, particularly the water hardness since LAS is precipitated at high Ca^{++} and Mg^{++} and is non-bioavailable.

Obviously, the large range of toxicity values makes it difficult to arrive at a single LAS concentration, below which there is reasonable assuredness that the ecosystem is safe. The Dutch government's aquatic risk assessment of LAS used a systematic approach to deal with large numbers of independent toxicity studies and arrived at a maximum permissible concentration of 0.25 mg/L [7]. This level is far above levels measured or calculated in surface waters, thus indicating a margin of environmental safety.

The structure of LAS, and presumably other alkyl aryl sulfonates, determines the degree of toxicity. The length of the alkyl group is one factor mentioned above. Generally, any structural change that results in a comparative increase in lipophilicity will show a concomitant increase in toxic response. The position of the phenyl group along the alkyl chain illustrates this principle. The internal phenyl isomers show less toxicity when compared to external (2-, 3-phenyl) isomers.

Alkyl Phenol Ethoxylates (APE)

$$R-\left\langle\bigcirc\right\rangle-O-(CH_2CH_2-O-)_n\,H$$

Where $R = C_8-C_{16}$ linear or branched alkyl chain, and $n = 1-30$. The structure above shows the *para* positional isomer which accounts for 90% or more of the ring substitution positions; however, meta and ortho isomers also occur at lesser frequency.

The most common APE is nonylphenol ethoxylate (NPE) containing around 9 EO and is prepared from the propylene trimer which produces a multitude of highly branched alkyl chains. A C_8 APE, called octylphenol ethoxylate (OPE) is prepared from the dimer of isobutylene to produce the 1,1,3,3-tetramethylbutyl chain. Cost and performance have been enduring properties of APE even though they are under environmental scrutiny. Although their use in household cleaners has diminished, they are still in great demand for industrial cleaning and uses in agriculture, plastics, textiles and paper.

There is a large body of information on APE related to environmental safety. Most of this is summarized in review articles [9, 13, 14, 16], and the greatest amount of information is on NPE because of the long history of questions on biodegradability. Primary biodegradation (biotransformations that result in loss of surfactant properties) of NPE hse been known to readily occur, and monitoring across sewage treatment plants has shown 70–97% removal. Analyses of effluents demonstrated that the parent molecules were transformed to mono- and di-ethoxylates (NPE$_1$, NPE$_2$) and 4-nonylphenol (NP) which have greater aquatic toxicity than the original, fully ethoxylated NPE. The problem is that the highly branched nonyl chain is resistant to biodegradation, at least in the residence time of sewage treatment and in conditions where cultures are not well-adapted or acclimated to NPE. The consequence was that investigators concluded that NPE was recalcitrant to ultimate biodegradation since NP, NPE$_1$, and NPE$_2$ appeared to accumulate, and worse, these intermediates were more toxic than the starting material. Side-by-side comparisons of linear APE with NPE consistently showed better biodegradation of the linear forms. Thus, NPE became targeted as environmentally suspect. Bad became worse when testing for endocrine disrupter activity showed that NP and the shorter ethoxylates like NPE$_1$ and NPE$_2$ exhibited estrogenic effects (albeit weak) in aquatic organisms, mammals and birds [49–53]. The future for NPE is uncertain.

APE Biodegradation. Primary biodegradation of APE, both linear and branched, occurs with proper time and acclimation of degraders. However, complete or ultimate biodegradation (mineralization) has not been consistently noted in biodegradation studies, and when mineralization is unequivocally observed, it is slower compared to many other surfactants. Even the linear APE show reduced rates of ultimate biodegradation.

The pathways for APE are predictable but not like that of AE [54, 55]. The intramolecular scission of AE to form the hydrophobe and polyethylene glycol (PEG) has not been observed in APE. It appears that the EO chain is shortened one EO unit at a time. The precise nature of this shortening is unclear. It could be oxidative attack of the terminal EO to the carboxylate followed by cleavage of the terminal C_2, or it could involve another ether scission mechanism recently proposed [56]. The proposed mechanism, shown below, is reminiscent of the proposed anaerobic AE biodegradative pathway and involves a hydroxyl shift in the terminal EO.

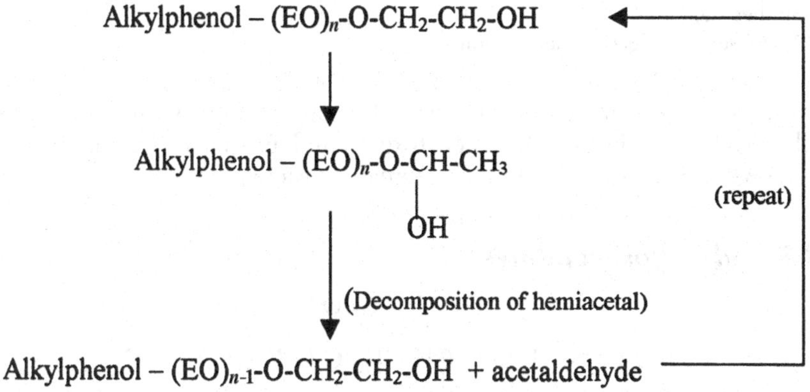

Alkylphenol – $(EO)_n$-O-CH_2-CH_2-OH

$\downarrow$

Alkylphenol – $(EO)_n$-O-CH-CH_3
$|$
OH   (repeat)

$|$ (Decomposition of hemiacetal)

$\downarrow$

Alkylphenol – $(EO)_{n-1}$-O-CH_2-CH_2-OH + acetaldehyde

Regardless of mechanism and groups of microbial degraders, the intermediate products invariably are NP, NPE_1, and NPE_2 which tend to accumulate. These intermediates partition readily into sludge and sediments which may make them less bioavailable and therefore more likely to resist further biodegradation. Further oxidation is possible, but studies are needed to elucidate the mechanisms. A carboxylate moiety at the terminus of the alkyl chain of NPE has been observed, and it is presumed that oxidative attack of the branched chain is feasible since many natural and inherently biodegradable aliphatic compounds contain branching.

APE Toxicity. Table 5 illustrates the increased toxicity of nonyl phenol and the 1 to 2 ethoxamer intermediates compared to the parent

Table 5.   Range of Toxicity Values for APE[a]

Taxonomic Group	Acute Toxicity, EC_{50} (mg/L)	$NOEC^b$ (mg/L)
Bacteria	20–800 (NPE_{4-30}) (growth inhibition)	—
Algae	5–>1000 (C_8–C_9 APE_{9-30}) 0.027–1.5 (NP) (growth inhibition)	—
Aquatic invertebrates	2.9–>100 (NPE_{9-12}) 0.043–3 (NP, NPE_1, NPE_2) (LC_{50})	10 (growth; NPE_9) 0.0067 (repro.; NP)
Fish	1.3–62 (C_8–C_9APE_{4-10}) 0.14–0.48 (NP)	2 (growth; NPE) 0.023 (growth; NP)
Other	<0.5 (NP; duckweed)	<10,000 (C_8–C_9APE; agric. plant tissue effects) <20 (NP; seedling growth)

[a] Studies cited in ref. 13
[b] No Observed Effects Concentration

surfactant with many ethoxamers. Toxicity studies with APE have observed a relationship between toxicity and EO chain length. Just as with AE, increasing EO number decreases toxicity.

Dialkyl Sulfosuccintates

$$O \qquad\qquad O$$
$$\parallel \qquad\qquad \parallel$$
$$R\text{-O-}C\text{-CH-CH}_2\text{-}C\text{-O-R}$$
$$\mid$$
$$SO_3^-$$

Where R = linear or branched alkyl groups, usually C_6, C_8, C_9.

The dialkyl sulfosuccinates are more of an industrial-use surfactant class rather than for use in detergents. Therefore, environmental data on these surfactants are meager. Early biodegradation data revealed an interesting property – primary biodegradation occurred in dibenzyl-, di-(2-ethylhexyl)-, di-(3,4,5-trimethylhexyl)-, and diisobutyl sulfosuccinates, but not in the di-(1,3-dimethyl)- or dicyclohexyl sulfosuccinates [57]. This indicates possible hindrance of hydrolytic cleavage when there is branching (e.g., methyl or saturated ring structure) at the number one carbon of the alcohol. Essentially the R group that showed no biodegradation is a secondary alcohol. More recent biodegradation studies of a dialkyl sulfosuccinate, comprised of linear, primary C_6–C_8 alcohols, exhibited good primary and ultimate biodegradation [58]. The proposed pathway had an initial hydrolysis of the ester linkage that is most distant from the sulfonate group. Subsequent omega oxidation and beta-oxidative chain shortening of the alkyl chain of the monoalkyl sulfosuccinate would occur, and eventually a second hydrolytic cleavage would form sulfosuccinic acid, which presumably is further metabolized.

Although the data are limited, the aquatic toxicity of dialkyl sulfosuccinates does not indicate unusual toxicity for a surfactant. The LC_{50} for rainbow trout is 28 mg/L [59].

Quaternary Ammonium Surfactants

$$R—N^+—R_2$$

with R_1 above and R_3 below the nitrogen.

Where the R groups may be methyl groups, linear or branched aliphatics, or aromatics.

Quaternary ammonium compounds or "quats" that are monoalky structures typically have a C_{12-16} alkyl chain and three methyl groups bonded to the quaternary nitrogen atom. Dialkyl quats have two alkyl groups and two methyl groups. Quats can also have aromatic structures such as a benzyl group as one of the R groups. Also, the quaternary nitrogen can be in a pyridine or an imidazole structure.

Quats are included in the list of surfactants for EOR, but technically they are used in other oil field operations, particularly in drilling muds. Their cationic nature, ability to emulsify oils plus potent germicidal activity make quats uniquely useful in mineral processing and oil applications. There is an abundance of environmental data on quats, not because of oil field operations, but rather because of their use as fabric softeners in detergents. The best known fabric softener and antistatic agent is ditallow dimethyl ammonium chloride (DTDMAC). Emulsifiers commonly have a N-alkyltrimethylammonium chloride or N-alkylimidazoline chloride configuration, and germicides, such as benzalkonium chloride, typically have a N-alkyldimethylbenzylammonium chloride structure. The fate and effects of these multifunctional surfactants are detailed in several reviews [9, 55, 60, 61].

Quats Biodegradation.   Laboratory testing to simulate wastewater treatment or river die-away generally comprises the data base on the fate of quats. Limited monitoring data of sewage treatment plants is also available. Removal in sewage treatment plants is expected to be 90% or greater, and although quats will sorb to solids and anionic surfactants, the removal mechanism is thought to be mostly biodegradation. One of the problems with testing germicidal quats is the establishment of test concentrations that do not produce a toxic response in the inoculum. Tests that use a low biomass can be completely inhibited by typical test concentrations of 10 mg/L. If this pitfall is avoided, the complete mineralization of quats can be demonstrated.

River water die-away and model stream testing has shown that detergent range quats degrade rapidly [62–68]. Monoalkyl, trimethyl quats degraded with a half-life of 2.7 days; however, the dialkyl, dimethyl varieties required much longer in river water without suspended solids. Subsequent work demonstrated that sorption on sediments improved biodegradability of the dialkyl quats possibly by substrate concentration and by maintaining an adaptive response of the attached microbial community to the quat. It was also shown that previous exposure to quats in model streams dramatically improved half lives 14–50 fold, presumably by ensuring an acclimated population. The same is true in soils. Subsurface sediments that were chronically exposed to detergent range quats showed the ability to support extensive mineralization of both monoalkyl and dialkyl quats; however, the dialkyl quats were still slower to degrade [17]. The overall picture for both water and soil environments is that if variables such as quat concentration, biomass, and acclimation are correct, then the quats will mineralize and will not accumulate in the receiving environmental compartment.

The pathways of quat biodegradation are straightforward: (1) the favored initial attack is fission of the C—N bond by a monooxygenase and (2) the alternative route is terminal (omega) oxidation of the alkyl portion(s) followed by shortening of the alkyl chain via beta oxidation. In the first or favored route a monoalkyl, trimethyl ammonium compound is cleaved to form alkanal and trimethylamine. The alkanal is further oxidized and catabolized by beta oxidation. The trimethylamine is broken down by sequential steps involving cleavage to formaldehyde + dimethylamine,then formaldehyde + methylamine and finally formalde-hyde + ammonia. Alternatively, the first steps can be attack of the C—N bond to release formaldehyde in consecutive steps until the longer alkylamine is cleaved to alkanal and ammonia. All of the steps involve monooxygenase activity. Therefore, the possibility for biodegradation in the absence of O_2 is very remote.

Quats Toxicity.   The structure of the quats has a big effect on the toxicity as evidenced by germicidal activity associated with the many structures of quats [66]. The ecotoxicological evaluation of quats should consider if the quat's function is as a germicide or as a surfactant, or as a fabric softener/antistatic agent. Studies on quats toxicity are designed from the viewpoint that these compounds are present at low concentra-

Table 6.   **Range of Toxicity Values of Quaternary Ammonium Surfactants**[a]

Taxonomic Group	Acute Toxicity, EC_{50} (mg/L)	$NOEC^b$ (mg/L)
Bacteria	(germicidal properties vary with quat)	3–40 (nitrification, respiration; monoalkyl quat)
Algae	0.1–>4 (DTDMAC) 0.03–0.6 (monoalkyl quats) (growth inhibition)	0.1–18 (photosynthesis; monoalkyl quats)
Aquatic invertebrates	1.2–5.8 (monoalkyl quat)	0.38 (DTDMAC)
Fish	0.4–~10	0.23
Other	200–>5000 mg/kg (rat oral toxicity)	—

[a] For all quats or as indicated; refs. 60, 67–70
[b] No Observed Effects Concentration

tions in surface waters, ostensibly from sewage that may or may not have been treated. The variables that affect toxic response of organisms in the receiving surface waters are the presence and level of acclimated biomass and the degree of suspended solids which sorb the quats and thus reduce toxicity. An aquatic risk assessment by Lewis and Wee [67] is a good example of these considerations.

Concluding Remarks

In general, the surfactants used in petroleum applications should not cause undue concern. The familiarity of human exposure to surfactants in the constant processes of bathing and dish washing coupled to the absence of observable acute or chronic toxicity problems has given us a sense that common surfactants are innocuous. We tend to extrapolate this "safe-to-use" concept to all terrestrial organisms and applications. However, this comfort zone with commercial surfactants should not be extended to situations where they enter surface waters, because surfactants exhibit considerable toxicity to aquatic organisms.

Surfactant usage in the petroleum industry will probably increase as new applications are found and older applications like surfactant flooding are implemented. The regulatory situation now is different than the 1970s, and the use of chemicals carries with them the need to understand the environmental fate and effects of these chemicals in normal applications and in accidental releases. Environmental risk assessment is a systematic, yet simple, process for doing this.

Risk assessments on surfactant usage require knowledge on biodegradation and toxicity of these chemicals. The foregoing review of studies on biodegradation of surfactant classes should help at least in understanding the basic principles and the types of data needed to make conclusions on persistence. There are underlying, common themes in surfactant biodegradation, such as mechanisms for degrading the alkyl chains that form the hydrophobes of all commercial surfactants. For example, regardless of surfactant class, the mechanism of omega oxidation followed by beta oxidation was prominent. Likewise, the hydrolytic cleavage of alkoxamers is a common theme. The microbial world is diverse and ubiquitous, yet the unity of biochemistry is evidenced in these common themes. It gives the investigator confidence in predicting the fate of surfactants in a variety of environmental compartments.

Toxicity of surfactants is also predictable. The wide range of toxicity values may seem confusing, but one must remember that surfactant structure can influence toxicity and that the standardized test methodology itself has many variables that affect toxicity values. Examples of the latter are length of testing, temperature, test water composition (e.g., hardness), species and the age of test organisms. In spite of these variables it is possible to make rationalizations and correlations and ultimately to arrive at sound judgements on the environmental safety of surfactants.

References

1. Roberts, D.W. *Sci. Total Environ.* **1991**, *109/110*, 557–568.
2. Ranney, M.W. In *Crude Oil Drilling Fluids*; Noyes Data Corp.: Park Ridge, New Jersey, 1979, pp 204–268.
3. DiStasio, J.I. In *Chemicals for Oil Field Operations*; Noyes Data Corp.: Park Ridge, New Jersey, 1981, pp 156–238.
4. Maddin, C.M. *First International Conference on Health, Safety and Environment, The Hague, The Netherlands, Nov. 1991*, Society of Petroleum Engineers, Richardson, Texas, 1991, paper SPE 23354.
5. *Environmental and Human Safety of Major Surfactants. Vol. I. Anionic Surfactants. Part 2. Alcohol Ethoxy Sulfates*, a final report to The Soap and Detergent Assoc., New York by Arthur D. Little, Inc. Cambridge, MA, 1991, Reference 65913.
6. *Environmental Data Review of Alkyl Ether Sulphates (AES)*, a report compiled by BKH Consulting Engineers, Delft, The Netherlands for the NVZ and European surfactant industry, October 1994.
7. Feijtel, T.C.J. ; van de Plassche, E.J. *Environmental Risk Characterization of 4 Major Surfactants Used in the Netherlands*, Report 79101–025, National Institute of Public Health and Dutch Soap Association, September, 1995.

8. Painter, H.A. In *Handbook of Environmental Chemistry, Vol. 3 Part F, Anthropogenic Compounds*; Hutzinger, O., Ed.; Springer-Verlag: Berlin, 1992; pp 1–88.
9. Swisher, R.D. *Surfactant Biodegradation*; Marcel Dekker: New York, 1987.
10. Steber, J.; Berger, H. In *Biodegradability of Surfactants*; Karsa, D.R; Porter, M.R., Eds.; Blackie Academic: London, 1995, pp 134–182.
11. Gilbert, P.A.; Pettigrew, R. *Int. J. Cosmet. Sci.* **1984**, *6*(4), 555–560.
12. Neubecker, T.A. *Environ. Sci. Technol.* **1985**, *19*(12), 1232–1236.
13. Talmadge, S.S. *Environmental and Human Safety of Major Surfactants; Alcohol Ethoxylates and Alkylphenol Ethoxylates*; Lewis Publishers: Boca Raton, 1994.
14. Holt, M.S.; Mitchell, G.C.; Watkinson, R.J. In *Handbook of Environmental Chemistry, Vol. 3 Part F, Anthropogenic Compounds*; Hutzinger, O., Ed.; Springer-Verlag: Berlin, 1992, pp 89–144.
15. *Environmental Data Review of Alcohol Ethoxylates (AE)*, a report compiled by BKH Consulting Engineers, Delft, The Netherlands for the NVZ and European surfactant industry, January 1994.
16. Balson, T.; Felix, M.S.B. In *Biodegradability of Surfactants*; Karsa, D.R.; Porter, M.R., Eds.; Blackie Academic: London, 1995, pp 204–230.
17. Federle, T.W.; Pastwa, G.M. *Groundwater* **1988**, *26*, 761–770.
18. Steber, J.; Wierich, P. *Water Res.* **1987**, *21*(6), 661–667.
19. White, G.F. *Pestic. Sci.* **1993**, *37*, 159–166.
20. White, G.F. *Microbiol. Rev.* **1996**, *60*, 216–232.
21. White, G.F, Higgins, T.P., John, D.M. *Proceedings of the International Symposium on Environmental Biotechnology*, Oostende, Belgium, 1997, pp 265–274.
22. Steber, J.; Wierich, P. *Appl. Environ. Microbiol.* **1985**, *49*, 530–537.
23. Wagener, S.; Schink, B. *Appl. Environ. Microbiol.* **1988**, *54*, 561–565.
24. Kravetz, L.; Salanitro, J.P.; Dorn, P.B.; Guin, K.F. *J. Am. Oil Chem. Soc.* **1991**, *68*, 610–618.
25. Naylor, C.G.; Castaldi, F.J.; Hayes, B.J. *J. Am. Oil Chem. Soc.* **1988**, *65*, 1669–1676.
26. Russel, G.L.; Britton, L.N. *Proceedings of the Annual Meeting, SETAC*, Nov. 1997.
27. Veith, G.D; Call, D.J.; Brooke, L.T. In *Aquatic Toxicology and Hazard Assessment: Sixth Symposium. ASTM STP 802*; Bishop, W.E.; Cardwell, R.D.; Heidolph, B.B., Eds., 1983, pp 90–97.
28. Wong, D.C.L.; Dorn, P.B.; Chai, E.Y. *Environ. Toxicol. Chem.* **1997**, *16*(9), 1970–1976.
29. Dorn, P.B; Rodgers, J.H., Jr.; Gillespie, W.B., Jr. *Environ. Toxicol. Chem.* **1997**, *16*(8), 1634–1645.
30. Feighner, G.C. In *Anionic Surfactants, Vol. 1*; Linfield, W.M., Ed.; Marcel Dekker, New York, 1976, pp 253–314.
31. *Environmental and Human Safety of Major Surfactants, Vol. 1. Anionic Surfactants, Part 1. Linear Alkylbenzene Sulfonates*, a final report to the Soap and Detergent Assoc. by Arthur D. Little, Cambridge, MA, Feb. 1991, reference 65913.

32. Painter, H.A.; Zabel, T.F. *Review of the Environmetal Safety of LAS*; a report prepared by WRc, Medmenham, UK, 1988.
33. *Environmental Fate and Behaviour of LAS; Literature Review*; a report prepared by BKH Consulting Engineers, Delft, The Netherlands, May 1993.
34. *The Use of Existing Toxicity Data for the Estimation of the Maximum Tolerable Environmental Concentration of Linear Alkyl Benzene Sulfonate; Part I: Main Report; Part II: Data Lists*; a report prepared by BKH Consulting Engineers, Delft, The Netherlands, May, 1993.
35. Britton, L.N. *J. Surf. and Det.* **1998**, *1*, 109–117.
36. McAvoy, D.C.; Eckhoff, W.S.; Rapaport, R.A. *Environ. Toxicol. Chem.* **1993**, *12*(6), 977–987.
37. McAvoy, D.C.; Dyer, S.D.; Fendinger, N.J.; Eckhoff. W.S.; Lawrence, D.L.; Begley W.M. *Environ. Toxicol. Chem.* **1998**, *17*(9), 1705–1711.
38. Waters, J.; Feijtel, T. *Chemosphere* **1995**, *30*, 1939–1956.
39. Tabor, C.F.; Barber, L.B. *Environ. Sci. Technol.* **1996**, *30*, 161–171.
40. Schöberl, P. *Tenside. Surf. Det.* **1997**, *34*(4), 233–237.
41. Zürrer, D.; Cook, A.M.; Leisinger, T. *Appl. Environ. Microbiol.* **1988**, *53*(7), 1459–1463.
42. Kertesz, M.A.; Kölbener, P.; Stockinger, H.; Beil, S.; Cook, A.M. *Appl. Environ. Microbiol.* **1994**, *60*(7), 2296–2303.
43. Denger, K.; Kertesz, M.A.; Vock, E.H.; Schön, R.; Mägli, A.; Cook, A.M. *Appl. Environ. Microbiol.* **1996**, *62*(5), 1526–1530.
44. Jimènez, L.; Breen, A.; Thomas, N.; Federle, T.W.; Sayler, G.S. *Appl. Environ. Microbiol.* **1991**, *57*, 1566–1569.
45. Cain, R.B. *Biochem. Soc. Trans.* **1987**, *15* (suppl.), 7S.
46. Schöberl, P. *Tenside Surf. Det.* **1989**, *26*, 86–94.
47. Trehy, M.L.; Gledhill, W.E.; Mieure, J.P.; Adamove. J.F.; Nielsen, A.M.; monoalkyl quats)Perkins, H.O.; Eckhoff, W.S. *Environ. Toxicol. Chem.* **1996**, *15*, 233–240.
48. Nielsen, A.M.; Britton, L.N.; Beall, C.E.; McCormick, T.P.; Russell, G.L. *Environ. Sci. Technol.* **1997**, *31*(12), 3397–3404.
49. Jobling, S.; Sumpter, J.P. *Aquat. Toxicol.* **1993**, *27*, 361–372.
50. White, R.; Jobling, S.; Hoare, S.A.; Sumpter, J.P.; Parker, M.G. *Endocrinology* **1994**, *135*, 175–183.
51. Jobling, S.; Sheahan, D.; Osborne, J.A.; Matthiessen, P.; Sumpter, J.P. *Environ. Toxicol. Chem.* **1996**, *15*, 194–202.
52. Soto, A.M.; Justicia, H.; Wray, J.W.; Sonenschein, C. *Environ. Health Perspect.* **1991**, *92*, 167–173.
53. Bicknell, R.J.; Herbison, A.E.; Sumpter, J.P. *J. Steroid Biochem. Mol. Biol.* **1995**, *54*, 7–9.
54. White, G.F. *Pestic. Sci.* **1993**, *37*, 159–166.
55. Van Ginkel, C.G. *Biodegradation* **1996**, *7*, 151–164.
56. John, D.M.; White, G.F. *J. Bacteriol.* **1998**, *180*, 4332–4338.
57. Hammerton, C. *Proc. Soc. Water Treat. Exam.* **1956**, *5*, 145–174.
58. Hales, S.G. *Environ. Toxicol. Chem.* **1993**, *12*, 1821–1828.
59. Goodrich, M.S.; Melancon, M.J.; Davies, R.A.; Lech, J.J. *Water Res.* **1991**, *25*, 119–125.

60. Boethling, R.S; Lynch, D.G. In *The Handbook of Environmental Chemistry, Vol. 3, Part F, Anthropogenic Compounds, Detergents*; Hutzinger, O., Ed.; Springer-Verlag: Berlin, 1992, pp 145–177.
61. Van Ginkel, C.G. In *Biodegradability of Surfactants*; Karsa, D.R.; Porter, M.R., Eds.; Blackie Academic: London, 1995, pp 183–203.
62. Larson, R.J.; Perry, R.L. *Water Res.* **1981**, *15*, 697–702.
63. Larson, R.J.; Vashon, R.D. *Dev. Indust. Microbiol.* **1983**, *24*, 425–434.
64. Ventullo, R.M.; Larson, R.J. *Appl. Environ. Microbiol.* **1986**, *51*, 356–536.
65. Larson, R.J.; Bishop, W.E. *Soap/Cosmetics/Chem. Spec.* **1988**, *64*, 58.
66. Lawrence, C.A. In *Cationic Surfactants*; Jungermann, E., Ed.; Marcel Dekker, New York, 1970, 491.
67. Lewis, M.A.; Wee, W.T. *Environ. Toxicol. Chem.* **1983**, *2*, 105–108.
68. Lewis, M.A.; Hamm, B.G. *Water Res.* **1986**, *20*, 1575.
69. Woltering, D.M.; Larson, R.J.; Hopping, W.D.; Jamieson, R.A.; de Oude, N.T. *Tenside Det. Surf.* **1987**, *24*, 286.
70. Knauf, W. *Tenside. Surf. Det.* **1973**, *10*, 251.

RECEIVED for review October 13, 1998. ACCEPTED revised manuscript January 6, 1999.

GLOSSARY AND INDEXES

14

Glossary of Surfactant Terminology

Laurier L. Schramm

Petroleum Recovery Institute, 100, 3512 – 33rd St. NW, Calgary, AB,
Canada T2L 2A6 and University of Calgary, Dept. of Chemistry, 2500
University Drive NW, Calgary, AB, Canada T2N 1N4

Preface

In the 200 years since Thomas Graham founded the discipline of colloid
science, a vast number of terms have come to be associated with colloid
and interface science and, in particular, with the sub-discipline of
surfactant science. In addition to the fundamental science, there is a
great diversity of occurrences and properties of surfactants in industry
and in everyday life. This chapter provides brief explanations for the most
important terms that may be encountered in a study of the fundamental
principles, experimental investigations, and petroleum industry-related
applications of surfactant science. Specific literature citations are given
when the sources for further information are particularly useful or
unique. For terms drawn from fundamental colloid and interface science,
much reliance was placed on the recommendations of the IUPAC
Commission on Colloid and Surface Chemistry [1]. For more comprehensive dictionaries and glossaries of terms in colloid and interface
science, see references [2–7].

Terms

Acid Number.   *See* Total Acid Number.

ACN.   Alkane carbon number, *see* Equivalent Alkane Carbon Number.

Activator.   Any agent that may be used in froth flotation to enhance
selectively the effectiveness of collectors for certain mineral components.
See also Froth Flotation.

Active Surfactant.   The primary surfactant in a detergent formulation. *See also* Detergent.

Adhesion.   The attachment of one phase to another.

Admicelle.   *See* Hemimicelle.

Adsolubilization.   A surface analog of micellar solubilization in which adsorbed surfactant bilayers (admicelles) absorb solutes from solution. Example: the partitioning of sparingly soluble organic molecules from water into admicelles. *See* reference [8]

Adsorbate.   A substance that becomes adsorbed at the interface or into the interfacial layer of another material, or adsorbent. *See* Adsorption.

Adsorbent.   The substrate material onto which a substance is adsorbed. *See* Adsorption.

Adsorption.   The increase in quantity of a component at an interface or in an interfacial layer. In most usage it is positive, but it can be negative (depletion); in this sense negative adsorption is a different process from desorption. Adsorption may also denote the process of components accumulating at an interface.

Adsorption Isotherm.   The mathematical or experimental relationship between the equilibrium quantity of a material adsorbed and the composition of the bulk phase, at constant temperature. The adsorption isobar is the analogous relationship for constant pressure, and the adsorption isostere is the analogous relationship for constant volume.

Aerated Emulsion.   A foam in which the liquid consists of two phases in the form of an emulsion. Also termed foam emulsion. Example: whipped cream consists of air bubbles dispersed in cream, which is an emulsion. *See also* Foam.

Aerating Agent.   *See* Foaming Agent.

Aerosol.   Colloidal dispersions of liquids or solids in a gas. Distinctions are made among aerosols of liquid droplets (fog, cloud, drizzle, mist, rain, spray) and aerosols of solid particles (fume and dust). *See* reference [3].

Ageing.   The properties of many colloidal systems may change with time in storage. Ageing in crude oils may refer to changes in composition due to oxidation, precipitation of components, bacterial action, or evaporation of low-boiling components. Ageing in emulsions or foams may refer to any of aggregation, coalescence, creaming or chemical changes. Aged emulsions and foams frequently have larger droplet or bubble sizes.

Aggregate.   A group of species, usually droplets, bubbles, particles or molecules, that are held together in some way. A micelle can be considered to be an aggregate of surfactant molecules or ions.

Aggregation. The process of forming a group of droplets, bubbles, particles, or molecules that are held together in some way. This process is sometimes referred to interchangeably as coagulation or flocculation, although in some usage these terms have discinct meanings. The reverse process is termed deflocculation or peptization.

Aggregation Number. The number of surfactant molecules or ions composing a micelle. Example: the aggregation number for dodecyl sulfate ions in water is about 70.

Air Drilling Fluid. Air when used as an oil and gas well drilling fluid. An air drilling fluid may contain a small amount of water, in which case a more specific term is mist drilling fluid. If the water also contains a foaming agent (surfactant), then the more specific term is foam drilling fluid. Gases other than air are sometimes used, such as nitrogen or natural gas. *See also* Foam Drilling Fluid, Stable Foam, Stiff Foam.

Alcohol Resisting Aqueous Film Forming Foam. (AFFF-AR) A fire extinguishing foam formulated specifically for alcohol, polar solvent, and hydrocarbon fires. *See also* Fluoroprotein Foam, Film Forming Fluoroprotein Foam, Aqueous Film Forming Foam.

Alkane Carbon Number. (ACN) *See* Equivalent Alkane Carbon Number.

Amphipathic. Having both lyophilic and lyophobic groups (properties) in the same molecule, as in the case of surfactants. Also referred to as being amphiphilic.

Amphiphilic. *See* Amphipathic.

Amphoteric Surfactant. A surfactant molecule for which the ionic character of the polar group depends on solution pH. For example, lauramidopropyl betaine $C_{11}H_{23}CONH(CH_2)_3N^+(CH_3)_2CH_2COO^-$ is positively charged at low pH but is electrically neutral, having both positive and negative charges at intermediate pH. Other combinations are possible and some amphoteric surfactants are negatively charged at high pH. *See also* Zwitterionic Surfactant.

Ancillaries. The non-surface active, complementary components in a detergent formulation. *See also* Detergent.

Anionic Surfactant. A surfactant molecule that can dissociate to yield a surfactant ion whose polar group is negatively charged. Example: sodium dodecyl sulfate, $CH_3(CH_2)_{11}SO_4^- Na^+$.

Anti-Bubbles.   A dispersion of liquid-in-gas-in-liquid wherein a droplet of liquid is surrounded by a thin layer of gas that in turn is surrounded by bulk liquid. Example: in an air–aqueous surfactant solution system this would be designated as water-in-air-in-water, or W/A/W. A liquid–liquid analogy can be drawn with the structures of multiple emulsions. *See also* reference [9], Fluid Film.

Antielectrostatic Agent.   A surfactant formulation that may be applied to a fabric or fibres to reduce the buildup of static electricity. Examples: alkyl sulfonates and alkyl phosphates.

Antifoaming Agent.   Any substance that acts to reduce the stability of a foam; it may also act to prevent foam formation. Terms such as antifoamer or foam inhibitor are used to specify the prevention of foaming, and terms such as defoamer or foam breaker are used to specify the reduction or elimination of foam stability. Example: poly(dimethylsiloxane)s, $(CH_3)_3SiO[(CH_3)_2SiO]_xR$, where R represents any of a number of organic functional groups. Antifoamers may act by any of a number of mechanisms.

Antiredeposition Agent.   A component in a detergent formulation that acts to help prevent redeposition of dispersed dirt or grease. Example: carboxymethylcellulose. *See also* reference [4], Detergent.

Antistatic Agent.   *See* Antielectrostatic Agent.

Antonow's Rule.   An empirical rule for the estimation of interfacial tension between two liquids as the difference between the surface tensions of each liquid. Even for pure liquids this rule is seldom very accurate.

A/O/W.   An abbreviation for a fluid film of oil between air and water. Usually designated W/O/A. *See* Fluid Film.

Aphrons.   *See* Microgas Emulsions.

Apolar.   Description applied to materials or surfaces that have no polar nature.

Aqueous Film Forming Foam.   (AFFF) A fire extinguishing foam based on blended hydrocarbon and fluorocarbon surfactants. Used as a rapidly spreading foam on hydrocarbon fires. *See also* Fluoroprotein Foam, Film Forming Fluoroprotein Foam, Alcohol Resisting Aqueous Film Forming Foam.

Areal Elasticity.   *See* Film Elasticity.

Asphaltene.   A high-molecular-mass, polyaromatic component of some crude oils that also has high sulfur, nitrogen, oxygen, and metal contents.

In practical work asphaltenes are usually defined operationally by using a standardized separation scheme. One such scheme defines asphaltenes as those components of a crude oil or bitumen that are soluble in toluene but insoluble in *n*-pentane.

Association Colloid. A dispersion of colloidal-sized aggregates of small molecules; it is lyophilic. Example: micelles of surfactant molecules or ions in water.

A/W/A. An abbreviation for a fluid film of water in air. *See* Fluid Film.

A/W/O. An abbreviation for a fluid film of water between air and oil phases. Also termed pseudoemulsion film. Usually designated O/W/A. *See* Fluid Film.

Bancroft's Rule. An empirical generalization that predicts that the continuous phase in an emulsion will be the phase in which the emulsifying agent is most soluble. An extension for solid particles acting as emulsifying agents predicts that the continuous phase will be the phase that preferentially wets the solid particles. *See also* Hydrophile–Lipophile Balance.

Beaker Test. *See* Bottle Test.

Bicontinuous System. A two-phase system in which both phases are continuous phases. For example, a possible structure for middle-phase microemulsions is one in which both oil and water phases are continuous throughout the microemulsion phase. *See also* Middle-Phase Microemulsion.

Bilayer. *See* Bimolecular Film.

Biliquid Foam. A concentrated emulsion of one liquid dispersed in another liquid.

Bimolecular Film. A membrane that separates two aqueous phases and is composed of two layers of polar organic molecules, such as surfactants or lipids, that are oriented with their hydrocarbon groups in the two molecular layers towards each other and the polar groups facing the respective aqueous phases. *See also* Vesicle.

Birefringent. A material that has different refractive indices in different directions. Example: liquid crystals.

Black Film. Fluid films yield interference colors in reflected white light that are characteristic of their thickness. At a thickness of about 0.1 μm the films appear white and are termed silver films. At reduced thicknesses they first become grey and then black (black films). There are two kinds of thin equilibrium (black) films: those that correspond to a primary minimum in interaction energy, typically at thicknesses of about 5 nm (Newton black films), and those that correspond to a secondary minimum, typically at thicknesses of about 30 nm (common black films).

Blender Test.   An empirical test in which an amount of potential foaming agent is added into a blender containing a specified volume of liquid to be foamed. After blending at a specified speed and for some specified time, the blending is halted and the extent (volume) of foam produced is measured both immediately and after a period of time of quiescent standing. There are many variations of this test. *See also* Bottle Test.

Bottle Test.   *Emulsions.*   An empirical test in which varying amounts of a potential demulsifier or coagulant are added into a series of tubes or bottles containing subsamples of an emulsion or other dispersion that is to be broken or coagulated. After some specified time the extent of phase separation and appearance of the interface separating the phases are noted. There are many variations of this test. For emulsions, in addition to the demulsifier, a diluent may be added to reduce viscosity. In the centrifuge test, centrifugal force may be added to speed up the phase separation. Other synonyms include jar test, beaker test.
Foams.   An empirical test in which an amount of potential foaming agent (or even defoaming agent) is added into a bottle containing a specified volume of liquid to be foamed. After shaking the bottle in a specific manner and for some specified time, the shaking is halted and the extent (volume) of foam produced is measured both immediately and after a period of time of quiescent standing. There are many variations of this test. *See also* Blender Test.

Bubble Point.   The gas pressure at which gas bubbles are generated and evolved from a liquid.

Builder.   A chemical compound added into detergent formulations to aid oil emulsification (by raising pH) and to complex and solubilize hardness ions. Example: sodium tripolyphosphate.

Calculation of Phase Inversion in Concentrated Emulsions. (CAPICO) A system in which potential cosmetic emulsion ingredients are numerically categorized so that one may calculate their influence on the phase inversion temperature of a formulated emulsion. See reference [10].

CAPICO.   *See* Calculation of Phase Inversion in Concentrated Emulsions.

Capillary Flow.   Liquid flow in response to a difference in pressures across curved interfaces. *See also* Capillary Pressure.

Capillary Forces.   The interfacial forces acting among oil, water, and solid in a capillary or in a porous medium. These determine the pressure difference (capillary pressure) across an oil–water interface in the capillary or in a pore. Capillary forces are largely responsible for oil entrapment under typical petroleum reservoir conditions.

Capillary Number.   A dimensionless ratio of viscous to capillary forces. It is used to provide a measure of the magnitude of forces that trap residual oil in a porous medium.

Capillary Pressure.   The pressure difference across an interface between two phases. When the interface is contained in a capillary, it is sometimes referred to as the suction pressure. In petroleum reservoirs it is the local pressure difference across the oil–water interface in a pore contained in a porous medium.

Capillary Ripples.   Surface or interfacial waves caused by perturbations of an interface. Where the perturbations are caused by mechanical means (e.g., barrier motion) the transverse waves are known as capillary ripples or Laplace waves, and the longitudinal waves are known as Marangoni waves. The characteristics of these waves depend on the surface tension and the surface elasticity.

Capillary Rise.   The surface tension-driven process by which a liquid rises in a capillary.

Capillary Waves.   *See* Capillary Ripples.

Cationic Surfactant.   A surfactant molecule that can dissociate to yield a surfactant ion whose polar group is positively charged. Example: cetyltrimethylammonium bromide, $CH_3(CH_2)_{15}N^+(CH_3)_3Br^-$.

CCC.   *See* Critical Coagulation Concentration.

Centrifuge Test.   *See* Bottle Test.

Charge Density.   In colloidal systems, the quantity of charge at an interface, expressed per unit area.

Charge of the Micelle.   *See* Micellar Charge.

Charge Reversal.   The process by which a charged substance takes on a new charge of the opposite sign. Such a change can be brought about by any of oxidation, reduction, dissociation, ion exchange, or adsorption. Example: the adsorption of cationic surfactant molecules onto negatively charged clay particles can exceed that required for charge neutralization and cause charge reversal.

Chocolate Mousse Emulsion.   A name frequently used to refer to the W/O emulsions of high water content that are formed when crude oils are spilled on the oceans. The name reflects the color and very viscous consistency of these emulsions. It has also been applied to other petroleum emulsions of similar appearance.

Clotted Soap.   *See* Middle Soap.

Cloud Point.   The transition temperature above which a nonionic surfactant or wax loses some of its water solubility and becomes ineffective as a surfactant. The originally transparent surfactant solution becomes cloudy because of the separation of a surfactant-rich phase. *See also* Coacervation.

cmc.   *See* Critical Micelle Concentration.

Coacervation.   When a lyophilic colloid loses stability, a separation into two liquid phases may occur. This process is termed coacervation. The phase that is more concentrated in the colloid is the coacervate, and the other phase is the equilibrium solution. *See also* Cloud Point.

Coactive Surfactant.   The secondary surfactant(s) in a detergent formulation. *See also* Detergent.

Coadsorption.   The adsorption of more than one species simultaneously.

Coagulation.   *See* Aggregation.

Coalescence.   The merging of two or more dispersed species into a single one. Coalescence reduces the total number of dispersed species and also the total interfacial area between phases. In emulsions and foams coalescence can lead to the separation of a macrophase, in which case the emulsion or foam is said to break.

Cohesion.   The tendency of a body of a substance to resist being mechanically pulled apart.

Collapse Pressure.   The film pressure required to cause a surface or interfacial monomolecular film to compress to an area that will no longer support a monolayer of adsorbed species; thus it will distort and collapse.

Collector.   A surfactant used in froth flotation to adsorb onto solid particles, make them hydrophobic, and thus facilitate their attachment to gas bubbles. *See also* Froth Flotation.

Colloidal.   A state of subdivision in which the particles, droplets, or bubbles dispersed in another phase have at least one dimension between about 1 and 1000 nm. A colloidal dispersion is a system in which colloidal species are dispersed in a continuous phase of different composition or state.

Colloidal Electrolyte.   An electrolyte that dissociates to yield ions at least one of which is of colloidal or near-colloidal size. Example: ionic surfactant micelles.

Colloidal Gas Aphrons.   *See* Microgas Emulsions.

Common Black Film.   *See* Black Film.

Complex Coacervation.   The process of coacervation when caused by the interaction of oppositely charged colloids.

Compressional Modulus.   *See* Film Elasticity.

Contact Angle.   When two immiscible fluids are in contact with a solid, the angle formed between the solid surface and the tangent to the fluid–fluid interface intersecting the three-phase contact point is termed the contact angle. By convention, if one of the fluids is water then the contact angle is measured through the water phase; otherwise, the contact angle is usually measured through the most dense phase. Distinctions may be made among advancing, receding, or equilibrium contact angles.

Continuous Phase.   In a colloidal dispersion, the phase in which another phase of particles, droplets, or bubbles is dispersed. Sometimes referred to as the external phase. Continuous phase is the opposite of dispersed phase. *See also* Dispersed Phase.

Cosurfactant.   A surfactant that may be added to a system to enhance the effectiveness of another surfactant. The term cosurfactant has also been improperly used to describe non-surface-active species that enhance a surfactant's effectiveness, such as an alcohol or a builder.

Counterions.   In systems containing large ionic species (colloidal ions, membrane surfaces, etc.), counterions are those that, compared to the large ions, have low molecular mass and opposite charge sign. For example, clay particles are usually negatively charged and are naturally associated with exchangeable counterions such as sodium and calcium.

Critical Coagulation Concentration.   (CCC) The electrolyte concentration that marks the onset of coagulation of dispersed species. The CCC is very system-specific, although the variation in CCC with electrolyte composition has been empirically generalized. *See also* Schulze–Hardy Rule.

Critical Film Thickness.   A fluid film may thin to a narrow range of film thicknesses within which it either becomes metastable to thickness changes (equilibrium film) or else ruptures. Persistent foams comprise fluid films at their critical film thickness.

Critical Micelle Concentration.   (cmc) The surfactant concentration above which molecular aggregates, termed micelles, begin to form. In practice a narrow range of surfactant concentrations represents the transition from a solution in which only single, unassociated surfactant molecules (monomers) are present to a solution containing micelles.

Critical Surface Tension of Wetting.   The minimum, or transition, surface tension of a liquid for which it will no longer exhibit complete wetting of a solid. This value is usually taken to be characteristic of a given solid and is sometimes used as an estimate of the solid's surface tension. *See also* Hydrophobic Index.

Critical Temperature.   In adsorption, the transition temperature at which a monolayer no longer exhibits the properties of a condensed state.

Critical Thickness.   *See* Critical Film Thickness.

Cuff-Layer Emulsion.   *See* Interface Emulsion.

Curd Soap (Fibres).   *See* Soap Curd.

Deaeration.   The removal of the gas phase from a dispersion. Example: some nonaqueous foams (made from bitumen or heavy crude oils) are very viscous and are deaerated by processes such as contacting with steam in cascading froth, countercurrent steam-flow vessels.

Deflocculation.   The reverse of aggregation (or flocculation or coagulation). Peptization means the same thing.

Defoamer.   *See* Foam Breaker, Antifoaming Agent.

Degree of Association.   In micelles, this is the number of surfactant molecules in the micelle. *See* Aggregation Number.

Demulsifier.   Any agent added to an emulsion that causes or enhances the rate of breaking of the emulsion (separation into its constituent liquid phases). Demulsifiers may act by any of a number of different mechanisms, which usually include enhancing the rate of droplet coalescence.

Dense Nonaqueous-Phase Liquid.   (DNAPL) *See* Nonaqueous-Phase Liquid.

Depressant.   Any agent that may be used in froth flotation to selectively reduce the effectiveness of collectors for certain mineral components. *See also* Froth Flotation.

Desorption.   The process by which the amount of adsorbed material becomes reduced. That is, the converse of adsorption. Desorption is a different process from negative adsorption. *See also* Adsorption.

Detergency. The action of surfactants that causes or aids in the removal of foreign material from solid surfaces by adsorbing at interfaces and reducing the energy needed to effect the removal. The processes of removal by dissolution and removal by abrasion are not considered to be part of detergency. *See also* Detergent.

Detergent. A surfactant that has cleaning properties in dilute solutions. As commercial cleaning products, detergents are actually formulations containing a number of chemical components, including surfactants, builders, bleaches, brighteners, enzymes, opacifiers, and fragrances. In such formulations there is usually a principal surfactant, termed the main active surfactant, and secondary surfactant(s), termed the coactive surfactant(s). The non-surface-active components are termed ancillary components, or ancillaries.

Detergent Oil. A lubricating oil, formulated to contain surfactant, that has detergent properties in the sense that solid particles are dispersed and kept in suspension. Example: a detergent oil may be used in an internal combustion engine. *See* reference [4].

Dewetting. In antifoaming, the process by which a droplet or particle of antifoaming agent enters the gas–liquid interface and displaces some of the original liquid from the interface. The liquid is usually an aqueous phase, so the process is sometimes referred to as dewetting.

Dilational Elasticity. *See* Film Elasticity.

Discontinuous Phase. *See* Dispersed Phase.

Disjoining Pressure. The negative derivative with respect to distance of the Gibbs energy of interaction per unit area yields a force per unit area between colloidal species, termed the disjoining pressure. Example: in a thin liquid film, the disjoining pressure equals the pressure, beyond the external pressure, that has to be applied to the liquid in the film in order to maintain a given film thickness.

Dispersant. Any species that may be used to aid in the formation of a colloidal dispersion. Examples: dispersant for dyestuffs, dispersant for pigments. Often a surfactant, such as a fatty acid derivative.

Dispersed Phase. In a colloidal dispersion, the phase that is distributed, in the form of particles, droplets, or bubbles, in a second, immiscible phase that is continuous. Also referred to as the disperse, discontinuous, or internal phase. *See also* Continuous Phase.

Dispersing Agent. *See* Dispersant.

Dispersion.   In colloids, a dispersion is a system in which finely divided droplets, particles, or bubbles are distributed in another phase. As it is usually used, dispersion implies a distribution without dissolution. An emulsion is an example of a colloidal dispersion; *see also* Colloidal.

Dispersion Medium.   The continuous phase in a dispersion.

Dissolved-Gas Flotation.   *See* Froth Flotation.

DNAPL.   *See* Nonaqueous-Phase Liquid.

Draves Wetting Test.   A method for comparing the wetting power of surfactants in which one measures the time required for complete wetting of a sample of material placed at the surface of a surfactant solution, under specified test conditions. Different systems are compared in terms of their wetting times. *See also* Wetting.

Duplex Film.   Any film that is thick enough for each of its two interfaces to be independent of each other and exhibit their own interfacial tensions. A duplex film is thus thicker than a monomolecular film.

Dynamic Foam Test.   Any of several methods for assessing foam stability in which one measures the steady-state foam volume generated under given conditions of gas flow, and shearing or shaking. *See also* Foaminess.

EACN.   *See* Equivalent Alkane Carbon Number.

Elasticity.   The ability of a material to change its physical dimensions when a force is applied to it, and then restore its original size and shape when the force is removed. *See also* Film Elasticity.

Elasticity Number.   A dimensionless quantity characterizing the surface-tension gradient in a thinning foam film.

Emulsifier.   Any agent that acts to stabilize an emulsion. The emulsifier may make it easier to form an emulsion and to provide stability against aggregation and possibly against coalescence. Emulsifiers are frequently but not necessarily surfactants.

Emulsion.   A dispersion of droplets of one liquid in another, immiscible liquid, in which the droplets are of colloidal or near-colloidal sizes. The term emulsion may also be used to refer to colloidal dispersions of liquid crystals in a liquid. *See also* Macroemulsion, Miniemulsion, Microemulsion.

Emulsion Test.   In general, emulsion tests range from simple identifications of emulsion presence and volume through to detailed component analyses. The term emulsion test frequently refers simply to the determination of sediments in an emulsion or oil sample.

Entering Coefficient. A measure of the tendency for an insoluble agent to penetrate, or "enter", an interface (usually gas–liquid or liquid–liquid). Entering is thermodynamically favored if the entering coefficient is greater than zero. When equilibria at the interfaces are not achieved instantaneously, reference is made to the initial and final (equilibrium) entering coefficient. *See also* Spreading Coefficient.

Equilibrium Film. *See* Fluid Film.

Equivalent Alkane Carbon Number. (EACN) Each surfactant, or surfactant mixture, in a reference series will produce a minimum interfacial tension (IFT) when measured against a different *n*-alkane. For any crude oil or oil component, a minimum IFT will be observed against one of the reference surfactants. The EACN for the crude oil refers to the *n*-alkane that would yield minimum IFT against that reference surfactant. The EACN thus allows predictions to be made about the interfacial tension behavior of a crude oil in the presence of surfactant. *See* references [*11, 12*]

Evanescent Foam. A transient foam that has no thin-film persistence and is therefore very unstable. Such foams exist only where new bubbles can be created faster than existing bubbles rupture. Examples: air bubbles blown rapidly into pure water; the foam created when a champagne bottle is opened.

Excluded Volume. The volume in a system, or near an interface, that is not accessible to molecules or dispersed species because of the presence of other species in that volume. *See also* Free Volume.

Expansion Factor. In foaming, the ratio of foam volume produced to the volume of liquid used to make the foam. Also termed the expansion ratio.

External Phase. *See* Continuous Phase.

Fatty Acid Soaps. A class of surfactants comprising the salts of aliphatic carboxylic acids having hydrocarbon chains of between 6 and 20 carbon atoms. Fatty acid soaps are no longer restricted to molecules having their origins in natural fats and oils.

Fatty Alcohol Surfactants. The class of primary alcohol surfactants having hydrocarbon chains of between 6 and 20 carbon atoms. Fatty alcohol surfactants are no longer restricted to molecules having their origins in natural fats and oils.

FFFP. *See* Film Forming Fluoroprotein Foam.

Film.   Any layer of material that covers a surface and is thin enough to not be significantly influenced by gravitational forces. *See also* Monolayer Adsorption, Duplex Film.

Film Balance.   A shallow trough that is filled with a liquid, and on top of which is placed material that may form a monolayer. The surface area available can be adjusted by moveable barriers, and, by means of a float, any surface pressure thus created can be measured. Also called Langmuir film balance, Langmuir trough, hydrophil balance, and Pockels–Langmuir–Adam–Wilson–McBain trough or PLAWM trough.

Film Compressibility.   The ratio of relative area change to differential change in surface tension. *See also* Film Elasticity.

Film Drainage.   The drainage of liquid from a lamella of liquid separating droplets or bubbles of another phase (i.e., in a foam or emulsion). Also termed thin-film drainage. *See also* Fluid Film.

Film Elasticity.   The differential change in surface tension with relative change in area. Also termed surface elasticity, dilational elasticity, areal elasticity, compressional modulus, surface dilational modulus, or modulus of surface elasticity. For fluid films the surface tension of one surface is used. The Gibbs film (surface) elasticity is the equilibrium value. If the surface tension is dynamic (time-dependent) in character then, for nonequilibrium values, the term Marangoni film (surface) elasticity is used. The compressibility of a film is the inverse of the film elasticity.

Film Element.   Any small, homogeneous region of a thin film. The film element includes the interfaces.

Film Flotation Technique.   *See* Hydrophobic Index.

Film Forming Fluoroprotein Foam.   (FFFP) A fire extinguishing foam based on very low surface tension producing fluouroprotein surfactants. Used as a rapidly spreading foam on hydrocarbon fires. *See also* Fluoroprotein Foam, Aqueous Film Forming Foam, Alcohol Resisting Aqueous Film Forming Foam.

Film Pressure.   The pressure, in two dimensions, exerted by an adsorbed monolayer. It is formally equal to the difference between the surface tension of pure solvent and that of the solution of adsorbing solute. It can be measured by using the film balance. *See also* Film Balance.

Film Tension.   An expression of surface tension applied to thin liquid films that have two equivalent surfaces. The film tension is twice the surface tension.

Film Water. In soil science, the film of water that remains, surrounding soil particles, after drainage. This layer may range from several to hundreds of molecules thick and comprises water of hydration plus water trapped by capillary forces.

Flocculation. *See* Aggregation. The products of the flocculation process are referred to as flocs or floccules.

Flotation. *See* Froth Flotation.

Fluid Film. A thin-fluid phase, usually of thickness less than about 1 μm. Such films may be specified by abbreviations similar to those used for emulsions, such as A/W/A, for a water film in air, or W/O/W for an oil film in water. There may be thicknesses at which such a film is stable or metastable to thickness changes (equilibrium films). Otherwise fluid films may be distinguished by rapid (mobile film) or slow (rigid film) thickness changes. *See also* Black Film.

Fluoroprotein Foam. (FP) A fire extinguishing foam based on fluoroprotein surfactants. *See also* Film Forming Fluoroprotein Foam, Aqueous Film Forming Foam, Alcohol Resisting Aqueous Film Forming Foam.

Foam. A dispersion of gas bubbles in a liquid, in which at least one dimension falls within the colloidal size range. Thus a foam typically contains either very small bubble sizes or, more commonly, quite large gas bubbles separated by thin liquid films. The thin liquid films are called lamellae (or laminae). Sometimes distinctions are drawn as follows. Concentrated foams, in which liquid films are thinner than the bubble sizes and the gas bubbles are polyhedral, are termed polyederschaum. Low-concentration foams, in which the liquid films have thicknesses on the same scale or larger than the bubble sizes and the bubbles are approximately spherical, are termed gas emulsions, gas dispersions, or kugelschaum.

Foam Booster. *See* Foaming Agent.

Foam Breaker. Any agent that acts to reduce or eliminate foam stability. Also termed defoamer. A more general term is antifoaming agent. *See also* Antifoaming Agent.

Foam Drainage. The drainage of liquid from liquid lamellae separating bubbles in a foam. *See also* Fluid Film.

Foam Drilling Fluid. A drilling fluid comprising air, water and a foaming agent (surfactant). These travel into a well as a mist, then change into a foam before returning up the annulus. *See also* Air Drilling Fluid, Stable Foam, Stiff Foam.

Foam Emulsion.   *See* Aerated Emulsion.

Foamer.   *See* Foaming Agent.

Foam Flooding.   In enhanced oil recovery, the process in which a foam is made to flow through an underground reservoir. The foam, which may be either generated on the surface and injected or generated in situ, is used to increase the drive fluid viscosity and improve its sweep efficiency. In refinery distillation and fractionation towers, the occurrence of foams which can carry liquid into regions of the towers intended for vapour.

Foam Fractionation.   A separation method in which a component of a liquid that is preferentially adsorbed at the liquid–gas interface is removed by foaming the liquid and collecting the foam produced. Foaming surfactants can be separated in this manner.

Foaminess.   A measure of the persistence of a foam (the time an average bubble exists before bursting). Ideally independent of the apparatus and procedure used, and characteristic of the foaming solution being tested. In practice these ideals have not been achieved but some approaches to determining foaminess using dynamic foam stability tests have been reviewed by Bikerman [13]. *See also* Dynamic Foam Test.

Foaming Agent.   Any agent that acts to stabilize a foam. The foaming agent may make it easier to form a foam or provide stability against coalescence. Foaming agents are usually surfactants. Also termed foam booster, whipping agent, and aerating agent.

Foaming Power.   *See* Increase of Volume upon Foaming.

Foam Inhibitor.   Any agent that acts to prevent foaming. Also termed foam preventative. A more general term is antifoaming agent. *See also* Antifoaming Agent.

Foam Number.   A relative drainage rate test for foams in which a foam is formed in a vessel and thereafter the remaining foam volume determined as a function of time. The foam number is the volume of bulk liquid that has separated after a specified time interval, expressed as a percentage of the original volume of liquid foamed.

Foamover.   In an industrial process vessel, unwanted foam may occasionally build up to such an extent that it becomes carried out the top of the vessel ("foamover") and on to the next part of the process. This carry over of foam and any entrained material that comes with it is frequently detrimental to other parts of a processing operation.

Foam Preventative.   *See* Foam Inhibitor, Antifoaming Agent.

Foam Quality. The gas volume fraction in a foam. Expressed as a percentage this is sometimes referred to as Mitchell foam quality. In three-phase systems other measures are used.

Foam Stability. *See* Foaminess.

Foam Texture. The bubble size distribution in a foam. For foams in porous media, it may be expressed in terms of the length scale of foam bubbles as compared to that for the spaces confining the foam. When the length scale of the confining space is comparable to or less than the length scale of the foam bubbles, the foam is sometimes termed lamellar foam, to distinguish it from the opposite case, termed bulk foam.

FP. *See* Fluoroprotein Foam.

Free Volume. The volume in a system, or near an interface, that is available and not occupied by other molecules or dispersed species. *See also* Excluded Volume.

Free Water. The readily separated, nonemulsified water associated with a practical water-in-oil emulsion.

Froth. A type of foam in which solid particles are also dispersed in the liquid (in addition to the gas bubbles). The solid particles may even be the stabilizing agent. The term froth is sometimes used to refer simply to a concentrated foam, but this usage is not preferred.

Frother. *See* Frothing Agent.

Froth Flotation. A separation process utilizing flotation, in which particulate matter becomes attached to gas (foam) bubbles. The flotation process produces a product layer of concentrated particles in foam termed froth. Variations include dissolved-gas flotation, in which gas is dissolved in water that is added to a colloidal dispersion. As microbubbles come out of solution they attach to and float the colloidal species.

Frothing Agent. Any agent that acts to stabilize a froth. May make it easier to form a froth and provide stability against coalescence. Frothing agents are usually surfactants. Analogous to foaming agent.

Gas Aphrons. *See* Microgas Emulsions.

Gas Dispersion. *See* Foam, Gas Emulsion.

Gas Emulsion. "Wet" foams in which the liquid lamellae have thicknesses on the same scale or larger than the bubble sizes. Typically in these cases the gas bubbles have spherical rather than polyhedral shape. Other synonyms include gas dispersion and kugelschaum. If the bubbles are very small and have a significant lifetime, the term microfoam is sometimes used. In petroleum production the term is used to specify crude oil that contains a small volume fraction of dispersed gas. *See also* Foam.

Gel Foam.   A foam which, in addition to the stabilizing surfactants, contains polymer and a cross-linking agent. The foam is first generated as a polymer-thickened foam, and after a delay period, gels. *See also* Stiff Foam.

Gibbs Effect.   The decrease in surface or interfacial tension that occurs as surfactant concentration increases towards the critical micelle concentration.

Gibbs Elasticity.   *See* Film Elasticity.

Gibbs–Marangoni Effect.   The effect in thin liquid films and foams whereby stretching an interface causes the surface excess surfactant concentration to decrease, hence surface tension to increase (Gibbs effect); the surface tension gradient thus created causes liquid to flow toward the stretched region, thus providing both a "healing" force and also a resisting force against further thinning (Marangoni effect). Sometimes referred to simply as the Marangoni effect.

Gibbs Ring.   *See* Plateau Border.

Gibbs Surface Elasticity.   *See* Film Elasticity.

Gibbs Surface Excess.   The excess amount of a component actually present in a system over that present in a reference system of the same volume as the real system, and in which the bulk concentrations in the two phases remain uniform up to the Gibbs dividing surface.

Half-Colloid.   *See* Lyophilic Colloid.

Half-Micelle.   *See* Hemimicelle.

Head Group.   The lyophilic functional group in a surfactant molecule. In aqueous systems the polar group of a surfactant. *See also* Surfactant, Surfactant Tail.

Hemimicelle.   An aggregate of adsorbed surfactant molecules that may form, distinct from monolayer formation, the enhanced adsorption being due to hydrophobic interactions between surfactant tails. Hemimicelles (half-micelles) have been considered to have the form of surface aggregates, or of a second adsorption layer with reversed orientation, somewhat like a bimolecular film. For bilayer surfactant aggregates, the term admicelles has also been used (references [8, 14]). Admicellar chromatography, adsolubilization, and admicellar catalysis make use of media bearing admicelles. *See also* Solloids.

Heterodisperse.   A colloidal dispersion in which the dispersed species (droplets, particles, etc.) do not all have the same size. Subcategories are paucidisperse (few sizes) and polydisperse (many sizes). *See also* Monodisperse.

HIOC.   *See* Hydrophobic Ionogenic Organic Compound.

HLB Scale.   *See* Hydrophile–Lipophile Balance.

HLB Temperature.   *See* Phase Inversion Temperature.

Humic Substances.   Polyaromatic and polyelectrolytic organic acids of high molecular mass (about 800–4000 g/mol or higher) that occur in natural water bodies, soils, and sediments. Significantly aromatic, these acids may have an appreciable aliphatic component, and may be surface-active (reference [15]). Humic substances are operationally divided into humic acids and fulvic acids on the basis of solubility: humic acids are water-soluble above pH 2 but water-insoluble below pH 2; fulvic acids are water-soluble at all pH levels.

Hydrophile–Lipophile Balance.   (HLB scale) An empirical scale categorizing surfactants in terms of their tendencies to be mostly oil-soluble or water-soluble, hence their tendencies to promote W/O or O/W emulsions, respectively. *See also* Phase Inversion Temperature.

Hydrophilic.   A qualitative term referring to the water-preferring nature of a species (atom, molecule, droplet, particle, etc.). For emulsions hydrophilic usually means that a species prefers the aqueous phase over the oil phase. In this example hydrophilic has the same meaning as oleophobic, but such is not always the case.

Hydrophobic.   A qualitative term referring to the water-avoiding nature of a species (atom, molecule, droplet, particle, etc.). For emulsions hydrophobic usually means that a species prefers the oil phase over the aqueous phase. In this example hydrophobic has the same meaning as oleophilic, but such is not always the case. A functional group of a molecule that is not very water-soluble is referred to as a hydrophobe.

Hydrophobic Bonding.   The attraction between hydrophobic species in water that arises from the fact that the solvent–solvent interactions are more favorable than the solvent–solute interactions.

Hydrophobic Effect.   The partitioning of a substance from an aqueous phase into (or onto) another phase due to its hydrophobicity. Often characterized by an octanol–water partitioning coefficient. *See also* Solvent-Motivated Sorption.

Hydrophobic Index.   An empirical measure of the relative wetting preference of very small solid particles. In one test method, solid particles of narrow size range are placed on the surfaces of a number of alcohol/ water solutions of decreasing surface tension. The percentage alcohol content at which the particles just begin to become hydrophilic and sink is the hydrophobic index; the corresponding solvent surface-tension value is taken as the critical surface tension of wetting. The technique is also referred to as the film-flotation technique (reference [16]) or sink-float method. *See also* Critical Surface Tension of Wetting.

Hydrophobic Interaction.   *See* Hydrophobic Bonding.

Hydrophobic Ionogenic Organic Compound.   (HIOC) An organic compound that is capable of ionizing, depending upon the solution pH. Upon ionization the properties of the molecule change and its sorption and subsurface migration (in the environment) vary accordingly.

Hydrosol.   A dispersion of very small diameter species in water or in aqueous solution. Dispersions of finely divided oil droplets in aqueous solution are sometimes referred to as oil hydrosols.

Hydrotrope.   Any species that enhances the solubility of another. Example: hydrotropes such as alkyl aryl sulfonates (e.g., toluene sulfonate) are added to detergent formulations to raise the cloud point.

Imbibition.   The displacement of a nonwetting phase by a wetting phase in a porous medium or a gel; the reverse of drainage.

Immersional Wetting.   The process of wetting when a solid (or liquid) that is initially in contact with gas becomes completely covered by an immiscible liquid phase. *See also* Wetting, Spreading Wetting.

Increase of Volume upon Foaming.   In foaming, 100 times the ratio of gas volume to liquid volume in a foam. Also termed the foaming power.

Induced Gas Flotation.   *See* Froth Flotation.

Initial Knockdown Capability.   *See* Knockdown Capability.

Interface.   The boundary between two immiscible phases, sometimes including a thin layer at the boundary within which the properties of one bulk phase change over to become the properties of the other bulk phase. An interfacial layer of finite specified thickness may be defined. When one of the phases is a gas, the term surface is frequently used.

Interface Emulsion.   An emulsion occurring between oil and water phases in a process separation or treatment apparatus. Such emulsions may have a high solids content and are frequently very viscous. In this case the term interface is used in a macroscopic sense and refers to a bulk phase separating two other bulk phases of higher and lower density. Other terms: cuff layer, pad layer, or rag layer emulsions.

Interfacial Film. A thin layer of material positioned between two immiscible phases, usually liquids, whose composition is different from either of the bulk phases.

Interfacial Layer. The layer at an interface that contains adsorbed species. Also termed the surface layer.

Interfacial Tension. *See* Surface Tension.

Interferometry. An experimental technique in which a beam of light is reflected from a film. Light reflected from the front and back surfaces of the film travels different distances and produces interference phenomena, a study of which allows calculation of the film thickness.

Intermicellar Liquid. An older term for the continuous (external) phase in micellar dispersions. *See also* Continuous Phase, Micelle.

Internal Phase. *See* Dispersed Phase.

Inverse Micelle. A micelle that is formed in a nonaqueous medium, thus having the surfactants' hydrophilic groups oriented inward away from the surrounding medium.

Inversion. The process by which one type of emulsion is converted to another, as when an O/W emulsion is transformed into a W/O emulsion, and vice versa. Inversion can be accomplished by a wide variety of physical and chemical means.

Invert Emulsion. A water-in-oil emulsion. This is different from the term reverse emulsion used in the petroleum field.

Invert-Oil Mud. An emulsion drilling fluid (mud) of the water-in-oil (W/O) type, and having a high water content. *See also* Oil-Base Mud, Oil Mud.

Ion Exchange. Adsorption of an ionic species that is accompanied by the simultaneous desorption of an equivalent charge quantity of other ionic species. Ion-exchange media can have specific selectivities. *See also* Sorbent-Motivated Sorption.

Ionic Strength. A measure of electrolyte concentration.

Iridescent Layers. *See* Schiller Layers.

Isodisperse. *See* Monodisperse.

Klevens Constants. The empirical parameters in an equation advanced by Klevens for predicting the critical micelle concentrations (cmc) of surfactants in terms of the number of carbon atoms in the hydrocarbon chain. Klevens constants for numerous surfactants are tabulated in references [17, 18].

Knockdown Capability.   A measure of the effectiveness of a defoamer. First, a column of foam is generated in a foam stability apparatus and the foam height is recorded. A measured amount of defoamer is added, and the reduction in foam height over a specified time period is noted. The knockdown capability is the reduction in foam height. There are many variations of this test. Sometimes referred to as initial knockdown capability.

Knockout Drops.   Demulsifier that may be used to enhance the separation of oil from water and solids in an emulsion. Also termed slugging compound.

Krafft Point.   The temperature above which the solubility of a surfactant increases sharply (micelles begin to be formed). In practice a narrow range of temperatures represents the transition from a solution in which only single, unassociated surfactant molecules (monomers) or ions (ionomers) can be present, up to a given solubility limit, to a solution that can contain micelles and thus allow much more surfactant to remain in solution in preference to precipitating. Numerous tabulations are given in references [17, 18]. In the soap industry the Krafft point is sometimes defined as the temperature at which a transparent soap solution becomes cloudy upon cooling (reference [4]).

Kugelschaum.   *See* Gas Emulsion.

Lamella.   *See* Foam.

Lamella Number.   A dimensionless parameter used to predict the likelihood that a combination of capillary suction in plateau borders and the influence of mechanical shear will cause an oil phase to become emulsified and imbibed into foam lamellae flowing in porous media (reference [19]).

Lamina.   *See* Foam.

Langmuir–Blodgett Film.   A film of molecules that is deposited onto a solid surface by repeatedly passing the solid through a monolayer of molecules at a gas–liquid interface. Each pass deposits an additional monolayer on the solid.

Langmuir Trough.   *See* Film Balance.

Laplace Flow.   *See* Capillary Flow.

Laplace Waves.   *See* Capillary Ripples.

Lather.   A foam produced by mechanical agitation on a solid surface. Example: the mechanical generation of shaving foam (lather) on a wet bar of soap.

Launderometer. A specialized machine used to perform a standardized test method for measuring the effectiveness of detergents. *See also* Detergent, Detergency.

Lens. A nonspreading droplet of liquid at an interface is said to form a lens. The lens is thick enough for its shape to be significantly influenced by gravitational forces.

Light Nonaqueous-Phase Liquid. (LNAPL) *See* Nonaqueous-Phase Liquid.

Limiting Capillary Pressure. For foam flow in porous media the maximum capillary pressure that can be attained by simply increasing the fraction of gas flow. Foams flowing at steady-state do so at or near this limiting capillary pressure. In the limiting capillary pressure regime the steady-state saturations remain essentially constant.

Line Tension. Where three phases meet there may exist a line tension (force) along the three-phase junction. For a lens of material at the interface between two other immiscible phases, the three-phase contact junction takes the form of a circle along which the line tension acts.

Lipid. Long-chain aliphatic hydrocarbons and derivatives originating in living cells. Some lipids, such as fatty acids, are also surfactants. Simple lipids tend to be hydrocarbon-soluble but not water-soluble. Examples: fatty acids, fats, waxes.

Lipid Bilayer. *See* Bimolecular Film.

Lipid Film. A thin film of oil in water in which the film is stabilized by lipids. The term is used even though the film is not a film of lipid. *See also* Fluid Film.

Lipophile. That part of a molecule that is organic-liquid-preferring in nature.

Lipophilic. The (usually fatty) organic-liquid-preferring nature of a species. Depending on the circumstances may also be a synonym for oleophilic. *See also* Hydrophile–Lipophile Balance.

Lipophobe. That part of a molecule that is organic-liquid-avoiding in nature.

Lipophobic. The (usually fatty) organic-liquid-avoiding nature of a species. Depending on the circumstances may also be a synonym for oleophobic.

Liposome. *See* Vesicle.

Liquid Aerosol. *See* Aerosol.

Liquid-Crystalline Phase.   *See* Mesomorphic Phase.

LNAPL.   *See* Nonaqueous-Phase Liquid.

Loose Emulsion.   A relatively unstable, easy-to-break emulsion. *See also* Tight Emulsion.

Lower-Phase Microemulsion.   A microemulsion that has a high water content and is stable while in contact with a bulk oil phase, and in laboratory tube or bottle tests tends to be situated at the bottom of the tube, underneath the oil phase. For chlorinated organic liquids, which are more dense than water, the oil will be the bottom phase rather than the top. *See also* Microemulsion, Winsor-Type Emulsions.

Lundelius Rule.   An expression for the inverse relation between solubility and the extent of adsorption of a species.

Lyocratic.   A dispersion stabilized principally by solvation forces. Example: the stability of aqueous biocolloid systems can be explained in terms of hydration and steric stabilization.

Lyophilic.   General term referring to the continuous-medium- (or solvent-) preferring nature of a species. *See* Hydrophilic.

Lyophilic Colloid.   An older term used to refer to single-phase colloidal dispersions. Examples: polymer and micellar solutions. Other synonyms no longer in use: semicolloid or half-colloid.

Lyophobic.   General term referring to the continuous-medium- (or solvent-) avoiding nature of a species. *See* Hydrophobic.

Lyophobic Colloid.   An older term used to refer to two-phase colloidal dispersions. Examples: suspensions, foams, emulsions.

Lyophobic Mesomorphic Phase.   *See* Mesomorphic Phase.

Lyoschizophrenic Surfactant.   A surfactant in a two-phase system whose behavior indicates a lack of preference for solubility in one phase or the other (reference [3]).

Lyotropic Liquid Crystals.   *See* Mesomorphic Phase.

Macroemulsion.   *See* Emulsion. The term macroemulsion is sometimes employed to identify emulsions having droplet sizes greater than a specified value, or alternatively, simply to distinguish an emulsion from the microemulsion or micellar emulsion types.

Macroion.   A charged colloidal species whose electric charge is attributable to the presence at the surface of ionic functionalities.

Main Active. The primary surfactant in a detergent formulation. *See also* Detergent.

Marangoni Effect. In surfactant-stabilized fluid films, any stretching in the film causes a local decrease in the interfacial concentration of adsorbed surfactant. This decrease causes the local interfacial tension to increase (Gibbs effect), which in turn acts in opposition to the original stretching force. With time the original interfacial concentration of surfactant is restored. The time-dependent restoring force is referred to as the Marangoni effect and is a mechanism for foam and emulsion stabilization. The combination of Gibbs and Marangoni effects is properly referred to as the Gibbs–Marangoni effect, but is frequently referred to simply as the Marangoni effect.

Marangoni Elasticity. *See* Film Elasticity, Marangoni Effect.

Marangoni Flow. Liquid flow in response to a gradient in surface or interfacial tension. *See* Marangoni Effect.

Marangoni Surface Elasticity. *See* Film Elasticity, Marangoni Effect.

Marangoni Waves. *See* Capillary Ripples.

Meniscus. The uppermost surface of a column of a liquid. The meniscus may be either convex or concave depending on the balance of gravitational and surface or interfacial tension forces acting on the liquid.

Mesomorphic Phase. A phase consisting of anisometric molecules or particles that are aligned in one or two directions but randomly arranged in other directions. Such a phase is also commonly referred to as a liquid-crystalline phase or simply a liquid crystal. The mesomorphic phase is in the nematic state if the molecules are oriented in one direction, and in the smectic state if oriented in two directions. Mesomorphic phases are also sometimes distinguished on the basis of whether their physical properties are mostly determined by interactions with surfactant and solvent (lyotropic liquid crystals) or by temperature (thermotropic liquid crystals). *See also* Neat Soap.

Micellar Aggregation Number. *See* Aggregation Number.

Micellar Catalysis. Catalytic reactions conducted in a surfactant solution in which micelles play a role in catalyzing the reaction. Typically the micelles either solubilize needed reactant(s) or they provide a medium of intermediate polarity to enhance the rate of a reaction.

Micellar Charge. The net charge of surfactant ions in a micelle including any counterions bound to the micelle.

Micellar Emulsion.   An emulsion that forms spontaneously and has extremely small droplet sizes (< 10 nm). Such emulsions are thermodynamically stable and are sometimes referred to as microemulsions.

Micellar Mass.   The mass of a micelle. For ionic surfactants this value includes the surfactant ions and their counterions.

Micellar Solubilization.   *See* Solubilization.

Micellar Weight.   *See* Micellar Mass.

Micelle.   An aggregate of surfactant molecules or ions in solution. Such aggregates form spontaneously at sufficiently high surfactant concentration, above the critical micelle concentration. The micelles typically contain from tens to hundreds of molecules and are of colloidal dimensions. If more than one kind of surfactant forms the micelles, they are referred to as mixed micelles. If a micelle becomes larger than usual as a result of either the incorporation of solubilized molecules or the formation of a mixed micelle, then the term swollen micelle is applied. *See also* Critical Micelle Concentration, Inverse Micelle.

Microemulsion.   A special kind of stabilized emulsion in which the dispersed droplets are extremely small (< 100 nm) and the emulsion is thermodynamically stable. These emulsions are transparent and may form spontaneously. In some usage a lower size limit of about 10 nm is implied in addition to the upper limit; *see also* Micellar Emulsion. In some usage the term microemulsion is reserved for a Winsor type IV system (water, oil, and surfactants all in a single phase). *See also* Winsor-Type Emulsions.

Microencapsulation.   The protection of a chemical species by containing it in small droplets, particles, or bubbles covered by a protective coating. Example: the encapsulation of liquid within vesicles.

Microfoam.   *See* Gas Emulsion.

Microgas Emulsions.   A kind of foam in which the gas bubbles have an unusually thick stabilizing film and exist clustered together as opposed to either separated, nearly spherical bubbles or the more concentrated, system-filling polyhedral bubbles. A microgas emulsion will cream to form a separate phase from water. Also termed aphrons or colloidal gas aphrons.

Middle-Phase Microemulsion.   A microemulsion that has high oil and water contents and is stable while in contact with either bulk oil or bulk water phases. This stability may be due to a bicontinuous structure in which both oil and water phases are continuous at the same time. In laboratory tube or bottle tests involving samples containing unemulsified oil and water, a middle-phase microemulsion will tend to be situated between the two phases. *See also* Bicontinuous System, Winsor-Type Emulsions.

Middle Soap.   A mesomorphic (liquid-crystal) phase of soap micelles, oriented in a hexagonal array of cylinders. Middle soap contains a similar or lower proportion of soap (e.g., 50%) as opposed to water. Middle soap is in contrast to neat soap, which contains more soap than water and is also a mesomorphic phase, but has a lamellar structure rather than a hexagonal array of cylinders. Also termed clotted soap. *See also* references [1, 4], Neat Soap.

Miniemulsion.   *See* Emulsion. The term miniemulsion is sometimes used to distinguish an emulsion from the microemulsion or micellar emulsion types. Thus a miniemulsion would contain droplet sizes greater than 100 nm and less than 1000 nm or some other specified upper size limit.

Mist Drilling Fluid.   *See* Air Drilling Fluid.

Mobile Film.   *See* Fluid Film.

Modulus of Surface Elasticity.   *See* Film Elasticity.

Monodisperse.   A colloidal dispersion in which all the dispersed species (droplets, particles, etc.) have the same size. Otherwise, the system is heterodisperse (paucidisperse or polydisperse).

Monolayer Adsorption.   Adsorption in which a first or only layer of molecules becomes adsorbed at an interface. In monolayer adsorption, all of the adsorbed molecules will be in contact with the surface of the adsorbent. The adsorbed layer is termed a monolayer or monomolecular film.

Monolayer Capacity.   In chemisorption, the amount of adsorbate needed to satisfy all available adsorption sites. For physisorption, the amount of adsorbate needed to cover the surface of the adsorbent with a complete monolayer.

Monomolecular Film.   *See* Monolayer Adsorption.

Mousse Emulsion.   *See* Chocolate Mousse Emulsion.

Multilayer Adsorption.   Adsorption in which the adsorption space contains more than a single layer of molecules; therefore, not all adsorbed molecules will be in contact with the surface of the adsorbent. *See also* Monolayer Adsorption.

Multiple Emulsion.   An emulsion in which the dispersed droplets themselves contain even more finely dispersed droplets of a separate phase. Thus, there may occur oil-dispersed-in-water-dispersed-in-oil

(O/W/O) and water-dispersed-in-oil-dispersed-in-water (W/O/W) multiple emulsions. These are sometimes called three-phase emulsions, triple-phase emulsions, or simply triple emulsions. More complicated multiple emulsions such as O/W/O/W and W/O/W/O are also possible.

Myelin Cylinders.   Long-chain polar compounds, above their solubility limit, may interact with surfactants to form mixed micelles that separate (as a coacervate) in the form of cylinders. These are termed myelin cylinders or myelinic figures. They are usually quite viscous and may be birefringent.

NAPL.   *See* Nonaqueous-Phase Liquid.

Neat Soap.   A mesomorphic (liquid-crystal) phase of soap micelles, oriented in a lamellar structure. Neat soap contains more soap (e.g., 75%) than water. Neat soap is in contrast to middle soap, which contains less soap than water and is also a mesomorphic phase, but has a hexagonal array of cylinders rather than a lamellar structure. *See also* reference [1].

Negative Adsorption.   *See* Adsorption.

Nelson-Type Emulsions.   Several types of phase behavior occur in microemulsions; they are denoted as Nelson type II −, type II +, and type III. These designations refer to equilibrium phase behaviors and distinguish, for example, the number of phases that may be in equilibrium and the nature of the continuous phase. *See also* reference [20]. Winsor-type emulsions are similarly identified, but with different type numbers.

Nematic State.   *See* Mesomorphic Phase.

Neumann's Triangle.   At the junction where three phases meet, three vectors representing the forces of interfacial tension among pairs of phases can be drawn. At equilibrium, the sum of these vectors of Neumann's triangle will equal zero.

Newton Black Film.   *See* Black Film.

Nitrified Foam.   A slang term used in some industries to denote foams in which nitrogen is the gas phase.

Nonaqueous-Phase Liquid.   (NAPL) Any liquid other than water. In environmental fields the term commonly refers to hydrocarbon liquids less dense than water (light nonaqueous-phase liquid, LNAPL), or chlorinated hydrocarbons that are more dense than water (dense nonaqueous-phase liquid, DNAPL). Example: 1,1,1-trichloroethane is a DNAPL.

Nonionic Surfactant. A surfactant molecule whose polar group is not electrically charged. Example: poly(oxyethylene) alcohol, $C_nH_{2n+1}(OCH_2CH_2)_mOH$.

Nonwetting. *See* Wetting.

Octanol–Water Partition Coefficient. The partitioning coefficient of a compound between octanol and water, that is, between specific nonpolar and polar phases. Used as an indication of the tendency of a compound to partition between oil and water phases. A variety of empirical equations estimate such partitioning of a compound on the basis of its octanol–water partition coefficient. *See also* Solvent-Motivated Sorption.

Oil-Base Mud. An emulsion drilling fluid (mud) of the water-dispersed-in-oil (W/O) type having a low water content. *See also* Oil Mud, Invert-Oil Mud.

Oil Colour. A qualitative test for the presence of emulsified water in an oil. Emulsified water droplets tend to impart a hazy appearance to the oil.

Oil Emulsion. An emulsion having an oil as the continuous phase.

Oil-Emulsion Mud. An emulsion drilling fluid (mud) of the oil-dispersed-in-water (O/W) type. *See also* Oil Mud.

Oil Hydrosol. An oil-in-water (O/W) emulsion in which the oil droplets are very small and the volume fraction of oil is also very small. The emulsion terminology is preferable. *See also* Hydrosol.

Oil Mud. An emulsion drilling fluid (mud) of the water-dispersed-in-oil (W/O) type. A mud of low water content is referred to as an oil-base mud, and a mud of high water content is referred to as an invert-oil mud. *See also* Oil-Emulsion Mud.

Oleophilic. The oil-preferring nature of a species. A synonym for lipophilic. *See also* Hydrophobic.

Oleophobic. The oil-avoiding nature of a species. A synonym for lipophobic. *See also* Hydrophilic.

O/O. Abbreviation for an oil-dispersed-in-oil emulsion in which one oil is polar and the other is not. Example: an emulsion of ethylene glycol in a liquid alkane.

Opacifiers. Agents that make a liquid appear more opaque, or pearlescent. For example, polystyrene latex is added to liquid detergents formulated for dishwashing or shampooing to give them a flat opaque appearance. *See also* Detergent.

Optimum Salinity. In microemulsions, the salinity for which the mixing of oil with a surfactant solution produces a middle-phase microemulsion containing an oil-to-water ratio of 1. In micellar enhanced oil recovery processes, extremely low interfacial tensions result, and oil recovery tends to be maximized when this condition is satisfied.

Oriented-Wedge Theory. An empirical generalization used to predict which phase in an emulsion will be continuous and which dispersed. It is based on a physical picture in which emulsifiers are considered to have a wedge shape and will favor adsorbing at an interface such that most efficient packing is obtained, that is, with the narrow ends pointed toward the centers of the droplets. A useful starting point, but there are many exceptions. *See also* Bancroft's Rule, Hydrophile–Lipophile Balance.

Orthokinetic Aggregation. The process of aggregation induced by hydrodynamic motions such as stirring, sedimentation, or convection. Orthokinetic aggregation is distinguished from perikinetic aggregation, the latter being caused by Brownian motions.

O/W. Abbreviation for an oil-dispersed-in-water emulsion.

O/W/A. Abbreviation for a fluid film of water between oil and air. *See* Fluid Film.

O/W/O. In emulsions, an abbreviation for an oil-in-water-in-oil multiple emulsion. The water droplets have oil droplets dispersed within them, and the water droplets themselves are dispersed in oil forming the continuous phase. In fluid films, an abbreviation for a thin fluid film of water in an oil phase. *See also* Fluid Film.

Pad Layer Emulsion. *See* Interface Emulsion.

Palisade Layer. In a micelle, the region of water molecules of hydration postulated to lie between relatively water-free hydrocarbon chains at the center of the micelle and the exposed, fully hydrated polar groups at the micelle surface.

Paraffin-Chain Salts. An older term for ionic surfactants.

Paucidisperse. A colloidal dispersion in which the dispersed species (droplets, particles, etc.) have a few different sizes. Paucidisperse is a category of heterodisperse systems. *See also* Monodisperse.

Peptization. The dispersion of an aggregated (coagulated or flocculated) system. Deflocculation means the same thing.

Perikinetic Aggregation. The process of aggregation when induced by Brownian motions. Perikinetic aggregation is distinguished from orthokinetic aggregation, the latter being caused by hydrodynamic motions such as sedimentation or convection.

Perrin Black Film.   A Newton black film. *See* Black Film.

Phase Diagram.   A graphical representation of the equilibrium relationships between phases in a system. For multicomponent systems, and considering varying temperatures, more than a simple two-dimensional phase diagram will be required.

Phase Inversion Temperature.   (PIT) The temperature at which the hydrophilic and oleophilic natures of a surfactant are in balance. As temperature is increased through the PIT, a surfactant will change from promoting one kind of emulsion, such as O/W, to another, such as W/O. Also termed the HLB temperature.

Phase Map.   *See* Phase Diagram.

Phase Ratio.   In emulsions phase ratio refers to the ratio between internal phase and continuous phase. Phase ratios are dimensionless, but the units used should be specified because mass ratios and volume ratios are commonly used.

Phospholipid.   Esters of phosphoric acid that contain fatty acid(s), an alcohol, and a nitrogen-containing base. *See also* Lipid.

Phospholipid Bilayer.   *See* Bimolecular Film.

Pickering Emulsion.   An emulsion stabilized by fine particles. The particles form a close-packed structure at the oil–water interface, with significant mechanical strength, which provides a barrier to coalescence.

PIT.   *See* Phase Inversion Temperature.

Plateau Border.   The region of transition at which thin fluid films are connected to other thin films or mechanical supports such as solid surfaces. For example, in foams plateau borders form the regions of liquid situated at the junction of liquid lamellae. Sometimes referred to as a Gibbs ring or Gibbs–Plateau Border.

PLAWM Trough.   Pockels–Langmuir–Adam–Wilson–McBain trough. *See* Film Balance.

Pockels Point.   When surfactant molecules are added into a system and form an insoluble film at an interface, surface tension does not decrease very strongly until enough is added to form a complete monolayer. The transition point is termed the Pockels point and corresponds to a surface area occupied per molecule of about 20 Å^2 for soaps.

Polar Group.   *See* Head Group.

Polar Substance. A substance having different, usually opposite, characteristics at two locations within it. Example: a permanent dipole. Increasing polarity generally increases solubility in water.

Polydisperse. A colloidal dispersion in which the dispersed species (droplets, particles, etc.) have a wide range of sizes. Polydisperse is a category of heterodisperse systems. *See also* Monodisperse.

Polyederschaum. *See* Foam.

Polymer-Thickened Foam. A foam which, in addition to the stabilizing surfactants, contains polymer. Polymer-thickened foams are formulated to produce increased stability and viscosity. *See also* Gel Foam, Stiff Foam.

Pour Point. The lowest temperature at which an emulsion, oil, surfactant solution, or other material will flow under a standardized set of test conditions.

Probe Molecule. Any species that is soluble in micelles and can be readily detected and measured. Example: pyrene solubilized in micelles can be a reporter probe for its environment through fluorescence spectroscopy.

Protected Lyophobic Colloids. *See* Sensitization.

Protective Colloid. A colloidal species that adsorbs onto and acts to "protect" the stability of another colloidal system. The term refers specifically to the protecting colloid and only indirectly to the protected colloid. Example: when a lyophilic colloid such as gelatin acts to protect another colloid in a dispersion by conferring steric stabilization.

Pseudoemulsion Film. A fluid film of an aqueous phase (water) between air and oil phases. These are usually designated O/W/A or A/W/O. *See also* Fluid Film.

Pseudophase Diagram. A phase diagram for a system in which there are more phases present than are allowed to vary in the diagram. A pseudophase diagram is thus only one of several that are needed to completely describe a system.

Pseudosolution. *See* Colloidal.

Quats. Quaternary ammonium compounds. These are cationic surfactants if the compound contains a hydrocarbon chain of sufficient length. Example: cetyltrimethylammonium bromide, $CH_3(CH_2)_{15}N^+(CH_3)_3Br^-$. *See also* Cationic Surfactant.

Rag Layer Emulsion. *See* Interface Emulsion.

Rayleigh–Taylor Instability. The instability of an interface between two fluids of different densities caused by the acceleration of the less dense fluid toward the more dense fluid.

Relative Molar Mass of Micelles. The mass of 1 mole of micelles (actually it is the mass of 1 mole of micelles divided by the mass of $\frac{1}{12}$ mole of ^{12}C). Synonyms include relative micellar mass and relative micellar weight.

Repeptization. Peptization, usually by dilution, of a once-stable dispersion that was aggregated (coagulated or flocculated) by the addition of electrolyte.

Reverse Emulsion. A petroleum industry term used to denote an oil-in-water emulsion (most wellhead emulsions are W/O). Reverse emulsion is the opposite from the meaning of the term invert emulsion. *See also* Invert Emulsion.

Reverse Micelles. Synonym for the dispersed phase in a water-in-oil type microemulsion. Here the surfactant heads, or polar groups, associate closely to minimize interaction with the oil phase. This close association can happen when they orient themselves inside water droplets, and it also allows the surfactant tails, or hydrocarbon groups, to stabilize the water droplets by orienting toward or into the oil.

Rigid Film. *See* Fluid Film.

Ross Foam. Foam produced from a binary or ternary solution under conditions at which its temperature and composition approach (but do not reach) the point of phase separation into separate immiscible liquid phases.

Ross-Miles Test. A method for assessing foam stability in which one measures the rate of collapse of a (static) column of foam that has been generated by allowing a certain quantity of foaming solution to fall a specified distance into a separate volume of the same solution contained in a vessel.

Rupture. *See* Fluid Film.

Salinity Requirement. *See* Optimum Salinity.

Salt Curve. A graphical representation of the viscosity of a system versus salt concentration. This curve can be an important characteristic of formulated systems in which viscosity control is necessary, such as in shampoo formulas.

Salting In. *Solutions.* When the addition of electrolyte to a solution causes an increase in the solubility of a solute. *See also* Salting Out.

Surfactants.   When the addition of electrolyte to a solution of nonionic surfactant causes the critical micelle concentration to increase. Also, addition of electrolyte to an ionic surfactant solution in a multiphase system can drive surfactant from the oil phase into the aqueous phase. *See also* Salting Out.

Salting Out.   *Solutions.*   When the addition of electrolyte to a solution causes a decrease in the solubility of a specified solute. *See also* Salting In.

Surfactants.   When the addition of electrolyte to a solution of nonionic surfactant causes the critical micelle concentration to decrease. Also, addition of electrolyte to an ionic surfactant solution in a multiphase system can drive surfactant from the aqueous phase into the oil phase. *See also* Salting In.

Emulsions.   The process of demulsification by the addition of electrolyte.

Saponification.   The reaction of a fat or a fatty acid with a base to produce soap.

Schiller Layers.   The layers of particles that may be formed during sedimentation such that the distances between layers are on the order of the wavelength of light, leading to iridescent, or Schiller layers.

Schulze–Hardy Rule.   An empirical rule summarizing the general tendency of the critical coagulation concentration (CCC) of a suspension, an emulsion, or other dispersion to vary inversely with the sixth power of the counterion charge number of added electrolyte. *See also* Critical Coagulation Concentration.

SEAR.   *See* Surfactant Enhanced Aquifer Remediation

Semi-Colloid.   *See* Lyophilic Colloid.

Sensitization.   The process in which small amounts of added hydrophilic colloidal material make a hydrophobic colloid more sensitive to coagulation by electrolyte. Example: the addition of polyelectrolyte to an oil-in-water emulsion to promote demulsification by salting out.

Septum.   In general, any dividing wall between two cavities. Example: the thin liquid films (lamellae) between bubbles in a foam.

Silicone Oil.   Any of a variety of silicon-containing polymer solutions. An example is a linear poly(dimethylsiloxane): $HO[(CH_3)_2SiO]_nH$.

Silver Film.   *See* Black Film.

Sink-Float Method.   *See* Hydrophobic Index.

Slugging Compound.   *See* Knockout Drops.

Smectic State. *See* Mesomorphic Phase.

Soap. A surface-active fatty acid salt containing at least eight carbon atoms. The term is no longer restricted to fatty acid salts originating from natural fats and oils. *See also* Surfactant.

Soap Curd. A mixture of soap crystals in a saturated solution in which the soap crystals produce a gel-like consistency. The soap crystals in this case are referred to as curd-fibers. Soap curd is not a mesomorphic (liquid-crystal) phase.

Soap Film. A thin film of water in air that is stabilized by surfactant. The term is used even though the film is not a film of soap and even where the surfactant is not a soap. *See also* Fluid Film.

Sol. A colloidal dispersion. In some usage the term sol is used to distinguish dispersions in which the dispersed-phase species are of very small size so that the dispersion appears transparent.

Solloids. Surface colloids. Colloidal-sized aggregates of surfactant and/or polymer species adsorbed on a surface. Used as a more general term than admicelles or hemimicelles. *See* reference [21].

Solubilizate. The solute whose solubility is increased in the process of solubilization.

Solubilization. The process by which the solubility of a solute is increased by the presence of another solute. Micellar solubilization refers to the incorporation of a solute (solubilizate) into or on micelles of another solute to thereby increase the solubility of the first solute.

Solubilizing Agent. Any product that may be used to aid in the solubilization of a species. Examples: solubilizing agents for dyestuffs or pigments. Often a surfactant, such as a fatty acid derivative. *See also* Dispersant.

Solvent-Motivated Sorption. Sorption that occurs as a result of the hydrophobicity of a compound. Accumulation of the compound at the interface or in the other phase is not due to its affinity for that phase so much as to its disaffinity for the initial phase. Such sorption occurs for organic contaminants in the environment. This kind of sorption can often be related to the octanol–water partitioning coefficient.

Sorbate. A substance that becomes sorbed into an interface or another material or both. *See also* Sorption.

Sorbent. The substrate into which or onto which a substance is sorbed or both. *See also* Sorption.

Sorbent-Motivated Sorption.   Sorption that occurs as a result of the affinity of the surface for a particular compound. Example: ion exchange. *See also* Solvent-Motivated Sorption.

Sorption.   A term used in a general sense to refer to either or both of the processes of adsorption and absorption.

Spreading.   The tendency of a liquid to flow and form a film coating an interface, usually a solid or immiscible liquid surface, in an attempt to minimize interfacial free energy. Such a liquid forms a zero contact angle as measured through itself.

Spreading Coefficient.   A measure of the tendency for a liquid to spread over a surface (usually of another liquid). It is thermodynamically favored if the spreading coefficient is greater than zero. For the spreading of liquid on a solid a modified equation, without the solid/gas tension, is sometimes used. Other usages of the concept have involved terms such as the spreading parameter or wetting power (reference [22]), or spreading tension (reference [4]). *See also* Entering Coefficient.

Spreading Pressure.   *See* Surface Pressure.

Spreading Wetting.   The process of wetting in which a liquid, already in contact with a solid (or second, immiscible liquid) surface, spreads over the solid surface, thereby increasing the interfacial area of contact between them. The spreading is thermodynamically favored when the spreading coefficient is positive. *See also* Immersional Wetting, Spreading Coefficient, Wetting.

Spread Layer.   The interfacial layer formed by an adsorbate when it becomes essentially completely adsorbed out of the bulk phase(s). If the layer is known to be one molecule thick, then the term spread monolayer is used.

Stable Foam.   An oil and gas well drilling fluid foam that contains film-stabilizing additives, such as polymers or clays, and is pre-formed at the surface. *See also* Air Drilling Fluid, Foam Drilling Fluid, Stiff Foam.

Stiff Foam.   An oil and gas well drilling fluid foam that contains film-stabilizing additives, such as polymers or clays, is pre-formed at the surface, and is more viscous than stable foam, having sufficient carrying capacity to remove drill cuttings from large diameter holes. Also termed gel foam. *See also* Air Drilling Fluid, Foam Drilling Fluid, Stable Foam.

Stratified Film.   A fluid film in which several thicknesses can exist simultaneously and can persist for a significant amount of time. *See also* Fluid Film.

Substrate. A material that provides a surface or interface at which adsorption or other phenomena take place. It may be simply an underlying or supporting phase.

Suction Pressure. *See* Capillary Pressure.

Surface. *See* Interface.

Surface-Active Agent. *See* Surfactant.

Surface Area. The area of a surface or interface, especially that between a dispersed and a continuous phase. The specific surface area is the total surface area divided by the mass of the appropriate phase.

Surface Colloids. *See* Solloids.

Surface Coverage. The ratio of the amount of adsorbed material to the monolayer capacity. The definition is the same for either of monolayer and multilayer adsorption.

Surface Dilational Modulus. *See* Film Elasticity.

Surface Dilational Viscosity. *See* Surface Viscosity.

Surface Elasticity. *See* Film Elasticity.

Surface Layer. *See* Interfacial Layer.

Surface of Tension. An imaginary boundary, having no thickness, at which surface or interfacial tension acts.

Surface Phenomena. Any phenomena whose effects are manifested at a surface separating two phases.

Surface Pressure. Actually an analog of pressure; the force per unit length exerted on a real or imaginary barrier separating an area of liquid or solid that is covered by a spreading substance from a clean area on the same liquid or solid. Also referred to as spreading pressure.

Surface Tension. The contracting force per unit length around the perimeter of a surface is usually referred to as surface tension if the surface separates gas from liquid or solid phases, and interfacial tension if the surface separates two nongaseous phases. Although not strictly defined the same way, surface tension can be expressed in units of energy per unit surface area. For practical purposes surface tension is frequently taken to reflect the change in surface free energy per unit increase in surface area. *See also* Surface Work.

Surface Viscosity. The two-dimensional analog of viscosity acting along the interface between two immiscible fluids. Also called interfacial viscosity. In fact, there are two kinds of surface viscosity: surface shear viscosity and surface dilational viscosity (or surface dilatational viscosity). Surface shear viscosity is the component that is analogous to three-dimensional shear viscosity: the rate of yielding of a layer of fluid due to an applied stress. Surface dilational viscosity relates to the rate of area expansion and is expressed as the local gradient in surface tension per change in relative area per unit time. Any shear rate dependence (non-Newtonian behavior) falls under the subject of surface rheology. Although usually termed surface viscosity or rheology, especially when one fluid is a gas, the more general terminology is surface or interfacial rheology.

Surface Work.   The work required to increase the area of the surface of tension. Under reversible, isothermal conditions the surface work (per unit surface area) equals the equilibrium, or static, surface tension.

Surfactant.   Any substance that lowers the surface or interfacial tension of the medium in which it is dissolved. The substance does not have to be completely soluble and may lower surface or interfacial tension by spreading over the interface. Soaps (fatty acid salts containing at least eight carbon atoms) are surfactants. Detergents are surfactants, or surfactant mixtures, whose solutions have cleaning properties. Also referred to as surface-active agents or tensides. The term surfactant was originally a trademark of the General Aniline and Film Corp., and later released to the public domain [23]. The term paraffin-chain salts was used in the older literature [24]. In some usage surfactants are defined as molecules capable of associating to form micelles.

Surfactant Effectiveness.   The surface excess concentration of surfactant corresponding to saturation of the surface or interface. Example: one indicator of effectiveness is the maximum reduction in surface or interfacial tension achievable by a surfactant. This term has a different meaning from surfactant efficiency. See references [17, 18].

Surfactant Efficiency.   The equilibrium solution surfactant concentration needed to achieve a specified level of adsorption at an interface. Example: one such measure of efficiency is the surfactant concentration needed to reduce the surface or interfacial tension by 20 mN/m from the value of the pure solvent(s). This term has a different meaning from surfactant effectiveness. See references [17, 18].

Surfactant Enhanced Aquifer Remediation.   (SEAR) A remediation technology based on reservoir chemical flooding principles (micellar solubilization and/or low interfacial tension flooding) and applied to the treatment of NAPL-contaminated soils.

Surfactant Macromonomers.   Hydrophobic monomers that also have surfactant character, also termed surfomers. Example: nonylphenoxypoly(etheroxy)ethyl acrylate. They are copolymerized with acrylamide to form hydrophobically associating polymers. *See* reference [25].

Surfactant Tail. The lyophobic portion of a surfactant molecule. It is commonly a hydrocarbon chain containing eight or more carbon atoms. *See also* Head Group.

Surfomers. *See* Surfactant Macromonomers.

Suspending Power. The ability of a detergent or detergent component to keep foreign material away from the solid material from which it has been removed in order to prevent redeposition. *See also* Detergency, Detergent.

Suspension. A system of solid particles dispersed in a liquid. Suspensions were previously referred to as suspensoids, meaning suspension colloids. Aside from the obvious definition of a colloidal suspension, a number of operational definitions are common in industry, such as any dispersed matter that can be removed by a 0.45 μm nominal pore size filter.

Swollen Micelle. *See* Micelle.

Syndet. A synthetic detergent, as opposed to a soap.

Tall Oil. Fatty and resinous carboxylic acids obtained from the sulfate process used to obtain cellulose from softwood trees.

TAN. *See* Total Acid Number.

Tenside. *See* Surfactant.

Thermocapillary Diffusion. Temperature induced Marangoni flow. The movement of suspended drops or bubbles when subjected to a temperature gradient, due to the resulting surface/interfacial tension gradient.

Thermotropic Liquid Crystals. Also called Thermotropic Mesomorphic Phase. *See* Mesomorphic Phase.

Thin Film. *See* Fluid Film.

Thin-Film Drainage. *See* Film Drainage, Fluid Film.

Three-Phase Emulsion. *See* Multiple Emulsion.

Tight Emulsion. A practically stable emulsion, as opposed to a loose emulsion.

Total Acid Number. (TAN) The acid number expresses the amount of base (potassium hydroxide) that will react with a given amount of material in a standardized titration procedure. A large acid number indicates a high concentration of acids in the original material, usually including natural surfactant precursors. A commonly measured property of crude oils.

Traube's Rule. A generalization for homologous series of organic compounds of type $R(CH_2)_nX$, that for each incremental CH_2 group the concentration of molecules required to produce a specified surface tension decreases by a factor of about 3. In adsorption Traube's rule is that a polar adsorbent will preferentially adsorb the most polar component from a nonpolar solution, and conversely, a nonpolar adsorbent will preferentially adsorb the least polar component from a polar solution.

Triple Emulsion. Also called Triple-Phase Emulsion. *See* Multiple Emulsion.

Turbidity. The property of dispersions that causes a reduction in the transparency of the continuous phase due to light scattering and absorption. Turbidity is a function of the size and concentration of the dispersed species. The turbidity coefficient is simply the extinction coefficient in the Beer–Lambert equation for absorbance when light scattering rather than absorbance proper is being studied (hence turbidimetry).

Upper-Phase Microemulsion. A microemulsion with a high oil content that is stable while in contact with a bulk water phase and in laboratory tube or bottle tests tends to be situated at the top of the tube above the water phase. For chlorinated organic liquids, which are more dense than water, the oil will be the top phase rather than the bottom. *See also* Microemulsion, Winsor Type Emulsions.

van der Waals Adsorption. An older term now replaced by physical adsorption, or physisorption.

Vesicle. A droplet that is stabilized by the presence at its surface of a lipid bimolecular film (bilayer) or series of concentric bilayers. Also termed liposome. *See also* Bimolecular Film.

Viscoelastic. A liquid (or solid) with both viscous and elastic properties. A viscoelastic liquid will deform and flow under the influence of an applied shear stress, but when the stress is removed the liquid will slowly recover from some of the deformation.

Wet Oil. An oil containing free water or emulsified water.

Wettability. A qualitative term referring to the water- or oil-preferring nature of surfaces, such as mineral surfaces. Wettability may be determined by direct measurement of contact angles, or inferred from measurements of fluid imbibition or relative permeabilities. Several conventions for describing wettability values exist. *See also* Contact Angle, Wettability Index, Wetting.

Wettability Index. A measure of wettability based on the U.S. Bureau of Mines (USBM) wettability test in which the wettability index (W) is determined as the logarithm of the ratio of areas under the capillary pressure curves for both increasing and decreasing saturation of the wetting phase. Complete oil-wetting occurs for $W = -\infty$ (in practice about -1.5), and complete water-wetting occurs for $W = \infty$ (in practice about 1.0). Another wettability index is derived from the Amott–Harvey test. *See also* reference [26], Wettability.

Wetting. A general term referring to one or more of the following specific kinds of wetting: adhesional wetting, spreading wetting, and immersional wetting. Frequently used to denote that the contact angle between a liquid and a solid is essentially zero and there is spontaneous spreading of the liquid over the solid. Nonwetting, on the other hand, is frequently used to denote the case where the contact angle is greater than $90°$ so that the liquid rolls up into droplets. *See also* Draves Wetting Test, Contact Angle, Wettability.

Wetting Power. *See* Spreading Coefficient.

Wetting Tension. The work done on a system during the process of immersional wetting, expressed per unit area of the phase being immersionally wetted. *See also* Immersional Wetting.

Whipping Agent. *See* Foaming Agent.

Wicking. The flow of liquid into a porous medium due to capillary forces.

Winsor-Type Emulsions. Several categories of microemulsions that refer to equilibrium phase behaviors and distinguish, for example, the number of phases that may be in equilibrium and the nature of the continuous phase. *See* reference [27]. They are denoted as Winsor Type I (oil-in-water), Type II (water-in-oil), Type III (most of the surfactant is in a middle phase with oil and water), and Type IV (water, oil, and surfactant are all present in a single phase). The Winsor Type III system is sometimes referred to as a middle-phase microemulsion, and the Type IV system is often referred to simply as a microemulsion. An advantage of the Winsor category system is that it is independent of the density of the oil phase and may lead to less ambiguity than do the lower-phase or upper-phase microemulsion type terminology. Nelson-type emulsions are similarly identified, but with different type numbers.

W/O. Abbreviation for a water-dispersed-in-oil emulsion.

W/O/A.   Abbreviation for a thin fluid film of oil between water and air phases. *See also* Fluid Film.

W/O/W.   In multiple emulsions, a water-in-oil-in-water multiple emulsion. Here the oil droplets have water droplets dispersed within them, and the oil droplets themselves are dispersed in water forming the continuous phase. In fluid films, a thin fluid film of oil in a water phase. *See also* Fluid Film.

Young's Equation.   A fundamental relationship giving the balance of forces at a point of three-phase contact in terms of surface and interfacial tensions and the contact angle.

Young–Laplace Equation.   The fundamental relationship giving the pressure difference across a curved interface in terms of the surface or interfacial tension and the principal radii of curvature. Also referred to as the equation of capillarity.

Zisman Plot.   A plot of the cosine of contact angle, between a solid of interest and a series of liquids, versus the surface tensions of those liquids. The surface tension extrapolated to zero contact angle is the critical surface tension of wetting of the solid.

Zwitterionic Surfactant.   A surfactant molecule that contains both negatively and positively charged groups. Example: lauramidopropylbetaine, $C_{11}H_{23}CONH(CH_2)_3N^+(CH_3)_2CH_2COO^-$, at neutral and alkaline solution pH. *See also* Amphoteric Surfactant.

Acknowledgments

The author gratefully acknowledges the financial and other support provided by the Natural Sciences and Engineering Research Council of Canada and the Petroleum Recovery Institute.

References

1. *Manual of Symbols and Terminology for Physicochemical Quantities and Units*; Appendix II; Prepared by IUPAC Commission on Colloid and Surface Chemistry; Butterworths: London, 1972. *See also* the additions in *Pure Appl. Chem.* **1983**, 55, 931–941; *Pure Appl. Chem.* **1985**, 57, 603–619.
2. Schramm, L.L. *The Language of Colloid and Interface Science*; American Chemical Society: Washington, DC, 1993.
3. Becher, P. *Dictionary of Colloid and Surface Science*; Dekker: New York, 1990
4. Carrrière, G. *Dictionary of Surface Active Agents, Cosmetics and Toiletries*; Elsevier: Amsterdam, 1978.

5. Schramm, L.L. In *Emulsions: Fundamentals and Applications in the Petroleum Industry*; Schramm, L.L., Ed.; American Chemical Society: Washington, DC, 1992, pp 385–405.
6. Schramm, L.L. In *Foams: Fundamentals and Applications in the Petroleum Industry*; Schramm, L.L., Ed.; American Chemical Society: Washington, DC, 1994, pp 487–535.
7. Schramm, L.L. In *Suspensions: Fundamentals and Applications in the Petroleum Industry*; Schramm, L.L., Ed.; American Chemical Society: Washington, DC, 1996, pp 727–783.
8. O'Haver, J.H.; Harwell, J.H. In *Surfactant Adsorption and Surface Solubilization*; American Chemical Society: Washington, DC, 1995, pp 49–66.
9. Isenberg, C. *The Science of Soap Films and Soap Bubbles*; Tieto Ltd.: Clevedon, United Kingdom, 1978, p 101.
10. Wadle, A.; Tesmann, H.; Leonard, M.; Förster, T. In *Surfactants in Cosmetics*, 2nd Edn; Rieger, M.M.; Rhein, L.D., Eds.; Dekker: New York, 1997, pp 207–224.
11. Cash, L.; Cayias, J.L.; Fournier, G.; Macallister, D.; Schares, T.; Schechter, R.S.; Wade, W.H. *J. Colloid Interface Sci.* **1977**, *59*, 39–44.
12. Cayias, J.L.; Schechter, R.S.; Wade, W.H. *Soc. Petrol. Eng. J.* **1976**, *16*, 351–357.
13. Bikerman, J.J. *Foams*; Springer-Verlage: New York, 1973.
14. Harwell, J.H.; Hoskins, J.C.; Schechter, R.S.; Wade, W.H. *Langmuir* **1985**, *1(2)*, 251–262.
15. *Surface and Colloid Chemistry in Natural Waters and Water Treatment*; Beckett, R., Ed.; Plenum Press: New York, 1990.
16. Fuerstenau, D.W.; Williams, M.C. *Colloids Surf.* **1987**, *22*, 87–91.
17. Myers, D. *Surfactant Science and Technology*; VCH: New York, 1988.
18. Rosen, M.J. *Surfactants and Interfacial Phenomena*; Wiley: New York, 1978.
19. Schramm, L.L.; Novosad, J.J. *Colloids Surf.* **1990**, *46*, 21–43.
20. Nelson, R.C. *Chem. Eng. Prog.* **1989**, *March*, 50–57.
21. Somasundaran, P.; Krishnakumar, S.; Kunjappu, J.T. In *Surfactant Adsorption and Surface Solubilization*; American Chemical Society: Washington, DC, 1995; pp 104–137.
22. Ross, S.; Becher, P. *J. Colloid Interface Sci.* **1992**, *149*, 575–579.
23. Stevens, C.E. In *Kirk-Othmer Encyclopedia of Chemical Technology*, 2nd Edn; Wiley: New York, Vol. 19, 1969, pp 507–593.
24. Adam, N.K. *Physical Chemistry*, Oxford University Press: Oxford, 1956, pp 577–635.
25. Schulz, D.N.; Kaladas, J.J.; Maurer, J.J.; Bock, J.; Pace, S.J.; Schulz, W.W. *Polymer* **1987**, *28*, 2110–2115.
26. Anderson, W.G. *J. Petrol. Technol.* **1986**, *38(12)*, 1246–1262.
27. Winsor, P.A. *Solvent Properties of Amphiphilic Compounds*; Butterworths: London, 1954.

RECEIVED for review July 10, 1998. ACCEPTED revised manuscript December 23, 1998.

AUTHOR INDEX

AFFILIATION INDEX

SUBJECT INDEX